MyMathLab (MML) is a series of online courses that accompany Pearson's textbooks in mathematics and statistics. Since 2001, MML has helped over five million students succeed at maths at more than 1,850 colleges and universities. MyMathLab engages students in active learning — it is modular, self-paced, accessible anywhere with internet access, and adaptable to each student's learning style — and lecturers can easily customise MyMathLab to meet their students' needs.

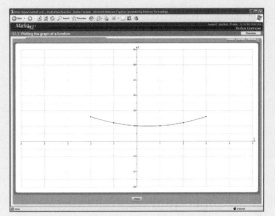

Your purchase of the fifth edition of *Foundation Maths* includes a free Student Access Code. You will need this access code to register.

If you did not purchase a new textbook and your lecturer requires you to enrol in MyMathLab, you may purchase online access. Go to **www.mymathlab.com/global** and follow the links to purchasing online.

Visit the *Foundation Maths,* fifth edition Companion Website at **www.pearsoned.co.uk/croft** to find valuable **student** learning material including:

- Student support pack containing one-page summaries of key topics, complete with examples, exercises and answers
- Extra end-of-chapter questions with answers
- Videos of selected exercises and examples

Other Pearson mathematics titles by the same authors:

- *Mathematics for Engineers*, Third edition, ISBN 9780273725497
- *Introduction to Engineering Mathematics*, ISBN 9780201624427
- *Engineering Mathematics: A Foundation for Electronic, Electrical, Communications and Systems Engineers*, Third edition, ISBN 9780130268587

Foundation Maths

Fifth edition

Anthony Croft
Loughborough University

Robert Davison
De Montfort University, Leicester

Prentice Hall
is an imprint of

Harlow, England • London • New York • Boston • San Francisco • Toronto • Sydney • Singapore • Hong Kong
Tokyo • Seoul • Taipei • New Delhi • Cape Town • Madrid • Mexico City • Amsterdam • Munich • Paris • Milan

Pearson Education Limited
Edinburgh Gate
Harlow
Essex CM20 2JE
England

and Associated Companies throughout the world

Visit us on the World Wide Web at:
www.pearsoned.co.uk

First published 1995
Third edition 2003
Fourth edition 2006
Fifth edition published 2010

© Pearson Education Limited 1995, 2010

ISBN: 978-0-273-72940-2

British Library Cataloguing-in-Publication Data
A catalogue record for this book is available from the British Library

Library of Congress Cataloging-in-Publication Data
Croft, Tony, 1957-
Foundation maths / Anthony Croft, Robert Davison. —5th ed.
     p. cm.
Includes index.
ISBN 978-0-273-72940-2 (pbk.)
1. Mathematics. I. Davison, Robert, II. Title.
QA37.3.C76 2010
510—dc22
                          2009049066

10 9 8 7
14 13

Typeset in 10/12.5 Times by 73
Printed by Ashford Colour Press Ltd, Gosport, UK

The publisher's policy is to use paper manufactured from sustainable forests.

Contents

Supporting resources

Visit **www.pearsoned.co.uk/croft** to find valuable online resources

Companion Website for students
- Student support pack containing one-page summaries of key topics, complete with examples, exercises and answers
- Extra end-of-chapter questions with answers
- Videos of selected exercises and examples

For instructors
- Solutions Manual containing solutions to the test and assignment exercises at the end of each chapter
- PowerPoint slides featuring figures from the book and key points from the book with their related worked examples

MyMathLab
- Online homework and assessment system
- Unlimited question practice due to algorithmically generated questions
- Personalised feedback
- Access to electronic pages of the textbook

For more information please contact your local Pearson Education sales representative or visit **www.pearsoned.co.uk/croft**

Preface

Today, a huge variety of disciplines require their students to have knowledge of certain mathematical tools in order to appreciate the quantitative aspects of their subjects. At the same time, higher education institutions have widened access so that there is much greater variety in the pre-university mathematical experiences of the student body. Some students are returning to education after many years in the workplace or at home bringing up families.

Foundation Maths has been written for those students in higher education who have not specialised in mathematics at A or AS level. It is intended for non-specialists who need some but not a great deal of mathematics as they embark upon their courses of higher education. It is likely to be especially useful to those students embarking upon a Foundation Degree with mathematical content. It takes students from around the lower levels of GCSE to a standard which will enable them to participate fully in a degree or diploma course. It is ideally suited for those studying marketing, business studies, management, science, engineering, social science, geography, combined studies and design. It will be useful for those who lack confidence and need careful, steady guidance in mathematical methods. Even for those whose mathematical expertise is already established, the book will be a helpful revision and reference guide. The style of the book also makes it suitable for those who wish to engage in self-study or distance learning.

We have tried throughout to adopt an informal, user-friendly approach and have described mathematical processes in everyday language. Mathematical ideas are usually developed by example rather than by formal proof. This reflects our experience that students learn better from examples than from abstract development. Where appropriate, the examples contain a great deal of detail so that the student is not left wondering how one stage of a calculation leads to the next. In *Foundation Maths*, objectives are clearly stated at the beginning of each chapter, and key points and formulae are highlighted throughout the book. Self-assessment questions are provided at the end of most sections. These test understanding of important features in the section and answers are given at the back of the book. These are followed by exercises; it is essential that these are attempted as the only way to develop competence and understanding is through practice. Solutions to these exercises are given at the back of the book and should be consulted only after the exercises have been attempted. A further set of test and assignment exercises is given at the end of each chapter. These are provided so that the tutor can set regular assignments or tests throughout the course. Solutions to these are not provided. Feedback from students who have used earlier editions of this book indicates that they have found the style and pace of the book helpful in their study of mathematics at university.

In order to keep the size of the book reasonable we have endeavoured to include topics which we think are most important, cause the most problems for students, and have the widest applicability. We have started the book with materials on arithmetic including whole numbers, fractions and decimals. This is followed by several chapters which gradually introduce important and commonly used topics in algebra. There follows chapters on sets, number bases and logic, collectively known as discrete mathematics. The remaining chapters introduce functions, trigonometry, vectors, matrices, statistics, probability and calculus. These will be found useful in the courses previously listed.

The best strategy for those using the book would be to read through each section, carefully studying all of the worked examples and solutions. Many of these solutions develop important results needed later in the book. It is then a good idea to cover up the solution and try to work the example again independently. It is only by doing the calculation that the necessary techniques will be mastered. At the end of each section the self-assessment questions should be attempted. If these cannot be answered then the previous few pages should be worked through again in order to find the answers in the text, before checking with answers given at the back of the book. Finally, the exercises should be attempted and, again, answers should be checked regularly with those given at the back of the book.

This new edition is enhanced by video clips (see **www.pearsoned.co.uk/ croft**) in which we, the authors, work through some algebraic examples and exercises taken from the book, pointing out techniques and key points. The icon next to an exercise signifies that there is a corresponding video clip. VIDEO

A further enhancement to this edition is the inclusion of MyMathLab, a tried and tested online self-study and assessment system that can add greatly to your understanding of mathematics. We encourage you to use it.

We hope that you find *Foundation Maths* useful and wish you the very best of luck.

Anthony Croft, Robert Davison 2010

Custom publishing

Custom publishing allows academics to pick and choose content from one or more texts for their course and combine it into a definitive course text.

Here are some common examples of Custom solutions which have helped over 500 courses across Europe:

- Different chapters from across our publishing imprints combined into one book
- Lecturer's own material combined together with textbook chapters or published in a separate booklet
- Third party cases and articles that you are keen for your students to read as part of the course
- Or any combination of the above

The Pearson Education Custom text published for your course is professionally produced and bound – just as you would expect from a normal Pearson Education text. You can even choose your own cover design and add your university logo. Since many of our titles have online resources accompanying them we can even build a Custom website that matches your course text.

Whatever your choice, our dedicated Editorial and Production teams will work with you throughout the process, right until you receive a copy of your Custom text.

The flexibility of Custom publishing allows you to include additional material for your students, either with more lower level maths to refresh the basics, or with some more advanced topics to cover additional topics. You could also choose to remove some of the earlier or later chapters to fully tailor the book to your course. Extending the student access period of this book's online MyMathLab course has also proved to be a popular custom option.

Here is a list of possible subject areas that Pearson Education provides material for that you might want to take additional chapters from:

- Access Level Mathematics
- Engineering Mathematics
- Statistics
- Calculus
- Linear Algebra

For full details on any of our publications, please visit: www.pearsoned.co.uk

If, once you have thought about your course, you feel Custom publishing might benefit you, please do get in contact. However minor, or major the change, we can help you out.

You can contact us at: www.pearsoncustom.co.uk or via your local representative at: www.pearsoned.co.uk/replocator

Guided tour

Key points highlight important results
that need to be referred to easily
or remembered. ——————————————————————————→

Worked examples are included throughout the book
to reinforce student learning and to illustrate a step-
by-step approach to solving mathematical problems.

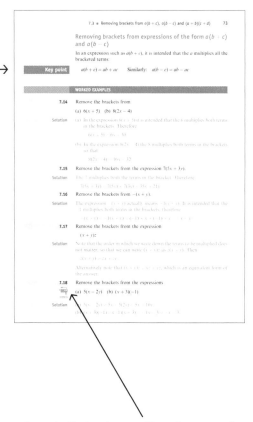

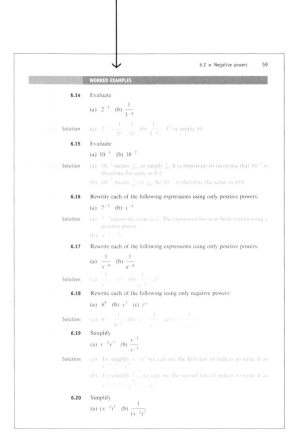

Icons indicate where a **video** of an example
or exercise can be found on the companion
website at **www.pearsoned.co.uk/croft**

Self-assessment questions are provided at the end of most sections to test understanding of important parts of the section. Answers are given at the back of the book.

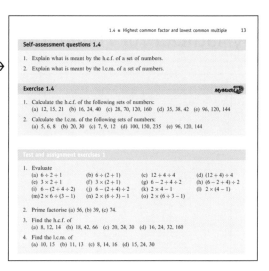

Exercises provide a key opportunity to develop competence and understanding through practice. Answers are given at the back of the book.

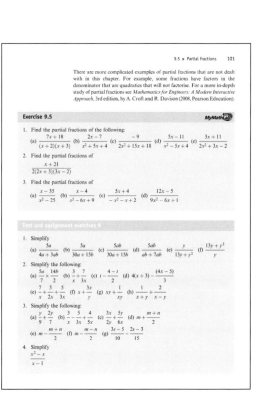

Test and assignment exercises (with answers provided in a separate Lecturers' Manual) allow lecturers and tutors to set regular assignments or tests throughout the course.

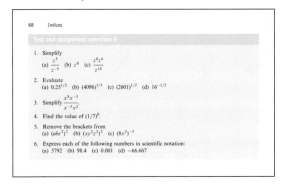

MyMathLab for *Foundation Maths*

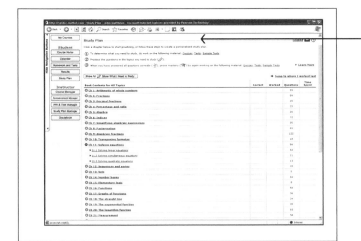

The **study plan** allows students unlimited practice on chosen areas of the textbook.

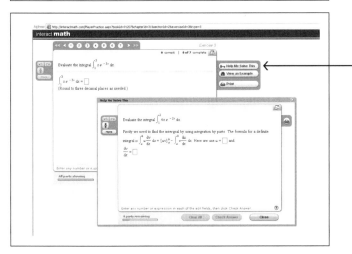

Feedback is given and **help** is provided to enable students to work through solutions step by step.

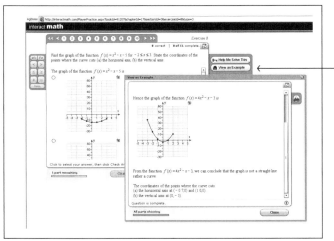

Detailed **examples** of questions are provided to aid understanding and learning.

List of videos

The following table lists the videos which accompany selected exercises and examples in the book. You can view the videos at **www.pearsoned.co.uk/croft**

Name	Reference
Substitution of a value into a quadratic expression	Exercise 5.3 Q13
Simplification of expressions requiring use of the first law of indices	Exercise 6.1 Q8
Simplification of expressions requiring use of the second and third laws of indices	Exercise 6.1 Q10
Simplification of expressions with negative powers	Exercise 6.2 Q4
Removing the brackets from expressions 1	Example 7.18
Removing the brackets from expressions 2	Example 7.24
Factorising a quadratic expression 1	Example 8.6
Factorising a quadratic expression 2	Example 8.12
Simplifying an algebraic fraction 1	Example 9.4
Simplifying an algebraic fraction 2	Example 9.8
Simplifying the product of two algebraic fractions	Example 9.17
Simplifying products and quotients of algebraic fractions	Exercise 9.3 Q4
Adding algebraic fractions 1	Example 9.24
Adding algebraic fractions 2	Example 9.25
An example of partial fractions	Example 9.27
Another example of partial fractions	Example 9.28
Transposition of a formula	Example 10.7
Solving simultaneous equations by elimination	Example 11.6
Solving a quadratic equation by factorisation	Example 11.10
Solving a quadratic equation using a formula	Example 11.15

$+$	plus
$-$	minus
$\pm$	plus or minus
$\times$	multiply by
$\cdot$	multiply by
$\div$	divide by
$=$	is equal to
$\equiv$	is identically equal to
$\approx$	is approximately equal to
$\neq$	is not equal to
$>$	is greater than
$\geqslant$	is greater than or equal to
$<$	is less than
$\leqslant$	is less than or equal to
$\in$	is a member of set
$\mathcal{E}$	universal set
$\cap$	intersection
$\cup$	union
$\emptyset$	empty set
$\bar{A}$	complement of set A
$\subseteq$	subset
$\mathbb{R}$	all real numbers
$\mathbb{R}^{+}$	all numbers greater than 0
$\mathbb{R}^{-}$	all numbers less than 0
$\mathbb{Z}$	all integers
$\mathbb{N}$	all positive integers

$\mathbb{Q}$ rational numbers

Π irrational numbers

$\therefore$ therefore

∞ infinity

e the base of natural logarithms (2.718...)

$\ln$ natural logarithm

$\log$ logarithm to base 10

$\sum$ sum of terms

$\int$ integral

$\frac{dy}{dx}$ derivative of y with respect to x

π 'pi' ≈ 3.14159

$\neg$ negation (not)

$\wedge$ conjunction (and)

$\vee$ disjunction (or)

$\rightarrow$ implication

Arithmetic of whole numbers

1

Objectives: This chapter:

- explains the rules for adding, subtracting, multiplying and dividing positive and negative numbers
- explains what is meant by an integer
- explains what is meant by a prime number
- explains what is meant by a factor
- explains how to prime factorise an integer
- explains the terms 'highest common factor' and 'lowest common multiple'

1.1 Addition, subtraction, multiplication and division

Arithmetic is the study of numbers and their manipulation. A clear and firm understanding of the rules of arithmetic is essential for tackling everyday calculations. Arithmetic also serves as a springboard for tackling more abstract mathematics such as algebra and calculus.

The calculations in this chapter will involve mainly whole numbers, or **integers** as they are often called. The **positive integers** are the numbers

$1, 2, 3, 4, 5 \ldots$

and the **negative integers** are the numbers

$\ldots -5, -4, -3, -2, -1$

The dots ($\ldots$) indicate that this sequence of numbers continues indefinitely. The number 0 is also an integer but is neither positive nor negative.

To find the **sum** of two or more numbers, the numbers are added together. To find the **difference** of two numbers, the second is subtracted from the first. The **product** of two numbers is found by multiplying the numbers together. Finally, the **quotient** of two numbers is found by dividing the first number by the second.

WORKED EXAMPLE

1.1   (a) Find the sum of 3, 6 and 4.

(b) Find the difference of 6 and 4.

(c) Find the product of 7 and 2.

(d) Find the quotient of 20 and 4.

Solution   (a) The sum of 3, 6 and 4 is

$$3 + 6 + 4 = 13$$

(b) The difference of 6 and 4 is

$$6 - 4 = 2$$

(c) The product of 7 and 2 is

$$7 \times 2 = 14$$

(d) The quotient of 20 and 4 is $\frac{20}{4}$, that is 5.

When writing products we sometimes replace the sign $\times$ by '·' or even omit it completely. For example, $3 \times 6 \times 9$ could be written as $3·6·9$ or $(3)(6)(9)$.

On occasions it is necessary to perform calculations involving negative numbers. To understand how these are added and subtracted consider Figure 1.1, which shows a number line.

Figure 1.1
The number line

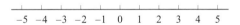

Any number can be represented by a point on the line. Positive numbers are on the right-hand side of the line and negative numbers are on the left. From any given point on the line, we can add a positive number by moving that number of places to the right. For example, to find the sum $5 + 3$, start at the point 5 and move 3 places to the right, to arrive at 8. This is shown in Figure 1.2.

Figure 1.2
To add a positive number, move that number of places to the right

To subtract a positive number, we move that number of places to the left. For example, to find the difference $5 - 7$, start at the point 5 and move 7 places to the left to arrive at -2. Thus $5 - 7 = -2$. This is shown in Figure 1.3. The result of finding $-3 - 4$ is also shown to be -7.

Figure 1.3
To subtract a positive number, move that number of places to the left

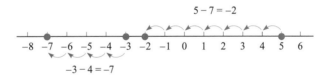

To add or subtract a negative number, the motions just described are reversed. So, to add a negative number, we move to the left. To subtract a negative number we move to the right. The result of finding $2 + (-3)$ is shown in Figure 1.4.

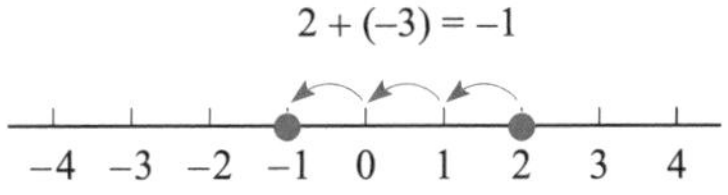

Figure 1.4
Adding a negative number involves moving to the left

We see that $2 + (-3) = -1$. Note that this is the same as the result of finding $2 - 3$, so that adding a negative number is equivalent to subtracting a positive number.

Figure 1.5
Subtracting a negative number involves moving to the right

The result of finding $5 - (-3)$ is shown in Figure 1.5.
We see that $5 - (-3) = 8$. This is the same as the result of finding $5 + 3$, so subtracting a negative number is equivalent to adding a positive number.

Key point

Adding a negative number is equivalent to subtracting a positive number.
Subtracting a negative number is equivalent to adding a positive number.

WORKED EXAMPLE

1.2 Evaluate (a) $8 + (-4)$, (b) $-15 + (-3)$, (c) $-15 - (-4)$.

Solution (a) $8 + (-4)$ is equivalent to $8 - 4$, that is 4.

(b) Because adding a negative number is equivalent to subtracting a positive number we find $-15 + (-3)$ is equivalent to $-15 - 3$, that is -18.

(c) $-15 - (-4)$ is equivalent to $-15 + 4$, that is -11.

When we need to multiply or divide negative numbers, care must be taken with the **sign** of the answer; that is, whether the result is positive or negative. The following rules apply for determining the sign of the answer when multiplying or dividing positive and negative numbers.

Key point

$$(\text{positive}) \times (\text{positive}) = \text{positive}$$
$$(\text{positive}) \times (\text{negative}) = \text{negative}$$
$$(\text{negative}) \times (\text{positive}) = \text{negative}$$
$$(\text{negative}) \times (\text{negative}) = \text{positive}$$

and

$$\frac{\text{positive}}{\text{positive}} = \text{positive}$$

$$\frac{\text{positive}}{\text{negative}} = \text{negative}$$

$$\frac{\text{negative}}{\text{positive}} = \text{negative}$$

$$\frac{\text{negative}}{\text{negative}} = \text{positive}$$

WORKED EXAMPLE

1.3 Evaluate

(a) $3 \times (-2)$ (b) $(-1) \times 7$ (c) $(-2) \times (-4)$ (d) $\dfrac{12}{(-4)}$ (e) $\dfrac{-8}{4}$ (f) $\dfrac{-6}{-2}$

Solution

(a) We have a positive number, 3, multiplied by a negative number, -2, and so the result will be negative:

$$3 \times (-2) = -6$$

(b) $(-1) \times 7 = -7$

(c) Here we have two negative numbers being multiplied and so the result will be positive:

$$(-2) \times (-4) = 8$$

(d) A positive number, 12, divided by a negative number, -4, gives a negative result:

$$\frac{12}{-4} = -3$$

(e) A negative number, -8, divided by a positive number, 4, gives a negative result:

$$\frac{-8}{4} = -2$$

(f) A negative number, -6, divided by a negative number, -2, gives a positive result:

$$\frac{-6}{-2} = 3$$

Self-assessment questions 1.1

1. Explain what is meant by an integer, a positive integer and a negative integer.

2. Explain the terms sum, difference, product and quotient.

3. State the sign of the result obtained after performing the following calculations:
    (a) $(-5) \times (-3)$ (b) $(-4) \times 2$ (c) $\frac{7}{-2}$ (d) $\frac{-8}{-4}$.

Exercise 1.1

MyMathLab

1. Without using a calculator, evaluate each of the following:
    (a) $6 + (-3)$ (b) $6 - (-3)$
    (c) $16 + (-5)$ (d) $16 - (-5)$
    (e) $27 - (-3)$ (f) $27 - (-29)$
    (g) $-16 + 3$ (h) $-16 + (-3)$
    (i) $-16 - 3$ (j) $-16 - (-3)$
    (k) $-23 + 52$ (l) $-23 + (-52)$
    (m) $-23 - 52$ (n) $-23 - (-52)$

2. Without using a calculator, evaluate
    (a) $3 \times (-8)$ (b) $(-4) \times 8$ (c) $15 \times (-2)$
    (d) $(-2) \times (-8)$ (e) $14 \times (-3)$

3. Without using a calculator, evaluate
    (a) $\frac{15}{-3}$ (b) $\frac{21}{7}$ (c) $\frac{-21}{7}$ (d) $\frac{-21}{-7}$ (e) $\frac{21}{-7}$
    (f) $\frac{-12}{2}$ (g) $\frac{-12}{-2}$ (h) $\frac{12}{-2}$

4. Find the sum and product of (a) 3 and 6, (b) 10 and 7, (c) 2, 3 and 6.

5. Find the difference and quotient of (a) 18 and 9, (b) 20 and 5, (c) 100 and 20.

1.2 The BODMAS rule

When evaluating numerical expressions we need to know the order in which addition, subtraction, multiplication and division are carried out. As a simple example, consider evaluating $2 + 3 \times 4$. If the addition is carried out first we get $2 + 3 \times 4 = 5 \times 4 = 20$. If the multiplication is carried out first we get $2 + 3 \times 4 = 2 + 12 = 14$. Clearly the order of carrying out numerical operations is important. The BODMAS rule tells us the order in which we must carry out the operations of addition, subtraction, multiplication and division.

Key point

BODMAS stands for

Brackets ( )	First priority
Of $\times$	Second priority
Division $\div$	Second priority
Multiplication $\times$	Second priority
Addition $+$	Third priority
Subtraction $-$	Third priority

This is the order of carrying out arithmetical operations, with bracketed expressions having highest priority and subtraction and addition having the lowest priority. Note that 'Of', 'Division' and 'Multiplication' have equal priority, as do 'Addition' and 'Subtraction'. 'Of' is used to show multiplication when dealing with fractions: for example, find $\frac{1}{2}$ of 6 means $\frac{1}{2} \times 6$.

   If an expression contains only multiplication and division, we evaluate by working from left to right. Similarly, if an expression contains only addition and subtraction, we also evaluate by working from left to right.

WORKED EXAMPLES

1.4   Evaluate

(a) $2 + 3 \times 4$ (b) $(2 + 3) \times 4$

Solution   (a) Using the BODMAS rule we see that multiplication is carried out first. So

$$2 + 3 \times 4 = 2 + 12 = 14$$

(b) Using the BODMAS rule we see that the bracketed expression takes priority over all else. Hence

$$(2 + 3) \times 4 = 5 \times 4 = 20$$

1.5   Evaluate

(a) $4 - 2 \div 2$ (b) $1 - 3 + 2 \times 2$

Solution   (a)   Division is carried out before subtraction, and so

$$4 - 2 \div 2 = 4 - \frac{2}{2} = 3$$

(b)   Multiplication is carried out before subtraction or addition:

$$1 - 3 + 2 \times 2 = 1 - 3 + 4 = 2$$

1.6   Evaluate

(a) $(12 \div 4) \times 3$ (b) $12 \div (4 \times 3)$

Solution   Recall that bracketed expressions are evaluated first.

(a) $(12 \div 4) \times 3 = \left(\dfrac{12}{4}\right) \times 3 = 3 \times 3 = 9$

(b) $12 \div (4 \times 3) = 12 \div 12 = 1$

Example 1.6 shows the importance of the position of brackets in an expression.

Self-assessment questions 1.2

1. State the BODMAS rule used to evaluate expressions.

2. The position of brackets in an expression is unimportant. True or false?

Exercise 1.2   *MyMathLab*

1. Evaluate the following expressions:
   (a) $6 - 2 \times 2$ (b) $(6 - 2) \times 2$
   (c) $6 \div 2 - 2$ (d) $(6 \div 2) - 2$
   (e) $6 - 2 + 3 \times 2$ (f) $6 - (2 + 3) \times 2$
   (g) $(6 - 2) + 3 \times 2$ (h) $\dfrac{16}{-2}$ (i) $\dfrac{-24}{-3}$
   (j) $(-6) \times (-2)$ (k) $(-2)(-3)(-4)$

2. Place brackets in the following expressions to make them correct:
   (a) $6 \times 12 - 3 + 1 = 55$
   (b) $6 \times 12 - 3 + 1 = 68$
   (c) $6 \times 12 - 3 + 1 = 60$
   (d) $5 \times 4 - 3 + 2 = 7$
   (e) $5 \times 4 - 3 + 2 = 15$
   (f) $5 \times 4 - 3 + 2 = -5$

1.3 Prime numbers and factorisation

A **prime number** is a positive integer, larger than 1, which cannot be expressed as the product of two smaller positive integers. To put it another way, a prime number is one that can be divided exactly only by 1 and itself.

For example, $6 = 2 \times 3$, so 6 can be expressed as a product of smaller numbers and hence 6 is not a prime number. However, 7 is prime. Examples of prime numbers are 2, 3, 5, 7, 11, 13, 17, 19, 23. Note that 2 is the only even prime.

Factorise means 'write as a product'. By writing 12 as 3×4 we have factorised 12. We say 3 is a **factor** of 12 and 4 is also a factor of 12. The way in which a number is factorised is not unique: for example, 12 may be expressed as 3×4 or 2×6. Note that 2 and 6 are also factors of 12.

When a number is written as a product of prime numbers we say the number has been **prime factorised**.

To prime factorise a number, consider the technique used in the following examples.

WORKED EXAMPLES

1.7   Prime factorise the following numbers:

(a)   12   (b)   42   (c)   40   (d)   70

Solution   (a)   We begin with 2 and see whether this is a factor of 12. Clearly it is, so we write

$$12 = 2 \times 6$$

Now we consider 6. Again 2 is a factor so we write

$$12 = 2 \times 2 \times 3$$

All the factors are now prime, that is the prime factorisation of 12 is $2 \times 2 \times 3$.

(b)   We begin with 2 and see whether this is a factor of 42. Clearly it is and so we can write

$$42 = 2 \times 21$$

Now we consider 21. Now 2 is not a factor of 21, so we examine the next prime, 3. Clearly 3 is a factor of 21 and so we can write

$$42 = 2 \times 3 \times 7$$

All the factors are now prime, and so the prime factorisation of 42 is $2 \times 3 \times 7$.

(c)   Clearly 2 is a factor of 40,

$$40 = 2 \times 20$$

Clearly 2 is a factor of 20,

$$40 = 2 \times 2 \times 10$$

Again 2 is a factor of 10,

$$40 = 2 \times 2 \times 2 \times 5$$

All the factors are now prime. The prime factorisation of 40 is $2 \times 2 \times 2 \times 5$.

(d) Clearly 2 is a factor of 70,

$$70 = 2 \times 35$$

We consider 35: 2 is not a factor, 3 is not a factor, but 5 is:

$$70 = 2 \times 5 \times 7$$

All the factors are prime. The prime factorisation of 70 is $2 \times 5 \times 7$.

1.8 Prime factorise 2299.

Solution We note that 2 is not a factor and so we try 3. Again 3 is not a factor and so we try 5. This process continues until we find the first prime factor. It is 11:

$$2299 = 11 \times 209$$

We now consider 209. The first prime factor is 11:

$$2299 = 11 \times 11 \times 19$$

All the factors are prime. The prime factorisation of 2299 is $11 \times 11 \times 19$.

Self-assessment questions 1.3

1. Explain what is meant by a prime number.

2. List the first 10 prime numbers.

3. Explain why all even numbers other than 2 cannot be prime.

Exercise 1.3

1. State which of the following numbers are prime numbers:
   (a) 13 (b) 1000 (c) 2 (c) 29 (d) $\frac{1}{2}$

2. Prime factorise the following numbers:
   (a) 26 (b) 100 (c) 27 (d) 71 (e) 64 (f) 87 (g) 437 (h) 899

3. Prime factorise the two numbers 30 and 42. List any prime factors which are common to both numbers.

1.4 Highest common factor and lowest common multiple

Highest common factor

Suppose we prime factorise 12. This gives $12 = 2 \times 2 \times 3$. From this prime factorisation we can deduce all the factors of 12:

> 2 is a factor of 12
> 3 is a factor of 12
> $2 \times 2 = 4$ is a factor of 12
> $2 \times 3 = 6$ is a factor of 12

Hence 12 has factors 2, 3, 4 and 6, in addition to the obvious factors of 1 and 12.

Similarly we could prime factorise 18 to obtain $18 = 2 \times 3 \times 3$. From this we can list the factors of 18:

> 2 is a factor of 18
> 3 is a factor of 18
> $2 \times 3 = 6$ is a factor of 18
> $3 \times 3 = 9$ is a factor of 18

The factors of 18 are 1, 2, 3, 6, 9 and 18. Some factors are common to both 12 and 18. These are 2, 3 and 6. These are **common factors** of 12 and 18. The highest common factor of 12 and 18 is 6.

The highest common factor of 12 and 18 can be obtained directly from their prime factorisation. We simply note all the primes common to both factorisations:

$$12 = 2 \times 2 \times 3 \qquad 18 = 2 \times 3 \times 3$$

Common to both is 2×3. Thus the highest common factor is $2 \times 3 = 6$. Thus 6 is the highest number that divides exactly into both 12 and 18.

Key point

Given two or more numbers the **highest common factor** (h.c.f.) is the largest (highest) number that is a factor of all the given numbers.
The highest common factor is also referred to as the **greatest common divisor** (g.c.d).

WORKED EXAMPLES

1.9   Find the h.c.f. of 12 and 27.

Solution   We prime factorise 12 and 27:

$$12 = 2 \times 2 \times 3 \qquad 27 = 3 \times 3 \times 3$$

Common to both is 3. Thus 3 is the h.c.f. of 12 and 27. This means that 3 is the highest number that divides both 12 and 27.

1.10   Find the h.c.f. of 28 and 210.

Solution   The numbers are prime factorised:

$$28 = 2 \times 2 \times 7$$
$$210 = 2 \times 3 \times 5 \times 7$$

The factors that are common are identified: a 2 is common to both and a 7 is common to both. Hence both numbers are divisible by $2 \times 7 = 14$. Since this number contains all the common factors it is the highest common factor.

1.11   Find the h.c.f. of 90 and 108.

Solution   The numbers are prime factorised:

$$90 = 2 \times 3 \times 3 \times 5$$
$$108 = 2 \times 2 \times 3 \times 3 \times 3$$

The common factors are 2, 3 and 3 and so the h.c.f. is $2 \times 3 \times 3$, that is 18. This is the highest number that divides both 90 and 108.

1.12   Find the h.c.f. of 12, 18 and 20.

Solution   Prime factorisation yields

$$12 = 2 \times 2 \times 3 \qquad 18 = 2 \times 3 \times 3 \qquad 20 = 2 \times 2 \times 5$$

There is only one factor common to all three numbers: it is 2. Hence 2 is the h.c.f. of 12, 18 and 20.

Lowest common multiple

Suppose we are given two or more numbers and wish to find numbers into which all the given numbers will divide. For example, given 4 and 6 we see that they both divide exactly into 12, 24, 36, 48, 60 and so on. The smallest number into which they both divide is 12. We say 12 is the **lowest common multiple** of 4 and 6.

Key point   The lowest common multiple (l.c.m.) of a set of numbers is the smallest (lowest) number into which all the given numbers will divide exactly.

WORKED EXAMPLE

1.13   Find the l.c.m. of 6 and 10.

Solution   We seek the smallest number into which both 6 and 10 will divide exactly. There are many numbers into which 6 and 10 will divide, for example 60,

120, 600, but we are seeking the smallest such number. By inspection, the smallest such number is 30. Thus the l.c.m. of 6 and 10 is 30.

A more systematic method of finding the l.c.m. involves the use of prime factorisation.

WORKED EXAMPLES

1.14   Find the l.c.m. of 15 and 20.

Solution   As a first step, the numbers are prime factorised:

$$15 = 3 \times 5 \qquad 20 = 2 \times 2 \times 5$$

Since 15 must divide into the l.c.m., then the l.c.m. must contain the factors of 15, that is 3×5. Similarly, as 20 must divide into the l.c.m., then the l.c.m. must also contain the factors of 20, that is $2 \times 2 \times 5$. The l.c.m. is the smallest number that contains both of these sets of factors. Note that the l.c.m. will contain only 2s, 3s and 5s as its prime factors. We now need to determine how many of these particular factors are needed.

To determine the l.c.m. we ask 'How many factors of 2 are required?', 'How many factors of 3 are required?', 'How many factors of 5 are required?'

The highest number of 2s occurs in the factorisation of 20. Hence the l.c.m. requires two factors of 2. Consider the number of 3s required. The highest number of 3s occurs in the factorisation of 15. Hence the l.c.m. requires one factor of 3. Consider the number of 5s required. The highest number of 5s is 1 and so the l.c.m. requires one factor of 5. Hence the l.c.m. is $2 \times 2 \times 3 \times 5 = 60$.

Hence 60 is the smallest number into which both 15 and 20 will divide exactly.

1.15   Find the l.c.m. of 20, 24 and 25.

Solution   The numbers are prime factorised:

$$20 = 2 \times 2 \times 5 \qquad 24 = 2 \times 2 \times 2 \times 3 \qquad 25 = 5 \times 5$$

By considering the prime factorisations of 20, 24 and 25 we see that the only primes involved are 2, 3 and 5. Hence the l.c.m. will contain only 2s, 3s and 5s.

Consider the number of 2s required. The highest number of 2s required is three from factorising 24. The highest number of 3s required is one, again from factorising 24. The highest number of 5s required is two, found from factorising 25. Hence the l.c.m. is given by

$$\text{l.c.m.} = 2 \times 2 \times 2 \times 3 \times 5 \times 5 = 600$$

Hence 600 is the smallest number into which 20, 24 and 25 will all divide exactly.

Self-assessment questions 1.4

1. Explain what is meant by the h.c.f. of a set of numbers.

2. Explain what is meant by the l.c.m. of a set of numbers.

Exercise 1.4

1. Calculate the h.c.f. of the following sets of numbers:
   (a) 12, 15, 21   (b) 16, 24, 40   (c) 28, 70, 120, 160   (d) 35, 38, 42   (e) 96, 120, 144

2. Calculate the l.c.m. of the following sets of numbers:
   (a) 5, 6, 8   (b) 20, 30   (c) 7, 9, 12   (d) 100, 150, 235   (e) 96, 120, 144

Test and assignment exercises 1

1. Evaluate
   (a) $6 \div 2 + 1$
   (b) $6 \div (2 + 1)$
   (c) $12 + 4 \div 4$
   (d) $(12 + 4) \div 4$
   (e) $3 \times 2 + 1$
   (f) $3 \times (2 + 1)$
   (g) $6 - 2 + 4 \div 2$
   (h) $(6 - 2 + 4) \div 2$
   (i) $6 - (2 + 4 \div 2)$
   (j) $6 - (2 + 4) \div 2$
   (k) $2 \times 4 - 1$
   (l) $2 \times (4 - 1)$
   (m) $2 \times 6 \div (3 - 1)$
   (n) $2 \times (6 \div 3) - 1$
   (o) $2 \times (6 \div 3 - 1)$

2. Prime factorise (a) 56, (b) 39, (c) 74.

3. Find the h.c.f. of
   (a) 8, 12, 14   (b) 18, 42, 66   (c) 20, 24, 30   (d) 16, 24, 32, 160

4. Find the l.c.m. of
   (a) 10, 15   (b) 11, 13   (c) 8, 14, 16   (d) 15, 24, 30

Fractions

2

Objectives: This chapter:

- explains what is meant by a fraction
- defines the terms 'improper fraction', 'proper fraction' and 'mixed fraction'
- explains how to write fractions in different but equivalent forms
- explains how to simplify fractions by cancelling common factors
- explains how to add, subtract, multiply and divide fractions

2.1 Introduction

The arithmetic of fractions is very important groundwork that must be mastered before topics in algebra such as formulae and equations can be understood. The same techniques that are used to manipulate fractions are used in these more advanced topics. You should use this chapter to ensure that you are confident at handling fractions before moving on to algebra. In all the examples and exercises it is important that you should carry out the calculations without the use of a calculator.

Fractions are numbers such as $\frac{1}{2}, \frac{3}{4}, \frac{11}{8}$ and so on. In general a fraction is a number of the form $\frac{p}{q}$, where the letters p and q represent whole numbers or integers. The integer q can never be zero because it is never possible to divide by zero.

In any fraction $\frac{p}{q}$ the number p is called the **numerator** and the number q is called the **denominator**.

Key point

$$\text{fraction} = \frac{\text{numerator}}{\text{denominator}} = \frac{p}{q}$$

Suppose that p and q are both positive numbers. If p is less than q, the fraction is said to be a **proper fraction**. So $\frac{1}{2}$ and $\frac{3}{4}$ are proper fractions since the numerator is less than the denominator. If p is greater than or equal to q, the fraction is said to be **improper**. So $\frac{11}{8}$, $\frac{7}{4}$ and $\frac{3}{3}$ are all improper fractions.

If either of p or q is negative, we simply ignore the negative sign when determining whether the fraction is proper or improper. So $-\frac{3}{5}$, $\frac{-7}{21}$ and $\frac{4}{-21}$ are proper fractions, but $\frac{3}{-3}$, $\frac{-8}{2}$ and $-\frac{11}{2}$ are improper.

Note that all proper fractions have a value less than 1.

The denominator of a fraction can take the value 1, as in $\frac{3}{1}$ and $\frac{7}{1}$. In these cases the result is a whole number, 3 and 7.

Self-assessment questions 2.1

1. Explain the terms (a) fraction, (b) improper fraction, (c) proper fraction. In each case give an example of your own.

2. Explain the terms (a) numerator, (b) denominator.

Exercise 2.1

1. Classify each of the following as proper or improper:
   (a) $\frac{9}{17}$ (b) $\frac{-9}{17}$ (c) $\frac{8}{8}$ (d) $-\frac{7}{8}$ (e) $\frac{110}{77}$

2.2 Expressing a fraction in equivalent forms

Given a fraction, we may be able to express it in a different form. For example, you will know that $\frac{1}{2}$ is equivalent to $\frac{2}{4}$. Note that multiplying both numerator and denominator by the same number leaves the value of the fraction unchanged. So, for example,

$$\frac{1}{2} = \frac{1 \times 2}{2 \times 2} = \frac{2}{4}$$

We say that $\frac{1}{2}$ and $\frac{2}{4}$ are **equivalent fractions**. Although they might look different, they have the same value.

Similarly, given the fraction $\frac{8}{12}$ we can divide both numerator and denominator by 4 to obtain

$$\frac{8}{12} = \frac{8/4}{12/4} = \frac{2}{3}$$

so $\frac{8}{12}$ and $\frac{2}{3}$ have the same value and are equivalent fractions.

| Key point | Multiplying or dividing both numerator and denominator of a fraction by the same number produces a fraction having the same value, called an equivalent fraction. |

A fraction is in its **simplest form** when there are no factors common to both numerator and denominator. For example, $\frac{5}{12}$ is in its simplest form, but $\frac{3}{6}$ is not since 3 is a factor common to both numerator and denominator. Its simplest form is the equivalent fraction $\frac{1}{2}$.

To express a fraction in its simplest form we look for factors that are common to both the numerator and denominator. This is done by prime factorising both of these. Dividing both the numerator and denominator by any common factors removes them but leaves an equivalent fraction. This is equivalent to cancelling any common factors. For example, to simplify $\frac{4}{6}$ we prime factorise to produce

$$\frac{4}{6} = \frac{2 \times 2}{2 \times 3}$$

Dividing both numerator and denominator by 2 leaves $\frac{2}{3}$. This is equivalent to cancelling the common factor of 2.

WORKED EXAMPLES

2.1 Express $\frac{24}{36}$ in its simplest form.

Solution We seek factors common to both numerator and denominator. To do this we prime factorise 24 and 36:

Prime factorisation has been described in §1.3.

$$24 = 2 \times 2 \times 2 \times 3 \qquad 36 = 2 \times 2 \times 3 \times 3$$

The factors $2 \times 2 \times 3$ are common to both 24 and 36 and so these may be cancelled. Note that only common factors may be cancelled when simplifying a fraction. Hence

Finding the highest common factor (h.c.f.) of two numbers is detailed in §1.4.

$$\frac{24}{36} = \frac{\not{2} \times \not{2} \times 2 \times \not{3}}{\not{2} \times \not{2} \times \not{3} \times 3} = \frac{2}{3}$$

In its simplest form $\frac{24}{36}$ is $\frac{2}{3}$. In effect we have divided 24 and 36 by 12, which is their h.c.f.

2.2 Express $\frac{49}{21}$ in its simplest form.

Solution Prime factorising 49 and 21 gives

$$49 = 7 \times 7 \qquad 21 = 3 \times 7$$

Their h.c.f. is 7. Dividing 49 and 21 by 7 gives

$$\frac{49}{21} = \frac{7}{3}$$

Hence the simplest form of $\frac{49}{21}$ is $\frac{7}{3}$.

Before we can start to add and subtract fractions it is necessary to be able to convert fractions into a variety of equivalent forms. Work through the following examples.

WORKED EXAMPLES

2.3   Express $\frac{3}{4}$ as an equivalent fraction having a denominator of 20.

Solution   To achieve a denominator of 20, the existing denominator must be multiplied by 5. To produce an equivalent fraction both numerator and denominator must be multiplied by 5, so

$$\frac{3}{4} = \frac{3 \times 5}{4 \times 5} = \frac{15}{20}$$

2.4   Express 7 as an equivalent fraction with a denominator of 3.

Solution   Note that 7 is the same as the fraction $\frac{7}{1}$. To achieve a denominator of 3, the existing denominator must be multiplied by 3. To produce an equivalent fraction both numerator and denominator must be multiplied by 3, so

$$7 = \frac{7}{1} = \frac{7 \times 3}{1 \times 3} = \frac{21}{3}$$

Self-assessment questions 2.2

1. All integers can be thought of as fractions. True or false?

2. Explain the use of h.c.f. in the simplification of fractions.

3. Give an example of three fractions that are equivalent.

Exercise 2.2

1. Express the following fractions in their simplest form:
   (a) $\frac{18}{27}$ (b) $\frac{12}{20}$ (c) $\frac{15}{45}$ (d) $\frac{25}{80}$ (e) $\frac{15}{60}$
   (f) $\frac{90}{200}$ (g) $\frac{15}{20}$ (h) $\frac{2}{18}$ (i) $\frac{16}{24}$ (j) $\frac{30}{65}$
   (k) $\frac{12}{21}$ (l) $\frac{100}{45}$ (m) $\frac{6}{9}$ (n) $\frac{12}{16}$ (o) $\frac{13}{42}$
   (p) $\frac{13}{39}$ (q) $\frac{11}{33}$ (r) $\frac{14}{30}$ (s) $-\frac{12}{16}$ (t) $\frac{11}{-33}$
   (u) $\frac{-14}{-30}$

2. Express $\frac{3}{4}$ as an equivalent fraction having a denominator of 28.

3. Express 4 as an equivalent fraction with a denominator of 5.

4. Express $\frac{5}{12}$ as an equivalent fraction having a denominator of 36.

5. Express 2 as an equivalent fraction with a denominator of 4.

6. Express 6 as an equivalent fraction with a denominator of 3.

7. Express each of the fractions $\frac{2}{3}$, $\frac{5}{4}$ and $\frac{5}{6}$ as an equivalent fraction with a denominator of 12.

8. Express each of the fractions $\frac{4}{9}$, $\frac{1}{2}$ and $\frac{5}{6}$ as an equivalent fraction with a denominator of 18.

9. Express each of the following numbers as an equivalent fraction with a denominator of 12:
   (a) $\frac{1}{2}$ (b) $\frac{3}{4}$ (c) $\frac{5}{2}$ (d) 5 (e) 4 (f) 12

2.3 Addition and subtraction of fractions

To add and subtract fractions we first rewrite each fraction so that they all have the same denominator. This is known as the **common denominator**. The denominator is chosen to be the lowest common multiple of the original denominators. Then the numerators only are added or subtracted as appropriate, and the result is divided by the common denominator.

WORKED EXAMPLES

2.5   Find $\frac{2}{3} + \frac{5}{4}$.

Solution   The denominators are 3 and 4. The l.c.m. of 3 and 4 is 12. We need to express both fractions with a denominator of 12.

Finding the lowest common multiple (l.c.m.) is detailed in §1.4.

To express $\frac{2}{3}$ with a denominator of 12 we multiply both numerator and denominator by 4. Hence $\frac{2}{3}$ is the same as $\frac{8}{12}$. To express $\frac{5}{4}$ with a denominator of 12 we multiply both numerator and denominator by 3. Hence $\frac{5}{4}$ is the same as $\frac{15}{12}$. So

$$\frac{2}{3} + \frac{5}{4} = \frac{8}{12} + \frac{15}{12} = \frac{8 + 15}{12} = \frac{23}{12}$$

2.6   Find $\frac{4}{9} - \frac{1}{2} + \frac{5}{6}$.

Solution   The denominators are 9, 2 and 6. Their l.c.m. is 18. Each fraction is expressed with 18 as the denominator:

$$\frac{4}{9} = \frac{8}{18} \qquad \frac{1}{2} = \frac{9}{18} \qquad \frac{5}{6} = \frac{15}{18}$$

Then

$$\frac{4}{9} - \frac{1}{2} + \frac{5}{6} = \frac{8}{18} - \frac{9}{18} + \frac{15}{18} = \frac{8 - 9 + 15}{18} = \frac{14}{18}$$

The fraction $\frac{14}{18}$ can be simplified to $\frac{7}{9}$. Hence

$$\frac{4}{9} - \frac{1}{2} + \frac{5}{6} = \frac{7}{9}$$

2.7   Find $\frac{1}{4} - \frac{5}{9}$.

Solution   The l.c.m. of 4 and 9 is 36. Each fraction is expressed with a denominator of 36. Thus

$$\frac{1}{4} = \frac{9}{36} \qquad \text{and} \qquad \frac{5}{9} = \frac{20}{36}$$

Then

$$\frac{1}{4} - \frac{5}{9} = \frac{9}{36} - \frac{20}{36}$$

$$= \frac{9 - 20}{36}$$

$$= \frac{-11}{36}$$

$$= -\frac{11}{36}$$

Consider the number $2\frac{3}{4}$. This is referred to as a **mixed fraction** because it contains a whole number part, 2, and a fractional part, $\frac{3}{4}$. We can convert this mixed fraction into an improper fraction as follows. Recognise that 2 is equivalent to $\frac{8}{4}$, and so $2\frac{3}{4}$ is $\frac{8}{4} + \frac{3}{4} = \frac{11}{4}$.

The reverse of this process is to convert an improper fraction into a mixed fraction. Consider the improper fraction $\frac{11}{4}$. Now 4 divides into 11 twice leaving a remainder of 3; so $\frac{11}{4} = 2$ remainder 3, which we write as $2\frac{3}{4}$.

WORKED EXAMPLE

2.8   (a)   Express $4\frac{2}{5}$ as an improper fraction.

(b)   Find $4\frac{2}{5} + \frac{1}{3}$.

Solution   (a) $4\frac{2}{5}$ is a mixed fraction. Note that $4\frac{2}{5}$ is equal to $4 + \frac{2}{5}$. We can write 4 as the equivalent fraction $\frac{20}{5}$. Therefore

$$4\frac{2}{5} = \frac{20}{5} + \frac{2}{5}$$

$$= \frac{22}{5}$$

(b) $4\frac{2}{5} + \frac{1}{3} = \frac{22}{5} + \frac{1}{3}$

$$= \frac{66}{15} + \frac{5}{15}$$

$$= \frac{71}{15}$$

Self-assessment question 2.3

1. Explain the use of l.c.m. when adding and subtracting fractions.

Exercise 2.3

1. Find
   (a) $\frac{1}{4} + \frac{2}{3}$ (b) $\frac{3}{5} + \frac{5}{3}$ (c) $\frac{12}{14} - \frac{2}{7}$

   (d) $\frac{3}{7} - \frac{1}{2} + \frac{2}{21}$ (e) $1\frac{1}{2} + \frac{4}{9}$

   (f) $2\frac{1}{4} - 1\frac{1}{3} + \frac{1}{2}$ (g) $\frac{10}{15} - 1\frac{2}{5} + \frac{8}{3}$

   (h) $\frac{9}{10} - \frac{7}{16} + \frac{1}{2} - \frac{2}{5}$

2. Find
   (a) $\frac{7}{8} + \frac{1}{3}$ (b) $\frac{1}{2} - \frac{3}{4}$ (c) $\frac{3}{5} + \frac{2}{3} + \frac{1}{2}$

   (d) $\frac{3}{8} + \frac{1}{3} + \frac{1}{4}$ (e) $\frac{2}{3} - \frac{4}{7}$

   (f) $\frac{1}{11} - \frac{1}{2}$ (g) $\frac{3}{11} - \frac{5}{8}$

3. Express as improper fractions:

   (a) $2\frac{1}{2}$ (b) $3\frac{2}{3}$ (c) $10\frac{1}{4}$ (d) $5\frac{2}{7}$

   (e) $6\frac{2}{9}$ (f) $11\frac{1}{3}$ (g) $15\frac{1}{2}$ (h) $13\frac{3}{4}$

   (i) $12\frac{1}{11}$ (j) $13\frac{2}{3}$ (k) $56\frac{1}{2}$

4. Without using a calculator express these improper fractions as mixed fractions:

   (a) $\frac{10}{3}$ (b) $\frac{7}{2}$ (c) $\frac{15}{4}$ (d) $\frac{25}{6}$

2.4 Multiplication of fractions

The product of two or more fractions is found by multiplying their numerators to form a new numerator, and then multiplying their denominators to form a new denominator.

WORKED EXAMPLES

2.9 Find $\frac{4}{9} \times \frac{3}{8}$.

Solution The numerators are multiplied: $4 \times 3 = 12$. The denominators are multiplied: $9 \times 8 = 72$. Hence

$$\frac{4}{9} \times \frac{3}{8} = \frac{12}{72}$$

This may now be expressed in its simplest form:

$$\frac{12}{72} = \frac{1}{6}$$

Hence

$$\frac{4}{9} \times \frac{3}{8} = \frac{1}{6}$$

An alternative, but equivalent, method is to cancel any factors common to both numerator and denominator at the outset:

$$\frac{4}{9} \times \frac{3}{8} = \frac{4 \times 3}{9 \times 8}$$

A factor of 4 is common to the 4 and the 8. Hence

$$\frac{4 \times 3}{9 \times 8} = \frac{1 \times 3}{9 \times 2}$$

A factor of 3 is common to the 3 and the 9. Hence

$$\frac{1 \times 3}{9 \times 2} = \frac{1 \times 1}{3 \times 2} = \frac{1}{6}$$

2.10   Find $\frac{12}{25} \times \frac{2}{7} \times \frac{10}{9}$.

Solution   We cancel factors common to both numerator and denominator. A factor of 5 is common to 10 and 25. Cancelling this gives

$$\frac{12}{25} \times \frac{2}{7} \times \frac{10}{9} = \frac{12}{5} \times \frac{2}{7} \times \frac{2}{9}$$

A factor of 3 is common to 12 and 9. Cancelling this gives

$$\frac{12}{5} \times \frac{2}{7} \times \frac{2}{9} = \frac{4}{5} \times \frac{2}{7} \times \frac{2}{3}$$

There are no more common factors. Hence

$$\frac{12}{25} \times \frac{2}{7} \times \frac{10}{9} = \frac{4}{5} \times \frac{2}{7} \times \frac{2}{3} = \frac{16}{105}$$

2.11   Find $\frac{3}{4}$ of $\frac{5}{9}$.

Recall that 'of' means multiply.

Solution $\frac{3}{4}$ of $\frac{5}{9}$ is the same as $\frac{3}{4} \times \frac{5}{9}$. Cancelling a factor of 3 from numerator and denominator gives $\frac{1}{4} \times \frac{5}{3}$, that is $\frac{5}{12}$. Hence $\frac{3}{4}$ of $\frac{5}{9}$ is $\frac{5}{12}$.

2.12   Find $\frac{5}{6}$ of 70.

Solution   We can write 70 as $\frac{70}{1}$. So

$$\frac{5}{6} \text{ of } 70 = \frac{5}{6} \times \frac{70}{1} = \frac{5}{3} \times \frac{35}{1} = \frac{175}{3} = 58\frac{1}{3}$$

2.13   Find $2\frac{7}{8} \times \frac{2}{3}$.

Solution   In this example the first fraction is a mixed fraction. We convert it to an improper fraction before performing the multiplication. Note that $2\frac{7}{8} = \frac{23}{8}$. Then

$$\frac{23}{8} \times \frac{2}{3} = \frac{23}{4} \times \frac{1}{3}$$

$$= \frac{23}{12}$$

$$= 1\frac{11}{12}$$

Self-assessment question 2.4

1. Describe how to multiply fractions together.

Exercise 2.4

1. Evaluate

    (a) $\frac{2}{3} \times \frac{6}{7}$ (b) $\frac{8}{15} \times \frac{25}{32}$ (c) $\frac{1}{4} \times \frac{8}{9}$

    (d) $\frac{16}{17} \times \frac{34}{48}$ (e) $2 \times \frac{3}{5} \times \frac{5}{12}$

    (f) $2\frac{1}{3} \times 1\frac{1}{4}$ (g) $1\frac{3}{4} \times 2\frac{1}{2}$

    (h) $\frac{3}{4} \times 1\frac{1}{2} \times 3\frac{1}{2}$

2. Evaluate

    (a) $\frac{2}{3}$ of $\frac{3}{4}$ (b) $\frac{4}{7}$ of $\frac{21}{30}$

    (c) $\frac{9}{10}$ of 80   (d) $\frac{6}{7}$ of 42

3. Is $\frac{3}{4}$ of $\frac{12}{15}$ the same as $\frac{12}{15}$ of $\frac{3}{4}$?

4. Find

    (a) $-\frac{1}{3} \times \frac{5}{7}$ (b) $\frac{3}{4} \times -\frac{1}{2}$

    (c) $\left(-\frac{5}{8}\right) \times \frac{8}{11}$ (d) $\left(-\frac{2}{3}\right) \times \left(-\frac{15}{7}\right)$

5. Find

    (a) $5\frac{1}{2} \times \frac{1}{2}$ (b) $3\frac{3}{4} \times \frac{1}{3}$

    (c) $\frac{2}{3} \times 5\frac{1}{9}$ (d) $\frac{3}{4} \times 11\frac{1}{2}$

6. Find

    (a) $\frac{3}{5}$ of $11\frac{1}{4}$ (b) $\frac{2}{3}$ of $15\frac{1}{2}$

    (c) $\frac{1}{4}$ of $-8\frac{1}{3}$

2.5 Division by a fraction

To divide one fraction by another fraction, we invert the second fraction and then multiply. When we invert a fraction we interchange the numerator and denominator.

WORKED EXAMPLES

2.14 Find $\frac{6}{25} \div \frac{2}{5}$.

Solution We invert $\frac{2}{5}$ to obtain $\frac{5}{2}$. Multiplication is then performed. So

$$\frac{6}{25} \div \frac{2}{5} = \frac{6}{25} \times \frac{5}{2} = \frac{3}{25} \times \frac{5}{1} = \frac{3}{5} \times \frac{1}{1} = \frac{3}{5}$$

2.15 Evaluate (a) $1\frac{1}{3} \div \frac{8}{3}$, (b) $\frac{20}{21} \div \frac{5}{7}$.

Solution (a) First we express $1\frac{1}{3}$ as an improper fraction:

$$1\frac{1}{3} = 1 + \frac{1}{3} = \frac{3}{3} + \frac{1}{3} = \frac{4}{3}$$

So we calculate

$$\frac{4}{3} \div \frac{8}{3} = \frac{4}{3} \times \frac{3}{8} = \frac{4}{8} = \frac{1}{2}$$

Hence

$$1\frac{1}{3} \div \frac{8}{3} = \frac{1}{2}$$

(b) $\frac{20}{21} \div \frac{5}{7} = \frac{20}{21} \times \frac{7}{5} = \frac{4}{21} \times \frac{7}{1} = \frac{4}{3}$

Self-assessment question 2.5

1. Explain the process of division by a fraction.

Exercise 2.5

MyMathLab

1. Evaluate

   (a) $\dfrac{3}{4} \div \dfrac{1}{8}$

   (b) $\dfrac{8}{9} \div \dfrac{4}{3}$

   (c) $\dfrac{-2}{7} \div \dfrac{4}{21}$

   (d) $\dfrac{9}{4} \div 1\dfrac{1}{2}$

   (e) $\dfrac{5}{6} \div \dfrac{5}{12}$

   (f) $\dfrac{99}{100} \div 1\dfrac{4}{5}$

   (g) $3\dfrac{1}{4} \div 1\dfrac{1}{8}$

   (h) $\left(2\dfrac{1}{4} \div \dfrac{3}{4} \right) \times 2$

   (i) $2\dfrac{1}{4} \div \left(\dfrac{3}{4} \times 2 \right)$

   (j) $6\dfrac{1}{4} \div 2\dfrac{1}{2} + 5$

   (k) $6\dfrac{1}{4} \div \left(2\dfrac{1}{2} + 5 \right)$

Test and assignment exercises 2

1. Evaluate

   (a) $\dfrac{3}{4} + \dfrac{1}{6}$

   (b) $\dfrac{2}{3} + \dfrac{3}{5} - \dfrac{1}{6}$

   (c) $\dfrac{5}{7} - \dfrac{2}{3}$

   (d) $2\dfrac{1}{3} - \dfrac{9}{10}$

   (e) $5\dfrac{1}{4} + 3\dfrac{1}{6}$

   (f) $\dfrac{9}{8} - \dfrac{7}{6} + 1$

   (g) $\dfrac{5}{6} - \dfrac{5}{3} + \dfrac{5}{4}$

   (h) $\dfrac{4}{5} + \dfrac{1}{3} - \dfrac{3}{4}$

2. Evaluate

   (a) $\dfrac{4}{7} \times \dfrac{21}{32}$

   (b) $\dfrac{5}{6} \times \dfrac{8}{15}$

   (c) $\dfrac{3}{11} \times \dfrac{20}{21}$

   (d) $\dfrac{9}{14} \times \dfrac{8}{18}$

   (e) $\dfrac{5}{4} \div \dfrac{10}{13}$

   (f) $\dfrac{7}{16} \div \dfrac{21}{32}$

   (g) $\dfrac{-24}{25} \div \dfrac{51}{50}$

   (h) $\dfrac{45}{81} \div \dfrac{25}{27}$

3. Evaluate the following expressions using the BODMAS rule:

   (a) $\dfrac{1}{2} + \dfrac{1}{3} \times 2$

   (b) $\dfrac{3}{4} \times \dfrac{2}{3} + \dfrac{1}{4}$

   (c) $\dfrac{5}{6} \div \dfrac{2}{3} + \dfrac{3}{4}$

   (d) $\left(\dfrac{2}{3} + \dfrac{1}{4} \right) \div 4 + \dfrac{3}{5}$

   (e) $\left(\dfrac{4}{3} - \dfrac{2}{5} \times \dfrac{1}{3} \right) \times \dfrac{1}{4} + \dfrac{1}{2}$

   (f) $\dfrac{3}{4}$ of $\left(1 + \dfrac{2}{3} \right)$

   (g) $\dfrac{2}{3}$ of $\dfrac{1}{2} + 1$

   (h) $\dfrac{1}{5} \times \dfrac{2}{3} + \dfrac{2}{5} \div \dfrac{4}{5}$

4. Express in their simplest form:

   (a) $\dfrac{21}{84}$

   (b) $\dfrac{6}{80}$

   (c) $\dfrac{34}{85}$

   (d) $\dfrac{22}{143}$

   (e) $\dfrac{69}{253}$

Decimal fractions

3

Objectives: This chapter:

- revises the decimal number system
- shows how to write a number to a given number of significant figures
- shows how to write a number to a given number of decimal places

3.1 Decimal numbers

Consider the whole number 478. We can regard it as the sum

$$400 + 70 + 8$$

In this way we see that, in the number 478, the 8 represents eight ones, or 8 units, the 7 represents seven tens, or 70, and the number 4 represents four hundreds or 400. Thus we have the system of hundreds, tens and units familiar from early years in school. All whole numbers can be thought of in this way.

When we wish to deal with proper fractions and mixed fractions, we extend the hundreds, tens and units system as follows. A **decimal point**, '.', marks the end of the whole number part, and the numbers that follow it, to the right, form the fractional part.

A number immediately to the right of the decimal point, that is in the **first decimal place**, represents tenths, so

$$0.1 = \frac{1}{10}$$

$$0.2 = \frac{2}{10} \quad \text{or} \quad \frac{1}{5}$$

$$0.3 = \frac{3}{10} \quad \text{and so on}$$

Note that when there are no whole numbers involved it is usual to write a zero in front of the decimal point, thus, .2 would be written 0.2.

WORKED EXAMPLE

3.1 Express the following decimal numbers as proper fractions in their simplest form

(a) 0.4   (b) 0.5   (c) 0.6

Solution The first number after the decimal point represents tenths.

(a) $0.4 = \frac{4}{10}$, which simplifies to $\frac{2}{5}$

(b) $0.5 = \frac{5}{10}$ or simply $\frac{1}{2}$

(c) $0.6 = \frac{6}{10} = \frac{3}{5}$

Frequently we will deal with numbers having a whole number part and a fractional part. Thus

$$5.2 = 5 \text{ units} + 2 \text{ tenths}$$

$$= 5 + \frac{2}{10}$$

$$= 5 + \frac{1}{5}$$

$$= 5\frac{1}{5}$$

Similarly,

$$175.8 = 175\frac{8}{10} = 175\frac{4}{5}$$

Numbers in the second position after the decimal point, or the **second decimal place**, represent hundredths, so

$$0.01 = \frac{1}{100}$$

$$0.02 = \frac{2}{100} \quad \text{or} \quad \frac{1}{50}$$

$$0.03 = \frac{3}{100} \quad \text{and so on}$$

Consider 0.25. We can think of this as

$$0.25 = 0.2 + 0.05$$

$$= \frac{2}{10} + \frac{5}{100}$$

$$= \frac{25}{100}$$

We see that 0.25 is equivalent to $\frac{25}{100}$, which in its simplest form is $\frac{1}{4}$.

In fact we can regard any numbers occupying the first two decimal places as hundredths, so that

$$0.25 = \frac{25}{100} \qquad \text{or simply} \quad \frac{1}{4}$$

$$0.50 = \frac{50}{100} \qquad \text{or} \quad \frac{1}{2}$$

$$0.75 = \frac{75}{100} = \frac{3}{4}$$

WORKED EXAMPLES

3.2   Express the following decimal numbers as proper fractions in their simplest form:

(a) 0.35   (b) 0.56   (c) 0.68

Solution   The first two decimal places represent hundredths:

(a) $0.35 = \frac{35}{100} = \frac{7}{20}$

(b) $0.56 = \frac{56}{100} = \frac{14}{25}$

(c) $0.68 = \frac{68}{100} = \frac{17}{25}$

3.3   Express 37.25 as a mixed fraction in its simplest form.

Solution $37.25 = 37 + 0.25$

$$= 37 + \frac{25}{100}$$

$$= 37 + \frac{1}{4}$$

$$= 37 \frac{1}{4}$$

Numbers in the third position after the decimal point, or **third decimal place**, represent thousandths, so

$$0.001 = \frac{1}{1000}$$

$$0.002 = \frac{2}{1000} \quad \text{or} \quad \frac{1}{500}$$

$$0.003 = \frac{3}{1000} \quad \text{and so on}$$

In fact we can regard any numbers occupying the first three positions after the decimal point as thousandths, so that

$$0.356 = \frac{356}{1000} \quad \text{or} \quad \frac{89}{250}$$

$$0.015 = \frac{15}{1000} \quad \text{or} \quad \frac{3}{200}$$

$$0.075 = \frac{75}{1000} = \frac{3}{40}$$

WORKED EXAMPLE

3.4   Write each of the following as a decimal number:

(a) $\frac{3}{10} + \frac{7}{100}$ (b) $\frac{8}{10} + \frac{3}{1000}$

Solution

(a) $\frac{3}{10} + \frac{7}{100} = 0.3 + 0.07 = 0.37$

(b) $\frac{8}{10} + \frac{3}{1000} = 0.8 + 0.003 = 0.803$

You will normally use a calculator to add, subtract, multiply and divide decimal numbers. Generally the more decimal places used, the more accurately we can state a number. This idea is developed in the next section.

Self-assessment questions 3.1

1. State which is the largest and which is the smallest of the following numbers:
   23.001, 23.0, 23.00001, 23.0008, 23.01

2. Which is the largest of the following numbers?
   0.1, 0.02, 0.003, 0.0004, 0.00005

Exercise 3.1

1. Express the following decimal numbers as proper fractions in their simplest form:
   (a) 0.7   (b) 0.8   (c) 0.9

2. Express the following decimal numbers as proper fractions in their simplest form:
   (a) 0.55   (b) 0.158   (c) 0.98
   (d) 0.099

3. Express each of the following as a mixed fraction in its simplest form:
   (a) 4.6   (b) 5.2   (c) 8.05   (d) 11.59
   (e) 121.09

4. Write each of the following as a decimal number:
   (a) $\frac{6}{10} + \frac{9}{100} + \frac{7}{1000}$ (b) $\frac{8}{100} + \frac{3}{1000}$
   (c) $\frac{17}{1000} + \frac{5}{10}$

3.2 Significant figures and decimal places

The accuracy with which we state a number often depends upon the context in which the number is being used. The volume of a petrol tank is usually given to the nearest litre. It is of no practical use to give such a volume to the nearest cubic centimetre.

When writing a number we often give the accuracy by stating the **number of significant figures** or the **number of decimal places** used. These terms are now explained.

Significant figures

Suppose we are asked to write down the number nearest to 857 using at most two non-zero digits, or numbers. We would write 860. This number is nearer to 857 than any other number with two non-zero digits. We say that 857 to 2 **significant figures** is 860. The words 'significant figures' are usually abbreviated to s.f. Because 860 is larger than 857 we say that the 857 has been **rounded up** to 860.

To write a number to three significant figures we can use no more than three non-zero digits. For example, the number closest to 1784 which has no more than three non-zero digits is 1780. We say that 1784 to 3 significant figures is 1780. In this case, because 1780 is less than 1784 we say that 1784 has been **rounded down** to 1780.

WORKED EXAMPLES

3.5 Write down the number nearest to 86 using only one non-zero digit. Has 86 been rounded up or down?

Solution   The number 86 written to one significant figure is 90. This number is nearer to 86 than any other number having only one non-zero digit. 86 has been rounded up to 90.

3.6   Write down the number nearest to 999 which uses only one non-zero digit.

Solution   The number 999 to one significant figure is 1000. This number is nearer to 999 than any other number having only one non-zero digit.

We now explain the process of writing to a given number of significant figures.

When asked to write a number to, say, three significant figures, 3 s.f., the first step is to look at the first four digits. If asked to write a number to two significant figures we look at the first three digits and so on. We always look at one more digit than the number of significant figures required.

For example, to write 6543.19 to 2 s.f. we would consider the number 6540.00; the digits 3, 1 and 9 are effectively ignored. The next step is to round up or down. If the final digit is a 5 or more then we round up by increasing the previous digit by 1. If the final digit is 4 or less we round down by leaving the previous digit unchanged. Hence when considering 6543.19 to 2 s.f., the 4 in the third place means that we round down to 6500.

To write 23865 to 3 s.f. we would consider the number 23860. The next step is to increase the 8 to a 9. Thus 23865 is rounded up to 23900.

Zeros at the beginning of a number are ignored. To write 0.004693 to 2 s.f. we would first consider the number 0.00469. Note that the zeros at the beginning of the number have not been counted. We then round the 6 to a 7, producing 0.0047.

The following examples illustrate the process.

WORKED EXAMPLES

3.7   Write 36.482 to 3 s.f.

Solution   We consider the first four digits, that is 36.48. The final digit is 8 and so we round up 36.48 to 36.5. To 3 s.f. 36.482 is 36.5.

3.8   Write 1.0049 to 4 s.f.

Solution   To write to 4 s.f. we consider the first five digits, that is 1.0049. The final digit is a 9 and so 1.0049 is rounded up to 1.005.

3.9   Write 695.3 to 2 s.f.

Solution   We consider 695. The final digit is a 5 and so we round up. We cannot round up the 9 to a 10 and so the 69 is rounded up to 70. Hence to 2 s.f. the number is 700.

3.10   Write 0.0473 to 1 s.f.

Solution   We do not count the initial zeros and consider 0.047. The final digit tells us to round up. Hence to 1 s.f. we have 0.05.

3.11   A number is given to 2 s.f. as 67.

(a)   What is the maximum value the number could have?
(b)   What is the minimum value the number could have?

Solution   (a)   To 2 s.f. 67.5 is 68. Any number just below 67.5, for example 67.49 or 67.499, to 2 s.f. is 67. Hence the maximum value of the number is 67.4999....

(b)   To 2 s.f. 66.4999... is 66. However, 66.5 to 2 s.f. is 67. The minimum value of the number is thus 66.5.

Decimal places

When asked to write a number to 3 decimal places (3 d.p.) we consider the first 4 decimal places, that is numbers after the decimal point. If asked to write to 2 d.p. we consider the first 3 decimal places and so on. If the final digit is 5 or more we round up, otherwise we round down.

WORKED EXAMPLES

3.12   Write 63.4261 to 2 d.p.

Solution   We consider the number to 3 d.p., that is 63.426. The final digit is 6 and so we round up 63.426 to 63.43. Hence 63.4261 to 2 d.p. is 63.43.

3.13   Write 1.97 to 1 d.p.

Solution   In order to write to 1 d.p. we consider the number to 2 d.p., that is we consider 1.97. The final digit is a 7 and so we round up. The 9 cannot be rounded up and so we look at 1.9. This can be rounded up to 2.0. Hence 1.97 to 1 d.p. is 2.0. Note that it is crucial to write 2.0 and not simply 2, as this shows that the number is written to 1 d.p.

3.14   Write −6.0439 to 2 d.p.

Solution   We consider −6.043. As the final digit is a 3 the number is rounded down to −6.04.

Self-assessment questions 3.2

1. Explain the meaning of 'significant figures'.
2. Explain the process of writing a number to so many decimal places.

Exercise 3.2

1. Write to 3 s.f.
   (a) 6962 (b) 70.406 (c) 0.0123
   (d) 0.010991 (e) 45.607 (f) 2345

2. Write 65.999 to
   (a) 4 s.f. (b) 3 s.f. (c) 2 s.f.
   (d) 1 s.f. (e) 2 d.p. (f) 1 d.p.

3. Write 9.99 to
   (a) 1 s.f. (b) 1 d.p.

4. Write 65.4555 to
   (a) 3 d.p. (b) 2 d.p. (c) 1 d.p.
   (d) 5 s.f. (e) 4 s.f. (f) 3 s.f. (g) 2 s.f.
   (h) 1 s.f.

Test and assignment exercises 3

1. Express the following numbers as proper fractions in their simplest form:
   (a) 0.74 (b) 0.96 (c) 0.05 (d) 0.25

2. Express each of the following as a mixed fraction in its simplest form:
   (a) 2.5 (b) 3.25 (c) 3.125 (d) 6.875

3. Write each of the following as a decimal number:
   (a) $\frac{3}{10} + \frac{1}{100} + \frac{7}{1000}$ (b) $\frac{5}{1000} + \frac{9}{100}$ (c) $\frac{4}{1000} + \frac{9}{10}$

4. Write 0.09846 to (a) 1 d.p, (b) 2 s.f., (c) 1 s.f.

5. Write 9.513 to (a) 3 s.f., (b) 2 s.f., (c) 1 s.f.

6. Write 19.96 to (a) 1 d.p., (b) 2 s.f., (c) 1 s.f.

Percentage and ratio

4

Objectives:

This chapter:

- explains the terms 'percentage' and 'ratio'
- shows how to perform calculations using percentages and ratios
- explains how to calculate the percentage change in a quantity

4.1 Percentage

In everyday life we come across percentages regularly. During sales periods shops offer discounts – for example, we might hear expressions like 'everything reduced by 50%'. Students often receive examination marks in the form of percentages – for example, to achieve a pass grade in a university examination, a student may be required to score at least 40%. Banks and building societies charge interest on loans, and the interest rate quoted is usually given as a percentage, for example 4.75%. Percentages also provide a way of comparing two or more quantities. For example, suppose we want to know which is the better mark: 40 out of 70, or 125 out of 200? By expressing these marks as percentages we will be able to answer this question.

Consequently an understanding of what a percentage is, and an ability to perform calculations involving percentages, are not only useful in mathematical applications, but also essential life skills.

Most calculators have a percentage button and we will illustrate the use of this later in the chapter. However, be aware that different calculators work in different and often confusing ways. Misleading results can be obtained if you do not know how to use your calculator correctly. So it is better if you are not over-reliant on your calculator and instead understand the principles behind percentage calculations.

Fundamentally, a **percentage** is a fraction whose denominator is 100. In fact you can think of the phrase 'per cent' meaning 'out of 100'. We use the symbol % to represent a percentage, as earlier. The following three fractions all have a denominator of 100, and are expressed as percentages as shown:

$$\frac{17}{100} \quad \text{may be expressed as} \quad 17\%$$

$$\frac{50}{100} \quad \text{may be expressed as} \quad 50\%$$

$$\frac{3}{100} \quad \text{may be expressed as} \quad 3\%$$

WORKED EXAMPLE

4.1   Express $\frac{19}{100}$, $\frac{35}{100}$ and $\frac{17.5}{100}$ as percentages.

Solution   All of these fractions have a denominator of 100. So it is straightforward to write down their percentage form:

$$\frac{19}{100} = 19\% \qquad \frac{35}{100} = 35\% \qquad \frac{17.5}{100} = 17.5\%$$

Sometimes it is necessary to convert a fraction whose denominator is not 100, for example $\frac{2}{5}$, into a percentage. This could be done by expressing the fraction as an equivalent fraction with denominator 100, as was explained in Section 2.2 on page 15. However, with calculators readily available, the calculation can be done as follows.

We can use the calculator to divide the numerator of the fraction by the denominator. The answer is then multiplied by 100. The resulting number is the required percentage. So, to convert $\frac{2}{5}$ we perform the following key strokes:

$$2 \div 5 \times 100 = 40$$

and so $\frac{2}{5} = 40\%$. You should check this now using your own calculator,

Key point   To convert a fraction to a percentage, divide the numerator by the denominator, multiply by 100 and then label the result as a percentage.

WORKED EXAMPLES

4.2   Convert $\frac{5}{8}$ into a percentage.

Solution   Using the method described above we find

$$5 \div 8 \times 100 = 62.5$$

Labelling the answer as a percentage, we see that $\frac{5}{8}$ is equivalent to 62.5%.

4.3   Bill scores $\frac{13}{17}$ in a test. In a different test, Mary scores $\frac{14}{19}$. Express the scores as percentages, and thereby make a comparison of the two marks.

Solution   Use your calculator to perform the division and then multiply the result by 100.

Bill's score: $13 \div 17 \times 100 = 76.5$ (1 d.p.)

Mary's score: $14 \div 19 \times 100 = 73.7$ (1 d.p.)

So we see that Bill scores 76.5% and Mary scores 73.7%. Notice that in these percentage forms it is easy to compare the two marks. We see that Bill has achieved the higher score. Making easy comparisons like this is one of the reasons why percentages are used so frequently.

We have seen that percentages are fractions with a denominator of 100, so that, for example, $\frac{19}{100} = 19\%$. Sometimes a fraction may be given not as a numerator divided by a denominator, but in its decimal form. For example, the decimal form of $\frac{19}{100}$ is 0.19. To convert a decimal fraction into a percentage we simply multiply by 100. So

$$0.19 = 0.19 \times 100\% = 19\%$$

Key point   To convert a decimal fraction to a percentage, multiply by 100 and then label the result as a percentage.

We may also want to reverse the process. Frequently in business calculations involving formulae for interest it is necessary to express a percentage in its decimal form. To convert a percentage to its equivalent decimal form we divide the percentage by 100. Alternatively, using a calculator, input the percentage and press the % button, to convert the percentage to its decimal form.

WORKED EXAMPLE

4.4   Express 50% as a decimal.

Solution   We divide the percentage by 100:

$$50 \div 100 = 0.5$$

So 50% is equivalent to 0.5. To see why this is the case, remember that 'per cent' literally means 'out of 100' so 50% means 50 out of 100, or $\frac{50}{100}$, or in its simplest form 0.5.

Alternatively, using a calculator, the key strokes

50 %

should give 0.5. Check whether you can do this on your calculator.

Key point	To convert a percentage to its equivalent decimal fraction form, divide by 100.

WORKED EXAMPLE

4.5 Express 17.5% as a decimal.

Solution We divide the percentage by 100:

$$17.5 \div 100 = 0.175$$

So 17.5% is equivalent to 0.175. Now check you can obtain the same result using the percentage button on your calculator.

Some percentages appear so frequently in everyday life that it is useful to learn their fraction and decimal fraction equivalent forms.

Key point	$10\% = 0.1 = \frac{1}{10}$ $25\% = 0.25 = \frac{1}{4}$
	$50\% = 0.5 = \frac{1}{2}$ $75\% = 0.75 = \frac{3}{4}$ $100\% = 1$

Recall from Section 1.2 that 'of' means multiply.

We are often asked to calculate a percentage of a quantity: for example, find 17.5% of 160 or 10% of 95. Such calculations arise when finding discounts on prices. Since $17.5\% = \frac{17.5}{100}$ we find

$$17.5\% \text{ of } 160 = \frac{17.5}{100} \times 160 = 28$$

and since $10\% = 0.1$ we may write

$$10\% \text{ of } 95 = 0.1 \times 95 = 9.5$$

Alternatively, the percentage button on a calculator can be used: check you can use your calculator correctly by verifying

Because finding 10% of a quantity is equivalent to dividing by 10, it is easy to find 10% by moving the decimal point one place to the left.

17.5 % × 160 = 28

10 % × 95 = 9.5

WORKED EXAMPLES

4.6   Calculate 27% of 90.

Solution   Using a calculator

$$27 \boxed{\%} \times 90 = 24.3$$

4.7   Calculate 100% of 6.

Solution $100 \boxed{\%} \times 6 = 6$

Observe that 100% of a number is simply the number itself.

4.8   A deposit of £750 increases by 9%. Calculate the resulting deposit.

Solution   We use a calculator to find 9% of 750. This is the amount by which the deposit has increased. Then

$$9 \boxed{\%} \times 750 = 67.50$$

The deposit has increased by £67.50. The resulting deposit is therefore $750 + 67.5 = £817.50$.

Alternatively we may perform the calculation as follows. The original deposit represents 100%. The deposit increases by 9% to 109% of the original. So the resulting deposit is 109% of £750:

$$109 \boxed{\%} \times 750 = £817.50$$

4.9   A television set is advertised at £315. The retailer offers a 10% discount. How much do you pay for the television?

Solution $10\% \text{ of } 315 = 31.50$

The discount is £31.50 and so the cost is $315 - 31.5 = £283.50$.

Alternatively we can note that since the discount is 10%, then the selling price is 90% of the advertised price:

$$90 \boxed{\%} \times 315 = 283.50$$

Performing the calculation in the two ways will increase your understanding of percentages and serve as a check.

When a quantity changes, it is sometimes useful to calculate the **percentage change**. For example, suppose a worker earns £14,500 in the current year, and last year earned £13,650. The actual amount earned has changed by 14,500 − 13,650 = £850. The percentage change is calculated from the formula:

Key point

$$\text{percentage change} = \frac{\text{change}}{\text{original value}} \times 100 = \frac{\text{new value} - \text{original value}}{\text{original value}} \times 100$$

If the change is positive, then there has been an increase in the measured quantity. If the change is negative, then there has been a decrease in the quantity.

WORKED EXAMPLES

4.10 A worker's earnings increase from £13,650 to £14,500. Calculate the percentage change.

Solution

$$\text{percentage change} = \frac{\text{new value} - \text{original value}}{\text{original value}} \times 100$$

$$= \frac{14,500 - 13,650}{13,650} \times 100$$

$$= 6.23$$

The worker's earnings increased by 6.23%.

4.11 A microwave oven is reduced in price from £149.95 to £135. Calculate the percentage change in price.

Solution

$$\text{percentage change} = \frac{\text{new value} - \text{original value}}{\text{original value}} \times 100$$

$$= \frac{135 - 149.95}{149.95} \times 100$$

$$= -9.97$$

The negative result is indicative of the price decrease. The percentage change in price is approximately –10%.

Self-assessment question 4.1

1. Give one reason why it is sometimes useful to express fractions as percentages.

Exercise 4.1

1. Calculate 23% of 124.

2. Express the following as percentages:

    (a) $\dfrac{9}{11}$ (b) $\dfrac{15}{20}$ (c) $\dfrac{9}{10}$ (d) $\dfrac{45}{50}$ (e) $\dfrac{75}{90}$

3. Express $\frac{13}{12}$ as a percentage.

4. Calculate 217% of 500.

5. A worker earns £400 a week. She receives a 6% increase. Calculate her new weekly wage.

6. A debt of £1200 is decreased by 17%. Calculate the remaining debt.

7. Express the following percentages as decimals:

    (a) 50% (b) 36% (c) 75%
    (d) 100% (e) 12.5%

8. A compact disc player normally priced at £256 is reduced in a sale by 20%. Calculate the sale price.

9. A bank deposit earns 7.5% interest in one year. Calculate the interest earned on a deposit of £15,000.

10. The cost of a car is increased from £6950 to £7495. Calculate the percentage change in price.

11. During a sale, a washing machine is reduced in price from £525 to £399. Calculate the percentage change in price.

4.2 Ratio

Ratios are simply an alternative way of expressing fractions. Consider the problem of dividing £200 between two people, Ann and Bill, in the ratio of $7:3$. This means that Ann receives £7 for every £3 that Bill receives. So every £10 is divided as £7 to Ann and £3 to Bill. Thus Ann receives $\frac{7}{10}$ of the money. Now $\frac{7}{10}$ of £200 is $\frac{7}{10} \times 200 = 140$. So Ann receives £140 and Bill receives £60.

WORKED EXAMPLE

4.12   Divide 170 in the ratio $3:2$.

Solution   A ratio of $3 : 2$ means that every 5 parts are split as 3 and 2. That is, the first number is $\frac{3}{5}$ of the total; the second number is $\frac{2}{5}$ of the total. So

$$\frac{3}{5} \text{ of } 170 = \frac{3}{5} \times 170 = 102$$

$$\frac{2}{5} \text{ of } 170 = \frac{2}{5} \times 170 = 68$$

The number is divided into 102 and 68.

Note from Worked Example 4.12 that to split a number in a given ratio we first find the total number of parts. The total number of parts is found by adding the numbers in the ratio. For example, if the ratio is given as $m : n$, the total number of parts is $m + n$. Then these $m + n$ parts are split into two with the first number being $\frac{m}{m+n}$ of the total, and the second number being $\frac{n}{m+n}$ of the total. Compare this with Worked Example 4.12.

WORKED EXAMPLE

4.13   Divide 250 cm in the ratio $1 : 3 : 4$.

Solution   Every 8 cm is divided into 1 cm, 3 cm and 4 cm. Thus the first length is $\frac{1}{8}$ of the total, the second length is $\frac{3}{8}$ of the total, and the third length is $\frac{4}{8}$ of the total:

$$\frac{1}{8} \text{ of } 250 = \frac{1}{8} \times 250 = 31.25$$

$$\frac{3}{8} \text{ of } 250 = \frac{3}{8} \times 250 = 93.75$$

$$\frac{4}{8} \text{ of } 250 = \frac{4}{8} \times 250 = 125$$

The 250 cm length is divided into 31.25 cm, 93.75 cm and 125 cm.

Ratios can be written in different ways. The ratio $3 : 2$ can also be written as $6 : 4$. This is clear if we note that $6 : 4$ is a total of 10 parts split as $\frac{6}{10}$ and $\frac{4}{10}$ of the total. Since $\frac{6}{10}$ is equivalent to $\frac{3}{5}$, and $\frac{4}{10}$ is equivalent to $\frac{2}{5}$, we see that $6 : 4$ is equivalent to $3 : 2$.

Generally, any ratio can be expressed as an equivalent ratio by multiplying or dividing each term in the ratio by the same number. So,

for example,

$$5 : 3 \text{ is equivalent to } 15 : 9$$

and

$$\frac{3}{4} : 2 \text{ is equivalent to } 3 : 8$$

WORKED EXAMPLES

4.14 Divide a mass of 380 kg in the ratio $\frac{3}{4} : \frac{1}{5}$.

Solution It is simpler to work with whole numbers, so first of all we produce an equivalent ratio by multiplying each term, first by 4, and then by 5, to give

$$\frac{3}{4} : \frac{1}{5} = 3 : \frac{4}{5} = 15 : 4$$

Note that this is equivalent to multiplying through by the lowest common multiple of 4 and 5.

So dividing 380 kg in the ratio $\frac{3}{4} : \frac{1}{5}$ is equivalent to dividing it in the ratio $15 : 4$.

Now the total number of parts is 19 and so we split the 380 kg mass as

$$\frac{15}{19} \times 380 = 300$$

and

$$\frac{4}{19} \times 380 = 80$$

The total mass is split into 300 kg and 80 kg.

4.15 Bell metal, which is a form of bronze, is used for casting bells. It is an alloy of copper and tin. To manufacture bell metal requires 17 parts of copper to every 3 parts of tin.

(a) Express this requirement as a ratio.

(b) Express the amount of tin required as a percentage of the total.

(c) If the total amount of tin in a particular casting is 150 kg, find the amount of copper.

Solution (a) Copper and tin are needed in the ratio $17 : 3$.

(b) $\frac{3}{20}$ of the alloy is tin. Since $\frac{3}{20} = 15\%$ we find that 15% of the alloy is tin.

(c)   A mass of 150 kg of tin makes up 15% of the total. So 1% of the total would have a mass of 10 kg. Copper, which makes up 85%, will have a mass of 850 kg.

Self-assessment question 4.2

1. Dividing a number in the ratio $2:3$ is the same as dividing it in the ratio $10:15$. True or false?

Exercise 4.2

1. Divide 180 in the ratio $8:1:3$.

2. Divide 930 cm in the ratio $1:1:3$.

3. A 6 m length of wood is cut in the ratio $2:3:4$. Calculate the length of each piece.

4. Divide 1200 in the ratio $1:2:3:4$.

5. A sum of £2600 is divided between Alan, Bill and Claire in the ratio of $2\frac{3}{4}:1\frac{1}{2}:2\frac{1}{4}$. Calculate the amount that each receives.

6. A mass of 40 kg is divided into three portions in the ratio $3:4:8$. Calculate the mass of each portion.

7. Express the following ratios in their simplest forms:
   (a) $12:24$ (b) $3:6$ (c) $3:6:12$
   (d) $\frac{1}{3}:7$

8. A box contains two sizes of nails. The ratio of long nails to short nails is $2:7$. Calculate the number of each type if the total number of nails is 108.

Test and assignment exercises 4

1. Express as decimals
   (a) 8%   (b) 18%   (c) 65%

2. Express as percentages
   (a) $\dfrac{3}{8}$ (b) $\dfrac{79}{100}$ (c) $\dfrac{56}{118}$

3. Calculate 27.3% of 1496.

4. Calculate 125% of 125.

5. Calculate 85% of 0.25.

6. Divide 0.5 in the ratio $2:4:9$.

7. A bill totals £234.5 to which is added tax at 17.5%. Calculate the amount of tax to be paid.

8. An inheritance is divided between three people in the ratio $4:7:2$. If the least amount received is £2300 calculate how much the other two people received.

9. Divide 70 in the ratio of $0.5 : 1.3 : 2.1$.

10. Divide 50% in the ratio $2 : 3$.

11. The temperature of a liquid is reduced from 39 °C to 35 °C. Calculate the percentage change in temperature.

12. A jacket priced at £120 is reduced by 30% in a sale. Calculate the sale price of the jacket.

13. The price of a car is reduced from £7250 to £6450. Calculate the percentage change in price.

14. The population of a small town increases from 17296 to 19437 over a five-year period. Calculate the percentage change in population.

15. A number, X, is increased by 20% to form a new number Y. Y is then decreased by 20% to form a third number Z. Express Z in terms of X.

Algebra

5

Objectives

This chapter:

- explains what is meant by 'algebra'
- introduces important algebraic notations
- explains what is meant by a 'power' or 'index'
- illustrates how to evaluate an expression
- explains what is meant by a 'formula'

5.1 What is algebra?

In order to extend the techniques of arithmetic so that they can be more useful in applications we introduce letters or **symbols** to represent quantities of interest. For example, we may choose the capital letter I to stand for the *interest rate* in a business calculation, or the lower case letter t to stand for the *time* in a scientific calculation, and so on. The choice of which letter to use for which quantity is largely up to the user, although some conventions have been developed. Very often the letters x and y are used to stand for arbitrary quantities. **Algebra** is the body of mathematical knowledge that has been developed to manipulate symbols. Some symbols take fixed and unchanging values, and these are known as **constants**. For example, suppose we let the symbol b stand for the boiling point of water. This is fixed at $100\,°C$ and so b is a constant. Some symbols represent quantities that can vary, and these are called **variables**. For example, the velocity of a car might be represented by the symbol v, and might vary from 0 to 100 kilometres per hour.

Algebraic notation

In algebraic work particular attention must be paid to the type of symbol used, so that, for example, the symbol T is quite different from the symbol t.

Table 5.1
The Greek alphabet

A	α	alpha	I	ι	iota	P	ρ	rho
B	β	beta	K	κ	kappa	Σ	σ	sigma
Γ	γ	gamma	Λ	λ	lambda	T	τ	tau
Δ	δ	delta	M	μ	mu	Y	υ	upsilon
E	ε	epsilon	N	ν	nu	Φ	ϕ	phi
Z	ζ	zeta	Ξ	ξ	xi	X	χ	chi
H	η	eta	O	o	omicron	Ψ	ψ	psi
Θ	θ	theta	Π	π	pi	Ω	ω	omega

Your scientific calculator is pre-programmed with the value of π. Check that you can use it.

Usually the symbols chosen are letters from the English alphabet although we frequently meet Greek letters. You may already be aware that the Greek letter 'pi', which has the symbol π, is used in the formula for the area of a circle, and is equal to the constant $3.14159\ldots$. In many calculations π can be approximated by $\frac{22}{7}$. For reference the full Greek alphabet is given in Table 5.1.

Another important feature is the position of a symbol in relation to other symbols. As we shall see in this chapter, the quantities xy, x^y, y^x and x_y all can mean quite different things. When a symbol is placed to the right and slightly higher than another symbol it is referred to as a **superscript**. So the quantity x^y contains the superscript y. Likewise, if a symbol is placed to the right and slightly lower than another symbol it is called a **subscript**. The quantity x_1 contains the subscript 1.

The arithmetic of symbols

Addition $(+)$ If the letters x and y stand for two numbers, their **sum** is written as $x + y$. Note that $x + y$ is the same as $y + x$ just as $4 + 7$ is the same as $7 + 4$.

Subtraction $(-)$ The quantity $x - y$ is called the **difference** of x and y, and means the number y subtracted from the number x. Note that $x - y$ is not the same as $y - x$, in the same way that $5 - 3$ is different from $3 - 5$.

Multiplication $(\times)$ Five times the number x is written $5 \times x$, although when multiplying the $\times$ sign is sometimes replaced with '$\cdot$', or is even left out altogether. This means that $5 \times x$, $5 \cdot x$ and $5x$ all mean five times the number x. Similarly $x \times y$ can be written $x \cdot y$ or simply xy. When multiplying, the order of the symbols is not important, so that xy is the same as yx just as 5×4 is the same as 4×5. The quantity xy is also known as the **product** of x and y.

Division $(\div)$ $x \div y$ means the number x divided by the number y. This is also written x/y. Here the order is important and x/y is quite different from y/x. An expression involving one symbol divided by another is known as an **algebraic fraction**. The top line is called the **numerator** and the bottom line is called the **denominator**. The quantity x/y is known as the **quotient** of x and y.

A quantity made up of symbols together with $+$, $-$, $\times$ or $\div$ is called an **algebraic expression**. When evaluating an algebraic expression the BODMAS rule given in Chapter 1 applies. This rule reminds us of the correct order in which to evaluate an expression.

Self-assessment questions 5.1

1. Explain what you understand by the term 'algebra'.

2. If m and n are two numbers, explain what is meant by mn.

3. What is an algebraic fraction? Explain the meaning of the terms 'numerator' and 'denominator'.

4. What is the distinction between a superscript and a subscript?

5. What is the distinction between a variable and a constant?

5.2 Powers or indices

Frequently we shall need to multiply a number by itself several times, for example $3 \times 3 \times 3$, or $a \times a \times a \times a$.

To abbreviate such quantities a new notation is introduced. $a \times a \times a$ is written a^3, pronounced 'a cubed'. The superscript 3 is called a **power** or **index** and the letter a is called the **base**. Similarly $a \times a$ is written a^2, pronounced 'a squared' or 'a raised to the power 2'.

The calculator button x^y is used to find powers of numbers.

Most calculators have a button marked x^y, which can be used to evaluate expressions such as 2^8, 3^{11} and so on. Check to see whether your calculator can do these by verifying that $2^8 = 256$ and $3^{11} = 177147$. Note that the plural of index is **indices**.

As a^2 means $a \times a$, and a^3 means $a \times a \times a$, then we interpret a^1 as simply a. That is, any number raised to the power 1 is itself.

Key point   Any number raised to the power 1 is itself, that is $a^1 = a$.

WORKED EXAMPLES

5.1   In the expression 3^8 identify the index and the base.

Solution   In the expression 3^8, the index is 8 and the base is 3.

5.2   Explain what is meant by y^5.

Solution y^5 means $y \times y \times y \times y \times y$.

5.3   Explain what is meant by $x^2 y^3$.

Solution x^2 means $x \times x$; y^3 means $y \times y \times y$. Therefore $x^2 y^3$ means $x \times x \times y \times y \times y$.

5.4   Evaluate 2^3 and 3^4.

Solution 2^3 means $2 \times 2 \times 2$, that is 8. Similarly 3^4 means $3 \times 3 \times 3 \times 3$, that is 81.

5.5   Explain what is meant by 7^1.

Solution   Any number to the power 1 is itself, that is 7^1 is simply 7.

5.6   Evaluate 10^2 and 10^3.

Solution 10^2 means 10×10 or 100. Similarly 10^3 means $10 \times 10 \times 10$ or 1000.

5.7   Use indices to write the expression $a \times a \times b \times b \times b$ more compactly.

Solution $a \times a$ can be written a^2; $b \times b \times b$ can be written b^3. Therefore $a \times a \times b \times b \times b$ can be written as $a^2 \times b^3$ or simply $a^2 b^3$.

5.8   Write out fully $z^3 y^2$.

Solution $z^3 y^2$ means $z \times z \times z \times y \times y$. Note that we could also write this as $zzzyy$.

We now consider how to deal with expressions involving not only powers but other operations as well. Recall from §5.1 that the BODMAS rule tells us the order in which operations should be carried out, but the rule makes no reference to powers. In fact, powers should be given higher priority than any other operation and evaluated first. Consider the expression -4^2. Because the power must be evaluated first -4^2 is equal to -16. On the other hand $(-4)^2$ means $(-4) \times (-4)$ which is equal to $+16$.

WORKED EXAMPLES

5.9   Simplify (a) -5^2,   (b) $(-5)^2$.

Solution   (a) The power is evaluated first. Noting that $5^2 = 25$, we see that $-5^2 = -25$.

Recall that when a negative number is multiplied by another negative number the result is positive.

(b) $(-5)^2$ means $(-5) \times (-5) = +25$.

Note how the brackets can significantly change the meaning of an expression.

5.10   Explain the meanings of $-x^2$ and $(-x)^2$. Are these different?

| Solution | In the expression $-x^2$ it is the quantity x that is squared, so that $-x^2 = -(x \times x)$. On the other hand $(-x)^2$ means $(-x) \times (-x)$, which equals $+x^2$. The two expressions are not the same. |

Following the previous two examples we emphasise again the importance of the position of brackets in an expression.

Self-assessment questions 5.2

1. Explain the meaning of the terms 'power' and 'base'.

2. What is meant by an index?

3. Explain the distinction between $(xyz)^2$ and xyz^2.

4. Explain the distinction between $(-3)^4$ and -3^4.

Exercise 5.2

MyMathLab

1. Evaluate the following without using a calculator: 2^4, $(\frac{1}{2})^2$, 1^8, 3^5 and 0^3.

2. Evaluate 10^4, 10^5 and 10^6 without using a calculator.

3. Use a calculator to evaluate 11^4, 16^8, 39^4 and 1.5^7.

4. Write out fully (a) a^4b^2c and (b) xy^2z^4.

5. Write the following expressions compactly using indices:
   (a) $xxxyyx$ (b) $xxyyzzz$
   (c) $xyzxyz$ (d) $abccba$

6. Using a calculator, evaluate
   (a) 7^4 (b) 7^5 (c) $7^4 \times 7^5$ (d) 7^9
   (e) 8^3 (f) 8^7 (g) $8^3 \times 8^7$ (h) 8^{10}

Can you spot a rule for multiplying numbers with powers?

7. Without using a calculator, find $(-3)^3$, $(-2)^2$, $(-1)^7$ and $(-1)^4$.

8. Use a calculator to find $(-16.5)^3$, $(-18)^2$ and $(-0.5)^5$.

9. Without using a calculator find
   (a) $(-6)^2$ (b) $(-3)^2$ (c) $(-4)^3$
   (d) $(-2)^3$
   Carefully compare your answers with the results of finding -6^2, -3^2, -4^3 and -2^3.

5.3 Substitution and formulae

Substitution means replacing letters by actual numerical values.

WORKED EXAMPLES

5.11   Find the value of a^4 when $a = 3$.

Solution a^4 means $a \times a \times a \times a$. When we **substitute** the number 3 in place of the letter a we find 3^4 or $3 \times 3 \times 3 \times 3$, that is 81.

5.12   Find the value of $a + 7b + 3c$ when $a = 1$, $b = 2$ and $c = 3$.

Solution   Letting $b = 2$ we note that $7b = 14$. Letting $c = 3$ we note that $3c = 9$. Therefore, with $a = 1$,
$$a + 7b + 3c = 1 + 14 + 9 = 24$$

5.13   If $x = 4$, find the value of (a) $8x^3$ and (b) $(8x)^3$.

Solution   (a)   Substituting $x = 4$ into $8x^3$ we find $8 \times 4^3 = 8 \times 64 = 512$.

(b)   Substituting $x = 4$ into $(8x)^3$ we obtain $(32)^3 = 32768$. Note that the use of brackets makes a significant difference to the result.

5.14   Evaluate mk, mn and nk when $m = 5$, $n = -4$ and $k = 3$.

Solution $mk = 5 \times 3 = 15$. Similarly $mn = 5 \times (-4) = -20$ and $nk = (-4) \times 3 = -12$.

5.15   Find the value of $-7x$ when (a) $x = 2$ and (b) $x = -2$.

Solution   (a)   Substituting $x = 2$ into $-7x$ we find -7×2, which equals -14.

(b)   Substituting $x = -2$ into $-7x$ we find -7×-2, which equals 14.

5.16   Find the value of x^2 when $x = -3$.

Solution   Because x^2 means $x \times x$, its value when $x = -3$ is -3×-3, that is $+9$.

5.17   Find the value of $-x^2$ when $x = -3$.

Solution   Recall that a power is evaluated first. So $-x^2$ means $-(x \times x)$. When $x = -3$, this evaluates to $-(-3 \times -3) = -9$.

5.18   Find the value of $x^2 + 3x$ when (a) $x = 2$, (b) $x = -2$.

Solution   (a)   Letting $x = 2$ we find
$$x^2 + 3x = (2)^2 + 3(2) = 4 + 6 = 10$$

(b)   Letting $x = -2$ we find
$$x^2 + 3x = (-2)^2 + 3(-2) = 4 - 6 = -2$$

5.19   Find the value of $\frac{3x^2}{4} + 5x$ when $x = 2$.

Solution   Letting $x = 2$ we find

$$\frac{3x^2}{4} + 5x = \frac{3(2)^2}{4} + 5(2)$$

$$= \frac{12}{4} + 10$$

$$= 13$$

5.20   Find the value of $\frac{x^3}{4}$ when $x = 0.5$.

Solution   When $x = 0.5$ we find

$$\frac{x^3}{4} = \frac{0.5^3}{4} = 0.03125$$

A **formula** is used to relate two or more quantities. You may already be familiar with the common formula used to find the area of a rectangle:

area $=$ length $\times$ breadth

In symbols, writing A for area, l for length and b for breadth we have

$A = l \times b$ or simply $A = lb$

If we are now given particular numerical values for l and b we can use this formula to find A.

WORKED EXAMPLES

5.21   Use the formula $A = lb$ to find A when $l = 10$ and $b = 2.5$.

Solution   Substituting the values $l = 10$ and $b = 2.5$ into the formula $A = lb$ we find $A = 10 \times 2.5 = 25$.

5.22   The formula $V = IR$ is used by electrical engineers. Find the value of V when $I = 12$ and $R = 7$.

Solution   Substituting $I = 12$ and $R = 7$ in $V = IR$ we find $V = 12 \times 7 = 84$.

5.23   Use the formula $y = x^2 + 3x + 4$ to find y when $x = -2$.

Solution   Substituting $x = -2$ into the formula gives

$$y = (-2)^2 + 3(-2) + 4 = 4 - 6 + 4 = 2$$

Self-assessment question 5.3

1. What is the distinction between an algebraic expression and a formula?

Exercise 5.3

1. Evaluate $3x^2y$ when $x = 2$ and $y = 5$.

2. Evaluate $8x + 17y - 2z$ when $x = 6$, $y = 1$ and $z = -2$.

3. The area A of a circle is found from the formula $A = \pi r^2$, where r is the length of the radius. Taking π to be 3.142 find the areas of the circles whose radii, in centimetres, are (a) $r = 10$, (b) $r = 3$, (c) $r = 0.2$.

4. Evaluate $3x^2$ and $(3x)^2$ when $x = 4$.

5. Evaluate $5x^2$ and $(5x)^2$ when $x = -2$.

6. If $y = 4.85$ find
   (a) $7y$ (b) y^2 (c) $5y + 2.5$
   (d) $y^3 - y$

7. If $a = 12.8$, $b = 3.6$ and $c = 9.1$ find
   (a) $a + b + c$ (b) ab (c) bc (d) abc

8. If $C = \frac{5}{9}(F - 32)$, find C when $F = 100$.

9. Evaluate (a) x^2, (b) $-x^2$ and (c) $(-x)^2$, when $x = 7$.

10. Evaluate the following when $x = -2$:
    (a) x^2 (b) $(-x)^2$ (c) $-x^2$
    (d) $3x^2$ (e) $-3x^2$ (f) $(-3x)^2$

11. Evaluate the following when $x = -3$:
    (a) $\frac{x^2}{3}$ (b) $(-x)^2$ (c) $-\left(\frac{x}{3}\right)^2$
    (d) $4x^2$ (e) $-4x^2$ (f) $(-4x)^2$

12. Evaluate $x^2 - 7x + 2$ when $x = -9$.

13. Evaluate $2x^2 + 3x - 11$ when $x = -3$.

14. Evaluate $-x^2 + 3x - 5$ when $x = -1$.

15. Evaluate $-9x^2 + 2x$ when $x = 0$.

16. Evaluate $5x^2 + x + 1$ when (a) $x = 3$, (b) $x = -3$, (c) $x = 0$, (d) $x = -1$.

17. Evaluate $\frac{2x^2}{3} - \frac{x}{2}$ when
    (a) $x = 6$ (b) $x = -6$ (c) $x = 0$
    (d) $x = 1$

18. Evaluate $\frac{4x^2}{5} + 3$ when
    (a) $x = 0$ (b) $x = 1$ (c) $x = 5$
    (d) $x = -5$

19. Evaluate $\frac{x^3}{2}$ when
    (a) $x = -1$ (b) $x = 2$ (c) $x = 4$

20. Use the formula $y = \frac{x^3}{2} + 3x^2$ to find y when
    (a) $x = 0$ (b) $x = 2$ (c) $x = 3$
    (d) $x = -1$

21. If $g = 2t^2 - 1$, find g when
    (a) $t = 3$ (b) $t = 0.5$ (c) $t = -2$

22. In business calculations, the simple interest earned on an investment, I, is calculated from the formula $I = Prn$, where P is the amount invested, r is the interest rate and n is the number of time periods. Evaluate I when
    (a) $P = 15000$, $r = 0.08$ and $n = 5$
    (b) $P = 12500$, $r = 0.075$ and $n = 3$.

23. An investment earning 'compound interest' has a value, S, given by $S = P(1 + r)^n$, where P is the amount invested, r is the interest rate and n is the number of time periods. Calculate S when
    (a) $P = 8250$, $r = 0.05$ and $n = 15$
    (b) $P = 125000$, $r = 0.075$ and $n = 11$.

VIDEO

Test and assignment exercises 5

1. Using a calculator, evaluate 44^3, 0.44^2 and 32.5^3.

2. Write the following compactly using indices:

   (a) $xxxyyyy$ (b) $\dfrac{xxx}{yyyy}$ (c) a^2baab

3. Evaluate the expression $4x^3yz^2$ when $x = 2$, $y = 5$ and $z = 3$.

4. The circumference C of a circle that has a radius of length r is given by the formula $C = 2\pi r$. Find the circumference of the circle with radius 0.5 cm. Take $\pi = 3.142$.

5. Find (a) $21^2 - 16^2$, (b) $(21 - 16)^2$. Comment upon the result.

6. If $x = 4$ and $y = -3$, evaluate

   (a) xy (b) $\dfrac{x}{y}$ (c) $\dfrac{x^2}{y^2}$ (d) $\left(\dfrac{x}{y}\right)^2$

7. Evaluate $2x(x + 4)$ when $x = 7$.

8. Evaluate $4x^2 + 7x$ when $x = 9$.

9. Evaluate $3x^2 - 7x + 12$ when $x = -2$.

10. Evaluate $-x^2 - 11x + 1$ when $x = -3$.

11. The formula $I = V/R$ is used by engineers. Find I when $V = 10$ and $R = 0.01$.

12. Given the formula $A = 1/x$, find A when (a) $x = 1$, (b) $x = 2$, (c) $x = 3$.

13. From the formula $y = 1/(x^2 + x)$ find y when (a) $x = 1$, (b) $x = -1$, (c) $x = 3$.

14. Find the value of $(-1)^n$ (a) when n is an even natural number and (b) when n is an odd natural number. (A natural number is a positive whole number.)

15. Find the value of $(-1)^{n+1}$ (a) when n is an even natural number and (b) when n is an odd natural number.

Indices

Objectives: This chapter:

- states three laws used for manipulating indices
- shows how expressions involving indices can be simplified using the three laws
- explains the use of negative powers
- explains square roots, cube roots and fractional powers
- revises multiplication and division by powers of 10
- explains 'scientific notation' for representing very large and very small numbers

6.1 The laws of indices

Recall from Chapter 5 that an index is simply a power and that the plural of index is indices. Expressions involving indices can often be simplified if use is made of the **laws of indices**.

The first law

$$a^m \times a^n = a^{m+n}$$

In words, this states that if two numbers involving the same base but possibly different indices are to be multiplied together, their indices are added. Note that this law can be applied only if both bases are the same.

Key point   The first law: $a^m \times a^n = a^{m+n}$.

WORKED EXAMPLES

6.1   Use the first law of indices to simplify $a^4 \times a^3$.

Solution   Using the first law we have $a^4 \times a^3 = a^{4+3} = a^7$. Note that the same result could be obtained by actually writing out all the terms:

$$a^4 \times a^3 = (a \times a \times a \times a) \times (a \times a \times a) = a^7$$

6.2   Use the first law of indices to simplify $3^4 \times 3^5$.

Solution   From the first law $3^4 \times 3^5 = 3^{4+5} = 3^9$.

6.3   Simplify $a^4 a^7 b^2 b^4$.

Solution $a^4 a^7 b^2 b^4 = a^{4+7} b^{2+4} = a^{11} b^6$. Note that only those quantities with the same base can be combined using the first law.

The second law

$$\frac{a^m}{a^n} = a^{m-n}$$

In words, this states that if two numbers involving the same base but possibly different indices are to be divided, their indices are subtracted.

Key point	The second law: $\dfrac{a^m}{a^n} = a^{m-n}$.

WORKED EXAMPLES

6.4   Use the second law of indices to simplify $\frac{a^5}{a^3}$.

Solution   The second law states that we subtract the indices, that is

$$\frac{a^5}{a^3} = a^{5-3} = a^2$$

6.5   Use the second law of indices to simplify $\frac{3^7}{3^4}$.

Solution   From the second law, $\frac{3^7}{3^4} = 3^{7-4} = 3^3$.

6.6   Using the second law of indices, simplify $\frac{x^3}{x^3}$.

Solution   Using the second law of indices we have $\frac{x^3}{x^3} = x^{3-3} = x^0$. However, note that any expression divided by itself equals 1, and so $\frac{x^3}{x^3}$ must equal 1. We can conclude from this that any number raised to the power 0 equals 1.

Key point	Any number raised to the power 0 equals 1, that is $a^0 = 1$.

WORKED EXAMPLE

6.7   Evaluate (a) 14^0, (b) 0.5^0.

Solution   (a)   Any number to the power 0 equals 1 and so $14^0 = 1$.

(b)   Similarly, $0.5^0 = 1$.

The third law

$$(a^m)^n = a^{mn}$$

If a number is raised to a power, and the result is itself raised to a power, then the two powers are multiplied together.

Key point   The third law: $(a^m)^n = a^{mn}$.

WORKED EXAMPLES

6.8   Simplify $(3^2)^4$.

Solution   The third law states that the two powers are multiplied:

$$(3^2)^4 = 3^{2 \times 4} = 3^8$$

6.9   Simplify $(x^4)^3$.

Solution   Using the third law:

$$(x^4)^3 = x^{4 \times 3} = x^{12}$$

6.10   Remove the brackets from the expression $(2a^2)^3$.

Solution $(2a^2)^3$ means $(2a^2) \times (2a^2) \times (2a^2)$. We can write this as

$$2 \times 2 \times 2 \times a^2 \times a^2 \times a^2$$

or simply $8a^6$. We could obtain the same result by noting that both terms in the brackets, that is the 2 and the a^2, must be raised to the power 3, that is

$$(2a^2)^3 = 2^3(a^2)^3 = 8a^6$$

The result of the previous example can be generalised to any term of the form $(a^m b^n)^k$. To simplify such an expression we make use of the formula $(a^m b^n)^k = a^{mk} b^{nk}$.

Key point $(a^m b^n)^k = a^{mk} b^{nk}$

WORKED EXAMPLE

6.11   Remove the brackets from the expression $(x^2 y^3)^4$.

Solution   Using the previous result we find

$$(x^2 y^3)^4 = x^8 y^{12}$$

We often need to use several laws of indices in one example.

WORKED EXAMPLES

6.12 Simplify $\frac{(x^3)^4}{x^2}$.

Solution $(x^3)^4 = x^{12}$ using the third law of indices

So

$$\frac{(x^3)^4}{x^2} = \frac{x^{12}}{x^2} = x^{10}$$ using the second law

6.13 Simplify $(t^4)^2 (t^2)^3$.

Solution $(t^4)^2 = t^8,$ $(t^2)^3 = t^6$ using the third law

So

$$(t^4)^2 (t^2)^3 = t^8 t^6 = t^{14}$$ using the first law

Self-assessment questions 6.1

1. State the three laws of indices.
2. Explain what is meant by a^0.
3. Explain what is meant by x^1.

Exercise 6.1

1. Simplify
   (a) $5^7 \times 5^{13}$ (b) $9^8 \times 9^5$
   (c) $11^2 \times 11^3 \times 11^4$

2. Simplify
   (a) $\dfrac{15^3}{15^2}$ (b) $\dfrac{4^{18}}{4^9}$ (c) $\dfrac{5^{20}}{5^{19}}$

3. Simplify
   (a) $a^7 a^3$ (b) $a^4 a^5$ (c) $b^{11} b^{10} b$

4. Simplify
   (a) $x^7 \times x^8$ (b) $y^4 \times y^8 \times y^9$

5. Explain why the laws of indices cannot be used to simplify $19^8 \times 17^8$.

6. Simplify
    (a) $(7^3)^2$ (b) $(4^2)^8$ (c) $(7^9)^2$

7. Simplify $\dfrac{1}{(5^3)^8}$.

VIDEO

8. Simplify
    (a) $(x^2y^3)(x^3y^2)$ (b) $(a^2bc^2)(b^2ca)$

9. Remove the brackets from
    (a) $(x^2y^4)^5$ (b) $(9x^3)^2$ (c) $(-3x)^3$
    (d) $(-x^2y^3)^4$

10. Simplify
    (a) $\dfrac{(z^2)^3}{z^3}$ (b) $\dfrac{(y^3)^2}{(y^2)^2}$ (c) $\dfrac{(x^3)^2}{(x^2)^3}$

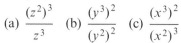
VIDEO

6.2 Negative powers

Sometimes a number is raised to a negative power. This is interpreted as follows:

$$a^{-m} = \frac{1}{a^m}$$

This can also be rearranged and expressed in the form

$$a^m = \frac{1}{a^{-m}}$$

Key point

$$a^{-m} = \frac{1}{a^m}, \qquad a^m = \frac{1}{a^{-m}}$$

For example,

$$3^{-2} \text{ means } \frac{1}{3^2}, \text{ that is } \frac{1}{9}$$

Similarly,

the number $\dfrac{1}{5^{-2}}$ can be written 5^2, or simply 25

To see the justification for this, note that because any number raised to the power 0 equals 1 we can write

$$\frac{1}{a^m} = \frac{a^0}{a^m}$$

Using the second law of indices to simplify the right-hand side we obtain $\frac{a^0}{a^m} = a^{0-m} = a^{-m}$ so that $\frac{1}{a^m}$ is the same as a^{-m}.

WORKED EXAMPLES

6.14   Evaluate

(a) 2^{-5} (b) $\dfrac{1}{3^{-4}}$

Solution   (a) $2^{-5} = \dfrac{1}{2^5} = \dfrac{1}{32}$ (b) $\dfrac{1}{3^{-4}} = 3^4$ or simply 81

6.15   Evaluate

(a) 10^{-1} (b) 10^{-2}

Solution   (a) 10^{-1} means $\frac{1}{10^1}$, or simply $\frac{1}{10}$. It is important to recognise that 10^{-1} is therefore the same as 0.1.

(b) 10^{-2} means $\frac{1}{10^2}$ or $\frac{1}{100}$. So 10^{-2} is therefore the same as 0.01.

6.16   Rewrite each of the following expressions using only positive powers:

(a) 7^{-3} (b) x^{-5}

Solution   (a) 7^{-3} means the same as $\frac{1}{7^3}$. The expression has now been written using a positive power.

(b) $x^{-5} = \frac{1}{x^5}$.

6.17   Rewrite each of the following expressions using only positive powers:

(a) $\dfrac{1}{x^{-9}}$ (b) $\dfrac{1}{a^{-4}}$

Solution   (a) $\dfrac{1}{x^{-9}} = x^9$ (b) $\dfrac{1}{a^{-4}} = a^4$

6.18   Rewrite each of the following using only negative powers:

(a) 6^8 (b) x^5 (c) z^a

Solution   (a) $6^8 = \dfrac{1}{6^{-8}}$ (b) $x^5 = \dfrac{1}{x^{-5}}$ (c) $z^a = \dfrac{1}{z^{-a}}$

6.19   Simplify

(a) $x^{-2}x^7$ (b) $\dfrac{x^{-3}}{x^{-5}}$

Solution   (a) To simplify $x^{-2}x^7$ we can use the first law of indices to write it as $x^{-2+7} = x^5$.

(b) To simplify $\dfrac{x^{-3}}{x^{-5}}$ we can use the second law of indices to write it as $x^{-3-(-5)} = x^{-3+5} = x^2$.

6.20   Simplify

(a) $(x^{-3})^5$ (b) $\dfrac{1}{(x^{-2})^2}$

Solution   (a) To simplify $(x^{-3})^5$ we can use the third law of indices and write it as $x^{-3 \times 5} = x^{-15}$. The answer could also be written as $\frac{1}{x^{15}}$.

(b) Note that $(x^{-2})^2 = x^{-4}$ using the third law. So $\dfrac{1}{(x^{-2})^2} = \dfrac{1}{x^{-4}}$. This could also be written as x^4.

Self-assessment question 6.2

1. Explain how the negative power in a^{-m} is interpreted.

Exercise 6.2

1. Without using a calculator express each of the following as a proper fraction:
    (a) 2^{-2} (b) 2^{-3} (c) 3^{-2} (d) 3^{-3}
    (e) 5^{-2} (f) 4^{-2} (g) 9^{-1} (h) 11^{-2}
    (i) 7^{-1}

2. Express each of the following as decimal fractions:
    (a) 10^{-1} (b) 10^{-2} (c) 10^{-6} (d) $\frac{1}{10^2}$
    (e) $\frac{1}{10^3}$ (f) $\frac{1}{10^4}$

3. Write each of the following using only a positive power:
    (a) x^{-4} (b) $\frac{1}{x^{-5}}$ (c) x^{-7} (d) y^{-2}
    (e) $\frac{1}{y^{-1}}$ (f) y^{-1} (g) y^{-2} (h) z^{-1}
    (i) $\frac{1}{z^{-1}}$

4. Simplify the following using the laws of indices and write your results using only positive powers:
    (a) $x^{-2}x^{-1}$ (b) $x^{-3}x^{-2}$ (c) x^3x^{-4}
    (d) $x^{-4}x^9$ (e) $\frac{x^{-2}}{x^{11}}$ (f) $(x^{-4})^2$
    (g) $(x^{-3})^3$ (h) $(x^2)^{-2}$

5. Simplify
    (a) $a^{13}a^{-2}$ (b) $x^{-9}x^{-7}$ (c) $x^{-21}x^2x$
    (d) $(4^{-3})^2$

6. Evaluate
    (a) 10^{-3} (b) 10^{-4} (c) 10^{-5}

7. Evaluate $4^{-8}/4^{-6}$ and $3^{-5}/3^{-8}$ without using a calculator.

6.3 Square roots, cube roots and fractional powers

Square roots

Consider the relationship between the numbers 5 and 25. We know that $5^2 = 25$ and so 25 is the square of 5. Equivalently we say that 5 is a **square root** of 25. The symbol $\sqrt[2]{}$, or simply $\sqrt{}$, is used to denote a square root and we write

$$5 = \sqrt{25}$$

We can picture this as follows:

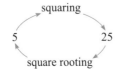

From this we see that taking the square root can be thought of as reversing the process of squaring.

We also note that

$$(-5) \times (-5) = (-5)^2$$
$$= 25$$

and so -5 is also a square root of 25. Hence we can write

$$-5 = \sqrt{25}$$

We can write both results together by using the 'plus or minus' sign $\pm$. We write

$$\sqrt{25} = \pm 5$$

In general, a **square root** of a number is a number that when squared gives the original number. Note that there are two square roots of any positive number but negative numbers possess no square roots.

Most calculators enable you to find square roots although only the positive value is normally given. Look for a $\sqrt{}$ or 'sqrt' button on your calculator.

WORKED EXAMPLE

6.21   (a)   Use your calculator to find $\sqrt{79}$ correct to 4 decimal places.

(b)   Check your answers are correct by squaring them.

Solution   (a)   Using the $\sqrt{}$ button on the calculator you should verify that

$$\sqrt{79} = 8.8882 \text{ (to 4 decimal places)}$$

The second square root is -8.8882. Thus we can write

$$\sqrt{79} = \pm 8.8882$$

(b)   Squaring either of the numbers ± 8.8882 we recover the original number, 79.

Cube roots

The **cube root** of a number is a number that when cubed gives the original number. The symbol for a cube root is $\sqrt[3]{}$. So, for example, since $2^3 = 8$ we can write $\sqrt[3]{8} = 2$.

We can picture this as follows:

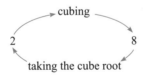

We can think of taking the cube root as reversing the process of cubing. As another example we note that $(-2)^3 = -8$ and hence $\sqrt[3]{-8} = -2$. All numbers, both positive and negative, possess a single cube root.

Your calculator may enable you to find a cube root. Look for a button marked $\sqrt[3]{}$. If so, check that you can use it correctly by verifying that

$$\sqrt[3]{46} = 3.5830$$

Fourth, fifth and other roots are defined in a similar way. For example, since

$$8^5 = 32768$$

we can write

$$\sqrt[5]{32768} = 8$$

Fractional powers

Sometimes fractional powers are used. The following example helps us to interpret a fractional power.

WORKED EXAMPLE

6.22   Simplify

(a) $x^{\frac{1}{2}}x^{\frac{1}{2}}$ (b) $x^{\frac{1}{3}}x^{\frac{1}{3}}x^{\frac{1}{3}}$

Use your results to interpret the fractional powers $\frac{1}{2}$ and $\frac{1}{3}$.

Solution   (a) Using the first law we can write

$$x^{\frac{1}{2}}x^{\frac{1}{2}} = x^{\frac{1}{2}+\frac{1}{2}} = x^1 = x$$

(b) Similarly,

$$x^{\frac{1}{3}}x^{\frac{1}{3}}x^{\frac{1}{3}} = x^{\frac{1}{3}+\frac{1}{3}+\frac{1}{3}} = x^1 = x$$

From (a) we see that

$$(x^{\frac{1}{2}})^2 = x$$

So when $x^{\frac{1}{2}}$ is squared, the result is x. Thus $x^{\frac{1}{2}}$ is simply the square root of x, that is

$$x^{\frac{1}{2}} = \sqrt{x}$$

Similarly, from (b)

$$(x^{\frac{1}{3}})^3 = x$$

and so $x^{\frac{1}{3}}$ is the cube root of x, that is

$$x^{\frac{1}{3}} = \sqrt[3]{x}$$

Key point	$x^{\frac{1}{2}} = \sqrt{x}, \qquad x^{\frac{1}{3}} = \sqrt[3]{x}$

More generally we have the following result:

Key point	$x^{\frac{1}{n}} = \sqrt[n]{x}$

Your scientific calculator will probably be able to find fractional powers. The button may be marked $x^{1/y}$ or $\sqrt[y]{x}$. Check that you can use it correctly by working through the following examples.

WORKED EXAMPLES

6.23 Evaluate to 3 decimal places, using a calculator:

(a) $3^{\frac{1}{4}}$ (b) $15^{1/5}$

Solution Use your calculator to obtain the following solutions:
(a) 1.316   (b) 1.719
Note in part (a) that although the calculator gives just a single fourth root, there is another, -1.316.

6.24 Evaluate $(81)^{1/2}$.

Solution $(81)^{1/2} = \sqrt{81} = \pm 9$.

6.25 Explain what is meant by the number $27^{1/3}$.

Solution $27^{1/3}$ can be written $\sqrt[3]{27}$, that is the cube root of 27. The cube root of 27 is 3, since $3 \times 3 \times 3 = 27$, and so $27^{1/3} = 3$. Note also that since $27 = 3^3$ we can write

$$(27)^{1/3} = (3^3)^{1/3} = 3^{(3 \times 1/3)} \quad \text{using the third law}$$
$$= 3^1 \quad = 3$$

The following worked example shows how we deal with negative fractional powers.

6.26 Explain what is meant by the number $(81)^{-1/2}$.

Solution Recall from our work on negative powers that $a^{-m} = 1/a^m$. Therefore we can write $(81)^{-1/2}$ as $1/(81)^{1/2}$. Now $81^{1/2} = \sqrt{81} = \pm 9$ and so

$$(81)^{-1/2} = \frac{1}{\pm 9} = \pm \frac{1}{9}$$

6.27   Write each of the following using a single index:

(a) $(5^2)^{\frac{1}{3}}$ (b) $(5^{-2})^{\frac{1}{3}}$

Solution   (a) Using the third law of indices we find

$$(5^2)^{\frac{1}{3}} = 5^{2 \times \frac{1}{3}} = 5^{\frac{2}{3}}$$

Note that $(5^2)^{\frac{1}{3}}$ is the cube root of 5^2, that is $\sqrt[3]{25}$ or 2.9240.

(b) Using the third law of indices we find

$$(5^{-2})^{\frac{1}{3}} = 5^{-2 \times \frac{1}{3}} = 5^{-\frac{2}{3}}$$

Note that there is a variety of equivalent ways in which this can be expressed, for example $\sqrt[3]{\frac{1}{5^2}}$ or $\sqrt[3]{\frac{1}{25}}$, or as $\frac{1}{5^{2/3}}$.

6.28   Write each of the following using a single index:

(a) $\sqrt{x^3}$ (b) $(\sqrt{x})^3$

Solution   (a) Because the square root of a number can be expressed as that number raised to the power $\frac{1}{2}$ we can write

$$\sqrt{x^3} = (x^3)^{\frac{1}{2}}$$

$$= x^{3 \times \frac{1}{2}} \qquad \text{using the third law}$$

$$= x^{\frac{3}{2}}$$

(b) $(\sqrt{x})^3 = (x^{\frac{1}{2}})^3$

$$= x^{\frac{3}{2}} \qquad \text{using the third law}$$

Note from this example that $\sqrt{x^3} = (\sqrt{x})^3$.

Note that by generalising the results of the two previous worked examples we have the following:

Key point $a^{\frac{m}{n}} = \sqrt[n]{a^m} = (\sqrt[n]{a})^m$

Self-assessment questions 6.3

1. Explain the meaning of the fractional powers $x^{1/2}$ and $x^{1/3}$.

2. What are the square roots of 100? Explain why the number -100 does not have any square roots.

Exercise 6.3

1. Evaluate
   (a) $64^{1/3}$ (b) $144^{1/2}$ (c) $16^{-1/4}$
   (d) $25^{-1/2}$ (e) $\dfrac{1}{32^{-1/5}}$

2. Simplify and then evaluate
   (a) $(3^{-1/2})^4$ (b) $(8^{1/3})^{-1}$

3. Write each of the following using a single index:
   (a) $\sqrt{8}$ (b) $\sqrt[3]{12}$ (c) $\sqrt[4]{16}$ (d) $\sqrt{13^3}$
   (e) $\sqrt[3]{4^7}$

4. Write each of the following using a single index:
   (a) $\sqrt{x}$ (b) $\sqrt[3]{y}$ (c) $\sqrt[2]{x^5}$ (d) $\sqrt[3]{5^7}$

6.4 Multiplication and division by powers of 10

To multiply and divide decimal fractions by powers of 10 is particularly simple. For example, to multiply 256.875 by 10 the decimal point is moved one place to the right, that is

$$256.875 \times 10 = 2568.75$$

To multiply by 100 the decimal point is moved two places to the right. So

$$256.875 \times 100 = 25687.5$$

To divide a number by 10, the decimal point is moved one place to the left. This is equivalent to multiplying by 10^{-1}. To divide by 100, the decimal point is moved two places to the left. This is equivalent to multiplying by 10^{-2}.

In general, to multiply a number by 10^n, the decimal point is moved n places to the right if n is a positive integer, and n places to the left if n is a negative integer. If necessary, additional zeros are inserted to make up the required number of digits. Consider the following example.

WORKED EXAMPLE

6.29 Without the use of a calculator, write down

(a) 75.45×10^3 (b) 0.056×10^{-2} (c) 96.3×10^{-3} (d) 0.00743×10^5

Solution

(a) The decimal point is moved three places to the right:
$75.45 \times 10^3 = 75450$. It has been necessary to include an additional zero to make up the required number of digits.

(b) The decimal point is moved two places to the left:
$0.056 \times 10^{-2} = 0.00056$.

(c) $96.3 \times 10^{-3} = 0.0963$.

(d) $0.00743 \times 10^5 = 743$.

Exercise 6.4

1. Without the use of a calculator write down:
   (a) 7.43×10^2 (b) 7.43×10^4 (c) 0.007×10^4 (d) 0.07×10^{-2}

2. Write each of the following as a multiple of 10^2:
   (a) 300 (b) 356 (c) 32 (d) 0.57

6.5 Scientific notation

It is often necessary to use very large numbers such as 65000000000 or very small numbers such as 0.000000001. **Scientific notation** can be used to express such numbers in a more concise form, which avoids writing very lengthy strings of numbers. Each number is written in the form

$a \times 10^n$

where a is usually a number between 1 and 10. We also make use of the fact that

$10 = 10^1, \qquad 100 = 10^2, \qquad 1000 = 10^3$ and so on

and also that

$$10^{-1} = \frac{1}{10} = 0.1, \qquad 10^{-2} = \frac{1}{100} = 0.01 \text{ and so on}$$

Then, for example,

the number 4000 can be written $4 \times 1000 = 4 \times 10^3$

Similarly

the number 68000 can be written $6.8 \times 10000 = 6.8 \times 10^4$

and

the number 0.09 can be written $9 \times 0.01 = 9 \times 10^{-2}$

Note that all three numbers have been written in the form $a \times 10^n$ where a lies between 1 and 10.

WORKED EXAMPLES

6.30   Express the following numbers in scientific notation:

(a) 54 (b) -276 (c) 0.3

Solution   (a) 54 can be written as 5.4×10, so in scientific notation we have 5.4×10^1.

(b) Negative numbers cause no problem: $-276 = -2.76 \times 10^2$.

(c) We can write 0.3 as 3×0.1 or 3×10^{-1}.

6.31   Write out fully the following numbers:

(a) 2.7×10^{-1} (b) 9.6×10^5 (c) -8.2×10^2

Solution   (a) $2.7 \times 10^{-1} = 0.27$.

(b) $9.6 \times 10^5 = 9.6 \times 100000 = 960000$.

(c) $-8.2 \times 10^2 = -8.2 \times 100 = -820$.

6.32   Simplify the expression $(3 \times 10^2) \times (5 \times 10^3)$.

Solution   The order in which the numbers are written down does not matter, and so we can write

$$(3 \times 10^2) \times (5 \times 10^3) = 3 \times 5 \times 10^2 \times 10^3 = 15 \times 10^5$$

Noting that $15 = 1.5 \times 10$ we can express the final answer in scientific notation:

$$15 \times 10^5 = 1.5 \times 10 \times 10^5 = 1.5 \times 10^6$$

Hence

$$(3 \times 10^2) \times (5 \times 10^3) = 1.5 \times 10^6$$

Self-assessment question 6.5

1. What is the purpose of using scientific notation?

Exercise 6.5

1. Express each of the following numbers in scientific notation:
    (a) 45   (b) 45000   (c) −450   (d) 90000000   (e) 0.15   (f) 0.00036   (g) 3.5
    (h) −13.2   (i) 1000000   (j) 0.0975   (k) 45.34

2. Write out fully the following numbers:
    (a) 3.75×10^2 (b) 3.97×10^1 (c) 1.875×10^{-1} (d) -8.75×10^{-3}

3. Simplify each of the following expressions, writing your final answer in scientific notation:
    (a) $(4 \times 10^3) \times (6 \times 10^4)$ (b) $(9.6 \times 10^4) \times (8.3 \times 10^3)$ (c) $(1.2 \times 10^{-3}) \times (8.7 \times 10^{-2})$
    (d) $\dfrac{9.37 \times 10^4}{6.14 \times 10^5}$ (e) $\dfrac{4.96 \times 10^{-2}}{9.37 \times 10^{-5}}$

1. Simplify

   (a) $\dfrac{z^5}{z^{-5}}$ (b) z^0 (c) $\dfrac{z^8 z^6}{z^{14}}$

2. Evaluate
   (a) $0.25^{1/2}$ (b) $(4096)^{1/3}$ (c) $(2601)^{1/2}$ (d) $16^{-1/2}$

3. Simplify $\dfrac{x^8 x^{-3}}{x^{-5} x^2}$.

4. Find the value of $(1/7)^0$.

5. Remove the brackets from
   (a) $(abc^2)^2$ (b) $(xy^2 z^3)^2$ (c) $(8x^2)^{-3}$

6. Express each of the following numbers in scientific notation:
   (a) 5792 (b) 98.4 (c) 0.001 (d) -66.667

Simplifying algebraic expressions

7

03

Objectives: This chapter:

- describes a number of ways in which complicated algebraic expressions can be simplified

7.1 Addition and subtraction of like terms

Like terms are multiples of the same quantity. For example, $3y$, $72y$ and $0.5y$ are all multiples of y and so are like terms. Similarly, $5x^2$, $-3x^2$ and $\frac{1}{2}x^2$ are all multiples of x^2 and so are like terms. xy, $17xy$ and $-91xy$ are all multiples of xy and are therefore like terms. Like terms can be collected together and added or subtracted in order to simplify them.

WORKED EXAMPLES

7.1 Simplify $3x + 7x - 2x$.

Solution All three terms are multiples of x and so are like terms. Therefore $3x + 7x - 2x = 8x$.

7.2 Simplify $3x + 2y$.

Solution $3x$ and $2y$ are not like terms. One is a multiple of x and the other is a multiple of y. The expression $3x + 2y$ cannot be simplified.

7.3 Simplify $x + 7x + x^2$.

Solution The like terms are x and $7x$. These can be simplified to $8x$. Then $x + 7x + x^2 = 8x + x^2$. Note that $8x$ and x^2 are not like terms and so this expression cannot be simplified further.

7.4 Simplify $ab + a^2 - 7b^2 + 9ab + 8b^2$.

Solution   The terms ab and $9ab$ are like terms. Similarly the terms $-7b^2$ and $8b^2$ are like terms. These can be collected together and then added or subtracted as appropriate. Thus

$$ab + a^2 - 7b^2 + 9ab + 8b^2 = ab + 9ab + a^2 - 7b^2 + 8b^2$$
$$= 10ab + a^2 + b^2$$

Exercise 7.1

1. Simplify, if possible,
    (a) $5p - 10p + 11q + 8q$ (b) $-7r - 13s + 2r + z$ (c) $181z + 13r - 2$
    (d) $x^2 + 3y^2 - 2y + 7x^2$ (e) $4x^2 - 3x + 2x + 9$

2. Simplify
    (a) $5y + 8p - 17y + 9q$ (b) $7x^2 - 11x^3 + 14x^2 + y^3$ (c) $4xy + 3xy + y^2$
    (d) $xy + yx$ (e) $xy - yx$

7.2 Multiplying algebraic expressions and removing brackets

Recall that when multiplying two numbers together the order in which we write them is irrelevant. For example, both 5×4 and 4×5 equal 20.

When multiplying three or more numbers together the order in which we carry out the multiplication is also irrelevant. By this we mean, for example, that when asked to multiply $3 \times 4 \times 5$ we can think of this as either $(3 \times 4) \times 5$ or as $3 \times (4 \times 5)$. Check for yourself that the result is the same, 60, either way.

It is also important to appreciate that $3 \times 4 \times 5$ could have been written as $(3)(4)(5)$.

It is essential that you grasp these simple facts about numbers in order to understand the algebra that follows. This is because identical rules are applied. Rules for determining the sign of the answer when multiplying positive and negative algebraic expressions are also the same as those used for multiplying numbers.

Key point   When multiplying

$$\text{positive} \times \text{positive} = \text{positive}$$
$$\text{positive} \times \text{negative} = \text{negative}$$
$$\text{negative} \times \text{positive} = \text{negative}$$
$$\text{negative} \times \text{negative} = \text{positive}$$

We introduce the processes involved in removing brackets using some simple examples.

WORKED EXAMPLES

7.5   Simplify $3(4x)$.

Solution   Just as with numbers $3(4x)$ could be written as $3 \times (4 \times x)$, and then as $(3 \times 4) \times x$, which evaluates to $12x$.
So $3(4x) = 12x$.

7.6   Simplify $5(3y)$.

Solution $5(3y) = 5 \times 3 \times y = 15y$.

7.7   Simplify $(5a)(3a)$.

Solution   Here we can write $(5a)(3a) = (5 \times a) \times (3 \times a)$. Neither the order in which we carry out the multiplications nor the order in which we write down the terms matters, and so we can write this as

$$(5a)(3a) = (5 \times 3)(a \times a)$$

As we have shown, it is usual to write numbers at the beginning of an expression. This simplifies to $15 \times a^2$, that is $15a^2$. Hence

$$(5a)(3a) = 15a^2$$

7.8   Simplify $4x^2 \times 7x^5$.

Solution   Recall that, when multiplying, the order in which we write down the terms does not matter. Therefore we can write

$$4x^2 \times 7x^5 = 4 \times 7 \times x^2 \times x^5$$

which equals $28x^{2+5} = 28x^7$.

7.9   Simplify $7(2b^2)$.

Solution $7(2b^2) = 7 \times (2 \times b^2) = (7 \times 2) \times b^2 = 14b^2$.

7.10   Simplify $(a) \times (-b)$

Solution   Here we have the product of a positive and a negative quantity. The result will be negative. We write

$$(a) \times (-b) = -ab$$

7.11   Explain the distinction between ab^2 and $(ab)^2$.

Solution ab^2 means $a \times b \times b$ whereas $(ab)^2$ means $(ab) \times (ab)$ which equals $a \times b \times a \times b$. The latter could also be written as a^2b^2.

7.12   Simplify (a) $(6z)(8z)$, (b) $(6z) + (8z)$, noting the distinction between the two results.

Solution   (a) $(6z)(8z) = 48z^2$.

(b) $(6z) + (8z)$ is the addition of like terms. This simplifies to $14z$.

7.13   Simplify (a) $(6x)(-2x)$, (b) $(-3y^2)(-2y)$.

Solution   (a) $(6x)(-2x)$ means $(6x) \times (-2x)$, which equals $-12x^2$.

(b) $(-3y^2)(-2y) = (-3y^2) \times (-2y) = 6y^3$.

Self-assessment questions 7.2

1. Two negative expressions are multiplied together. State the sign of the resulting product.

2. Three negative expressions are multiplied together. State the sign of the resulting product.

Exercise 7.2

1. Simplify each of the following:
    (a) $(4)(3)(7)$ (b) $(7)(4)(3)$ (c) $(3)(4)(7)$

2. Simplify
    (a) $5 \times (4 \times 2)$ (b) $(5 \times 4) \times 2$

3. Simplify each of the following:
    (a) $7(2z)$ (b) $15(2y)$ (c) $(2)(3)x$
    (d) $9(3a)$ (e) $(11)(5a)$ (f) $2(3x)$

4. Simplify each of the following:
    (a) $5(4x^2)$ (b) $3(2y^3)$ (c) $11(2u^2)$
    (d) $(2 \times 4) \times u^2$ (e) $(13)(2z^2)$

5. Simplify
    (a) $(7x)(3x)$ (b) $3a(7a)$ (c) $14a(a)$

6. Simplify
    (a) $5y(3y)$ (b) $5y + 3y$
    Explain why the two results are not the same.

7. Simplify the following:
    (a) $(abc)(a^2bc)$ (b) $x^2y(xy)$
    (c) $(xy^2)(xy^2)$

8. Explain the distinction, if any, between $(xy^2)(xy^2)$ and xy^2xy^2.

9. Explain the distinction, if any, between $(xy^2)(xy^2)$ and $(xy^2) + (xy^2)$.
    In both cases simplify the expressions.

10. Simplify
    (a) $(3z)(-7z)$ (b) $3z - 7z$

11. Simplify
    (a) $(-x)(3x)$ (b) $-x + 3x$

12. Simplify
    (a) $(-2x)(-x)$ (b) $-2x - x$

7.3   Removing brackets from $a(b+c)$, $a(b-c)$ and $(a+b)(c+d)$

Recall from your study of arithmetic that the expression $(5 - 4) + 7$ is different from $5 - (4 + 7)$ because of the position of the brackets. In order to simplify an expression it is often necessary to remove brackets.

Removing brackets from expressions of the form $a(b+c)$ and $a(b-c)$

In an expression such as $a(b+c)$, it is intended that the a multiplies all the bracketed terms:

Key point

$$a(b+c) = ab + ac \qquad \text{Similarly:} \quad a(b-c) = ab - ac$$

WORKED EXAMPLES

7.14   Remove the brackets from

(a) $6(x+5)$ (b) $8(2x-4)$

Solution   (a) In the expression $6(x+5)$ it is intended that the 6 multiplies both terms in the brackets. Therefore

$$6(x+5) = 6x + 30$$

(b) In the expression $8(2x-4)$ the 8 multiplies both terms in the brackets so that

$$8(2x-4) = 16x - 32$$

7.15   Remove the brackets from the expression $7(5x+3y)$.

Solution   The 7 multiplies both the terms in the bracket. Therefore

$$7(5x+3y) = 7(5x) + 7(3y) = 35x + 21y$$

7.16   Remove the brackets from $-(x+y)$.

Solution   The expression $-(x+y)$ actually means $-1(x+y)$. It is intended that the -1 multiplies both terms in the brackets, therefore

$$-(x+y) = -1(x+y) = (-1) \times x + (-1) \times y = -x - y$$

7.17   Remove the brackets from the expression

$$(x+y)z$$

Solution   Note that the order in which we write down the terms to be multiplied does not matter, so that we can write $(x+y)z$ as $z(x+y)$. Then

$$z(x+y) = zx + zy$$

Alternatively note that $(x+y)z = xz + yz$, which is an equivalent form of the answer.

7.18   Remove the brackets from the expressions

VIDEO

(a) $5(x-2y)$ (b) $(x+3)(-1)$

Solution   (a) $5(x-2y) = 5x - 5(2y) = 5x - 10y$.

(b) $(x+3)(-1) = (-1)(x+3) = -1x - 3 = -x - 3$.

7.19   Simplify $x + 8(x - y)$.

Solution   An expression such as this is simplified by first removing the brackets and then collecting together like terms. Removing the brackets we find

$$x + 8(x - y) = x + 8x - 8y$$

Collecting like terms we obtain $9x - 8y$.

7.20   Remove the brackets from

(a) $\frac{1}{2}(x + 2)$ (b) $\frac{1}{2}(x - 2)$ (c) $-\frac{1}{3}(a + b)$

Solution   (a) In the expression $\frac{1}{2}(x + 2)$ it is intended that the $\frac{1}{2}$ multiplies both the terms in the brackets. So

$$\frac{1}{2}(x + 2) = \frac{1}{2}x + \frac{1}{2}(2) = \frac{1}{2}x + 1$$

(b) Similarly,

$$\frac{1}{2}(x - 2) = \frac{1}{2}x - \frac{1}{2}(2) = \frac{1}{2}x - 1$$

(c) In the expression $-\frac{1}{3}(a + b)$ the term $-\frac{1}{3}$ multiplies both terms in the brackets. So

$$-\frac{1}{3}(a + b) = -\frac{1}{3}a - \frac{1}{3}b$$

Removing brackets from expressions of the form $(a + b)(c + d)$

In the expression $(a + b)(c + d)$ it is intended that the quantity $(a + b)$ multiplies both the c and the d in the second brackets. Therefore

$$(a + b)(c + d) = (a + b)c \quad + \quad (a + b)d$$

Each of these two terms can be expanded further to give

$$(a + b)c = ac + bc \qquad \text{and} \qquad (a + b)d = ad + bd$$

Therefore

Key point $$(a + b)(c + d) = ac + bc + ad + bd$$

WORKED EXAMPLES

7.21   Remove the brackets from $(3 + x)(2 + y)$.

Solution $(3 + x)(2 + y) = (3 + x)(2) + (3 + x)y$

$$= 6 + 2x + 3y + xy$$

7.22   Remove the brackets from $(x + 6)(x - 3)$.

Solution $(x + 6)(x - 3) = (x + 6)x + (x + 6)(-3)$

$$= x^2 + 6x - 3x - 18$$

$$= x^2 + 3x - 18$$

7.23   Remove the brackets from

(a) $(1 - x)(2 - x)$ (b) $(-x - 2)(2x - 1)$

Solution   (a) $(1 - x)(2 - x) = (1 - x)2 + (1 - x)(-x)$

$$= 2 - 2x - x + x^2$$

$$= 2 - 3x + x^2$$

(b) $(-x - 2)(2x - 1) = (-x - 2)(2x) + (-x - 2)(-1)$

$$= -2x^2 - 4x + x + 2$$

$$= -2x^2 - 3x + 2$$

7.24   Remove the brackets from the expression $3(x + 1)(x - 1)$.

VIDEO

Solution   First consider the expression $(x + 1)(x - 1)$:

$$(x + 1)(x - 1) = (x + 1)x + (x + 1)(-1)$$

$$= x^2 + x - x - 1$$

$$= x^2 - 1$$

Then $3(x + 1)(x - 1) = 3(x^2 - 1) = 3x^2 - 3$.

Exercise 7.3   MyMathLab

1. Remove the brackets from
    (a) $4(x + 1)$ (b) $-4(x + 1)$
    (c) $4(x - 1)$ (d) $-4(x - 1)$

2. Remove the brackets from the following
    expressions:
    (a) $5(x - y)$ (b) $19(x + 3y)$
    (c) $8(a + b)$ (d) $(5 + x)y$
    (e) $12(x + 4)$ (f) $17(x - 9)$
    (g) $-(a - 2b)$ (h) $\frac{1}{2}(2x + 1)$
    (i) $-3m(-2 + 4m + 3n)$

3. Remove the brackets and simplify the
    following:
    (a) $18 - 13(x + 2)$ (b) $x(x + y)$

4. Remove the brackets and simplify the
    following expressions:
    (a) $(x + 1)(x + 6)$ (b) $(x + 4)(x + 5)$
    (c) $(x - 2)(x + 3)$ (d) $(x + 6)(x - 1)$
    (e) $(x + y)(m + n)$ (f) $(4 + y)(3 + x)$
    (g) $(5 - x)(5 + x)$
    (h) $(17x + 2)(3x - 5)$

5. Remove the brackets and simplify the following expressions:
   (a) $(x+3)(x-7)$ (b) $(2x-1)(3x+7)$
   (c) $(4x+1)(4x-1)$
   (d) $(x+3)(x-3)$ (e) $(2-x)(3+2x)$

6. Remove the brackets and simplify the following expressions:
   (a) $\dfrac{1}{2}(x+2y)+\dfrac{7}{2}(4x-y)$

   (b) $\dfrac{3}{4}(x-1)+\dfrac{1}{4}(2x+8)$

7. Remove the brackets from
   (a) $-(x-y)$ (b) $-(a+2b)$
   (c) $-\dfrac{1}{2}(3p+q)$

8. Remove the brackets from $(x+1)(x+2)$. Use your result to remove the brackets from $(x+1)(x+2)(x+3)$.

Test and assignment exercises 7

1. Simplify
   (a) $7x^2+4x^2+9x-8x$ (b) $y+7-18y+1$ (c) $a^2+b^2+a^3-3b^2$

2. Simplify
   (a) $(3a^2b)\times(-a^3b^2c)$ (b) $\dfrac{x^3}{-x^2}$

3. Remove the brackets from
   (a) $(a+3b)(7a-2b)$ (b) $x^2(x+2y)$ (c) $x(x+y)(x-y)$

4. Remove the brackets from
   (a) $(7x+2)(3x-1)$ (b) $(1-x)(x+3)$ (c) $(5+x)x$ (d) $(8x+4)(7x-2)$

5. Remove the brackets and simplify
   (a) $3x(x+2)-7x^2$ (b) $-(2a+3b)(a+b)$ (c) $4(x+7)+13(x-2)$
   (d) $5(2a+5)-3(5a-2)$ (e) $\dfrac{1}{2}(a+4b)+\dfrac{3}{2}a$

Factorisation

8

Objectives: This chapter:

- explains what is meant by the 'factors' of an algebraic expression
- shows how an algebraic expression can be factorised
- shows how to factorise quadratic expressions

8.1 Factors and common factors

Recall from Chapter 1 that a number is **factorised** when it is written as a product. For example, 15 may be factorised into 3×5. We say that 3 and 5 are **factors** of 15. The number 16 can be written as 8×2 or 4×4, or even as 16×1, and so the factorisation may not be unique.

Algebraic expressions can also be factorised. Consider the expression $5x + 20y$. Both $5x$ and $20y$ have the number 5 common to both terms. We say that 5 is a **common factor**. Any common factors can be written outside a bracketed term. Thus $5x + 20y = 5(x + 4y)$. Removal of the brackets will result in the original expression and can always be used to check your answer. We see that factorisation can be thought of as reversing the process of removing brackets. Similarly, if we consider the expression $x^2 + 2x$, we note that both terms contain the factor x, and so $x^2 + 2x$ can be written as $x(x + 2)$. Hence x and $x + 2$ are both factors of $x^2 + 2x$.

WORKED EXAMPLES

8.1 Factorise $3x + 12$.

Solution The number 12 can be factorised as 3×4 so that 3 is a common factor of $3x$ and 12. We can write $3x + 12 = 3x + 3(4)$. Any common factors are written in front of the brackets and the contents of the brackets are adjusted accordingly. So

$$3x + 3(4) = 3(x + 4)$$

Note again that this answer can be checked by removing the brackets.

8.2   List the ways in which $15x^2$ can be written as a product of its factors.

Solution $15x^2$ can be written in many different ways. Some of these are $15x^2 \times 1$, $15x \times x$, $15 \times x^2$, $5x \times 3x$, $5 \times 3x^2$ and $3 \times 5x^2$.

8.3   Factorise $8x^2 - 12x$.

Solution   We can write $8x^2 - 12x = (4x)(2x) - (4x)3$ so that both terms contain the factor $4x$. This is placed at the front of the brackets to give

$$8x^2 - 12x = 4x(2x - 3)$$

8.4   What factors are common to the terms $5x^2$ and $15x^3$? Factorise $5x^2 + 15x^3$.

Solution   Both terms contain a factor of 5. Because x^3 can be written as $x^2 \times x$, both $5x^2$ and $15x^3$ contain a factor x^2. Therefore

$$5x^2 + 15x^3 = 5x^2 + (5x^2)(3x) = 5x^2(1 + 3x)$$

8.5   Factorise $6x + 3x^2 + 9xy$.

Solution   By careful inspection of all of the terms we see that $3x$ is a factor of each term. Hence

$$6x + 3x^2 + 9xy = 3x(2 + x + 3y)$$

Hence the factors of $6x + 3x^2 + 9xy$ are $3x$ and $2 + x + 3y$.

Self-assessment question 8.1

1.   Explain what is meant by 'factorising an expression'.

Exercise 8.1

1.   Remove the brackets from
      (a) $9(x + 3)$ (b) $-5(x - 2)$
      (c) $\dfrac{1}{2}(x + 1)$
      (d) $-(a - 3b)$ (e) $\dfrac{1}{2(x + y)}$
      (f) $\dfrac{x}{y(x - y)}$

2.   List all the factors of each of (a) $4x^2$, (b) $6x^3$.

3.   Factorise
      (a) $3x + 18$ (b) $3y - 9$ (c) $-3y - 9$
      (d) $-3 - 9y$ (e) $20 + 5t$ (f) $20 - 5t$
      (g) $-5t - 20$ (h) $3x + 12$ (i) $17t + 34$
      (j) $-36 + 4t$

4. Factorise
   (a) $x^4 + 2x$ (b) $x^4 - 2x$ (c) $3x^4 - 2x$
   (d) $3x^4 + 2x$ (e) $3x^4 + 2x^2$
   (f) $3x^4 + 2x^3$ (g) $17z - z^2$
   (h) $-xy + 3x$ (i) $-xy + 3y$
   (j) $x + 2xy + 3xyz$

5. Factorise
   (a) $10x + 20y$ (b) $12a + 3b$
   (c) $4x - 6xy$ (d) $7a + 14$

   (e) $10m - 15$ (f) $\dfrac{1}{5a + 35b}$

   (g) $\dfrac{1}{5a^2 + 35ab}$

6. Factorise
   (a) $15x^2 + 3x$ (b) $4x^2 - 3x$
   (c) $4x^2 - 8x$ (d) $15 - 3x^2$
   (e) $10x^3 + 5x^2 + 15x^2y$
   (f) $6a^2b - 12ab^2$
   (g) $16abc - 8ab^2 + 24bc$

8.2 Factorising quadratic expressions

Expressions of the form $ax^2 + bx + c$, where a, b and c are numbers, are called **quadratic expressions**. The numbers b or c may equal zero but a must not be zero. The number a is called the **coefficient** of x^2, b is the coefficient of x, and c is called the **constant term**.

We see that

$$2x^2 + 3x - 1, \quad x^2 + 3x + 2, \quad x^2 + 7 \quad \text{and} \quad 2x^2 - x$$

are all quadratic expressions.

Key point

An expression of the form $ax^2 + bx + c$, where a, b and c are numbers, is called a quadratic expression. The coefficient of x^2 is a, the coefficient of x is b, and the constant term is c.

To factorise such an expression means to express it as a product of two terms. For example, removing the brackets from $(x + 6)(x - 3)$ gives $x^2 + 3x - 18$ (see Worked Example 7.22). Reversing the process, $x^2 + 3x - 18$ can be factorised to $(x + 6)(x - 3)$. Not all quadratic expressions can be factorised in this way. We shall now explore how such factorisation is attempted.

Quadratic expressions where the coefficient of x^2 is 1

Consider the expression $(x + m)(x + n)$. Removing the brackets we find

$$(x + m)(x + n) = (x + m)x + (x + m)n$$

$$= x^2 + mx + nx + mn$$

$$= x^2 + (m + n)x + mn$$

Note that the coefficient of the x term is the sum $m + n$ and the constant term is the product mn. Using this information several quadratic expressions can be factorised by careful inspection. For example, suppose we wish to factorise $x^2 + 5x + 6$. We know that $x^2 + (m + n)x + mn$ can be factorised to $(x + m)(x + n)$. We seek values of m and n so that

$$x^2 + 5x + 6 = x^2 + (m + n)x + mn$$

Comparing the coefficients of x on both sides we require

$$5 = m + n$$

Comparing the constant terms on both sides we require

$$6 = mn$$

By inspection we see that $m = 3$ and $n = 2$ have this property and so

$$x^2 + 5x + 6 = (x + 3)(x + 2)$$

Note that the answer can be easily checked by removing the brackets again.

WORKED EXAMPLES

8.6   Factorise the quadratic expression $x^2 + 8x + 12$.

VIDEO

Solution   The factorisation of $x^2 + 8x + 12$ will be of the form $(x + m)(x + n)$. This means that mn must equal 12 and $m + n$ must equal 8. The two numbers must therefore be 2 and 6. So

$$x^2 + 8x + 12 = (x + 2)(x + 6)$$

Note again that the answer can be checked by removing the brackets.

8.7   Factorise $x^2 + 10x + 25$.

Solution   We try to factorise in the form $(x + m)(x + n)$. We require $m + n$ to equal 10 and mn to equal 25. If $m = 5$ and $n = 5$ this requirement is met. Therefore $x^2 + 10x + 25 = (x + 5)(x + 5)$. It is usual practice to write this as $(x + 5)^2$.

8.8   Factorise $x^2 - 121$.

Solution   In this example the x term is missing. We still attempt to factorise as $(x + m)(x + n)$. We require $m + n$ to equal 0 and mn to equal -121. Some thought shows that if $m = 11$ and $n = -11$ this requirement is met. Therefore $x^2 - 121 = (x + 11)(x - 11)$.

8.9   Factorise $x^2 - 5x + 6$.

Solution   We try to factorise in the form $(x + m)(x + n)$. We require $m + n$ to equal -5 and mn to equal 6. By inspection we see that if $m = -3$ and $n = -2$ this requirement is met. Therefore $x^2 - 5x + 6 = (x - 3)(x - 2)$.

Quadratic expressions where the coefficient of x^2 is not 1

These expressions are a little harder to factorise. All possible factors of the first and last terms must be found, and various combinations of these should be attempted until the required answer is found. This involves trial and error along with educated guesswork and practice.

WORKED EXAMPLES

8.10   Factorise, if possible, the expression $2x^2 + 11x + 12$.

Solution   The factors of the first term, $2x^2$, are $2x$ and x. The factors of the last term, 12, are

$$12, 1 \quad -12, -1 \quad 6, 2 \quad -6, -2 \quad \text{and} \quad 4, 3 \quad -4, -3$$

We can try each of these combinations in turn to find which gives us a coefficient of x of 11. For example, removing the brackets from

$$(2x + 12)(x + 1)$$

gives

$$(2x + 12)(x + 1) = (2x + 12)x + (2x + 12)(1)$$
$$= 2x^2 + 12x + 2x + 12$$
$$= 2x^2 + 14x + 12$$

which has an incorrect middle term. By trying further combinations it turns out that the only one producing a middle term of $11x$ is $(2x + 3)(x + 4)$ because

$$(2x + 3)(x + 4) = (2x + 3)(x) + (2x + 3)(4)$$
$$= 2x^2 + 3x + 8x + 12$$
$$= 2x^2 + 11x + 12$$

so that $(2x + 3)(x + 4)$ is the correct factorisation.

8.11   Factorise, if possible, $4x^2 + 6x + 2$.

Solution   Before we try to factorise this quadratic expression notice that there is a factor of 2 in each term so that we can write it as $2(2x^2 + 3x + 1)$. Now consider the quadratic expression $2x^2 + 3x + 1$. The factors of the first term, $2x^2$, are $2x$ and x. The factors of the last term, 1, are simply 1 and 1, or -1 and -1. We can try these combinations in turn to find which gives us a middle term of $3x$. Removing the brackets from $(2x + 1)(x + 1)$ gives $2x^2 + 3x + 1$, which has the correct middle term. Finally, we can write

$$4x^2 + 6x + 2 = 2(2x^2 + 3x + 1) = 2(2x + 1)(x + 1)$$

8.12   Factorise $6x^2 + 7x - 3$.

VIDEO

Solution   The first term may be factorised as $6x \times x$ and also as $3x \times 2x$. The factors of the last term are

$$3, -1 \quad\quad \text{and} \quad\quad -3, 1$$

We need to try each combination in turn to find which gives us a coefficient of x of 7. For example, removing the brackets from

$$(6x + 3)(x - 1)$$

gives

$$(6x + 3)(x - 1) = (6x + 3)x + (6x + 3)(-1)$$
$$= 6x^2 + 3x - 6x - 3$$
$$= 6x^2 - 3x - 3$$

which has an incorrect middle term. By trying further combinations it turns out that the only one producing a middle term of $7x$ is $(3x - 1)(2x + 3)$ because

$$(3x - 1)(2x + 3) = (3x - 1)2x + (3x - 1)(3)$$
$$= 6x^2 - 2x + 9x - 3$$
$$= 6x^2 + 7x - 3$$

The correct factorisation is therefore $(3x - 1)(2x + 3)$.

Until you have sufficient experience at factorising quadratic expressions you must be prepared to go through the process of trying all possible combinations until the correct answer is found.

Self-assessment question 8.2

1. Not all quadratic expressions can be factorised. Try to find an example of one such expression.

Exercise 8.2

MyMathLab Global

1. Factorise the following quadratic expressions:
   (a) $x^2 + 3x + 2$ (b) $x^2 + 13x + 42$
   (c) $x^2 + 2x - 15$ (d) $x^2 + 9x - 10$
   (e) $x^2 - 11x + 24$ (f) $x^2 - 100$
   (g) $x^2 + 4x + 4$ (h) $x^2 - 36$
   (i) $x^2 - 25$ (j) $x^2 + 10x + 9$
   (k) $x^2 + 8x - 9$ (l) $x^2 - 8x - 9$
   (m) $x^2 - 10x + 9$ (n) $x^2 - 5x$

2. Factorise the following quadratic expressions:
   (a) $2x^2 - 5x - 3$ (b) $3x^2 - 5x - 2$
   (c) $10x^2 + 11x + 3$ (d) $2x^2 + 12x + 16$
   (e) $2x^2 + 5x + 3$ (f) $3s^2 + 5s + 2$
   (g) $3z^2 + 17z + 10$ (h) $9x^2 - 36$
   (i) $4x^2 - 25$

3. (a) By removing the brackets show that

$$(x+y)(x-y) = x^2 - y^2$$

This result is known as the **difference of two squares**.

(b) Using the result in part (a) write down the factorisation of

(i) $16x^2 - 1$ (ii) $16x^2 - 9$

(iii) $25t^2 - 16r^2$

4. Factorise the following quadratic expressions:

(a) $x^2 + 3x - 10$ (b) $2x^2 - 3x - 20$
(c) $9x^2 - 1$ (d) $10x^2 + 14x - 12$
(e) $x^2 + 15x + 26$ (f) $-x^2 - 2x + 3$

5. Factorise
(a) $100 - 49x^2$ (b) $36x^2 - 25y^2$
(c) $\frac{1}{4} - 9v^2$ (d) $\frac{x^2}{y^2} - 4$

Test and assignment exercises 8

1. Factorise the following expressions:
(a) $7x + 49$ (b) $121x + 22y$ (c) $a^2 + ab$ (d) $ab + b^2$ (e) $ab^2 + ba^2$

2. Factorise the following quadratic expressions:
(a) $3x^2 + x - 2$ (b) $x^2 - 144$ (c) $s^2 - 5s + 6$ (d) $2y^2 - y - 15$

3. Factorise the following:
(a) $1 - x^2$ (b) $x^2 - 1$ (c) $9 - x^2$ (d) $x^2 - 81$ (e) $25 - y^2$

4. Factorise the denominators of the following expressions:

(a) $\dfrac{1}{x^2 + 6x}$ (b) $\dfrac{3}{s^2 + 3s + 2}$ (c) $\dfrac{3}{s^2 + s - 2}$ (d) $\dfrac{5}{x^2 + 11x + 28}$ (e) $\dfrac{x}{2x^2 - 17x - 9}$

Algebraic fractions

9

Objectives : This chapter:

- explains how to simplify algebraic fractions by cancelling common factors
- explains how algebraic fractions can be multiplied and divided
- explains how algebraic fractions can be added and subtracted
- explains how to express a fraction as the sum of its partial fractions

9.1 Introduction

Just as one whole number divided by another is a numerical fraction, so one algebraic expression divided by another is called an **algebraic fraction**.

$$\frac{x}{y} \qquad \frac{x^2 + y}{x} \qquad \frac{3x + 2}{7}$$

are all examples of algebraic fractions. The top line is known as the **numerator** of the fraction, and the bottom line is the **denominator**.

Rules for determining the sign of the answer when dividing positive and negative algebraic expressions are the same as those used for dividing numbers.

Key point

When dividing

$$\frac{\text{positive}}{\text{positive}} = \text{positive} \qquad\qquad \frac{\text{negative}}{\text{positive}} = \text{negative}$$

$$\frac{\text{positive}}{\text{negative}} = \text{negative} \qquad\qquad \frac{\text{negative}}{\text{negative}} = \text{positive}$$

Using these rules we see that an algebraic expression can often be written in different but equivalent forms. For example, note that

$$\frac{x}{-y} \text{ can be written as } -\frac{x}{y}$$

and that

$$\frac{-x}{y} \text{ can be written as } -\frac{x}{y}$$

and also that

$$\frac{-x}{-y} \text{ can be written as } \frac{x}{y}$$

9.2 Cancelling common factors

Cancellation of common factors was described in detail in §2.2.

Consider the numerical fraction $\frac{3}{12}$. To simplify this we factorise both the numerator and the denominator. Any factors which appear in both the numerator and the denominator are called **common factors**. These can be cancelled. For example,

$$\frac{3}{12} = \frac{1 \times 3}{4 \times 3} = \frac{1 \times \cancel{3}}{4 \times \cancel{3}} = \frac{1}{4}$$

The same process is applied when dealing with algebraic fractions.

WORKED EXAMPLES

9.1 For each pair of expressions, state which factors are common to both.

(a) $3xy$ and $6xz$ (b) xy and $5y^2$ (c) $3(x+2)$ and $(x+2)^2$
(d) $3(x-1)$ and $(x-1)(x+4)$

Solution (a) The expression $6xz$ can be written $(3)(2)xz$. We see that factors common to both this and $3xy$ are 3 and x.

(b) The expression $5y^2$ can be written $5(y)(y)$. We see that the only factor common to both this and xy is y.

(c) $(x+2)^2$ can be written $(x+2)(x+2)$. Thus $(x+2)$ is a factor common to both $(x+2)^2$ and $3(x+2)$.

(d) $3(x-1)$ and $(x-1)(x+4)$ have a common factor of $(x-1)$.

9.2   Simplify

$$\frac{18x^2}{6x}$$

Solution   First note that 18 can be factorised as 6×3. So there are factors of 6 and x in both the numerator and the denominator. Then common factors can be cancelled. That is,

$$\frac{18x^2}{6x} = \frac{(6)(3)x^2}{6x} = \frac{3x}{1} = 3x$$

Key point   When simplifying an algebraic fraction only factors common to both the numerator and denominator can be cancelled.

A fraction is expressed in its simplest form by factorising the numerator and denominator and cancelling any common factors.

WORKED EXAMPLES

9.3   Simplify

$$\frac{5}{25 + 15x}$$

Solution   First of all note that the denominator can be factorised as $5(5 + 3x)$. There is therefore a factor of 5 in both the numerator and denominator. So 5 is a common factor. This can be cancelled. That is,

$$\frac{5}{25 + 15x} = \frac{1 \times 5}{5(5 + 3x)} = \frac{1 \times \cancel{5}}{\cancel{5}(5 + 3x)} = \frac{1}{5 + 3x}$$

It is very important to note that the number 5 that has been cancelled is a common factor. It is incorrect to try to cancel terms that are not common factors.

9.4   Simplify

VIDEO

$$\frac{5x}{25x + 10y}$$

Solution   Factorising the denominator we can write

$$\frac{5x}{25x + 10y} = \frac{5x}{5(5x + 2y)}$$

We see that there is a common factor of 5 in both numerator and denominator that can be cancelled. Thus

$$\frac{5x}{25x + 10y} = \frac{\cancel{5}x}{\cancel{5}(5x + 2y)} = \frac{x}{5x + 2y}$$

Note that no further cancellation is possible. x is not a common factor because it is not a factor of the denominator.

9.5   Simplify

$$\frac{4x}{3x^2 + x}$$

Solution   Note that the denominator factorises to $x(3x + 1)$. Once both numerator and denominator have been factorised, any common factors are cancelled. So

$$\frac{4x}{3x^2 + x} = \frac{4x}{x(3x + 1)} = \frac{4\cancel{x}}{\cancel{x}(3x + 1)} = \frac{4}{3x + 1}$$

Note that the factor x is common to both numerator and denominator and so has been cancelled.

9.6   Simplify

$$\frac{x}{x^2 + 2x}$$

Solution   Note that the denominator factorises to $x(x + 2)$. Also note that the numerator can be written as $1 \times x$. So

$$\frac{x}{x^2 + 2x} = \frac{1 \times x}{x(x + 2)} = \frac{1 \times \cancel{x}}{\cancel{x}(x + 2)} = \frac{1}{x + 2}$$

9.7   Simplify

(a) $\dfrac{2(x - 1)}{(x + 3)(x - 1)}$ (b) $\dfrac{x - 4}{(x - 4)^2}$

Solution   (a) There is a factor of $(x - 1)$ common to both the numerator and denominator. This is cancelled to give

$$\frac{2(x - 1)}{(x + 3)(x - 1)} = \frac{2}{x + 3}$$

(b) There is a factor of $x - 4$ in both numerator and denominator. This is cancelled as follows:

$$\frac{x - 4}{(x - 4)^2} = \frac{1(x - 4)}{(x - 4)(x - 4)} = \frac{1}{x - 4}$$

9.8   Simplify

VIDEO

$$\frac{x+2}{x^2+3x+2}$$

Solution   The denominator is factorised and then any common factors are cancelled:

$$\frac{x+2}{x^2+3x+2} = \frac{1(x+2)}{(x+2)(x+1)} = \frac{1}{x+1}$$

9.9   Simplify

(a) $\dfrac{3x+xy}{x^2+5x}$ (b) $\dfrac{x^2-1}{x^2+3x+2}$

Solution   The numerator and denominator are both factorised and any common factors are cancelled:

(a) $\dfrac{3x+xy}{x^2+5x} = \dfrac{x(3+y)}{x^2+5x} = \dfrac{\cancel{x}(3+y)}{\cancel{x}(x+5)} = \dfrac{3+y}{5+x}$

(b) $\dfrac{x^2-1}{x^2+3x+2} = \dfrac{(x+1)(x-1)}{(x+1)(x+2)} = \dfrac{x-1}{x+2}$

Self-assessment questions 9.2

1. Explain why no cancellation is possible in the expression $\dfrac{3x}{3x+y}$.

2. Explain why no cancellation is possible in the expression $\dfrac{x+1}{x+3}$.

3. Explain why it is possible to perform a cancellation in the expression $\dfrac{x+1}{2x+2}$, and perform it.

Exercise 9.2   *MyMathLab* Global

1. Simplify

    (a) $\dfrac{9x}{3y}$ (b) $\dfrac{9x}{x^2}$ (c) $\dfrac{9xy}{3x}$ (d) $\dfrac{9xy}{3y}$

    (e) $\dfrac{9xy}{xy}$ (f) $\dfrac{9xy}{3xy}$

2. Simplify

    (a) $\dfrac{15x}{3y}$ (b) $\dfrac{15x}{5y}$ (c) $\dfrac{15xy}{x}$ (d) $\dfrac{15xy}{xy}$

    (e) $\dfrac{x^5}{-x^3}$ (f) $\dfrac{-y^3}{y^7}$ (g) $\dfrac{-y}{-y^2}$ (h) $\dfrac{-y^{-3}}{-y^4}$

3. Simplify the following algebraic fractions:

    (a) $\dfrac{4}{12+8x}$ (b) $\dfrac{5+10x}{5}$ (c) $\dfrac{2}{4+14x}$

    (d) $\dfrac{2x}{4+14x}$ (e) $\dfrac{2x}{2+14x}$ (f) $\dfrac{7}{49x+7y}$

    (g) $\dfrac{7y}{49x+7y}$ (h) $\dfrac{7x}{49x+7y}$

4. Simplify

(a) $\dfrac{15x + 3}{3}$ (b) $\dfrac{15x + 3}{3x + 6y}$ (c) $\dfrac{12}{4x + 8}$

(d) $\dfrac{12x}{4xy + 8x}$ (e) $\dfrac{13x}{x^2 + 5x}$

(f) $\dfrac{17y}{9y^2 + 4y}$

5. Simplify the following:

(a) $\dfrac{5}{15 + 10x}$ (b) $\dfrac{2x}{x^2 + 7x}$

(c) $\dfrac{2x + 8}{x^2 + 2x - 8}$ (d) $\dfrac{7ab}{a^2b^2 + 9ab}$

(e) $\dfrac{xy}{xy + x}$

6. Simplify

(a) $\dfrac{x - 4}{(x - 4)(x - 2)}$ (b) $\dfrac{2x - 4}{x^2 + x - 6}$

(c) $\dfrac{3x}{3x^2 + 6x}$ (d) $\dfrac{x^2 + 2x + 1}{x^2 - 2x - 3}$

(e) $\dfrac{2(x - 3)}{(x - 3)^2}$ (f) $\dfrac{x - 3}{(x - 3)^2}$

(g) $\dfrac{x - 3}{2(x - 3)^2}$ (h) $\dfrac{4(x - 3)}{2(x - 3)^2}$

(i) $\dfrac{x + 4}{2(x + 4)^2}$ (j) $\dfrac{x + 4}{2(x + 4)}$

(k) $\dfrac{2(x + 4)}{(x + 4)}$ (l) $\dfrac{(x + 4)(x - 3)}{x - 3}$

(m) $\dfrac{x + 4}{(x - 3)(x + 4)}$ (n) $\dfrac{x + 3}{x^2 + 7x + 12}$

(o) $\dfrac{x + 4}{2x + 8}$ (p) $\dfrac{x + 4}{2x + 9}$

9.3 Multiplication and division of algebraic fractions

To multiply two algebraic fractions together we multiply their numerators together and multiply their denominators together:

$$\frac{a}{b} \times \frac{c}{d} = \frac{a \times c}{b \times d}$$

Any common factors in the result should be cancelled.

WORKED EXAMPLES

9.10 Simplify

$$\frac{4}{5} \times \frac{x}{y}$$

Solution   We multiply the numerators together and multiply the denominators together. That is,

$$\frac{4}{5} \times \frac{x}{y} = \frac{4x}{5y}$$

9.11   Simplify

$$\frac{4}{x} \times \frac{3y}{16}$$

Solution   The numerators are multiplied together and the denominators are multiplied together. Therefore

$$\frac{4}{x} \times \frac{3y}{16} = \frac{4 \times 3y}{16x}$$

Because $16x = 4 \times 4x$, the common factor 4 can be cancelled. So

$$\frac{4 \times 3y}{16x} = \frac{\cancel{4} \times 3y}{\cancel{4} \times 4x} = \frac{3y}{4x}$$

9.12   Simplify

(a) $\dfrac{1}{2} \times x$ (b) $\dfrac{1}{2} \times (a + b)$

Solution   (a) Writing x as $\dfrac{x}{1}$ we can state

$$\frac{1}{2} \times x = \frac{1}{2} \times \frac{x}{1} = \frac{1 \times x}{2 \times 1} = \frac{x}{2}$$

(b) Writing $a + b$ as $\dfrac{a + b}{1}$ we can state

$$\frac{1}{2} \times (a + b) = \frac{1}{2} \times \frac{(a + b)}{1} = \frac{1 \times (a + b)}{2 \times 1} = \frac{a + b}{2}$$

9.13   Simplify

$$\frac{4x^2}{y} \times \frac{3x^3}{yz}$$

Solution   We multiply the numerators together and multiply the denominators together:

$$\frac{4x^2}{y} \times \frac{3x^3}{yz} = \frac{4x^2 \times 3x^3}{y \times yz} = \frac{12x^5}{y^2 z}$$

9.14   Simplify

$$5 \times \left(\frac{x-3}{25} \right)$$

Solution   This means

$$\frac{5}{1} \times \frac{x-3}{25}$$

which equals

$$\frac{5 \times (x-3)}{1 \times 25}$$

A common factor of 5 can be cancelled from the numerator and denominator to give

$$\frac{(x-3)}{5}$$

9.15   Simplify

$$-\frac{1}{5} \times \frac{3x-4}{8}$$

Solution   We can write

$$-\frac{1}{5} \times \frac{3x-4}{8} = -\frac{1 \times (3x-4)}{5 \times 8} = -\frac{3x-4}{40}$$

Note that the answer can also be expressed as $\frac{(4-3x)}{40}$ because

$$-\frac{3x-4}{40} = \frac{-1}{1} \times \frac{3x-4}{40} = \frac{-3x+4}{40} = \frac{4-3x}{40}$$

You should be aware from the last worked example that a solution can often be expressed in a number of equivalent ways.

WORKED EXAMPLES

9.16   Simplify

$$\frac{a}{a+b} \times \frac{b}{5a^2}$$

Solution

$$\frac{a}{a+b} \times \frac{b}{5a^2} = \frac{ab}{5a^2(a+b)}$$

Cancelling the common factor of a in numerator and denominator gives

$$\frac{b}{5a(a+b)}$$

9.17 Simplify

VIDEO

$$\frac{x^2+4x+3}{2x+8} \times \frac{x+4}{x+1}$$

Solution   Before multiplying the two fractions together we should try to factorise if possible so that common factors can be identified. By factorising, we can write the given expressions as

$$\frac{(x+1)(x+3)}{2(x+4)} \times \frac{x+4}{x+1} = \frac{(x+1)(x+3)(x+4)}{2(x+4)(x+1)}$$

Cancelling common factors this simplifies to just

$$\frac{x+3}{2}$$

Division is performed by inverting the second fraction and multiplying:

Key point

$$\frac{a}{b} \div \frac{c}{d} = \frac{a}{b} \times \frac{d}{c}$$

WORKED EXAMPLES

9.18 Simplify

$$\frac{10a}{b} \div \frac{a^2}{3b}$$

Solution   The second fraction is inverted and then multiplied by the first. That is,

$$\frac{10a}{b} \div \frac{a^2}{3b} = \frac{10a}{b} \times \frac{3b}{a^2} = \frac{30ab}{a^2b} = \frac{30}{a}$$

9.19 Simplify

$$\frac{x^2y^3}{z} \div \frac{y}{x}$$

Solution

$$\frac{x^2 y^3}{z} \div \frac{y}{x} = \frac{x^2 y^3}{z} \times \frac{x}{y} = \frac{x^3 y^3}{zy}$$

Any common factors in the result can be cancelled. So

$$\frac{x^3 y^3}{zy} = \frac{x^3 y^2}{z}$$

Self-assessment question 9.3

1. The technique of multiplying and dividing algebraic fractions is identical to that used for numbers. True or false?

Exercise 9.3

VIDEO

1. Simplify

   (a) $\dfrac{1}{2} \times \dfrac{y}{3}$ (b) $\dfrac{1}{3} \times \dfrac{z}{2}$ (c) $\dfrac{2}{5}$ of $\dfrac{1}{y}$

   (d) $\dfrac{2}{5}$ of $\dfrac{1}{x}$ (e) $\dfrac{3}{4}$ of $\dfrac{x}{y}$ (f) $\dfrac{3}{5} \times \dfrac{x^2}{y}$

   (g) $\dfrac{x}{y} \times \dfrac{3}{5}$ (h) $\dfrac{7}{8} \times \dfrac{x}{2y}$ (i) $\dfrac{1}{2} \times \dfrac{1}{2x}$

   (j) $\dfrac{1}{2} \times \dfrac{x}{2}$ (k) $\dfrac{1}{2} \times \dfrac{2}{x}$ (l) $\dfrac{1}{3} \times \dfrac{x}{3}$

   (m) $\dfrac{1}{3} \times \dfrac{3}{x}$ (n) $\dfrac{1}{3} \times \dfrac{1}{3x}$ (o) $\dfrac{1}{3} \times \dfrac{3x}{2}$

2. Simplify

   (a) $\dfrac{1}{2} \div \dfrac{x}{2}$ (b) $\dfrac{1}{2} \div \dfrac{2}{x}$ (c) $\dfrac{2}{x} \div \dfrac{2}{x}$

   (d) $\dfrac{x}{2} \div \dfrac{1}{2}$ (e) $\dfrac{2}{x} \div 2$ (f) $\dfrac{2}{x} \div \dfrac{1}{2}$

   (g) $\dfrac{3}{x} \div \dfrac{1}{2}$

3. Simplify the following:

   (a) $\dfrac{5}{4} \times \dfrac{a}{25}$ (b) $\dfrac{5}{4} \times \dfrac{a}{b}$ (c) $\dfrac{8a}{b^2} \times \dfrac{b}{16a^2}$

   (d) $\dfrac{9x}{3y} \times \dfrac{2x}{y^2}$ (e) $\dfrac{3}{5a} \times \dfrac{b}{a}$ (f) $\dfrac{1}{4} \times \dfrac{x}{y}$

   (g) $\dfrac{1}{3} \times \dfrac{x}{x+y}$ (h) $\dfrac{x-3}{x+4} \times \dfrac{1}{3x-9}$

4. Simplify the following:

   (a) $\dfrac{3}{x} \times \dfrac{xy}{z^3}$ (b) $\dfrac{(3+x)}{x} \div \dfrac{y}{x}$ (c) $\dfrac{4}{3} \div \dfrac{16}{x}$

   (d) $\dfrac{a}{bc^2} \times \dfrac{b^2 c}{a}$

5. Simplify

   (a) $\dfrac{x+2}{(x+5)(x+4)} \times \dfrac{x+5}{x+2}$

   (b) $\dfrac{x-2}{4} \div \dfrac{x}{16}$ (c) $\dfrac{12ab}{5ef} \div \dfrac{4ab^2}{f}$

   (d) $\dfrac{x+3y}{2x} \div \dfrac{y}{4x^2}$ (e) $\dfrac{3}{x} \times \dfrac{3}{y} \times \dfrac{1}{z}$

6. Simplify

   $$\dfrac{1}{x+1} \times \dfrac{2x+2}{x+3}$$

7. Simplify

   $$\dfrac{x+1}{x+2} \times \dfrac{x^2+6x+8}{x^2+4x+3}$$

9.4 Addition and subtraction of algebraic fractions

For revision of adding and subtracting fractions see §2.3.

The method is the same as that for adding or subtracting numerical fractions. Note that it is not correct simply to add or subtract the numerator and denominator. The lowest common denominator must first be found. This is the simplest expression that contains all original denominators as its factors. Each fraction is then written with this common denominator. The fractions can then be added or subtracted by adding or subtracting just the numerators, and dividing the result by the common denominator.

WORKED EXAMPLES

9.20 Add the fractions $\frac{3}{4}$ and $\frac{1}{x}$.

Solution We must find $\frac{3}{4} + \frac{1}{x}$. To do this we must first rewrite the fractions to ensure they have a common denominator. The common denominator is the simplest expression that has the given denominators as its factors. The simplest such expression is $4x$. We write

$$\frac{3}{4} \text{ as } \frac{3x}{4x} \qquad \text{and} \qquad \frac{1}{x} \text{ as } \frac{4}{4x}$$

Then

$$\frac{3}{4} + \frac{1}{x} = \frac{3x}{4x} + \frac{4}{4x}$$

$$= \frac{3x + 4}{4x}$$

No further simplification is possible.

9.21 Simplify

$$\frac{3}{x} + \frac{4}{x^2}$$

Solution The expression $\frac{3}{x}$ is rewritten as $\frac{3x}{x^2}$, which makes the denominators of both terms x^2 but leaves the value of the expression unaltered. Note that both the original denominators, x and x^2, are factors of the new denominator. We call this denominator the **lowest common denominator**. The fractions are then added by adding just the numerators. That is,

$$\frac{3x}{x^2} + \frac{4}{x^2} = \frac{3x + 4}{x^2}$$

9.22 Express $\dfrac{5}{a} - \dfrac{4}{b}$ as a single fraction.

Solution Both fractions are rewritten to have the same denominator. The simplest expression containing both a and b as its factors is ab. Therefore ab is the lowest common denominator. Then

$$\frac{5}{a} - \frac{4}{b} = \frac{5b}{ab} - \frac{4a}{ab} = \frac{5b - 4a}{ab}$$

9.23 Write $\dfrac{4}{x + y} - \dfrac{3}{y}$ as a single fraction.

Solution The simplest expression that contains both denominators as its factors is $(x + y)y$. We must rewrite each term so that it has this denominator:

$$\frac{4}{x + y} = \frac{4}{x + y} \times \frac{y}{y} = \frac{4y}{(x + y)y}$$

Similarly,

$$\frac{3}{y} = \frac{3}{y} \times \frac{x + y}{x + y} = \frac{3(x + y)}{(x + y)y}$$

The fractions are then subtracted by subtracting just the numerators:

$$\frac{4}{x + y} - \frac{3}{y} = \frac{4y}{(x + y)y} - \frac{3(x + y)}{(x + y)y} = \frac{4y - 3(x + y)}{(x + y)y}$$

which simplifies to

$$\frac{y - 3x}{(x + y)y}$$

9.24 Express as a single fraction

VIDEO

$$\frac{2}{x + 3} + \frac{5}{x - 1}$$

Solution The simplest expression having both $x + 3$ and $x - 1$ as its factors is

$$(x + 3)(x - 1)$$

This is the lowest common denominator. Each term is rewritten so that it has this denominator. Thus

$$\frac{2}{x + 3} = \frac{2(x - 1)}{(x + 3)(x - 1)} \qquad \text{and} \qquad \frac{5}{x - 1} = \frac{5(x + 3)}{(x + 3)(x - 1)}$$

Then

$$\frac{2}{x+3}+\frac{5}{x-1}=\frac{2(x-1)}{(x+3)(x-1)}+\frac{5(x+3)}{(x+3)(x-1)}$$

$$=\frac{2(x-1)+5(x+3)}{(x+3)(x-1)}$$

which simplifies to

$$\frac{7x+13}{(x+3)(x-1)}$$

9.25   Express as a single fraction

VIDEO

$$\frac{1}{x-4}+\frac{1}{(x-4)^2}$$

Solution   The simplest expression having $x-4$ and $(x-4)^2$ as its factors is $(x-4)^2$. Both fractions are rewritten with this denominator:

$$\frac{1}{x-4}+\frac{1}{(x-4)^2}=\frac{(x-4)}{(x-4)^2}+\frac{1}{(x-4)^2}$$

$$=\frac{x-4+1}{(x-4)^2}$$

$$=\frac{x-3}{(x-4)^2}$$

Self-assessment question 9.4

1. Explain what is meant by the 'lowest common denominator' and how it is found.

Exercise 9.4

1. Express each of the following as a single fraction:

   (a) $\dfrac{z}{2}+\dfrac{z}{3}$ (b) $\dfrac{x}{3}+\dfrac{x}{4}$ (c) $\dfrac{y}{5}+\dfrac{y}{25}$

2. Express each of the following as a single fraction:

   (a) $\dfrac{1}{2}+\dfrac{1}{x}$ (b) $\dfrac{1}{2}+x$ (c) $\dfrac{1}{3}+y$

   (d) $\dfrac{1}{3}+\dfrac{1}{y}$ (e) $8+\dfrac{1}{y}$

3. Express each of the following as a single fraction:

   (a) $\dfrac{5}{x}-\dfrac{1}{2}$ (b) $\dfrac{5}{x}+2$ (c) $\dfrac{3}{x}-\dfrac{1}{3}$

   (d) $\dfrac{x}{3}-\dfrac{1}{2}$ (e) $\dfrac{3}{x}+\dfrac{1}{3}$

4. Express each of the following as a single fraction:

(a) $\dfrac{3}{x} + \dfrac{4}{y}$ (b) $\dfrac{3}{x^2} + \dfrac{4y}{x}$ (c) $\dfrac{4ab}{x} + \dfrac{3ab}{2y}$

(d) $\dfrac{4xy}{a} + \dfrac{3xy}{2b}$ (e) $\dfrac{3}{x} - \dfrac{6}{2x}$ (f) $\dfrac{3x}{2y} - \dfrac{7y}{4x}$

(g) $\dfrac{3}{x+y} - \dfrac{2}{y}$ (h) $\dfrac{1}{a+b} - \dfrac{1}{a-b}$

(i) $2x + \dfrac{1}{2x}$ (j) $2x - \dfrac{1}{2x}$

5. Express each of the following as a single fraction:

(a) $\dfrac{x}{y} + \dfrac{3x^2}{z}$ (b) $\dfrac{4}{a} + \dfrac{5}{b}$

(c) $\dfrac{6x}{y} - \dfrac{2y}{x}$ (d) $3x - \dfrac{3x+1}{4}$

(e) $\dfrac{5a}{12} + \dfrac{9a}{18}$ (f) $\dfrac{x-3}{4} + \dfrac{3}{5}$

6. Express each of the following as a single fraction:

(a) $\dfrac{1}{x+1} + \dfrac{1}{x+2}$ (b) $\dfrac{1}{x-1} + \dfrac{2}{x+3}$

(c) $\dfrac{3}{x+5} + \dfrac{1}{x+4}$ (d) $\dfrac{1}{x-2} + \dfrac{3}{x-4}$

(e) $\dfrac{3}{2x+1} + \dfrac{1}{x+1}$ (f) $\dfrac{3}{1-2x} + \dfrac{1}{x}$

(g) $\dfrac{3}{x+1} + \dfrac{4}{(x+1)^2}$

(h) $\dfrac{1}{x-1} + \dfrac{1}{(x-1)^2}$

9.5 Partial fractions

We have seen how to add and/or subtract algebraic fractions to yield a single fraction. Section 9.4 illustrates the process with some examples.

Sometimes we wish to use the reverse of this process. That is, starting with a single fraction we wish to express it as the sum of two or more simpler fractions. Each of these simpler fractions is known as a **partial fraction** because it is part of the original fraction.

Let us refer to Worked Example 9.24. From this example we can see that

$$\frac{2}{x+3} + \frac{5}{x-1}$$

can be expressed as the single fraction, $\frac{7x+13}{(x+3)(x-1)}$. Worked Example 9.26 illustrates the process of starting with $\frac{7x+13}{(x+3)(x-1)}$ and finding its partial fractions, $\frac{2}{x+3}$ and $\frac{5}{x-1}$.

WORKED EXAMPLES

9.26 Express $\frac{7x+13}{(x+3)(x-1)}$ as its partial fractions.

Solution   The denominator has two factors: $x + 3$ and $x - 1$. It is these factors that determine the form of the partial fractions. Each factor in the denominator produces a partial fraction – this is an important point.

The factor $x + 3$ produces a partial fraction of the form $\frac{A}{x+3}$ where A is a constant. Similarly the factor $x - 1$ produces a partial fraction of the form $\frac{B}{x-1}$ where B is a constant. Hence we have

$$\frac{7x + 13}{(x + 3)(x - 1)} = \frac{A}{x + 3} + \frac{B}{x - 1} \tag{9.1}$$

We now need to find the values of A and B. By multiplying both sides of (9.1) by $(x + 3)(x - 1)$ we have

$$\frac{7x + 3}{(x + 3)(x - 1)} \times (x + 3)(x - 1) = \frac{A}{x + 3} \times (x + 3)(x - 1)$$

$$+ \frac{B}{x - 1} \times (x + 3)(x - 1) \tag{9.2}$$

By cancelling the common factors in each term, (9.2) simplifies to

$$7x + 13 = A(x - 1) + B(x + 3) \tag{9.3}$$

Note that (9.3) is true for *all* values of x. To find the values of A and B we can substitute into (9.3) any value of x we choose. We choose values of x that are most helpful and convenient. Let us choose x to be -3. When $x = -3$ is substituted into (9.3) we obtain

$$7(-3) + 13 = A(-3 - 1) + B(-3 + 3)$$

from which

$$-8 = A(-4)$$
$$A = 2$$

As you can see, the value of $x = -3$ was chosen in order to simplify (9.3) in such a way that the value of A could then be found.

Now we return to (9.3) and let $x = 1$. Then (9.3) simplifies to

$$7(1) + 13 = A(1 - 1) + B(1 + 3)$$

from which $B = 5$. Clearly we chose $x = 1$ in order to simplify (9.3) by eliminating the A term, thus allowing B to be found. Putting $A = 2$, $B = 5$ into (9.1) we have the partial fractions

$$\frac{7x + 13}{(x + 3)(x - 1)} = \frac{2}{x + 3} + \frac{5}{x - 1}$$

9.27   Express $\frac{5x+28}{x^2+7x+10}$ as partial fractions.

VIDEO

Solution   The factors of the denominator must first be found, that is $x^2 + 7x + 10$ must be factorised:

$$x^2 + 7x + 10 = (x + 2)(x + 5)$$

As the denominator has two factors, then there are two partial fractions. The factor $x + 2$ leads to a partial fraction $\frac{A}{x+2}$ and the factor $x + 5$ leads to a partial fraction $\frac{B}{x+5}$, where A and B are constants whose values have yet to be found.

So

$$\frac{5x + 28}{x^2 + 7x + 10} = \frac{5x + 28}{(x + 2)(x + 5)} = \frac{A}{x + 2} + \frac{B}{x + 5} \tag{9.4}$$

To find the values of A and B we multiply both sides of (9.4) by $(x + 2)(x + 5)$. After cancelling common factors in each term we have

$$5x + 28 = A(x + 5) + B(x + 2) \tag{9.5}$$

We now select convenient values of x to simplify (9.5) so that A and B can be found. We see that choosing $x = -2$ will simplify (9.5) so that A can be determined. With $x = -2$, (9.5) becomes

$$5(-2) + 28 = A(-2 + 5) + B(-2 + 2)$$
$$18 = 3A$$
$$A = 6$$

Returning to (9.5), we let $x = -5$ so that B can be found:

$$5(-5) + 28 = A(-5 + 5) + B(-5 + 2)$$
$$3 = -3B$$
$$B = -1$$

Putting the values of A and B into (9.4) yields the partial fractions

$$\frac{5x + 28}{x^2 + 7x + 10} = \frac{6}{x + 2} - \frac{1}{x + 5}$$

If a denominator has a repeated factor then a slight variation of the previous method is employed.

WORKED EXAMPLE

9.28   Find the partial fractions of $\frac{6x-5}{4x^2-4x+1}$

VIDEO

Solution   As in the previous example the denominator must first be factorised:

$$4x^2 - 4x + 1 = (2x - 1)(2x - 1) = (2x - 1)^2$$

We note that although there are two factors, they are identical (that is, a repeated factor). As there are two factors, then two partial fractions are generated. The partial fractions in such a case are of the form

$$\frac{A}{2x - 1} + \frac{B}{(2x - 1)^2}$$

where A and B are constants. Note that the denominators $(2x - 1)$ and $(2x - 1)^2$ are used. So

$$\frac{6x - 5}{4x^2 - 4x + 1} = \frac{6x - 5}{(2x - 1)^2} = \frac{A}{2x - 1} + \frac{B}{(2x - 1)^2} \tag{9.6}$$

We need to find the values of the constants A and B. Multiplying both sides of (9.6) by $(2x - 1)^2$ and then cancelling any common factors yields:

$$6x - 5 = A(2x - 1) + B \tag{9.7}$$

The left-hand side and right-hand side of (9.7) are equal for all values of x. We choose convenient values of x to help us find the values of A and B. By substituting $x = \frac{1}{2}$ into (9.7) it simplifies to

$$-2 = A(0) + B$$

and so $B = -2$. The value of $x = \frac{1}{2}$ was chosen as the factor $(2x - 1)$ then evaluated to 0 and the A term on the right-hand side became 0, allowing B to be found directly. We now need to find the value of A. It is impossible to eliminate the B term from (9.7) by any choice of an x value. Hence we simply choose any value of x that simplifies (9.7) considerably. Let us for example substitute $x = 0$ into (9.7) to obtain

$$-5 = A(-1) + B$$
$$A = B + 5$$

As we have already found the value of B to be -2, this is then substituted into the above equation to yield $A = 3$.

Substituting the values of A and B into (9.6) produces the partial fractions

$$\frac{6x - 5}{4x^2 - 4x + 1} = \frac{3}{2x - 1} - \frac{2}{(2x - 1)^2}$$

From Worked Examples 9.26, 9.27 and 9.28 we note the following:

Key point

When calculating partial fractions:

- the denominator must be factorised

- each factor of the denominator generates a partial fraction

- for repeated factors, the partial fractions are generated by the factor and the square of the factor

There are more complicated examples of partial fractions that are not dealt with in this chapter. For example, some fractions have factors in the denominator that are quadratics that will not factorise. For a more in-depth study of partial fractions see *Mathematics for Engineers: A Modern Interactive Approach*, 3rd edition, by A. Croft and R. Davison (2008, Pearson Education).

Exercise 9.5

1. Find the partial fractions of the following:

(a) $\dfrac{7x + 18}{(x + 2)(x + 3)}$ (b) $\dfrac{2x - 7}{x^2 + 5x + 4}$ (c) $\dfrac{-9}{2x^2 + 15x + 18}$ (d) $\dfrac{5x - 11}{x^2 - 5x + 4}$ (e) $\dfrac{3x + 11}{2x^2 + 3x - 2}$

2. Find the partial fractions of

$$\dfrac{x + 21}{2(2x + 3)(3x - 2)}$$

3. Find the partial fractions of

(a) $\dfrac{x - 35}{x^2 - 25}$ (b) $\dfrac{x - 4}{x^2 - 6x + 9}$ (c) $\dfrac{5x + 4}{-x^2 - x + 2}$ (d) $\dfrac{12x - 5}{9x^2 - 6x + 1}$

Test and assignment exercises 9

1. Simplify

(a) $\dfrac{5a}{4a + 3ab}$ (b) $\dfrac{5a}{30a + 15b}$ (c) $\dfrac{5ab}{30a + 15b}$ (d) $\dfrac{5ab}{ab + 7ab}$ (e) $\dfrac{y}{13y + y^2}$ (f) $\dfrac{13y + y^2}{y}$

2. Simplify the following:

(a) $\dfrac{5a}{7} \times \dfrac{14b}{2}$ (b) $\dfrac{3}{x} + \dfrac{7}{3x}$ (c) $t - \dfrac{4 - t}{2}$ (d) $4(x + 3) - \dfrac{(4x - 5)}{3}$

(e) $\dfrac{7}{x} + \dfrac{3}{2x} + \dfrac{5}{3x}$ (f) $x + \dfrac{3x}{y}$ (g) $xy + \dfrac{1}{xy}$ (h) $\dfrac{1}{x + y} + \dfrac{2}{x - y}$

3. Simplify the following:

(a) $\dfrac{y}{9} + \dfrac{2y}{7}$ (b) $\dfrac{3}{x} - \dfrac{5}{3x} + \dfrac{4}{5x}$ (c) $\dfrac{3x}{2y} + \dfrac{5y}{6x}$ (d) $m + \dfrac{m + n}{2}$

(e) $m - \dfrac{m + n}{2}$ (f) $m - \dfrac{m - n}{2}$ (g) $\dfrac{3s - 5}{10} - \dfrac{2s - 3}{15}$

4. Simplify

$$\dfrac{x^2 - x}{x - 1}$$

5. Find the partial fractions of

(a) $\dfrac{4x + 13}{(x + 2)(x + 7)}$ (b) $\dfrac{x + 4}{(x + 1)(x + 2)}$ (c) $\dfrac{x - 14}{(x + 4)(x - 5)}$ (d) $\dfrac{-x}{(3x + 1)(2x + 1)}$

(e) $\dfrac{6x - 13}{(2x + 3)(3x - 1)}$

6. Calculate the partial fractions of

(a) $\dfrac{8x + 19}{x^2 + 5x + 6}$ (b) $\dfrac{3x + 11}{x^2 + 9x + 20}$ (c) $\dfrac{3x + 7}{x^2 + 4x + 4}$ (d) $\dfrac{x - 6}{x^2 - 6x + 9}$ (e) $\dfrac{8x - 7}{4x^2 - 4x + 1}$

Transposing formulae

10

Objectives: This chapter:

■ explains how formulae can be rearranged or transposed

10.1 Rearranging a formula

In the formula for the area of a circle, $A = \pi r^2$, we say that A is the **subject** of the formula. The subject appears by itself on one side, usually the left, of the formula, and nowhere else. If we are asked to **transpose** the formula for r, then we must rearrange the formula so that r becomes the subject. The rules for transposing formulae are quite simple. Essentially whatever you do to one side you must also do to the whole of the other side.

Key point

To transpose a formula you may:

■ add the same quantity to both sides
■ subtract the same quantity from both sides
■ multiply or divide both sides by the same quantity
■ perform operations on both sides, such as 'square both sides', 'square root both sides' etc.

WORKED EXAMPLES

10.1 The circumference C of a circle is given by the formula $C = 2\pi r$. Transpose this to make r the subject.

Solution   The intention is to obtain r by itself on the left-hand side. Starting with $C = 2\pi r$ we must try to isolate r. Dividing both sides by 2π we find

$$C = 2\pi r$$

$$\frac{C}{2\pi} = \frac{2\pi r}{2\pi}$$

$$\frac{C}{2\pi} = r \qquad \text{by cancelling the common factor } 2\pi \text{ on the right}$$

Finally we can write $r = \frac{C}{2\pi}$ and the formula has been transposed for r.

10.2   Transpose the formula $y = 3(x + 7)$ for x.

Solution   We must try to obtain x on its own on the left-hand side. This can be done in stages. Dividing both sides by 3 we find

$$\frac{y}{3} = \frac{3(x + 7)}{3} = x + 7 \qquad \text{by cancelling the common factor 3}$$

Subtracting 7 from both sides then gives

$$\frac{y}{3} - 7 = x + 7 - 7$$

so that

$$\frac{y}{3} - 7 = x$$

Equivalently we can write $x = \frac{y}{3} - 7$, and the formula has been transposed for x.

Alternatively, again starting from $y = 3(x + 7)$ we could proceed as follows. Removing the brackets we obtain $y = 3x + 21$. Then, subtracting 21 from both sides,

$$y - 21 = 3x + 21 - 21 = 3x$$

Dividing both sides by 3 will give x on its own:

$$\frac{y - 21}{3} = x \qquad \text{so that} \qquad x = \frac{y - 21}{3}$$

Noting that

$$\frac{y - 21}{3} = \frac{y}{3} - \frac{21}{3} = \frac{y}{3} - 7$$

we see that this answer is equivalent to the expression obtained previously.

10.3   Transpose the formula $y - z = 3(x + 2)$ for x.

Solution   Dividing both sides by 3 we find

$$\frac{y - z}{3} = x + 2$$

Subtracting 2 from both sides gives

$$\frac{y - z}{3} - 2 = x + 2 - 2$$

$$= x$$

so that

$$x = \frac{y - z}{3} - 2$$

10.4   Transpose $x + xy = 7$, (a) for x, (b) for y.

Solution   (a) In the expression $x + xy$, note that x is a common factor. We can then write the given formula as $x(1 + y) = 7$. Dividing both sides by $1 + y$ will isolate x. That is,

$$\frac{x(1 + y)}{1 + y} = \frac{7}{1 + y}$$

Cancelling the common factor $1 + y$ on the left-hand side gives

$$x = \frac{7}{1 + y}$$

and the formula has been transposed for x.

(b) To transpose $x + xy = 7$ for y we first subtract x from both sides to give $xy = 7 - x$. Finally, dividing both sides by x gives

$$\frac{xy}{x} = \frac{7 - x}{x}$$

Cancelling the common factor of x on the left-hand side gives $y = \dfrac{7 - x}{x}$, which is the required transposition.

10.5   Transpose the formula $I = \frac{V}{R}$ for R.

Solution   Starting with $I = \frac{V}{R}$ we first multiply both sides by R. This gives

$$IR = \frac{V}{R} \times R = V$$

Then, dividing both sides by I gives

$$\frac{IR}{I} = \frac{V}{I}$$

so that $R = \frac{V}{I}$.

10.6   Make x the subject of the formula $y = \dfrac{4}{x - 7}$.

Solution   We must try to obtain x on its own on the left-hand side. It is often useful to multiply both sides of the formula by the same quantity in order to remove fractions. Multiplying both sides by $(x - 7)$ we find

$$(x - 7)y = (x - 7) \times \frac{4}{x - 7} = 4$$

Removing the brackets on the left-hand side we find

$$xy - 7y = 4 \qquad \text{so that} \qquad xy = 7y + 4$$

Finally, dividing both sides by y we obtain

$$\frac{xy}{y} = \frac{7y + 4}{y}$$

That is,

$$x = \frac{7y + 4}{y}$$

10.7   Transpose the formula

VIDEO

$$T = 2\pi \sqrt{\frac{l}{g}}$$

to make l the subject.

Solution   We must attempt to isolate l. We can do this in stages as follows. Dividing both sides by 2π we find

$$\frac{T}{2\pi} = \frac{2\pi\sqrt{\dfrac{l}{g}}}{2\pi} = \sqrt{\frac{l}{g}} \qquad \text{by cancelling the common factor } 2\pi$$

Recall from §6.3 that $(\sqrt{a})^2 = a$.

The square root sign over the l/g can be removed by squaring both sides. Recall that if a term containing a square root is squared then the square root will disappear:

$$\left(\frac{T}{2\pi}\right)^2 = \frac{l}{g}$$

Then, multiplying both sides by g we find

$$g\left(\frac{T}{2\pi}\right)^2 = l$$

Equivalently we have

$$l = g\left(\frac{T}{2\pi}\right)^2$$

and the formula has been transposed for l.

Self-assessment questions 10.1

1. Explain what is meant by the subject of a formula.

2. In what circumstances might you want to transpose a formula?

3. Explain what processes are allowed when transposing a formula.

Exercise 10.1

1. Transpose each of the following formulae to make x the subject:

    (a) $y = 3x$ (b) $y = \dfrac{1}{x}$ (c) $y = 7x - 5$

    (d) $y = \dfrac{1}{2}x - 7$ (e) $y = \dfrac{1}{2x}$

    (f) $y = \dfrac{1}{2x + 1}$ (g) $y = \dfrac{1}{2x} + 1$

    (h) $y = 18x - 21$ (i) $y = 19 - 8x$

2. Transpose $y = mx + c$, (a) for m, (b) for x, (c) for c.

3. Transpose the following formulae for x:

    (a) $y = 13(x - 2)$ (b) $y = x\left(1 + \dfrac{1}{x}\right)$

    (c) $y = a + t(x - 3)$

4. Transpose each of the following formulae to make the given variable the subject:

    (a) $y = 7x + 11$, for x (b) $V = IR$, for I

    (c) $V = \dfrac{4}{3}\pi r^3$, for r (d) $F = ma$, for m

5. Make n the subject of the formula $l = a + (n - 1)d$.

6. If $m = n + t\sqrt{x}$, find an expression for x.

7. Make x the subject of the following formulae:

(a) $y = 1 - x^2$　(b) $y = \dfrac{1}{1 - x^2}$

(c) $y = \dfrac{1 - x^2}{1 + x^2}$

Test and assignment exercises 10

1. Make x the subject of the following formulae:

(a) $y = 13x - 18$　(b) $y = \dfrac{13}{x + 18}$　(c) $y = \dfrac{13}{x} + 18$

2. Transpose the following formulae:

(a) $y = \dfrac{7 + x}{14}$ for x　(b) $E = \dfrac{1}{2}mv^2$ for v　(c) $8x - 13y = 12$ for y　(d) $y = \dfrac{x^2}{2g}$ for x

3. Make x the subject of the formula $v = k/\sqrt{x}$.

4. Transpose the formula $V = \pi r^2 h$ for h.

5. Make r the subject of the formula $y = H + Cr$.

6. Transpose the formula $s = ut + \dfrac{1}{2}at^2$ for a.

7. Transpose the formula $v^2 = u^2 + 2as$ for u.

8. Transpose the formula $k = pv^3$ for v.

Solving equations

Objectives: This chapter:

- explains what is meant by an equation and its solution
- shows how to solve linear, simultaneous and quadratic equations

An **equation** states that two quantities are equal, and will always contain an **unknown quantity** that we wish to find. For example, in the equation $5x + 10 = 20$ the unknown quantity is x. To **solve** an equation means to find all values of the unknown quantity that can be substituted into the equation so that the left side equals the right side. Each such value is called a **solution** or alternatively a **root** of the equation. In the example above the solution is $x = 2$ because when $x = 2$ is substituted both the left side and the right side equal 20. The value $x = 2$ is said to **satisfy** the equation.

11.1 Solving linear equations

A **linear** equation is one of the form $ax + b = 0$ where a and b are numbers and the unknown quantity is x. The number a is called the **coefficient** of x. The number b is called the **constant term**. For example, $3x + 7 = 0$ is a linear equation. The coefficient of x is 3 and the constant term is 7. Similarly, $-2x + 17.5 = 0$ is a linear equation. The coefficient of x is -2 and the constant term is 17.5. Note that the unknown quantity occurs only to the first power, that is as x, and not as x^2, x^3, $x^{1/2}$ etc. Linear equations may appear in other forms that may seem to be different but are nevertheless equivalent. Thus

$$4x + 13 = -7, \qquad 3 - 14x = 0 \qquad \text{and} \qquad 3x + 7 = 2x - 4$$

are all linear equations that could be written in the form $ax + b = 0$ if necessary.

An equation such as $3x^2 + 2 = 0$ is not linear because the unknown quantity occurs to the power 2. Linear equations are solved by trying to obtain the unknown quantity on its own on the left-hand side; that is, by making the unknown quantity the subject of the equation. This is done using the rules given for transposing formulae in Chapter 10. Consider the following examples.

WORKED EXAMPLE

11.1   Solve the equation $x + 10 = 0$.

Solution   We make x the subject by subtracting 10 from both sides to give $x = -10$. Therefore $x = -10$ is the solution. It can be easily checked by substituting into the original equation:

$$(-10) + 10 = 0 \qquad \text{as required}$$

Note from the last worked example that the solution should be checked by substitution to ensure it satisfies the given equation. If it does not then a mistake has been made.

WORKED EXAMPLES

11.2   Solve the equation $4x + 8 = 0$.

Solution   In order to find the unknown quantity x we attempt to make it the subject of the equation. Subtracting 8 from both sides we find

$$4x + 8 - 8 = 0 - 8 = -8$$

That is,

$$4x = -8$$

Then dividing both sides by 4 gives

$$\frac{4x}{4} = \frac{-8}{4}$$

so that $x = -2$. The solution of the equation $4x + 8 = 0$ is $x = -2$.

11.3   Solve the equation $5x + 17 = 4x - 3$.

Solution   First we collect together all terms involving x. This is done by subtracting $4x$ from both sides to remove this term from the right. This gives

$$5x - 4x + 17 = -3$$

That is, $x + 17 = -3$. To make x the subject of this equation we now subtract 17 from both sides to give $x = -3 - 17 = -20$. The solution of the equation $5x + 17 = 4x - 3$ is $x = -20$. Note that the answer can be easily checked by substituting $x = -20$ into the original equation and verifying that the left side equals the right side.

11.4   Solve the equation $\dfrac{x - 3}{4} = 1$.

Solution   We attempt to obtain x on its own. First note that if we multiply both sides of the equation by 4 this will remove the 4 in the denominator. That is,

$$4 \times \left(\frac{x - 3}{4} \right) = 4 \times 1$$

so that

$$x - 3 = 4$$

Finally, adding 3 to both sides gives $x = 7$.

Self-assessment questions 11.1

1. Explain what is meant by a root of an equation.

2. Explain what is meant by a linear equation.

3. State the rules that can be used to solve a linear equation.

4. You may think that a formula and an equation look very similar. Try to explain the distinction between a formula and an equation.

Exercise 11.1

1. Verify that the given values of x satisfy the given equations:
   (a) $x = 7$ satisfies $3x + 4 = 25$
   (b) $x = -5$ satisfies $2x - 11 = -21$
   (c) $x = -4$ satisfies $-x - 8 = -4$
   (d) $x = \frac{1}{2}$ satisfies $8x + 4 = 8$
   (e) $x = -\frac{1}{3}$ satisfies $27x + 8 = -1$
   (f) $x = 4$ satisfies $3x + 2 = 7x - 14$

2. Solve the following linear equations:
   (a) $3x = 9$ (b) $\dfrac{x}{3} = 9$ (c) $3t + 6 = 0$
   (d) $3x - 13 = 2x + 9$ (e) $3x + 17 = 21$
   (f) $4x - 20 = 3x + 16$

   (g) $5 - 2x = 2 + 3x$
   (h) $\dfrac{x + 3}{2} = 3$ (i) $\dfrac{3x + 2}{2} + 3x = 1$

3. Solve the following equations:
   (a) $5(x + 2) = 13$
   (b) $3(x - 7) = 2(x + 1)$
   (c) $5(1 - 2x) = 2(4 - 2x)$

4. Solve the following equations:
   (a) $3t + 7 = 4t - 2$
   (b) $3v = 17 - 4v$
   (c) $3s + 2 = 14(s - 1)$

5. Solve the following linear equations:
    (a) $5t + 7 = 22$ (b) $7 - 4t = -13$
    (c) $7 - 4t = 27$ (d) $5 = 14 - 3t$
    (e) $4x + 13 = -x + 25$
    (f) $\dfrac{x+3}{2} = \dfrac{x-3}{4}$ (g) $\dfrac{1}{3}x + 6 = \dfrac{1}{2}x + 2$
    (h) $\dfrac{1}{5}x + 7 = \dfrac{1}{3}x + 5$
    (i) $\dfrac{2x+4}{5} = \dfrac{x-3}{2}$ (j) $\dfrac{x-7}{8} = \dfrac{3x+1}{5}$

6. The following equations may not appear to be linear at first sight but they can all be rewritten in the standard form of a linear equation. Find the solution of each equation.
    (a) $\dfrac{1}{x} = 5$ (b) $\dfrac{1}{x} = \dfrac{5}{2}$ (c) $\dfrac{1}{x+1} = \dfrac{5}{2}$
    (d) $\dfrac{1}{x-1} = \dfrac{5}{2}$ (e) $\dfrac{1}{x} = \dfrac{1}{2x+1}$
    (f) $\dfrac{3}{x} = \dfrac{1}{2x+1}$ (g) $\dfrac{1}{x+1} = \dfrac{1}{3x+2}$
    (h) $\dfrac{3}{x+1} = \dfrac{2}{4x+1}$

11.2 Solving simultaneous equations

Sometimes equations contain more than one unknown quantity. When this happens there are usually two or more equations. For example, in the two equations

$$x + 2y = 14 \qquad 3x + y = 17$$

the unknowns are x and y. Such equations are called **simultaneous equations** and to solve them we must find values of x and y that satisfy both equations at the same time. If we substitute $x = 4$ and $y = 5$ into either of the two equations above we see that the equation is satisfied. We shall demonstrate how simultaneous equations can be solved by removing, or **eliminating**, one of the unknowns.

WORKED EXAMPLES

11.5 Solve the simultaneous equations

$$x + 3y = 14 \tag{11.1}$$
$$2x - 3y = -8 \tag{11.2}$$

Solution Note that if these two equations are added the unknown y is removed or eliminated:

$$
\begin{array}{r}
x + 3y = 14 \\
2x - 3y = -8 \\
\hline
3x \qquad = 6
\end{array}
\quad +
$$

so that $x = 2$. To find y we substitute $x = 2$ into either equation. Substituting into Equation 11.1 gives

$$2 + 3y = 14$$

Solving this linear equation will give y. We have

$$2 + 3y = 14$$

$$3y = 14 - 2 = 12$$

$$y = \frac{12}{3} = 4$$

Therefore the solution of the simultaneous equations is $x = 2$ and $y = 4$. Note that these solutions should be checked by substituting back into both given equations to check that the left-hand side equals the right-hand side.

11.6   Solve the simultaneous equations

VIDEO

$$5x + 4y = 7 \tag{11.3}$$

$$3x - y = 11 \tag{11.4}$$

Solution   Note that, in this example, if we multiply the first equation by 3 and the second by 5 we shall have the same coefficient of x in both equations. This gives

$$15x + 12y = 21 \tag{11.5}$$

$$15x - 5y = 55 \tag{11.6}$$

We can now eliminate x by subtracting Equation 11.6 from Equation 11.5, giving

$$15x + 12y = \quad 21$$
$$\underline{15x - 5y = \quad 55} \quad -$$
$$17y = -34$$

from which $y = -2$. In order to find x we substitute our solution for y into either of the given equations. Substituting into Equation 11.3 gives

$$5x + 4(-2) = 7 \quad \text{so that} \quad 5x = 15 \quad \text{or} \quad x = 3$$

The solution of the simultaneous equations is therefore $x = 3$, $y = -2$.

Exercise 11.2

1. Verify that the given values of x and y satisfy the given simultaneous equations.
   (a) $x = 7$, $y = 1$ satisfy $2x - 3y = 11$, $3x + y = 22$
   (b) $x = -7$, $y = 2$ satisfy $2x + y = -12$, $x - 5y = -17$
   (c) $x = -1$, $y = -1$ satisfy $7x - y = -6$, $x - y = 0$

2. Solve the following pairs of simultaneous equations:
    (a) $3x + y = 1, \quad 2x - y = 2$ (b) $4x + 5y = 21, \quad 3x + 5y = 17$
    (c) $2x - y = 17, \quad x + 3y = 12$ (d) $-2x + y = -21, \quad x + 3y = -14$
    (e) $-x + y = -10, \quad 3x + 7y = 20$ (f) $4x - 2y = 2, \quad 3x - y = 4$

3. Solve the following simultaneous equations:
    (a) $5x + y = 36, \ 3x - y = 20$ (b) $x - 3y = -13, \ 4x + 2y = -24$
    (c) $3x + y = 30, \ -5x + 3y = -50$ (d) $3x - y = -5, \ -7x + 3y = 15$
    (e) $11x + 13y = -24, \ x + y = -2$

11.3 Solving quadratic equations

A **quadratic equation** is an equation of the form $ax^2 + bx + c = 0$ where a, b and c are numbers and x is the unknown quantity we wish to find. The number a is the **coefficient** of x^2, b is the coefficient of x, and c is the **constant term**. Sometimes b or c may be zero, although a can never be zero. For example,

$$x^2 + 7x + 2 = 0 \qquad 3x^2 - 2 = 0 \qquad -2x^2 + 3x = 0 \qquad 8x^2 = 0$$

are all quadratic equations.

Key point

A quadratic equation has the form $ax^2 + bx + c = 0$ where a, b and c are numbers, and x represents the unknown we wish to find.

WORKED EXAMPLES

11.7 State the coefficient of x^2 and the coefficient of x in the following quadratic equations:

(a) $4x^2 + 3x - 2 = 0$ (b) $x^2 - 23x + 17 = 0$ (c) $-x^2 + 19 = 0$

Solution

(a) In the equation $4x^2 + 3x - 2 = 0$ the coefficient of x^2 is 4 and the coefficient of x is 3.

(b) In the equation $x^2 - 23x + 17 = 0$ the coefficient of x^2 is 1 and the coefficient of x is -23.

(c) In the equation $-x^2 + 19 = 0$ the coefficient of x^2 is -1 and the coefficient of x is 0, since there is no term involving just x.

11.8 Verify that both $x = -7$ and $x = 5$ satisfy the quadratic equation $x^2 + 2x - 35 = 0$.

Solution We substitute $x = -7$ into the left-hand side of the equation. This yields

$$(-7)^2 + 2(-7) - 35$$

That is,

$$49 - 14 - 35$$

This simplifies to zero, and so the left-hand side equals the right-hand side of the given equation. Therefore $x = -7$ is a solution.

Similarly, if $x = 5$ we find

$$(5^2) + 2(5) - 35 = 25 + 10 - 35$$

which also simplifies to zero. We conclude that $x = 5$ is also a solution.

Solution by factorisation

If the quadratic expression on the left-hand side of the equation can be factorised, solutions can be found using the method in the following examples.

WORKED EXAMPLES

11.9 Solve the quadratic equation $x^2 + 3x - 10 = 0$.

Solution The left-hand side of the equation can be factorised to give

$$x^2 + 3x - 10 = (x + 5)(x - 2) = 0$$

Whenever the product of two quantities equals zero, then one or both of these quantities must be zero. It follows that either $x + 5 = 0$ or $x - 2 = 0$ from which $x = -5$ and $x = 2$ are the required solutions.

11.10 Solve the quadratic equation $6x^2 + 5x - 4 = 0$.

VIDEO

Solution The left-hand side of the equation can be factorised to give

$$6x^2 + 5x - 4 = (2x - 1)(3x + 4) = 0$$

It follows that either $2x - 1 = 0$ or $3x + 4 = 0$ from which $x = \frac{1}{2}$ and $x = -\frac{4}{3}$ are the required solutions.

11.11 Solve the quadratic equation $x^2 - 8x = 0$.

Solution Factorising the left-hand side gives

$$x^2 - 8x = x(x - 8) = 0$$

from which either $x = 0$ or $x - 8 = 0$. The solutions are therefore $x = 0$ and $x = 8$.

11.12   Solve the quadratic equation $x^2 - 36 = 0$.

Solution   The left-hand side factorises to give

$$x^2 - 36 = (x + 6)(x - 6) = 0$$

so that $x + 6 = 0$ or $x - 6 = 0$. The solutions are therefore $x = -6$ and $x = 6$.

11.13   Solve the equation $x^2 = 81$.

Solution   Writing this in the standard form of a quadratic equation we obtain $x^2 - 81 = 0$. The left-hand side can be factorised to give

$$x^2 - 81 = (x - 9)(x + 9) = 0$$

from which $x - 9 = 0$ or $x + 9 = 0$. The solutions are then $x = 9$ and $x = -9$.

Solution of quadratic equations using the formula

When it is difficult or impossible to factorise the quadratic expression $ax^2 + bx + c$, solutions of a quadratic equation can be sought using the following formula:

Key point

If $ax^2 + bx + c = 0$, then $x = \dfrac{-b \pm \sqrt{b^2 - 4ac}}{2a}$

This formula gives possibly two solutions: one solution is obtained by taking the positive square root and the second solution by taking the negative square root.

WORKED EXAMPLES

11.14   Solve the equation $x^2 + 9x + 20 = 0$ using the formula.

Solution   Comparing the given equation with the standard form $ax^2 + bx + c = 0$ we see that $a = 1$, $b = 9$ and $c = 20$. These values are substituted into the formula:

$$x = \frac{-b \pm \sqrt{b^2 - 4ac}}{2a}$$

$$= \frac{-9 \pm \sqrt{81 - 4(1)(20)}}{(2)(1)}$$

$$= \frac{-9 \pm \sqrt{81 - 80}}{2}$$

$$= \frac{-9 \pm \sqrt{1}}{2}$$

$$= \frac{-9 \pm 1}{2}$$

$$= \begin{cases} -4 & \text{by taking the positive square root} \\ -5 & \text{by taking the negative square root} \end{cases}$$

The two solutions are therefore $x = -4$ and $x = -5$.

11.15   Solve the equation $2x^2 - 3x - 7 = 0$ using the formula.

VIDEO

Solution   In this example $a = 2$, $b = -3$ and $c = -7$. Care should be taken with the negative signs. Substituting these into the formula we find

$$x = \frac{-(-3) \pm \sqrt{(-3)^2 - 4(2)(-7)}}{2(2)}$$

$$= \frac{3 \pm \sqrt{9 + 56}}{4}$$

$$= \frac{3 \pm \sqrt{65}}{4}$$

$$= \frac{3 \pm 8.062}{4}$$

$$= \begin{cases} 2.766 & \text{by taking the positive square root} \\ -1.266 & \text{by taking the negative square root} \end{cases}$$

The two solutions are therefore $x = 2.766$ and $x = -1.266$.

If the values of a, b and c are such that $b^2 - 4ac$ is positive, the formula will produce two solutions known as **distinct real roots** of the equation. If $b^2 - 4ac = 0$ there will be a single root known as a **repeated root**. Some books refer to the equation having **equal roots**. If the equation is such that $b^2 - 4ac$ is negative, the formula requires us to find the square root of a negative number. In ordinary arithmetic this is impossible and we say that in such a case the quadratic equation does not possess any real roots.

However, it is still useful to be able to write down formally expressions for the roots. To do this mathematicians have developed a number system called **complex numbers**. When it is necessary to find the square root of a negative number, such as -4, we proceed as follows. We write

$$-4 = 4 \times -1$$

so that

$$\sqrt{-4} = \sqrt{4} \times \sqrt{-1} = 2 \times \sqrt{-1}$$

Now $\sqrt{-1}$ is not a real number, but we denote it by the symbol i and refer to it as an **imaginary number**. Using the symbol i enables us to write down $\sqrt{-4} = 2\mathrm{i}$. In a similar way we can write down the square root of any negative number. For example,

$$\sqrt{-9} = 3\mathrm{i}, \quad \sqrt{-16} = 4\mathrm{i} \quad \text{and} \quad \sqrt{-7} = \sqrt{7}\,\mathrm{i}$$

Then, a complex number is one which can have both a **real part** and an **imaginary part**. For example, the complex number $6 + 2\mathrm{i}$ has real part 6 and imaginary part 2. The imaginary part is always the number which multiplies the symbol i. We shall see how this helps us to solve the quadratic equation $x^2 - 6x + 10 = 0$ for which $b^2 - 4ac$ is negative in Example 11.17.

The quantity $b^2 - 4ac$ is called the **discriminant**, because it allows us to distinguish between the three possible cases. In summary:

Key point   Given

$$ax^2 + bx + c = 0$$

then

$b^2 - 4ac > 0$ two distinct real roots

$b^2 - 4ac = 0$ repeated (equal) root

$b^2 - 4ac < 0$ no real roots – there are two distinct complex roots

WORKED EXAMPLES

11.16   Use the formula to solve the equation $4x^2 + 4x + 1 = 0$.

Solution   In this example $a = 4$, $b = 4$ and $c = 1$. Applying the formula gives

$$x = \frac{-4 \pm \sqrt{4^2 - 4(4)(1)}}{(2)(4)}$$

$$= \frac{-4 \pm \sqrt{16 - 16}}{8}$$

$$= \frac{-4 \pm 0}{8} = -\frac{1}{2}$$

There is a single, repeated root $x = -\frac{1}{2}$. Note that $b^2 - 4ac = 0$.

11.17   Use the formula to solve the quadratic equation $x^2 - 6x + 10 = 0$.

Solution   We use the formula with $a = 1$, $b = -6$ and $c = 10$. We find

$$x = \frac{6 \pm \sqrt{(-6)^2 - 4(1)(10)}}{2}$$

$$= \frac{6 \pm \sqrt{36 - 40}}{2}$$

$$= \frac{6 \pm \sqrt{-4}}{2}$$

We are now faced with finding the square root of -4. We use the imaginary number i as explained on page 118. Writing -4 as 4×-1 we have $\sqrt{-4} = \sqrt{4} \times \sqrt{-1} = 2\mathrm{i}$. Then the solution of the equation becomes

$$x = \frac{6 \pm 2\mathrm{i}}{2} = 3 \pm \mathrm{i}$$

There are thus two roots, $x = 3 + \mathrm{i}$ and $x = 3 - \mathrm{i}$. These are complex roots with real part 3 and imaginary part ± 1.

For a fuller introduction to complex numbers refer to the companion text *Mathematics for Engineers*, 3rd edition (2008), by the same authors.

Self-assessment questions 11.3

1. Under what conditions will a quadratic equation possess distinct real roots?

2. Under what conditions will a quadratic equation possess a repeated root?

3. Under what conditions will a quadratic equation possess complex roots? What is it necessary to do in order to write down the roots in this case?

Exercise 11.3

1. Solve the following quadratic equations by factorisation:
    (a) $x^2 + x - 2 = 0$
    (b) $x^2 - 8x + 15 = 0$
    (c) $4x^2 + 6x + 2 = 0$
    (d) $x^2 - 6x + 9 = 0$
    (e) $x^2 - 81 = 0$
    (f) $x^2 + 4x + 3 = 0$
    (g) $x^2 + 2x - 3 = 0$
    (h) $x^2 + 3x - 4 = 0$
    (i) $x^2 + 6x + 5 = 0$
    (j) $x^2 - 12x + 35 = 0$
    (k) $x^2 + 12x + 35 = 0$
    (l) $2x^2 + x - 3 = 0$
    (m) $2x^2 - x - 6 = 0$
    (n) $2x^2 - 7x - 15 = 0$
    (o) $3x^2 - 2x - 1 = 0$
    (p) $9x^2 - 12x - 5 = 0$
    (q) $7x^2 + x = 0$
    (r) $4x^2 + 12x + 9 = 0$

2. Solve the following quadratic equations using the formula:
   (a) $3x^2 - 6x - 5 = 0$ (b) $x^2 + 3x - 77 = 0$ (c) $2x^2 - 9x + 2 = 0$
   (d) $x^2 + 3x - 4 = 0$ (e) $3x^2 - 3x - 4 = 0$ (f) $4x^2 + x - 1 = 0$
   (g) $x^2 - 7x - 3 = 0$ (h) $x^2 + 7x - 3 = 0$ (i) $11x^2 + x + 1 = 0$
   (j) $2x^2 - 3x - 7 = 0$

3. Solve the following quadratic equations:
   (a) $6x^2 + 13x + 6 = 0$ (b) $3t^2 + 13t + 12 = 0$ (c) $t^2 - 7t + 3 = 0$

Test and assignment exercises 11

1. Solve the following equations:
   (a) $13t - 7 = 2t + 5$ (b) $-3t + 9 = 13 - 7t$ (c) $-5t = 0$

2. Solve the following equations:
   (a) $4x = 16$ (b) $\dfrac{x}{12} = 9$ (c) $4x - 13 = 3$ (d) $4x - 14 = 2x + 8$
   (e) $3x - 17 = 4$ (f) $7 - x = 9 + 3x$ (g) $4(2 - x) = 8$

3. Solve the following equations:
   (a) $2y = 8$ (b) $5u = 14u + 3$ (c) $5 = 4I$ (d) $13i + 7 = 2i - 9$
   (e) $\dfrac{1}{4}x + 9 = 3 - \dfrac{1}{2}x$ (f) $\dfrac{4x + 2}{2} + 8x = 0$

4. Solve the following equations:
   (a) $3x^2 - 27 = 0$ (b) $2x^2 + 18x + 14 = 0$ (c) $x^2 = 16$
   (d) $x^2 - x - 72 = 0$ (e) $2x^2 - 3x - 44 = 0$ (f) $x^2 - 4x - 21 = 0$

5. The solutions of the equation $x^2 + 4x = 0$ are identical to the solutions of $3x^2 + 12x = 0$. True or false?

6. Solve the following simultaneous equations:
   (a) $x - 2y = -11$, $7x + y = -32$ (b) $2x - y = 2$, $x + 3y = 29$
   (c) $x + y = 19$, $-x + y = 1$

Sequences and series

12

12.1 Sequences

A **sequence** is a set of numbers written down in a specific order. For example,

$1, 3, 5, 7, 9$

$-1, -2, -3, -4$

are both sequences. There need not be an obvious rule relating the numbers in the sequence. For example,

$9, -11, \dfrac{1}{2}, 32.5$

is a sequence.

Each number in the sequence is called a **term** of the sequence. The number of terms in the first sequence above is five, and the number of terms in the second is four. It does not matter that some of the terms are the same, as in $1, 0, 1, 0, 1, 0, 1$.

Sometimes we use the symbol '...' to indicate that the sequence continues. For example, the sequence $1, 2, 3, \ldots, 20$ is the sequence of whole numbers from 1 to 20 inclusive. All of the sequences given above have a finite number of terms. They are known as **finite sequences**. Some sequences go on for ever, and these are called **infinite sequences**. To indicate that a sequence might go on for ever we can use the '...' notation. So when we write

$$1, 3, 5, 7, 9 \ldots$$

it can be assumed that this sequence continues indefinitely.

Very often you will be able to spot a rule that allows you to find the next term in a sequence. For example $1, 3, 5, 7, 9 \ldots$ is a sequence of odd integers and the next term is likely to be 11. The next term in the sequence $1, \frac{1}{2}, \frac{1}{4} \ldots$ might well be $\frac{1}{8}$.

Notation used for sequences

We use a subscript notation to refer to different terms in a sequence. For example, suppose we denote the sequence $1, 3, 5, 7, 9, 11$ by x. Then the first term can be labelled x_1, the second term x_2 and so on. That is,

$$x_1 = 1, \quad x_2 = 3, \quad x_3 = 5, \quad x_4 = 7 \qquad \text{and so on}$$

However, note that sometimes the first term in a sequence is labelled x_0, the second is labelled x_1 and so on.

The terms of a sequence can often be found by using a formula. Consider the examples below.

WORKED EXAMPLES

12.1   The terms of a sequence x are given by $x_k = 2k + 3$. Write down the terms x_1, x_2 and x_3.

Solution   If $x_k = 2k + 3$ then replacing k by 1 gives $x_1 = 2 \times 1 + 3 = 5$. Similarly $x_2 = 7$ and $x_3 = 9$.

12.2   The terms of a sequence x are given by $x_k = 4k^2$. Write down the terms x_0, x_1, x_2 and x_3.

Solution   If $x_k = 4k^2$ then replacing k by 0 gives $x_0 = 4 \times 0^2 = 0$. Similarly $x_1 = 4$, $x_2 = 16$ and $x_3 = 36$.

Self-assessment question 12.1

1. Explain what is meant by a finite sequence and an infinite sequence.

Exercise 12.1

1. Write down the first five terms of the sequences given by
    (a) $x_k = 3k$, starting from $k = 1$ (b) $x_k = 4^k$, starting from $k = 0$
    (c) $x_k = k/2$, starting from $k = 1$ (d) $x_k = 7k - 3$, starting from $k = 0$
    (e) $x_k = -k$, starting from $k = 0$

2. Write down the first four terms of the sequence given by $x_n = 2^n + 3^n$, starting from $n = 0$.

12.2 Arithmetic progressions

A particularly simple way of forming a sequence is to calculate each new term by *adding* a fixed amount to the previous term. For example, suppose the first term is 1 and we find subsequent terms by repeatedly adding 6. We obtain

$$1, 7, 13, 19 \ldots$$

Such a sequence is called an **arithmetic progression** or **arithmetic sequence**. The fixed amount that is added each time is a constant called the **common difference**.

WORKED EXAMPLES

12.3   Write down four terms of the arithmetic progression that has first term 10 and common difference 3.

Solution   The second term is found by adding the common difference 3 to the first term 10. This gives a second term of 13. Similarly to find the third term we add on another 3, and so on. This gives the sequence 10, 13, 16, 19 . . ..

12.4   Write down six terms of the arithmetic progression that has first term 5 and common difference -2.

Solution   Note that in this example the common difference is negative. To find each new term we must add -2. In other words we must subtract 2. This results in

$$5, 3, 1, -1, -3, -5$$

We shall now introduce a general notation for arithmetic progressions. Suppose we let the first term be a, and the common difference be d. Then the second term is $a + d$. The third term is $a + d + d$, that is $a + 2d$, and so on. We note the following result:

Key point

An arithmetic progression can be written

$$a, a + d, a + 2d, a + 3d \ldots$$

a is the **first term**, d is the **common difference**.

Study the following pattern:

the first term is a
the second term is $a + d$
the third term is $a + 2d$
the fourth term is $a + 3d$

and so on. This leads to the following formula for the nth term:

Key point

The nth term of an arithmetic progression is given by $a + (n - 1)d$.

WORKED EXAMPLE

12.5   Use the formula for the nth term to find the 10th term of an arithmetic progression with first term 3 and common difference 5.

Solution   Using the formula nth term $= a + (n - 1)d$ with $a = 3$, $d = 5$ and $n = 10$ we find

$$\text{10th term} = 3 + (10 - 1) \times 5 = 3 + 9 \times 5 = 48$$

The 10th term is therefore 48. Using the formula saves having to write out each term individually.

Self-assessment questions 12.2

1. What is meant by an arithmetic progression?

2. Give an example of a sequence that is an arithmetic progression.

3. Give an example of a sequence that is not an arithmetic progression.

Exercise 12.2

1. Write down the first five terms of the arithmetic progression with common difference 4 and first term 1.

2. Write down the first six terms of the arithmetic progression with common difference -1 and first term 3.

3. Find the 16th term of the arithmetic sequence with first term 2 and common difference 5.

4. Find the 23rd term of the arithmetic progression with first term 11 and common difference -3.

12.3 Geometric progressions

Another simple way of forming a sequence is to calculate each new term by *multiplying* the previous term by a fixed amount. For example, suppose the first term of a sequence is 2, and we find the second, third and fourth terms by repeatedly multiplying by 5. We obtain the sequence

2, 10, 50, 250

Such a sequence is called a **geometric sequence** or **geometric progression**. The fixed amount by which each term is multiplied is called the **common ratio**. If you are given the first term of a sequence, and the common ratio, it is easy to generate subsequent terms.

WORKED EXAMPLES

12.6 Find the first six terms of the geometric sequence with first term 3 and common ratio 2.

Solution The first term is 3. The second term is found by multiplying the first by the common ratio 2. So, the second term is 6. The third term is found by multiplying the second by 2, to give 12. We continue in this fashion to generate the sequence 3, 6, 12, 24, 48, 96.

12.7 Write down the first five terms of the geometric sequence with first term 5 and common ratio $\frac{2}{3}$.

Solution The second term is found by multiplying the first by $\frac{2}{3}$. So the second term is $5 \times \frac{2}{3} = \frac{10}{3}$. The third term is found by multiplying this by $\frac{2}{3}$. Continuing in this way we generate the sequence

$$5, \frac{10}{3}, \frac{20}{9}, \frac{40}{27}, \frac{80}{81}$$

12.8 A geometric progression is given by $1, \frac{1}{2}, \frac{1}{4} \ldots$. What is its common ratio?

Solution The first term is 1. We must ask by what number must the first term be multiplied to give us the second. The answer is clearly $\frac{1}{2}$. The common ratio is $\frac{1}{2}$. This can be checked by verifying that the third term is indeed $\frac{1}{4}$. More generally, the common ratio of a geometric sequence can be found by dividing any term by the previous term. Observe this in the earlier examples.

12.9 Write down the first six terms of the geometric sequence with first term 4 and common ratio -1.

Solution Note that in this example the common ratio is negative. This will have an interesting effect, as you will see. Starting with the first term and repeatedly multiplying by -1 gives the sequence $4, -4, 4, -4, 4, -4$. Notice that because the common ratio is negative, the terms of the sequence alternate in sign.

We now introduce a general notation that can be used to describe geometric sequences. Suppose the first term is a and the common ratio is r. Then the second term will be $a \times r$, or ar. The third term is found by multiplying again by the common ratio, to give $(ar) \times r = ar^2$. Continuing in this way we obtain the sequence $a, ar, ar^2, ar^3 \dots$.

Key point

A geometric progression can be written

$$a, ar, ar^2, ar^3 \dots$$

a is the **first term**, r is the **common ratio**.

Study the following pattern:

the first term is a
the second term is ar
the third term is ar^2
the fourth term is ar^3

and so on. This leads us to the following formula for the nth term:

Key point

The nth term of a geometric progression is given by ar^{n-1}.

WORKED EXAMPLE

12.10   Write down the seventh term of the geometric progression that has first term 2 and common ratio 3.

Solution   The nth term is ar^{n-1}. Here $a = 2$ and $r = 3$. The seventh term is $(2)(3)^{7-1} = 2 \times 3^6 = 2 \times 729 = 1458$.

Self-assessment questions 12.3

1. Explain what is meant by a geometric sequence.

2. Give one example of a sequence that is a geometric sequence and one that is not.

3. What can you say about the terms of a geometric progression when the common ratio is 1?

4. What can you say about the terms of a geometric progression when the common ratio is -1?

Exercise 12.3

1. Write down the first four terms of the geometric sequence with first term 8 and common ratio 3.

2. Write down the first four terms of the geometric progression with first term $\frac{1}{4}$ and common ratio $\frac{4}{5}$.

3. A geometric sequence has first term 4 and common ratio 2. Find (a) the 5th term and (b) the 11th term.

4. A geometric progression is given by $2, -1, \frac{1}{2}, -\frac{1}{4} \ldots$. What is its common ratio?

12.4 Infinite sequences

Some sequences continue indefinitely, and these are called **infinite sequences**. Recall that we can use the … notation to indicate this. Thus $1, 2, 3, 4, 5 \ldots$ is the infinite sequence of positive whole numbers.

It can happen that as we move along the sequence the terms get closer and closer to a fixed value. For example, consider the sequence

$$1, \frac{1}{2}, \frac{1}{3}, \frac{1}{4}, \frac{1}{5} \ldots$$

Notice that the terms are getting smaller and smaller. If we continue on for ever these terms approach the value 0. The sequence can be written in the abbreviated form $x_k = \frac{1}{k}$, for $k = 1, 2, 3 \ldots$. As k gets larger and larger, and approaches infinity, the terms of the sequence get closer and closer to zero. We say that '$\frac{1}{k}$ tends to zero as k tends to infinity', or alternatively 'as k tends to infinity, the **limit** of the sequence is zero'. We write this concisely as

$$\lim_{k \to \infty} \frac{1}{k} = 0$$

'lim' is an abbreviation for limit, so $\lim_{k \to \infty}$ means we must examine the behaviour of the sequence as k gets larger and larger. When a sequence possesses a limit it is said to **converge**. However, not all sequences possess a limit. The sequence defined by $x_k = 3k - 2$, which is $1, 4, 7, 10 \ldots$, is one such example. As k gets larger and larger so too do the terms of the sequence. This sequence is said to **diverge**.

WORKED EXAMPLE

12.11   (a) Write down the first four terms of the sequence $x_k = 3 + \frac{1}{k^2}$, $k = 1, 2, 3 \ldots$.

(b) Find, if possible, the limit of this sequence as k tends to infinity.

Solution   (a) The first four terms are given by

$$x_1 = 3 + \tfrac{1}{1} = 4, \quad x_2 = 3 + \tfrac{1}{4} = 3\tfrac{1}{4}, \quad x_3 = 3 + \tfrac{1}{9} = 3\tfrac{1}{9},$$
$$x_4 = 3 + \tfrac{1}{16} = 3\tfrac{1}{16}$$

(b) As more terms are included we see that x_k approaches 3 because the quantity $\frac{1}{k^2}$ becomes smaller and smaller. We can write

$$\lim_{k \to \infty} \left(3 + \frac{1}{k^2} \right) = 3$$

This sequence converges to the limit 3.

Self-assessment question 12.4

1. Explain what is meant by the limit of a convergent sequence.

Exercise 12.4

1. Find the limit of each of the following sequences, as k tends to infinity, if such a limit exists.
   (a) $x_k = k$, $k = 1, 2, 3 \ldots$
   (b) $x_k = k^2$, $k = 1, 2, 3 \ldots$
   (c) $x_k = \frac{100}{k}$, $k = 1, 2, 3 \ldots$
   (d) $x_k = \frac{1}{k^2}$, $k = 1, 2, 3 \ldots$
   (e) $x_k = k + 1$, $k = 1, 2, 3 \ldots$
   (f) $x_k = 2^k$, $k = 1, 2, 3 \ldots$
   (g) $x_k = (\tfrac{1}{2})^k$, $k = 1, 2, 3 \ldots$
   (h) $x_k = 7 + \frac{3}{k^2}$, $k = 1, 2, 3 \ldots$
   (i) $x_k = \frac{k+1}{k}$, $k = 1, 2, 3 \ldots$

12.5 Series and sigma notation

If the terms of a sequence are added the result is known as a **series**. For example, if we add the terms of the sequence $1, 2, 3, 4, 5$ we obtain the series

$$1 + 2 + 3 + 4 + 5$$

Clearly, a series is a **sum**. If the series contains a finite number of terms we are able to add them all up and obtain the sum of the series. If the series contains an infinite number of terms the situation is more complicated. An infinite series may have a finite sum, in which case it is said to converge. Alternatively, it may not have, and then it is said to diverge.

Sigma notation

Sigma notation, $\sum$, provides a concise and convenient way of writing long sums. The sum

$$1 + 2 + 3 + 4 + 5 + \cdots + 10 + 11 + 12$$

can be written very concisely using the capital Greek letter $\sum$ as

$$\sum_{k=1}^{k=12} k$$

The $\sum$ stands for a sum, in this case the sum of all the values of k as k ranges through all whole numbers from 1 to 12. Note that the lowermost and uppermost values of k are written at the bottom and top of the sigma sign respectively.

The lowermost value of k is commonly $k=1$ or $k=0$, but other values are certainly possible. Do take care when working with sigma notation to check the lowermost value of k. Sometimes the sigma notation itself is abbreviated. The '$k=$' part, written at the bottom and the top of the sigma sign, can be omitted. For example, $\sum_{k=1}^{k=6} 3k+1$ may sometimes be written as $\sum_{k=1}^{6} 3k+1$ or even more simply as $\sum_{1}^{6} 3k+1$.

WORKED EXAMPLES

12.12   Write out explicitly what is meant by

$$\sum_{k=1}^{k=5} k^3$$

Solution   We must let k range from 1 to 5, cube each value of k, and add the results:

$$\sum_{k=1}^{k=5} k^3 = 1^3 + 2^3 + 3^3 + 4^3 + 5^3$$

12.13   Express $\frac{1}{1} + \frac{1}{2} + \frac{1}{3} + \frac{1}{4}$ concisely using sigma notation.

Solution   Each term takes the form $\frac{1}{k}$, where k varies from 1 to 4. In sigma notation we could write this as

$$\sum_{k=1}^{k=4} \frac{1}{k}$$

12.14   Write the sum

$$x_1 + x_2 + x_3 + x_4 + \cdots + x_{19} + x_{20}$$

using sigma notation.

Solution   The sum may be written as

$$\sum_{k=1}^{k=20} x_k$$

There is nothing special about using the letter k. For example,

$$\sum_{n=1}^{n=7} n^2 \qquad \text{stands for} \qquad 1^2 + 2^2 + 3^2 + 4^2 + 5^2 + 6^2 + 7^2$$

We can also use a little trick to alternate the signs of the numbers between $+$ and $-$. Note that $(-1)^2 = 1$, $(-1)^3 = -1$ and so on.

WORKED EXAMPLES

12.15 Write out fully what is meant by $\sum_{k=1}^{4} (-1)^k 2^k$.

Solution When $k = 1$, $(-1)^k 2^k = (-1)^1 2^1 = -2$.

When $k = 2$, $(-1)^k 2^k = (-1)^2 2^2 = 4$.

When $k = 3$, $(-1)^k 2^k = (-1)^3 2^3 = -8$.

When $k = 4$, $(-1)^k 2^k = (-1)^4 2^4 = 16$.
So

$$\sum_{k=1}^{4} (-1)^k 2^k = -2 + 4 - 8 + 16$$

Note that the signs are alternating owing to the presence of $(-1)^k$ in the sum.

12.16 Write out fully what is meant by

$$\sum_{i=0}^{5} \frac{(-1)^{i+1}}{2i+1}$$

Solution Note that the lowermost value of i is $i = 0$.

When $i = 0$, $\dfrac{(-1)^{i+1}}{2i+1} = \dfrac{(-1)^1}{1} = -1$

When $i = 1$, $\dfrac{(-1)^{i+1}}{2i+1} = \dfrac{(-1)^2}{3} = \dfrac{1}{3}$

Substituting in values for $i = 2$, $i = 3$, $i = 4$ and $i = 5$ produces the sum

$$\sum_{i=0}^{5} \frac{(-1)^{i+1}}{2i+1} = -1 + \frac{1}{3} - \frac{1}{5} + \frac{1}{7} - \frac{1}{9} + \frac{1}{11}$$

Exercise 12.5

1. Write out fully what is meant by

    (a) $\displaystyle\sum_{i=1}^{i=5} i^2$ (b) $\displaystyle\sum_{k=1}^{4} (2k+1)^2$ (c) $\displaystyle\sum_{k=0}^{4} (2k+1)^2$

2. Write out fully what is meant by

    (a) $\displaystyle\sum_{i=1}^{4} \frac{i}{i+1}$ (b) $\displaystyle\sum_{n=0}^{3} \frac{n+1}{n+2}$

    Comment upon your answers.

3. Write out fully what is meant by

    $$\sum_{k=1}^{3} \frac{(-1)^k}{k}$$

12.6 Arithmetic series

If the terms of an arithmetic sequence are added, the result is known as an **arithmetic series**. For example, the arithmetic progression with five terms having first term 4 and common difference 5 is $4, 9, 14, 19, 24$. If these terms are added we obtain the arithmetic series $4 + 9 + 14 + 19 + 24$. It is easily verified that this has sum 70. If the series has a large number of terms then finding its sum by directly adding all the terms will be laborious. Fortunately there is a formula that enables us to find the sum of an arithmetic series.

Key point

The sum of the first n terms of an arithmetic series with first term a and common difference d is denoted by S_n and given by

$$S_n = \frac{n}{2}(2a + (n-1)d)$$

WORKED EXAMPLE

12.17 Find the sum of the first 10 terms of the arithmetic series with first term 3 and common difference 4.

Solution Use the formula $S_n = \frac{n}{2}(2a + (n-1)d)$ with $n = 10$, $a = 3$ and $d = 4$:

$$S_{10} = \frac{10}{2}(2 \times 3 + (10-1) \times 4) = 5(6 + 36) = 210$$

Self-assessment questions 12.6

1. Explain what is meant by an arithmetic series.

2. Write down the formula for the sum of the first n terms of an arithmetic series.

Exercise 12.6

1. Find the sum of the first 12 terms of the arithmetic series with first term 10 and common difference 8.

2. Find the sum of the first seven terms of the arithmetic series with first term -3 and common difference -2.

3. The sum of an arithmetic series is 270. The common difference is 1 and the first term is 4. Calculate the number of terms in the series.

4. The sum of the first 15 terms of an arithmetic series is 165. The common difference is 2. Calculate the first term of the series.

5. The sum of the first 13 terms of an arithmetic series is 0. The first term is 3. Calculate the common difference.

6. The first term of an arithmetic series is 16. The 30th term is 100. Calculate S_{30}.

7. Show that the sum of the first n terms of an arithmetic series, that is S_n, may be written as

$$S_n = \frac{n}{2}(\text{first term} + \text{last term})$$

12.7 Geometric series

If the terms of a geometric sequence are added, the result is known as a **geometric series**. For example, the geometric progression with five terms having first term 2 and common ratio 3 is 2, 6, 18, 54, 162. If these terms are added we obtain the geometric series $2 + 6 + 18 + 54 + 162$. It is easily verified that this has sum 242. There is a formula that enables us to find the sum of a geometric series:

Key point

The sum of the first n terms of a geometric series with first term a and common ratio r is denoted by S_n and given by

$$S_n = \frac{a(1 - r^n)}{1 - r} \qquad \text{provided } r \text{ is not equal to } 1$$

The formula excludes the use of $r = 1$ because in this case the denominator becomes zero, and division by zero is never allowed.

WORKED EXAMPLE

12.18 Find the sum of the first five terms of the geometric series with first term 2 and common ratio 3.

Solution Use the formula $S_n = \frac{a(1 - r^n)}{1 - r}$ with $n = 5$, $a = 2$ and $r = 3$:

$$S_5 = \frac{2(1 - 3^5)}{1 - 3} = \frac{2(1 - 243)}{-2} = 242$$

Self-assessment questions 12.7

1. Explain what is meant by a geometric series.

2. Write down the formula for the sum of the first n terms of a geometric series.

3. Explain why the formula is not valid when $r = 1$.

4. Suppose a geometric sequence has first term a and common ratio $r = 1$. We have seen that the formula given above for finding the sum of the first n terms is not valid. Deduce a valid formula.

Exercise 12.7

1. Find the sum of the first six terms of the geometric series with first term 2 and common ratio 3.

2. Find the sum of the first 12 terms of the geometric series with first term 10 and common ratio 4.

3. Find the sum of the first seven terms of the geometric series with first term 3 and common ratio -2.

12.8 Infinite geometric series

When the terms of an infinite sequence are added we obtain an infinite series. It may seem strange to try to add together an infinite number of terms but under some circumstances their sum is finite and can be found.

Consider the special case of an infinite geometric series for which the common ratio r lies between -1 and 1. In such a case the sum always exists, and its value can be found from the following formula:

The sum of an infinite number of terms of a geometric series is denoted by S_∞ and is given by

$$S_\infty = \frac{a}{1-r} \qquad \text{provided } -1 < r < 1$$

Note that if the common ratio is bigger than 1 or less than -1, that is $r > 1$ or $r < -1$, then the sum of an infinite geometric series cannot be found.

WORKED EXAMPLE

12.19   Find the sum of the infinite geometric series with first term 2 and common ratio $\frac{1}{3}$.

Solution   Using the formula $S_\infty = \frac{a}{1-r}$ with $a = 2$ and $r = \frac{1}{3}$,

$$S_\infty = \frac{2}{1 - \frac{1}{3}} = \frac{2}{\frac{2}{3}} = 3$$

Notice that we can only make use of the formula because the value of r lies between -1 and 1.

Exercise 12.8

1. Find the sum of the infinite geometric series with first term 1 and common ratio $\frac{2}{3}$.

2. Find the sum of the infinite geometric series with first term 1 and common ratio $-\frac{2}{3}$.

3. Find the sum of the infinite geometric series with first term 4 and common ratio $-\frac{3}{4}$.

4. Find the sum of the infinite geometric series with first term -8 and common ratio 0.25.

Test and assignment exercises 12

1. Write down the first five terms of the sequences given by (a) $x_k = 9k - 7$, (b) $x_k = k + \frac{1}{k}$, starting from $k = 1$.

2. State which of the following sequences are arithmetic, geometric or neither:
    (a) $6, 18, 54, 162 \ldots$ (b) $6, 7, 8, 9 \ldots$ (c) $1, -1, 1, -1 \ldots$ (d) $1, 3, 6, 8 \ldots$

3. An arithmetic progression has first term 4 and common difference $\frac{1}{2}$. State the 100th term.

4. An arithmetic series has first term 3 and common difference 2. Find the sum of the first 100 terms.

5. A geometric progression has first term 8 and common ratio 4. Find the ninth term.

6. A geometric series has first term 6 and common ratio 3. Find the sum of the first 10 terms.

7. A geometric series has first term 2 and common ratio $\frac{2}{5}$. Find the fifth term.

8. If $x_k = 2 - k$ evaluate $\sum_{k=1}^{4} x_k$.

9. Find the sum of the infinite geometric series $1 + \frac{1}{4} + \frac{1}{16} + \dots$.

10. Find the sum of the infinite geometric series $2 - \frac{2}{3} + \frac{2}{9} - \dots$.

Sets

13

Objectives: This chapter:

- introduces and explains terminology of set theory
- explains complement, intersection and union of sets
- shows how to represent sets using Venn diagrams

13.1 Terminology

A **set** is a collection of clearly defined objects, things or states. Sets are usually denoted by a capital letter and the objects contained within braces {...}. Here are some sets which we will refer to in the examples that follow:

$A = \{\text{all even numbers}\}$
$B = \{0, 1, 2, 3\}$
$C = \{\text{the days in the week}\}$
$D = \{\text{employees in a firm who have been absent more than 10 days in the past 12 months}\}$
$E = \{\text{on, off}\}$

Key point

A set is a collection of clearly defined objects, things or states.

As you can see, the objects in a set can be almost anything. The objects are usually called the **elements** of a set.

A is the set of all even numbers. These cannot be listed and so the elements are described. B is the set comprising the four digits 0, 1, 2 and 3. Since B has a small number of elements they can easily be listed. The elements of C are the days of the week: these could be listed or described as they are here. The elements of D are the names of employees with more than 10 days' absence in the past 12 months. Note that it is possible that D contains no names at all, that is it may be empty. E lists the two possible states of a simple switch.

The set A has an infinite number of elements and so is an example of an **infinite set**. In contrast, B has a finite number of elements and so is a **finite set**. Note that C, D and E are also finite sets.

Consider the sets

$$F = \{2, 4, 6, 8\} \quad G = \{4, 8, 6, 2\}$$

F and G contain the same elements although they are listed in a different order. Because F and G contain exactly the same elements they are **equal sets**.

Let H be defined by

$$H = \{2, 2, 4, 6, 6, 6, 8\}$$

Some elements of H have been repeated. Repeated elements of a set are ignored so we could equally well define H to be

$$H = \{2, 4, 6, 8\}$$

Note that F, G and H are all equal sets.

Key point

The order of the elements of a set is not important. There is no need to repeat elements.

Sometimes all the elements of one set are contained completely within another set. For example, all the elements of $F = \{2, 4, 6, 8\}$ are contained within the set $A = \{\text{all even numbers}\}$. We say that F is a **subset** of A and denote this by

$$F \subseteq A$$

Key point

If all the elements of a set X are also elements of a set Y then X is a subset of Y, denoted $X \subseteq Y$.

To denote that an element belongs to a particular set we use the symbol $\in$. This means 'is a member of' or 'belongs to'. Thus, for example $2 \in A$, $4 \in F$, $3 \in B$. If an element does not belong to a set we use $\notin$. For example,

$$1 \notin A, \quad 10 \notin F$$

As mentioned earlier, a set may have no elements. Such a set is said to be **empty** and this is denoted by $\emptyset$.

We now introduce the universal set. The set containing all the objects of interest in a particular context is called the **universal set**, denoted by $\mathcal{E}$. The universal set depends upon the particular context. For example, if we are interested only in whole numbers then $\mathcal{E}$ will be the set of whole numbers.

If we are concerned with decimal digits then

$$\mathcal{E} = \{0, 1, 2, 3, 4, 5, 6, 7, 8, 9\}$$

Given a set X and a universal set $\mathcal{E}$ we can define a new set, called the **complement** of X, denoted $\overline{X}$. The complement of X contains all the elements of the universal set that are not in X.

Key point   The complement of X, denoted $\overline{X}$, contains all the elements that are not in X.

The following examples illustrate the points discussed so far.

WORKED EXAMPLES

13.1   List the elements of the following sets:

A = {all even numbers between 4 and 13 inclusive}
B = {all integers between 3.4 and 7.6 inclusive}
C = {months of the year beginning with the letter J}

Solution A = {4, 6, 8, 10, 12}
B = {4, 5, 6, 7}
C = {January, June, July}

13.2   The sets X, Y and Z are defined by

X = {all positive odd numbers}
Y = {1, 2, 3, 4, 5}
Z = {all decimal digits}

State whether the following are true or false:

(a) X is a finite set

(b) Y is a finite set

(c) $Z \subseteq Y$

(d) $Y \subseteq Z$

(e) $Y \subseteq X$

(f) $Z \subseteq X$

(g) $3 \in Y$

(h) $4 \in Z$

(i) $6 \notin X$

Solution   (a) F   (b) T   (c) F   (d) T   (e) F   (f) F   (g) T   (h) T   (i) T

13.3   Given

$$\mathcal{E} = \{0, 1, 2, 3, 4, 5, 6, 7, 8, 9\}, \ X = \{2, 4, 6, 8\}, \ Y = \{1, 2, 3, 4, 5, 6, 7\}$$

state (a) $\overline{X}$, (b) $\overline{Y}$.

Solution   (a) $\overline{X} = \{0, 1, 3, 5, 7, 9\}$ (b) $\overline{Y} = \{0, 8, 9\}$

Self-assessment question 13.1

1. Explain what the following terms mean: (a) element of a set, (b) subset, (c) universal set, (d) equal sets, (e) complement of a set.

Exercise 13.1

1. $\mathcal{E} = \{\text{all positive integers}\}$, $X = \{\text{all positive even integers}\}$, $Y = \{10, 20, 50\}$, $Z = \{5, 10, 15, 20\}$. State whether the following are true or false:
   (a) $Y \subseteq X$ (b) $10 \in X$ (c) $15 \notin X$ (d) $Y \subseteq Z$ (e) $Z \subseteq X$

2. Given $\mathcal{E} = \{2, 4, 6, 8, 10, 12\}$, $X = \{2, 4, 6, 8\}$, $Y = \{6, 8, 10\}$, state (a) $\overline{X}$, (b) $\overline{Y}$.

13.2   Sets defined mathematically

Some sets are defined using mathematical terms. Suppose we wish the set S to contain all numbers that are divisible by both 4 and 7. We write

$$S = \{x : x \text{ is divisible by 4 and } x \text{ is divisible by 7}\}$$

We read this as 'S is the set comprising elements x, where x is divisible by 4 and also x is divisible by 7'.

Consider

$$R = \left\{x : x = \frac{a}{b} \text{ where } a \text{ is even and } b \text{ is odd}\right\}$$

Then R is the set whose elements are fractions where the numerator is even and the denominator is odd.

Intersection, ∩

We now introduce the intersection of two sets.

Sometimes, two sets have elements that are common to both. Consider

$$X = \{0, 1, 2, 3, 4, 5, 6, 7, 8, 9\}, \ Y = \{0, 3, 6, 9, 12, 15\}$$

Then X and Y have some common elements, namely 0, 3, 6 and 9. The set which comprises elements which are common to both X and Y is called the **intersection** of X and Y, denoted $X \cap Y$. So

$$X \cap Y = \{0, 3, 6, 9\}$$

In general

$$X \cap Y = \{x: x \in X \text{ and } x \in Y\}$$

This says that the intersection of X and Y contains elements, x, such that x belongs to X and x belongs to Y.

Key point   The intersection of X and Y contains the elements that are common to both sets.

WORKED EXAMPLE

13.4   With

$$X = \{0, 1, 2, 3, 4, 5, 6, 7, 8, 9\}, \ Y = \{0, 3, 6, 9, 12, 15\}, \ Z = \{2, 4, 6, 8, 10, 12, 14\}$$

state (a) $Y \cap Z$, (b) $X \cap Z$, (c) $X \cap (Y \cap Z)$.

Solution   (a) $Y \cap Z$ is the set of elements which are in both Y and Z:

$$Y \cap Z = \{6, 12\}$$

(b) $X \cap Z$ is the set of elements which are in both X and Z:

$$X \cap Z = \{2, 4, 6, 8\}$$

(c) $X \cap (Y \cap Z)$ is the set of elements in both X and $(Y \cap Z)$. Recall that $Y \cap Z$ was found in (a).

$$X \cap (Y \cap Z) = \{6\}$$

Sometimes there are no elements common to two given sets. In such a case then their intersection will not contain any elements, that is the intersection will be empty. The empty set is denoted by $\emptyset$. For example, if $R = \{1, 2, 3\}$ and $S = \{5, 6, 7\}$ then

$$R \cap S = \emptyset$$

because there are no elements common to both R and S. Such sets are said to be **disjoint**.

Key point   If $X \cap Y = \emptyset$, that is the intersection has no elements, then X and Y are disjoint sets.

Union, $\cup$

We now turn to the union of two sets. The **union** of two sets, X and Y, is the set that contains all the elements from both X and Y. The union is denoted by $X \cup Y$ and is formally defined by

$$X \cup Y = \{x : x \in X \text{ or } x \in Y \text{ or both}\}$$

Key point   The union of X and Y contains all the elements of both X and Y.

WORKED EXAMPLE

13.5   With

$$X = \{0, 1, 2, 3, 4, 5, 6, 7, 8, 9\}, \quad Y = \{0, 3, 6, 9, 12, 15\}, \quad Z = \{2, 4, 6, 8, 10, 12, 14\}$$

state (a) $X \cup Y$, (b) $X \cup Z$, (c) $Y \cup Z$.

Solution   (a) $X \cup Y$ comprises all the elements from X and Y. So

$$X \cup Y = \{0, 1, 2, 3, 4, 5, 6, 7, 8, 9, 12, 15\}$$

(b) $X \cup Z = \{0, 1, 2, 3, 4, 5, 6, 7, 8, 9, 10, 12, 14\}$

(c) $Y \cup Z = \{0, 2, 3, 4, 6, 8, 9, 10, 12, 14, 15\}$
Recall that there is no need to repeat elements in a set.

Self-assessment questions 13.2

1. Explain the terms (a) intersection of sets, (b) union of sets.

2. What is the meaning of the symbol '$\in$'?

Exercise 13.2

1. List the elements of the following finite sets:

    $A = \{x : x \text{ is a positive integer, greater than 5 and less than 11}\}$
    $B = \{x : x \text{ is odd and } x \text{ is greater than 5 and less than 20}\}$
    $C = \{x : x \text{ is between 10 and 50 inclusive and } x \text{ is divisible by 3 and } x \text{ is divisible by 5}\}$

2. Given $A = \{5, 6, 7, 9\}$, $B = \{0, 2, 4, 6, 8\}$ and $\mathcal{E} = \{0, 1, 2, 3, 4, 5, 6, 7, 8, 9\}$, list the elements of each of the following sets:

    (a) $\overline{A}$ (b) $\overline{B}$ (c) $\overline{\overline{A}}$ (d) $\overline{\overline{B}}$ (e) $\overline{A} \cup \overline{B}$ (f) $\overline{A} \cap \overline{B}$ (g) $A \cup B$ (h) $A \cap B$
    (i) $\overline{A \cup B}$ (j) $\overline{A \cap B}$
    What do you notice about (i) (e) and (j), (ii) (f) and (i)?

3. Given a universal set, a non-empty set A and its complement $\overline{A}$, state which of the following
    are true and which are false:
    (a) $A \cup \overline{A} = \mathcal{E}$ (b) $A \cap \mathcal{E} = \emptyset$ (c) $A \cap \overline{A} = \emptyset$ (d) $A \cap \overline{A} = \mathcal{E}$ (e) $A \cup \emptyset = \mathcal{E}$
    (f) $A \cup \emptyset = A$ (g) $A \cup \emptyset = \emptyset$ (h) $A \cap \emptyset = A$ (i) $A \cap \emptyset = \emptyset$ (j) $A \cup \mathcal{E} = \emptyset$
    (k) $A \cup \mathcal{E} = \mathcal{E}$

13.3 Venn diagrams

Sets can be represented diagrammatically using **Venn diagrams**. Union,
intersection and complement can all be represented on Venn diagrams.

The universal set, $\mathcal{E}$, is usually represented by a rectangle; other sets are
represented by circles. Figure 13.1 shows a universal set and a set A.

Figure 13.1
The universal set $\mathcal{E}$ and
the set A

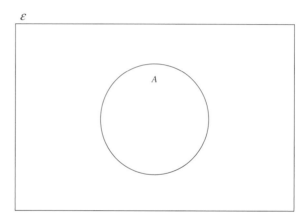

Figure 13.2
The complement of A,
that is $\overline{A}$, is represented
by the shaded area

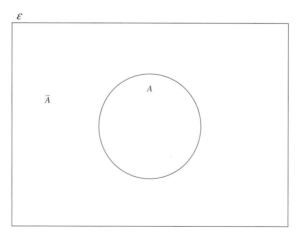

The shaded area in Figure 13.2 represents the complement of A, that is $\overline{A}$.

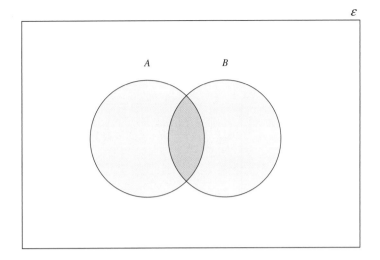

Figure 13.3
The deeply shaded area represents the set $A \cap B$

Figure 13.3 shows the universal set and two sets A and B. The deeply shaded area represents the intersection of A and B, that is $A \cap B$.

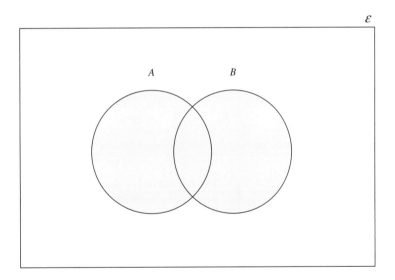

Figure 13.4
The set $A \cup B$

The shaded area of Figure 13.4 represents $A \cup B$.

WORKED EXAMPLE

13.6 A and B are intersecting sets. Represent on a Venn diagram (a) $\overline{A \cup B}$, (b) $\overline{A \cap B}$, (c) $\overline{A} \cup B$.

Solution   (a) $\overline{A \cup B}$ is the complement of $A \cup B$. It comprises all elements that are not in $A \cup B$. The shaded area in Figure 13.5(a) illustrates this.

Figure 13.5(a)
The shaded area represents $\overline{A \cup B}$

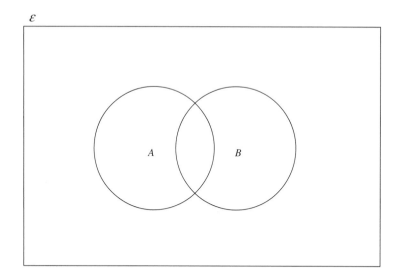

(b) $\overline{A \cap B}$ is the complement of $A \cap B$. It comprises all elements that are not in $A \cap B$ and this is represented by the shaded area in Figure 13.5(b).

Figure 13.5(b)
The shaded area represents $\overline{A \cap B}$

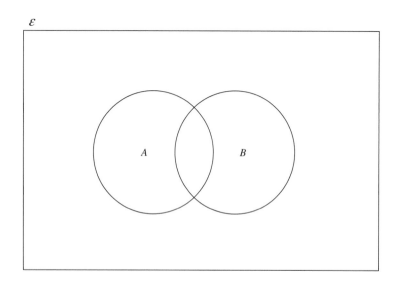

(c) $\overline{A} \cup B$ is the union of $\overline{A}$ and B. It comprises all elements in $\overline{A}$ and B. This is illustrated in Figure 13.5(c)

Figure 13.5(c)
The shaded area represents $\overline{A} \cup B$

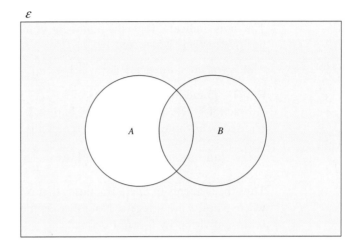

Exercise 13.3

MyMathLab

1. A and B are disjoint sets. Represent the following on a Venn diagram:
   (a) $\overline{A}$ (b) $\overline{A} \cap B$ (c) $\overline{A} \cap \overline{B}$

2. Given $\mathcal{E} = \{0, 1, 2, 3, 4, 5, 6, 7, 8, 9\}$, $A = \{2, 4, 6, 8\}$, $B = \{1, 3, 5, 7\}$ and $C = \{4, 5, 6, 7\}$ show A, B and C on a single Venn diagram.

13.4 Number sets

In this section we introduce some sets which relate to numbers. We use $\mathbb{R}$ to denote the set of all numbers from minus infinity to plus infinity. $\mathbb{R}$ includes integers and fractions, both positive and negative. Formally we say

$$\mathbb{R} = \{x : -\infty < x < \infty\}$$

where ∞ is the symbol for infinity.

There are several important subsets of $\mathbb{R}$. The set of all positive numbers is denoted by $\mathbb{R}^+$:

$$\mathbb{R}^+ = \{\text{all positive numbers}\} = \{x : x \text{ is greater than } 0\}$$

Similarly the set of all negative numbers is denoted by $\mathbb{R}^-$:

$$\mathbb{R}^- = \{\text{all negative numbers}\} = \{x : x \text{ is less than } 0\}$$

Integers (whole numbers) were introduced in Chapter 1. We now define sets where the elements are integers:

$$\mathbb{Z} = \{\text{all integers}\} = \{\ldots -3, -2, -1, 0, 1, 2, 3, \ldots\}$$

The dots indicate that the numbers continue indefinitely. The set of positive integers is denoted by $\mathbb{N}$:

$$\mathbb{N} = \{\text{positive integers}\} = \{1, 2, 3, 4, \ldots\}$$

$\mathbb{R}^+$, $\mathbb{R}^-$, $\mathbb{Z}$ and $\mathbb{N}$ are all subsets of $\mathbb{R}$.

Fractions were introduced in Chapter 2. Commonly fractions are also referred to as **rational numbers**. $\mathbb{Q}$ represents the set of all rational numbers:

$$\mathbb{Q} = \{\text{all rational numbers}\} = \{x : x = \frac{p}{q}, p \in \mathbb{Z}, q \in \mathbb{Z}, q \text{ is not } 0\}$$

Thus the elements of $\mathbb{Q}$ are fractions, that is they are numbers of the form $\frac{p}{q}$ where p and q are integers. Note there is the added condition that q is not 0, since division by 0 is not defined.

Any number that is not rational is called **irrational**. Examples of irrational numbers include $\sqrt{2}$ and π. Commonly we let Π denote the set of irrational numbers:

$$\Pi = \{\text{all irrational numbers}\}$$

Since any number is either rational or irrational then together they make up all possible numbers. This is expressed in set notation by

$$\mathbb{Q} \cup \Pi = \mathbb{R}$$

Note that no number is both rational and irrational so that the intersection of $\mathbb{Q}$ and Π is empty, that is

$$\mathbb{Q} \cap \Pi = \emptyset$$

Further work on sets relating to intervals appears in Chapter 17.

Self-assessment question 13.4

1. Explain what is meant by (a) rational number, (b) irrational number. Give two examples of each.

Test and assignment exercises 13

1. Given $A = \{17, 18, 19, 20, 21\}$, $B = \{\text{all odd numbers}\}$ and $C = \{\text{all even numbers}\}$, state
   (a) $A \cap B$ (b) $A \cap C$ (c) $B \cap C$ (d) $B \cup C$ (e) $A \cup B \cup C$

2. $A = \{0, 1, 2, 5, 9\}$, $B = \{1, 2, 7, 8, 9\}$ and $\mathcal{E} = \{\text{all decimal digits}\}$. List the elements of the following sets: (a) $A \cup B$, (b) $A \cap B$, (c) $\overline{A}$, (d) $\overline{B}$, (e) $\overline{A \cup B}$, (f) $\overline{A} \cap \overline{B}$, (g) $\overline{A \cap B}$, (h) $\overline{A} \cup \overline{B}$. What do you conclude from (e) and (f)? What do you conclude from (g) and (h)?

Figure 13.6
The sets A, B and C

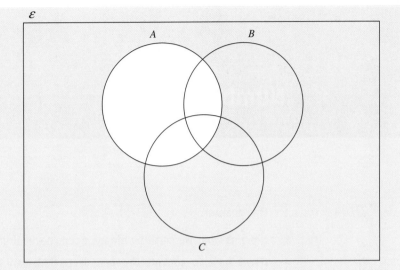

3. A, B and C are three sets such that A and B are not disjoint, A and C are not disjoint and B and C are not disjoint. No set is a subset of any other set. Figure 13.6 shows a Venn diagram of the three sets.

   Draw Venn diagrams, shading the following sets:
   (a) $A \cap B$ (b) $B \cup C$ (c) $B \cup (A \cap C)$ (d) $\overline{C}$ (e) $\overline{B \cap C}$ (f) $A \cap B \cap C$ (g) $\overline{B} \cap (A \cap C)$

4. The sets A, B and C are given by $A = \{1, 2, 3, 4, 5, 6\}$, $B = \{4, 6\}$, $C = \{3, 4, 5, 6, 7, 8, 9\}$. Illustrate A, B and C on a single Venn diagram.

Number bases

14

Objectives: This chapter:

- explains what we mean by binary, octal, decimal and hexadecimal numbers
- shows how to convert between decimal, binary, octal and hexadecimal numbers

14.1 The decimal system

The numbers that we commonly use are based on 10. For example, 253 can be written as

$$253 = 200 + 50 + 3$$
$$= 2(100) + 5(10) + 3(1)$$

Using our knowledge of powers we know that 100 can be written as 10^2, 10 can be written as 10^1 and 1 can be written as 10^0. So

$$253 = 2(100) + 5(10) + 3(1)$$
$$= 2(10^2) + 5(10^1) + 3(10^0)$$

from which it is clear why we refer to this as a 'base 10' number.

When we use 10 as a base we say we are writing in the **decimal system**. Note that in the decimal system there are 10 digits: 0, 1, 2, 3, 4, 5, 6, 7, 8, 9. You may recall the phrase 'hundreds, tens and units' and as we have already noted these are simply powers of 10. We denote numbers in base 10 with a small subscript, for example 5192_{10}:

$$5192_{10} = 5000 + 100 + 90 + 2$$
$$= 5(1000) + 1(100) + 9(10) + 2(1)$$
$$= 5(10^3) + 1(10^2) + 9(10^1) + 2(10^0)$$

This chapter introduces the concept of writing numbers in bases other than 10.

14.2 The binary system

A binary system uses base 2. A binary system has only two digits: 0 and 1. Numbers in base 2 are called **binary digits** or simply **bits** for short. Binary numbers are based on powers of 2. Binary numbers have important applications in computer science and electronic engineering.

Key point   The binary system is based upon powers of 2.

Converting from binary to decimal

Consider the binary number 110101_2. As the base is 2 this means that powers of 2 essentially replace powers of 10:

$$110101_2 = 1(2^5) + 1(2^4) + 0(2^3) + 1(2^2) + 0(2^1) + 1(2^0)$$
$$= 1(32) + 1(16) + 0(8) + 1(4) + 0(2) + 1(1)$$
$$= 32 + 16 + 4 + 1$$
$$= 53_{10}$$

Hence 110101_2 and 53_{10} are equivalent.

WORKED EXAMPLE

14.1   Convert (a) 1111_2, (b) 101010_2 to decimal.

Solution   (a) $$1111_2 = 1(2^3) + 1(2^2) + 1(2^1) + 1(2^0)$$
$$= 1(8) + 1(4) + 1(2) + 1(1)$$
$$= 8 + 4 + 2 + 1$$
$$= 15_{10}$$

(b) $$101010_2 = 1(2^5) + 0(2^4) + 1(2^3) + 0(2^2) + 1(2^1) + 0(2^0)$$
$$= 1(32) + 0 + 1(8) + 0 + 1(2) + 0$$
$$= 32 + 8 + 2$$
$$= 42_{10}$$

Converting decimal to binary

We now look at some examples of converting numbers in base 10 to numbers in base 2, that is from decimal to binary. We make use of Table 14.1, which shows various powers of 2 when converting from decimal to binary. Table 14.1 may be extended as necessary.

Table 14.1
Powers of 2

2^0	1	2^4	16	2^8	256
2^1	2	2^5	32	2^9	512
2^2	4	2^6	64	2^{10}	1024
2^3	8	2^7	128		

WORKED EXAMPLES

14.2   Convert 83_{10} to a binary number.

Solution   We need to express 83_{10} as the sum of a set of numbers each of which is a power of 2. From Table 14.1 we see that 64 is the highest number in the table that does not exceed the given number of 83. We write

$$83 = 64 + 19$$

We now focus on the 19. From Table 14.1, 16 is the highest number that does not exceed 19. So we write

$$19 = 16 + 3$$

giving

$$83 = 64 + 16 + 3$$

We now focus on the 3 and again using Table 14.1 we may write

$$83 = 64 + 16 + 2 + 1$$
$$= 2^6 + 2^4 + 2^1 + 2^0$$
$$= 1(2^6) + 0(2^5) + 1(2^4) + 0(2^3) + 0(2^2) + 1(2^1) + 1(2^0)$$
$$= 1010011_2$$

14.3   Express 200_{10} as a binary number.

Solution   From Table 14.1 we note that 128 is the highest number that does not exceed 200 so we write

$$200 = 128 + 72$$

Using Table 14.1 repeatedly we may write

$$200 = 128 + 72$$
$$= 128 + 64 + 8$$
$$= 2^7 + 2^6 + 2^3$$
$$= 1(2^7) + 1(2^6) + 0(2^5) + 0(2^4) + 1(2^3) + 0(2^2) + 0(2^1) + 0(2^0)$$
$$= 11001000_2$$

Another way to convert decimal numbers to binary numbers is to divide by 2 repeatedly and note the remainder. We rework the previous two examples using this method.

WORKED EXAMPLE

14.4 Convert the following decimal numbers to binary: (a) 83, (b) 200.

Solution (a) We divide by 2 repeatedly and note the remainder:

	Remainder
$83 \div 2 = 41$ r 1	1
$41 \div 2 = 20$ r 1	1
$20 \div 2 = 10$ r 0	0
$10 \div 2 = 5$ r 0	0
$5 \div 2 = 2$ r 1	1
$2 \div 2 = 1$ r 0	0
$1 \div 2 = 0$ r 1	1

To obtain the binary number we write out the remainder, working from the bottom one to the top one. This gives

$$83_{10} = 1010011_2$$

as before.

(b) We repeat the process by repeatedly dividing 200 by 2 and noting the remainder:

	Remainder
$200 \div 2 = 100$ r 0	0
$100 \div 2 = 50$ r 0	0
$50 \div 2 = 25$ r 0	0
$25 \div 2 = 12$ r 1	1

$$12 \div 2 = 6 \text{ r } 0 \qquad\qquad 0$$
$$6 \div 2 = 3 \text{ r } 0 \qquad\qquad 0$$
$$3 \div 2 = 1 \text{ r } 1 \qquad\qquad 1$$
$$1 \div 2 = 0 \text{ r } 1 \qquad\qquad 1$$

Reading the remainder column from the bottom to the top gives the required binary number:

$$200_{10} = 11001000_2$$

Self-assessment questions 14.2

1. Describe one way of converting decimal numbers to binary numbers.

2. What number is the binary system based upon?

3. How many different digits occur in the binary system?

Exercise 14.2

MyMathLab

1. Convert the following decimal numbers to binary numbers:
   (a) 19   (b) 36   (c) 100   (d) 796   (e) 5000

2. Convert the following binary numbers to decimal numbers:
   (a) 111   (b) 10101   (c) 111001   (d) 1110001   (e) 11111111

3. What is the highest decimal number that can be written in binary form using a maximum of (a) two binary digits, (b) three binary digits, (c) four binary digits, (d) five binary digits? Can you spot a pattern? (e) Write a formula for the highest decimal number that can be written using N binary digits.

4. Write the decimal number 0.5 in binary.

14.3 Octal system

Octal numbers use 8 as a base. The eight digits used in the octal system are 0, 1, 2, 3, 4, 5, 6 and 7. Octal numbers use powers of 8, just as decimal numbers use powers of 10 and binary numbers use powers of 2. The octal number 325_8 means

$$325_8 = 3(8^2) + 2(8^1) + 5(8^0)$$

| Key point | Octal numbers are based upon powers of 8. |

Converting octal numbers to decimal numbers

The following examples illustrate the technique.

| WORKED EXAMPLES |

14.5   Convert 325_8 to a decimal number.

Solution
$$325_8 = 3(8^2) + 2(8^1) + 5(8^0)$$
$$= 3(64) + 2(8) + 5(1)$$
$$= 192 + 16 + 5$$
$$= 213_{10}$$

14.6   Convert 7046_8 to a decimal number.

Solution
$$7046_8 = 7(8^3) + 0(8^2) + 4(8^1) + 6(8^0)$$
$$= 7(512) + 0 + 4(8) + 6(1)$$
$$= 3622_{10}$$

Converting decimal numbers to octal numbers

We now consider how to convert decimal numbers to octal numbers.

Table 14.2
Powers of 8

8^0	1
8^1	8
8^2	64
8^3	512
8^4	4096
8^5	32768

We begin by noting Table 14.2 where the values of powers of 8 are recorded. The following example illustrates how Table 14.2 is used to convert decimal numbers to octal numbers.

| WORKED EXAMPLES |

14.7   Convert 1001 to an octal number.

Solution   From Table 14.2 we note that the highest number that does not exceed 1001 is 512. So we write

$$1001 = 512 + 489$$

Looking at the 489, we see that 64 is the highest number that does not exceed 489. We note that

$$489 = 7(64) + 41$$

Finally, looking at 41, we note that 8 is the highest number in Table 14.2 that does not exceed 41. We note that

$$41 = 5(8) + 1$$

so we may write

$$
\begin{aligned}
1001 &= 512 + 489 \\
&= 512 + 7(64) + 41 \\
&= 512 + 7(64) + 5(8) + 1 \\
&= 1(8^3) + 7(8^2) + 5(8^1) + 1(8^0) \\
&= 1751_8
\end{aligned}
$$

As an alternative we can divide repeatedly by 8, noting the remainder:

	Remainder
$1001 \div 8 = 125 \text{ r } 1$	1
$125 \div 8 = 15 \text{ r } 5$	5
$15 \div 8 = 1 \text{ r } 7$	7
$1 \div 8 = 0 \text{ r } 1$	1

Reading up the remainder column gives the required octal number: 1751_8.

14.8   Convert 3726 to an octal number.

Solution   From Table 14.2 we see that 512 is the highest number that does not exceed 3726. We note that

$$3726 = 7(512) + 142$$

We focus on the 142. From Table 14.2 the highest number that does not exceed 142 is 64. We may write

$$142 = 2(64) + 14$$

Finally we note that $14 = 8 + 6$. So

$$
\begin{aligned}
3726 &= 7(512) + 2(64) + 8 + 6 \\
&= 7(8^3) + 2(8^2) + 1(8^1) + 6(8^0) \\
&= 7216_8
\end{aligned}
$$

We can also arrive at this result by successively dividing by 8 and noting the remainder:

	Remainder
$3726 \div 8 = 465 \text{ r } 6$	6
$465 \div 8 = 58 \text{ r } 1$	1
$58 \div 8 = 7 \text{ r } 2$	2
$7 \div 8 = 0 \text{ r } 7$	7

Reading up the remainder column gives the required number: 7216_8.

Self-assessment questions 14.3

1. How many digits are used in the octal system?

2. What number are octal numbers based upon?

Exercise 14.3

MyMathLab

1. Convert the following decimal numbers to octal numbers:
   (a) 971 (b) 2841 (c) 5014 (d) 10000 (e) 17926

2. Convert the following octal numbers to decimal numbers:
   (a) 73 (b) 1237 (c) 7635 (d) 6677 (e) 67765

3. What is the highest decimal number that can be represented by an octal number of
   (a) one digit, (b) two digits, (c) three digits, (d) four digits, (e) five digits, (f) N digits?

14.4 Hexadecimal system

Finally, we look at number systems which use 16 as a base. These systems are termed **hexadecimal**. There are 16 digits in the hexadecimal system: 0, 1, 2, 3, 4, 5, 6, 7, 8, 9, A, B, C, D, E, F. Notice that conventional decimal digits are insufficient to represent hexadecimal numbers and so additional 'digits', A, B, C, D, E, F, are included. Table 14.3 shows the equivalence between decimal and hexadecimal digits.

Table 14.3
Hexadecimal digits

Decimal	Hexadecimal
0	0
1	1
2	2
3	3
4	4
5	5
6	6
7	7
8	8
9	9
10	A
11	B
12	C
13	D
14	E
15	F

Converting from hexadecimal to decimal

The following example illustrates how to convert from hexadecimal to decimal.

WORKED EXAMPLE

14.9   Convert the following hexadecimal numbers to decimal numbers: (a) 93A, (b) F9B3.

Solution   (a)   Noting that hexadecimal numbers use base 16 we have

$$93A_{16} = 9(16^2) + 3(16^1) + A(16^0)$$
$$= 9(256) + 3(16) + 10(1)$$
$$= 2362_{10}$$

(b) $F9B3_{16} = F(16^3) + 9(16^2) + B(16^1) + 3(16^0)$
$$= 15(4096) + 9(256) + 11(16) + 3(1)$$
$$= 63923_{10}$$

Converting from decimal to hexadecimal

Table 14.4 gives useful information to help in the conversion from decimal to hexadecimal.

16^0	1
16^1	16
16^2	256
16^3	4096
16^4	65536

The following example illustrates how to convert from decimal to hexadecimal.

WORKED EXAMPLES

14.10 Convert 14397 to a hexadecimal number.

Solution From Table 14.4 the highest number that does not exceed 14397 is 4096. We write

$$14397 = 3(4096) + 2109$$

We now focus on the 2109. From Table 14.4, the highest number that does not exceed 2109 is 256:

$$2109 = 8(256) + 61$$

Finally, $61 = 3(16) + 13$. So we have

$$14397 = 3(4096) + 8(256) + 3(16) + 13$$
$$= 3(16^3) + 8(16^2) + 3(16^1) + 13$$

From Table 14.3 we see that 13_{10} is D in hexadecimal so we have

$$14397_{10} = 383D_{16}$$

As with other number bases that we have studied, we can convert decimal numbers by repeated division and noting the remainder. The previous example is reworked to illustrate this.

14.11 Convert 14397 to hexadecimal.

Solution We divide repeatedly by 16, noting the remainder:

	Remainder
$14397 \div 16 = 899 \text{ r } 13$	13
$899 \div 16 = 56 \text{ r } 3$	3
$56 \div 16 = 3 \text{ r } 8$	8
$3 \div 16 = 0 \text{ r } 3$	3

Recall that 13 in hexadecimal is D. Reading up the remainder column we have

$$14397_{10} = 383D_{16}$$

as before.

Self-assessment question 14.4

1. What number is the hexadecimal system based upon?

Exercise 14.4

1. Convert the following hexadecimal numbers to decimal numbers:
   (a) 91   (b) 6C   (c) A1B   (d) F9D4   (e) ABCD

2. Convert the following decimal numbers to hexadecimal numbers:
   (a) 160   (b) 396   (c) 5010   (d) 25000   (e) 1000000

3. Calculate the highest decimal number that can be represented by a hexadecimal number with
   (a) one digit   (b) two digits   (c) three digits   (d) four digits   (e) N digits

Test and assignment exercises 14

1. Convert the following decimal numbers into binary, octal and hexadecimal:
   (a) 63   (b) 200   (c) 371   (d) 693   (e) 3750

2. Convert the following numbers to hexadecimal numbers:
   (a) 1110111_2 (b) 671_8 (c) 9364_{10}

3. Convert the following hexadecimal numbers to binary, octal and decimal:
   (a) 9A   (b) 1BF   (c) DE   (d) ABC7   (e) FA1D

Elementary logic

15

Objectives : This chapter:

- explains the terms commonly used in logic
- explains negation, conjunction, disjunction and implication
- shows how to construct truth tables for logical expressions

15.1 Logic and propositions

A proposition is a statement that is either true (T) or false (F). Here are three examples of propositions:

1. One plus two is nine.
2. Shakespeare wrote *Hamlet*.
3. All cars run on petrol.

Here the first and third are false, the second is true. Some statements are not propositions as it is impossible to assign a truth value to them. For example, 'Maths is interesting' cannot unequivocally be said to be either true or false. Some people find maths interesting, others do not. Similarly the statements

Mike is a funny person.
It was warm yesterday.
Go home now.

cannot have a truth value assigned to them and hence are not propositions.

Key point   A proposition is a statement that is either true or false.

Self-assessment question 15.1

1. Explain what is meant by a proposition. Give two examples of statements which are propositions and two examples of statements which are not propositions.

Exercise 15.1

1. Determine which of the following are propositions:
   (a) The moon is made of cheese.
   (b) Swimming is a good exercise.
   (c) Rectangles have four sides.
   (d) Hand in your homework tomorrow.
   (e) Your homework is due in tomorrow.

15.2 Symbolic logic

Logic is the study of whether an argument is robust and leads to valid conclusions. It is used to help in the structuring of computer programs. In order to develop the theory of logic we now introduce a concise notation. A proposition is given a single letter to represent it. For example, we let M stand for the proposition 'The moon is made of cheese' and write

M: The moon is made of cheese.

The truth value of a proposition is either true (T) or false (F). We now introduce some symbols which act upon or join together propositions.

Negation

The **negation** of a proposition is the proposition that is true whenever the original proposition is false and false when the original is true. Negation is denoted by the symbol $\neg$.

Key point   Negation has the symbol $\neg$. For any proposition, P, $\neg P$ is true whenever P is false. $\neg P$ is false whenever P is true.

WORKED EXAMPLE

15.1   State the negation of

W: The word is made up of six characters.

Solution   The negation is $\neg W$ where

$\neg W$: The word is not made up of six characters.

Conjunction

Given any two propositions we can form their **conjunction**. If

W: The word is made up of six characters
D: The first character is a digit

then the conjunction of W and D, denoted $W \wedge D$, is

$W \wedge D$: The word is made up of six characters and the first character is a digit

Conjunction corresponds to 'and' in everyday English. Note that $W \wedge D$ is itself a proposition. Since it is made up of two simpler propositions it is called a **compound proposition**. The truth value of $W \wedge D$ depends upon the truth values of W and D. This will be explained more fully in the next section on truth tables.

Key point   Conjunction has the symbol $\wedge$. It corresponds to 'and' in English.

WORKED EXAMPLE

15.2   If

S: The switch is open
F: The current is flowing

state in words (a) $S \wedge F$, (b) $\neg S \wedge F$, (c) $S \wedge \neg F$.

Solution   (a)   The switch is open and the current is flowing.

(b)   The switch is not open and the current is flowing.

(c)   The switch is open and the current is not flowing.

Disjunction

The **disjunction** of W and D is denoted $W \vee D$. It corresponds to the word 'or' but we need to be careful as the conjunction has a different interpretation to the usual meaning of the word 'or'. $W \vee D$ is true whenever W is true or D is true or both are true. Consider the example:

R: A rectangle has four sides
C: A circle has five sides

Then $R \vee C$ is

$R \vee C$: A rectangle has four sides or a circle has five sides

The compound proposition, $R \vee C$, is true whenever R is true or C is true or both are true. In this example, since R is true then $R \vee C$ is true.

Key point   Disjunction has the symbol $\vee$. It corresponds to 'or' in English.

WORKED EXAMPLE

15.3   Referring to Worked Example 15.2 for the definitions of S and F, state in words (a) $S \vee F$, (b) $\neg(S \vee F)$.

Solution   (a) The switch is open or the current is flowing.

(b) Here the negation is applied to $(S \vee F)$ to give 'Neither the switch is open nor the current is flowing.' This could be reworded as 'The switch is not open and the current is not flowing.'

Implication

Consider the statement 'If I pass the maths exam then I will celebrate at the quiz night.'

A proposition such as this one, using the 'If ... then...' structure, is called a **conditional proposition**.

Let us define the simple propositions

P: I pass the maths exam
Q: I will celebrate at the quiz night

The compound proposition 'If P then Q' is denoted by $P \rightarrow Q$.

The truth value of the conditional proposition depends upon the truth value of the individual simple propositions. $P \rightarrow Q$ is sometimes read as 'P implies Q'.

Key point   We read $P \rightarrow Q$ as 'P implies Q' or 'If P ... then Q'.

WORKED EXAMPLE

15.4   Using S and F as in Worked Example 15.2, state symbolically

(a) If the switch is open then the current is flowing.
(b) If the switch is not open, then the current is flowing.
(c) If the current is flowing, then the switch is not open.

Solution   (a) $S \rightarrow F$

(b) $\neg S \rightarrow F$

(c) $F \rightarrow \neg S$

Self-assessment question 15.2

1. Explain the terms (a) conjunction, (b) implication, (c) disjunction.

Exercise 15.2

1. Given

    W: The word has less than seven characters
    C: The program compiles

    state in words
    (a) $\neg W$ (b) $\neg C$ (c) $W \rightarrow C$ (d) $W \vee \neg C$ (e) $C \wedge W$

2. Given

    D: The first character is a digit
    V: The word has six characters

    state symbolically
    (a) The first character is not a digit.
    (b) If the word has six characters then the first character is a digit.
    (c) The word does not have six characters or the first character is a digit.

15.3 Truth tables

Consider a compound proposition, that is one made up of simple propositions and the logic symbols $\neg$ (negation), $\wedge$ (conjunction), $\vee$ (disjunction) and $\rightarrow$ (implication). A truth table lists all possible combinations of truth values of the simple propositions together with the resulting truth value of the compound proposition.

Negation

Table 15.1 shows the truth table for negation.

Table 15.1
Truth table for negation

P	$\neg P$
T	F
F	T

Table 15.1 states that when a proposition P is T, the proposition $\neg P$ is F; when P is F then $\neg P$ is T. Consider a simple example:

P: The word has six characters

So

$\neg P$: The word does not have six characters

If P is true, that is the word in question does have six characters, then $\neg P$ must clearly be false. If P is false so that the word does not have six characters then $\neg P$ is true.

Conjunction

Table 15.2 shows the truth table for a conjunction, $P \wedge Q$.

Table 15.2
Truth table for a conjunction $P \wedge Q$

P	Q	$P \wedge Q$
T	T	T
T	F	F
F	T	F
F	F	F

The first two columns show the four possible combinations of truth values of P and Q. The conjunction, $P \wedge Q$, is true only when both P is true and Q is true. Otherwise the conjunction is false.

WORKED EXAMPLE

15.5   Construct a truth table for (a) $(\neg P) \wedge Q$, (b) $(\neg P) \wedge (\neg Q)$.

Solution   (a)   Table 15.3 shows the required truth table. The first two columns show the truth values of P and Q. There are four possible combinations of truth values for P and Q. The third column shows the truth value of $\neg P$. We put in this column as an intermediate step because it helps us to calculate the truth values in the final column.

Table 15.3
Truth table for $(\neg P) \wedge Q$

P	Q	$\neg P$	$(\neg P) \wedge Q$
T	T	F	F
T	F	F	F
F	T	T	T
F	F	T	F

Note that $(\neg P) \wedge Q$ is true only when both $(\neg P)$ is true and Q is true.

(b) Table 15.4 shows the truth table for $(\neg P) \wedge (\neg Q)$.

Table 15.4
Truth table for
$(\neg P) \wedge (\neg Q)$

P	Q	$\neg P$	$\neg Q$	$(\neg P) \wedge (\neg Q)$
T	T	F	F	F
T	F	F	T	F
F	T	T	F	F
F	F	T	T	T

Note that $(\neg P) \wedge (\neg Q)$ is true only when both $(\neg P)$ and $(\neg Q)$ are true.

Disjunction

Table 15.5 shows the truth table for the disjunction $P \vee Q$.

Table 15.5
Truth table for the
disjunction $P \vee Q$

P	Q	$P \vee Q$
T	T	T
T	F	T
F	T	T
F	F	F

Note that $P \vee Q$ is true whenever either P is true or Q is true or both P and Q are true. $P \vee Q$ is false only when both P is false and Q is also false.

WORKED EXAMPLE

15.6 Construct the truth table for (a) $P \vee (\neg Q)$, (b) $(\neg P) \vee (\neg Q)$.

Solution (a) Table 15.6 shows the required truth table.

Table 15.6
Truth table for $P \vee (\neg Q)$

P	Q	$\neg Q$	$P \vee (\neg Q)$
T	T	F	T
T	F	T	T
F	T	F	F
F	F	T	T

Note that $P \vee (\neg Q)$ is false only when both P and $(\neg Q)$ are false.

(b) Table 15.7 shows the truth table for $(\neg P) \vee (\neg Q)$.

Table 15.7
Truth table for $\neg P \vee \neg Q$

P	Q	$\neg P$	$\neg Q$	$\neg P \vee \neg Q$
T	T	F	F	F
T	F	F	T	T
F	T	T	F	T
F	F	T	T	T

Implication

Table 15.8 shows the truth table for the implication $P \rightarrow Q$.

Table 15.8
Truth table for
implication $P \rightarrow Q$

P	Q	$P \rightarrow Q$
T	T	T
T	F	F
F	T	T
F	F	T

The only time that $P \rightarrow Q$ is false occurs when P is true and Q is false.

WORKED EXAMPLE

15.7   Construct the truth table for (a) $(\neg P) \rightarrow Q$, (b) $P \rightarrow (\neg Q)$.

Solution   (a) Table 15.9 shows the truth table for $\neg P \rightarrow Q$.

Table 15.9
Truth table for $\neg P \rightarrow Q$

P	Q	$\neg P$	$\neg P \rightarrow Q$
T	T	F	T
T	F	F	T
F	T	T	T
F	F	T	F

(b)   Table 15.10 shows the truth table for $P \rightarrow \neg Q$.

Table 15.10
Truth table for $P \rightarrow \neg Q$

P	Q	$\neg Q$	$P \rightarrow \neg Q$
T	T	F	F
T	F	T	T
F	T	F	T
F	F	T	T

We have already met several examples of compound propositions. For example, $P \wedge Q$, $P \vee Q$, $P \rightarrow Q$ and $P \rightarrow \neg Q$ are all compound propositions. Here are a few more complex examples of compound propositions:

$$P \wedge (P \rightarrow \neg Q), \ (\neg Q \vee R) \wedge (\neg R \wedge P), \ [P \vee (\neg Q)] \rightarrow [Q \wedge (\neg P)]$$

Just as with arithmetic operations of $+, -, \times$ and $\div$, there is an order to the way in which the logical connectives are applied.

Negation ($\neg$) is applied first. Conjunction ($\wedge$) and disjunction ($\vee$) are applied next and are of equal strength. Implication ($\rightarrow$) is applied next.

Key point

In a logical expression:
$\neg$ is applied first
$\vee$ and $\wedge$ are applied next with equal strength
$\rightarrow$ is applied last.

For example, in the expression $\neg Q \vee R$, the not ($\neg$) is applied first and to the Q only. Then the disjunction ($\vee$) is applied to the $\neg Q$ and R. The expression could have been written as $(\neg Q) \vee R$. As another example consider $P \vee Q \rightarrow R \wedge Q$. The disjunction ($\vee$) is applied to the P and the Q, the conjunction ($\wedge$) is applied to the R and the Q, and then, finally, the implication ($\rightarrow$) is applied. The expression could have been written as: $(P \vee Q) \rightarrow (R \wedge Q)$. In some expressions it is imperative to use brackets, otherwise the expression is not defined. For example, if we write $P \wedge Q \vee R$, because 'or' and 'and' have equal strength it is not clear whether we mean $(P \wedge Q) \vee R$ or $P \wedge (Q \vee R)$.

The next example illustrates truth tables for some more complex logical expressions.

WORKED EXAMPLE

15.8   Draw up the truth table for

(a) $[(P \rightarrow Q) \wedge (Q \rightarrow R)] \rightarrow (P \rightarrow R)$
(b) $\neg R \wedge [S \wedge (\neg S \vee R)]$

Solution (a) Table 15.11 shows the required truth table. There are several intermediate columns included which aid the calculation of the truth values in the final column.

Table 15.11
Truth table for
$[(P \to Q) \land (Q \to R)] \to (P \to R)$

P	Q	R	$P \to Q$	$Q \to R$	$(P \to Q) \land (Q \to R)$
T	T	T	T	T	T
T	T	F	T	F	F
T	F	T	F	T	F
T	F	F	F	T	F
F	T	T	T	T	T
F	T	F	T	F	F
F	F	T	T	T	T
F	F	F	T	T	T

$P \to R$	$[(P \to Q) \land (Q \to R)] \to (P \to R)$
T	T
F	T
T	T
F	T
T	T
T	T
T	T
T	T

Note that the final logical expression is always true. Such an expression is called a **tautology**.

(b) Table 15.12 shows the truth table for $\neg R \land [S \land (\neg S \lor R)]$.

Table 15.12
Truth table for
$\neg R \land [S \land (\neg S \lor R)]$

S	R	$\neg S$	$\neg R$	$\neg S \lor R$	$S \land (\neg S \lor R)$	$\neg R \land [S \land (\neg S \lor R)]$
T	T	F	F	T	T	F
T	F	F	T	F	F	F
F	T	T	F	T	F	F
F	F	T	T	T	F	F

The final expression has a truth value which is always false. Such an expression is called a **contradiction**.

Self-assessment questions 15.3

1. Explain the terms (a) contradiction, (b) tautology.

2. A logical expression involves (a) four, (b) five, (c) N simple propositions. How many rows are needed in the truth tables?

Exercise 15.3

1. (a) Draw up the truth tables for   (i) $\neg(P \vee Q)$,   (ii) $\neg P \wedge \neg Q$,   (iii) $\neg(P \wedge Q)$,   (iv) $\neg P \vee \neg Q$.
   (b) Expressions with the same truth values are said to be **logically equivalent**. Which pairs of expressions in (a) are logically equivalent?

2. Draw up the truth tables for
   (a) $\neg(\neg P)$ (b) $(P \wedge Q) \vee R$ (c) $P \wedge (Q \vee R)$

3. (a) Compare the truth tables of $P \rightarrow Q$ and $\neg Q \rightarrow \neg P$.
   (b) State whether the expressions are logically equivalent (see Question 1).

Test and assignment exercises 15

1. Which of the following statements are propositions?

   (a) Mike has passed the maths examination.
   (b) Mike is really good at maths.
   (c) The exam was more difficult than the one last year.
   (d) The average mark was 53.
   (e) Pete worked hard but failed the exam.

2. Draw up the truth tables for the following:

   (a) $(P \wedge Q) \rightarrow P$
   (b) $P \wedge (Q \rightarrow P)$
   (c) $(\neg P \vee Q) \rightarrow Q$
   (d) $(\neg P \vee Q) \rightarrow \neg Q$
   (e) $\neg(P \vee Q) \rightarrow \neg P$

3. Draw up the truth tables for

   (a) $P \wedge (Q \wedge R)$
   (b) $P \vee (Q \vee R)$
   (c) $(P \rightarrow \neg R) \wedge (R \rightarrow Q)$
   (d) $(R \wedge \neg P) \rightarrow (P \vee \neg R)$
   (e) $[P \vee (P \rightarrow Q)] \wedge (\neg Q \rightarrow R)$

Functions

03

16

Objectives: This chapter:

- explains what is meant by a function
- describes the notation used to write functions
- explains the terms 'independent variable' and 'dependent variable'
- explains what is meant by a composite function
- explains what is meant by the inverse of a function

16.1 Definition of a function

A **function** is a rule that receives an input and produces an output. It is shown schematically in Figure 16.1. For example, the rule may be 'add 2 to the input'. If 6 is the input, then $6 + 2 = 8$ will be the output. If -5 is the

Figure 16.1
A function produces an output from an input

input, then $-5 + 2 = -3$ will be the output. In general, if x is the input then $x + 2$ will be the output. Figure 16.2 illustrates this function schematically.

Figure 16.2
The function adds 2 to the input

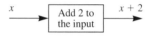

For a rule to be a function then it is crucial that only *one* output is produced for any given input.

| Key point | A function is a rule that produces a *single* output for any given input. |

The input to a function can usually take many values and so is called a **variable**. The output, too, varies depending upon the value of the input, and so is also a variable. The input is referred to as the **independent variable** because we are free to choose its value. The output is called the **dependent variable** because its value depends upon the value of the input.

16.2 Notation used for functions

We usually denote the input, the output and the function by letters or symbols. Commonly we use x to represent the input, y the output and f the function, although other letters will be used as well.

Consider again the example from §16.1. We let f be the function 'add 2 to the input' and we let x be the input. In mathematical notation we write

$$f : x \rightarrow x + 2$$

This means that the function f takes an input x and produces an output $x + 2$.

An alternative, but commonly used, notation is

$$f(x) = x + 2$$

The quantity $f(x)$ does not mean f times x but rather indicates that the function f acts on the quantity in the brackets. Because we also call the output y we can write $y = f(x) = x + 2$, or simply $y = x + 2$.

We could represent the same function using different letters. If h represents the function and t the input then we can write

$$h(t) = t + 2$$

WORKED EXAMPLES

16.1 A function multiplies the input by 4. Write down the function in mathematical notation.

Solution Let us call the function f and the input x. Then we have

$$f : x \rightarrow 4x \quad \text{or alternatively} \quad f(x) = 4x$$

If we call the output y, we can write $y = f(x) = 4x$, or simply $y = 4x$.

16.2 A function divides the input by 6 and then adds 3 to the result. Write the function in mathematical notation.

Solution   Let us call the function z and the input t. Then we have

$$z(t) = \frac{t}{6} + 3$$

16.3   A function f is given by the rule $f : x \rightarrow 9$, or alternatively as $f(x) = 9$. Describe in words what this function does.

Solution   Whatever the value of the input to this function, the output is always 9.

16.4   A function squares the input and then multiplies the result by 6. Write down the function using mathematical notation.

Solution   Let us call the function f, the input x and the output y. Then

$$y = f(x) = 6x^2$$

16.5   Describe in words what the following functions do:

(a) $h(x) = \dfrac{1}{x}$ (b) $g(t) = t + t^2$

Solution   (a) The function $h(x) = \dfrac{1}{x}$ divides 1 by the input.

(b) The function $g(t) = t + t^2$ adds the input to the square of the input.

Often we are given a function and need to calculate the output from a given input.

WORKED EXAMPLES

16.6   A function f is defined by $f(x) = 3x + 1$. Calculate the output when the input is (a) 4, (b) -1, (c) 0.

Solution   The function f multiplies the input by 3 and then adds 1 to the result.

(a) When the input is 4, the output is $3 \times 4 + 1 = 12 + 1 = 13$. We write

$$f(x = 4) = 3(4) + 1 = 12 + 1 = 13$$

or more simply

$$f(4) = 13$$

Note that 4 has been substituted for x in the formula for f.

(b) We require the output when the input is -1, that is $f(-1)$:

$$f(x = -1) = f(-1) = 3(-1) + 1 = -3 + 1 = -2$$

The output is -2 when the input is -1.

(c) We require the output when the input is 0, that is $f(0)$:

$$f(x = 0) = f(0) = 3(0) + 1 = 0 + 1 = 1$$

16.7 A function g is defined by $g(t) = 2t^2 - 1$. Find

(a) $g(3)$ (b) $g(0.5)$ (c) $g(-2)$

Solution (a) We obtain $g(3)$ by substituting 3 for t:

$$g(3) = 2(3)^2 - 1 = 2(9) - 1 = 17$$

(b) $g(0.5) = 2(0.5)^2 - 1 = 0.5 - 1 = -0.5$.
(c) $g(-2) = 2(-2)^2 - 1 = 8 - 1 = 7$.

16.8 A function h is defined by $h(x) = \dfrac{x}{3} + 1$. Find

(a) $h(3)$ (b) $h(t)$ (c) $h(\alpha)$ (d) $h(2\alpha)$ (e) $h(2x)$

Solution (a) If $h(x) = \dfrac{x}{3} + 1$ then $h(3) = \dfrac{3}{3} + 1 = 1 + 1 = 2$.

(b) The function h divides the input by 3 and then adds 1. We require the output when the input is t, that is we require $h(t)$. Now

$$h(t) = \frac{t}{3} + 1$$

since the input t has been divided by 3 and then 1 has been added to the result. Note that $h(t)$ is obtained by substituting t in place of x in $h(x)$.

(c) We require the output when the input is α. This is obtained by substituting α in place of x. We find

$$h(\alpha) = \frac{\alpha}{3} + 1$$

(d) We require the output when the input is 2α. We substitute 2α in place of x. This gives

$$h(2\alpha) = \frac{2\alpha}{3} + 1$$

(e) We require the output when the input is $2x$. We substitute $2x$ in place of x. That is,

$$h(2x) = \frac{2x}{3} + 1$$

16.9 Given $f(x) = x^2 + x - 1$ write expressions for

(a) $f(\alpha)$ (b) $f(x + 1)$ (c) $f(2t)$

Solution (a) Substituting α in place of x we obtain

$$f(\alpha) = \alpha^2 + \alpha - 1$$

(b) Substituting $x + 1$ for x we obtain

$$f(x + 1) = (x + 1)^2 + (x + 1) - 1$$
$$= x^2 + 2x + 1 + x + 1 - 1$$
$$= x^2 + 3x + 1$$

(c) Substituting $2t$ in place of x we obtain

$$f(2t) = (2t)^2 + 2t - 1 = 4t^2 + 2t - 1$$

$<$ is the symbol for less than.
$\leqslant$ is the symbol for less than or equal to.
$\geqslant$ is the symbol for greater than or equal to.

Sometimes a function uses different rules on different intervals. For example, we could define a function as

$$f(x) = \begin{cases} 3x & \text{when} & 0 \leqslant x \leqslant 4 \\ 2x + 6 & \text{when} & 4 < x < 5 \\ 9 & \text{when} & x \geqslant 5 \end{cases}$$

Here the function is defined in three 'pieces'. The value of x determines which part of the definition is used to evaluate the function. The function is said to be a **piecewise** function.

WORKED EXAMPLE

16.10   A piecewise function is defined by

$$y(x) = \begin{cases} x^2 + 1 & \text{when} & -1 \leqslant x \leqslant 2 \\ 3x & \text{when} & 2 < x \leqslant 6 \\ 2x + 1 & \text{when} & x > 6 \end{cases}$$

Evaluate
(a) $y(0)$ (b) $y(4)$ (c) $y(2)$ (d) $y(7)$

Solution   (a) We require the value of y when $x = 0$. Since 0 lies between -1 and 2 we use the first part of the definition, that is $y = x^2 + 1$. Hence

$$y(0) = 0^2 + 1 = 1$$

(b) We require y when $x = 4$. The second part of the definition must be used because x lies between 2 and 6. Therefore

$$y(4) = 3(4) = 12$$

(c) We require y when $x = 2$. The value $x = 2$ occurs in the first part of the definition. Therefore

$$y(2) = 2^2 + 1 = 5$$

(d) We require y when $x = 7$. The final part of the function must be used. Therefore

$$y(7) = 2(7) + 1 = 15$$

Self-assessment questions 16.2

1. Explain what is meant by a function.

2. Explain the meaning of the terms 'dependent variable' and 'independent variable'.

3. Given $f(x)$, is the statement '$f(1/x)$ means $1/f(x)$' true or false?

4. Give an example of a function $f(x)$ such that $f(2) = f(3)$, that is the outputs for the inputs 2 and 3 are identical.

Exercise 16.2

1. Describe in words each of the following functions:
   (a) $h(t) = 10t$ (b) $g(x) = -x + 2$
   (c) $h(t) = 3t^4$ (d) $f(x) = \dfrac{4}{x^2}$
   (e) $f(x) = 3x^2 - 2x + 9$ (f) $f(x) = 5$
   (g) $f(x) = 0$

2. Describe in words each of the following functions:
   (a) $f(t) = 3t^2 + 2t$ (b) $g(x) = 3x^2 + 2x$
   Comment upon your answers.

3. Write the following functions using mathematical notation:
   (a) The input is cubed and the result is divided by 12.
   (b) The input is added to 3 and the result is squared.
   (c) The input is squared and added to 4 times the input. Finally, 10 is subtracted from the result.
   (d) The input is squared and added to 5. Then the input is divided by this result.
   (e) The input is cubed and then 1 is subtracted from the result.

   (f) 1 is subtracted from the input and the result is squared.
   (g) Twice the input is subtracted from 7 and the result is divided by 4.
   (h) The output is always -13 whatever the value of the input.

4. Given the function $A(n) = n^2 - n + 1$ evaluate
   (a) $A(2)$ (b) $A(3)$ (c) $A(0)$
   (d) $A(-1)$

5. Given $y(x) = (2x - 1)^2$ evaluate
   (a) $y(1)$ (b) $y(-1)$ (c) $y(-3)$
   (d) $y(0.5)$ (e) $y(-0.5)$

6. The function f is given by $f(t) = 4t + 6$. Write expressions for
   (a) $f(t + 1)$ (b) $f(t + 2)$
   (c) $f(t + 1) - f(t)$ (d) $f(t + 2) - f(t)$

7. The function $f(x)$ is defined by $f(x) = 2x^2 - 3$. Write expressions for
   (a) $f(n)$ (b) $f(z)$ (c) $f(t)$ (d) $f(2t)$
   (e) $f\left(\dfrac{1}{z}\right)$ (f) $f\left(\dfrac{3}{n}\right)$ (g) $f(-x)$
   (h) $f(-4x)$ (i) $f(x + 1)$ (j) $f(2x - 1)$

8. Given the function $a(p) = p^2 + 3p + 1$ write an expression for $a(p+1)$. Verify that $a(p+1) - a(p) = 2p + 4$.

9. Sometimes the output from one function forms the input to another function. Suppose we have two functions: f given by $f(t) = 2t$, and h given by $h(t) = t + 1$. $f(h(t))$ means that t is input to h, and the output from h is input to f. Evaluate
(a) $f(3)$ (b) $h(2)$ (c) $f(h(2))$
(d) $h(f(3))$

10. The functions f and h are defined as in Question 9. Write down expressions for
(a) $f(h(t))$ (b) $h(f(t))$

11. A function is defined by
$$f(x) = \begin{cases} x & 0 \leqslant x < 1 \\ 2 & x = 1 \\ 1 & x > 1 \end{cases}$$
Evaluate
(a) $f(0.5)$ (b) $f(1.1)$ (c) $f(1)$

16.3 Composite functions

Sometimes we wish to apply two or more functions, one after the other. The output of one function becomes the input of the next function.

Suppose $f(x) = 2x$ and $g(x) = x + 3$. We note that the function $f(x)$ doubles the input while the function $g(x)$ adds 3 to the input. Now, we let the output of $g(x)$ become the input to $f(x)$. Figure 16.3 illustrates the position.

Figure 16.3
The output of g is the input of f

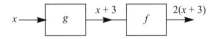

We have

$$g(x) = x + 3$$

$$f(x+3) = 2(x+3) = 2x + 6$$

Note that $f(x+3)$ may be written as $f(g(x))$. Referring to Figure 16.3 we see that the initial input is x and that the final output is $2x + 6$. The functions $g(x)$ and $f(x)$ have been combined. We call $f(g(x))$ a **composite function**. It is composed of the individual functions $f(x)$ and $g(x)$. In this example we have

$$f(g(x)) = 2x + 6$$

WORKED EXAMPLES

16.11   Given $f(x) = 2x$ and $g(x) = x + 3$ find the composite function $g(f(x))$.

Solution   The output of $f(x)$ becomes the input to $g(x)$. Figure 16.4 illustrates this.

Figure 16.4
The composite function
$g(f(x))$

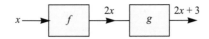

We see that

$$g(f(x)) = g(2x)$$

$$= 2x + 3$$

Note that in general $f(g(x))$ and $g(f(x))$ are different functions.

16.12   Given $f(t) = t^2 + 1$, $g(t) = \frac{3}{t}$ and $h(t) = 2t$ determine each of the following composite functions:

(a) $f(g(t))$ (b) $g(h(t))$ (c) $f(h(t))$ (d) $f(g(h(t)))$ (e) $g(f(h(t)))$

Solution   (a) $f(g(t)) = f\left(\dfrac{3}{t}\right) = \left(\dfrac{3}{t}\right)^2 + 1 = \dfrac{9}{t^2} + 1$

(b) $g(h(t)) = g(2t) = \dfrac{3}{2t}$

(c) $f(h(t)) = f(2t) = (2t)^2 + 1 = 4t^2 + 1$

(d) $f(g(h(t))) = f\left(\dfrac{3}{2t}\right)$ using (b)

$$= \left(\dfrac{3}{2t}\right)^2 + 1$$

$$= \dfrac{9}{4t^2} + 1$$

(e) $g(f(h(t))) = g(4t^2 + 1)$ using (c)

$$= \dfrac{3}{4t^2 + 1}$$

Self-assessment questions 16.3

1. Explain the term 'composite function'.

2. Give examples of functions $f(x)$ and $g(x)$ such that $f(g(x))$ and $g(f(x))$ are equal.

Exercise 16.3

1. Given $f(x) = 4x$ and $g(x) = 3x - 2$ find
    (a) $f(g(x))$ (b) $g(f(x))$

2. If $x(t) = t^3$ and $y(t) = 2t$ find
    (a) $y(x(t))$ (b) $x(y(t))$

3. Given $r(x) = \dfrac{1}{2x}$, $s(x) = 3x$ and

    $t(x) = x - 2$ find

    (a) $r(s(x))$ (b) $t(s(x))$ (c) $t(r(s(x)))$
    (d) $r(t(s(x)))$ (e) $r(s(t(x)))$

4. A function can be combined with itself. This is known as **self-composition**. Given $v(t) = 2t + 1$ find
    (a) $v(v(t))$ (b) $v(v(v(t)))$

5. Given $m(t) = (t + 1)^3$, $n(t) = t^2 - 1$ and $p(t) = t^2$ find
    (a) $m(n(t))$ (b) $n(m(t))$ (c) $m(p(t))$
    (d) $p(m(t))$ (e) $n(p(t))$ (f) $p(n(t))$
    (g) $m(n(p(t)))$ (h) $p(p(t))$ (i) $n(n(t))$
    (j) $m(m(t))$

16.4 The inverse of a function

Note that the symbol f^{-1} does not mean $\dfrac{1}{f}$.

We have described a function f as a rule which receives an input, say x, and generates an output, say y. We now consider the reversal of that process, namely finding a function which receives y as input and generates x as the output. If such a function exists it is called the **inverse function** of f. Figure 16.5 illustrates this schematically. The inverse of $f(x)$ is denoted by $f^{-1}(x)$.

Figure 16.5
The inverse of f reverses the effect of f

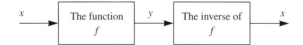

WORKED EXAMPLES

16.13   The functions f and g are defined by

$$f(x) = 2x \qquad g(x) = \frac{x}{2}$$

(a) Verify that f is the inverse of g.

(b) Verify that g is the inverse of f.

Solution   (a) The function g receives an input of x and generates an output of $x/2$; that is, it halves the input. In order to reverse the process, the inverse of g should receive $x/2$ as input and generate x as output. Now consider the function $f(x) = 2x$. This function doubles the input. Hence

$$f\left(\frac{x}{2}\right) = 2\left(\frac{x}{2}\right) = x$$

The function f has received $x/2$ as input and generated x as output. Hence f is the inverse of g. This is shown schematically in Figure 16.6.

Figure 16.6
The function f is the
inverse of g

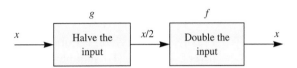

(b) The function f receives x as input and generates $2x$ as output. In order to reverse the process, the inverse of f should receive $2x$ as input and generate x as output. Now $g(x) = x/2$, that is the input is halved, and so

$$g(2x) = \frac{2x}{2} = x$$

Hence g is the inverse of f. This is shown schematically in Figure 16.7.

Figure 16.7
The function g is the
inverse of f

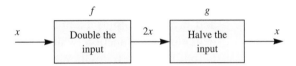

16.14   Find the inverse of the function $f(x) = 3x - 4$.

Solution   The function f multiplies the input by 3 and subtracts 4 from the result. To reverse the process, the inverse function, g say, must add 4 to the input and then divide the result by 3. Hence

$$g(x) = \frac{x + 4}{3}$$

16.15   Find the inverse of $h(t) = -\frac{1}{2}t + 5$.

Solution   The function h multiplies the input by $-\frac{1}{2}$ and then adds 5 to the result. Therefore the inverse function, g say, must subtract 5 from the input and then divide the result by $-\frac{1}{2}$. Hence

$$g(t) = \frac{(t - 5)}{-1/2} = -2(t - 5) = -2t + 10$$

There is an algebraic method of finding an inverse function that is often easier to apply. Suppose we wish to find the inverse of the function $f(x) = 6 - 2x$. We let

$$y = 6 - 2x$$

and then transpose this for x. This gives

$$x = \frac{6 - y}{2}$$

Finally, we interchange x and y to give $y = (6 - x)/2$. This is the required inverse function. To summarize these stages:

Key point   To find the inverse of $y = f(x)$,

- transpose the formula to make x the subject
- interchange x and y

The result is the required inverse function.

We shall meet some functions that do not have an inverse function. For example, consider the function $f(x) = x^2$. If 3 is the input, the output is 9. Now if -3 is the input, the output will also be 9 since $(-3)^2 = 9$. In order to reverse this process an inverse function would have to take an input of 9 and produce outputs of both 3 and -3. However, this contradicts the definition of a function, which states that a function must have only *one* output for a given input. We say that $f(x) = x^2$ does not have an inverse function.

Self-assessment questions 16.4

1. Explain what is meant by the inverse of a function.

2. Explain why the function $f(x) = 4x^4$ does not possess an inverse function.

Exercise 16.4

1. Find the inverse of each of the following functions:
   (a) $f(x) = 3x$ (b) $f(x) = \dfrac{x}{4}$
   (c) $f(x) = x + 1$ (d) $f(x) = x - 3$
   (e) $f(x) = 3 - x$ (f) $f(x) = 2x + 6$
   (g) $f(x) = 7 - 3x$ (h) $f(x) = \dfrac{1}{x}$
   (i) $f(x) = \dfrac{3}{x}$ (j) $f(x) = -\dfrac{3}{4x}$

2. Find the inverse, $f^{-1}(x)$, when $f(x)$ is given by
   (a) $6x$ (b) $6x + 1$ (c) $x + 6$
   (d) $\dfrac{x}{6}$ (e) $\dfrac{6}{x}$

3. Find the inverse, $g^{-1}(t)$, when $g(t)$ is given by

    (a) $3t + 1$ (b) $\dfrac{1}{3t + 1}$

    (c) t^3 (d) $3t^3$

    (e) $3t^3 + 1$ (f) $\dfrac{3}{t^3 + 1}$

4. The functions $g(t)$ and $h(t)$ are defined by

    $$g(t) = 2t - 1, \; h(t) = 4t + 3$$

    Find
    (a) the inverse of $h(t)$, that is $h^{-1}(t)$
    (b) the inverse of $g(t)$, that is $g^{-1}(t)$
    (c) $g^{-1}(h^{-1}(t))$ (d) $h(g(t))$
    (e) the inverse of $h(g(t))$
    What observations do you make from (c) and (e)?

Test and assignment exercises 16

1. Given $r(t) = t^2 - t/2 + 4$ evaluate

    (a) $r(0)$ (b) $r(-1)$ (c) $r(2)$ (d) $r(3.6)$ (e) $r(-4.6)$

2. A function is defined as $h(t) = t^2 - 7$.

    (a) State the dependent variable.   (b) State the independent variable.

3. Given the functions $a(x) = x^2 + 1$ and $b(x) = 2x + 1$ write expressions for

    (a) $a(\alpha)$ (b) $b(t)$ (c) $a(2x)$ (d) $b\left(\dfrac{x}{3}\right)$ (e) $a(x + 1)$ (f) $b(x + h)$

    (g) $b(x - h)$ (h) $a(b(x))$ (i) $b(a(x))$

4. Find the inverse of each of the following functions:

    (a) $f(x) = \pi - x$ (b) $h(t) = \dfrac{t}{3} + 2$ (c) $r(n) = \dfrac{1}{n}$ (d) $r(n) = \dfrac{1}{n - 1}$

    (e) $r(n) = \dfrac{2}{n - 1}$ (f) $r(n) = \dfrac{a}{n - b}$ where a and b are constants

5. Given $A(n) = n^2 + n - 6$ find expressions for
    (a) $A(n + 1)$ (b) $A(n - 1)$
    (c) $2A(n + 1) - A(n) + A(n - 1)$

6. Find $h^{-1}(x)$ when $h(x)$ is given by
    (a) $\dfrac{x + 1}{3}$ (b) $\dfrac{3}{x + 1}$ (c) $\dfrac{x + 1}{x}$ (d) $\dfrac{x}{x + 1}$

7. Given $v(t) = 4t - 2$ find
    (a) $v^{-1}(t)$ (b) $v(v(t))$

8. Given $h(x) = 9x - 6$ and $g(x) = \dfrac{1}{3x}$ find

   (a) $h(g(x))$ (b) $g(h(x))$

9. Given $f(x) = (x + 1)^2$, $g(x) = 4x$ and $h(x) = x - 1$ find

   (a) $f(g(x))$ (b) $g(f(x))$ (c) $f(h(x))$ (d) $h(f(x))$ (e) $g(h(x))$
   (f) $h(g(x))$ (g) $f(g(h(x)))$ (h) $g(h(f(x)))$

10. Given $x(t) = t^3$, $y(t) = \dfrac{1}{t+1}$ and $z(t) = 3t - 1$ find

    (a) $y(x(t))$ (b) $y(z(t))$ (c) $x(y(z(t)))$ (d) $z(y(x(t)))$

11. Write each of the following functions using mathematical notation:

    (a) The input is multiplied by 7.
    (b) Five times the square of the input is subtracted from twice the cube of the input.
    (c) The output is 6.
    (d) The input is added to the reciprocal of the input.
    (e) The input is multiplied by 11 and then 6 is subtracted from this. Finally, 9 is divided by this result.

12. Find the inverse of

$$f(x) = \frac{x + 1}{x - 1}$$

17

Graphs of functions

Objectives: This chapter:

- shows how coordinates are plotted on the $x-y$ plane
- introduces notation to denote intervals on the x axis
- introduces symbols to denote 'greater than' and 'less than'
- shows how to draw a graph of a function
- explains what is meant by the domain and range of a function
- explains how to use the graph of a function to solve equations
- explains how to solve simultaneous equations using graphs

In Chapter 16 we introduced functions and represented them by algebraic expressions. Another useful way of representing functions is pictorially, by means of **graphs**.

17.1 The $x-y$ plane

We introduce horizontal and vertical **axes** as shown in Figure 17.1. These axes intersect at a point O called the **origin**. The horizontal axis is used to represent the independent variable, commonly x, and the vertical axis is used to represent the dependent variable, commonly y. The region shown is then referred to as the $x-y$ plane.

A **scale** is drawn on both axes in such a way that at the origin $x = 0$ and $y = 0$. Positive x values lie to the right and negative x values to the left. Positive y values are above the origin, negative y values are below the origin. It is not essential that the scales on both axes are the same. Note that anywhere on the x axis the value of y must be zero. Anywhere on the y axis the value of x must be zero. Each point in the plane corresponds to a specific value of x and y. We usually call the x value the x **coordinate** and call the y

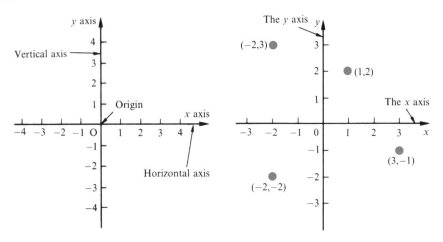

Figure 17.1
The x–y plane

Figure 17.2
Several points in the x–y plane

value the y **coordinate**. To refer to a specific point we give both its coordinates in brackets in the form (x, y), always giving the x coordinate first.

WORKED EXAMPLE

17.1   Draw the x–y plane and on it mark the points whose coordinates are $(1, 2)$, $(3, -1)$, $(-2, 3)$, $(-2, -2)$.

Solution   The plane is drawn in Figure 17.2, and the given points are indicated. Note that the first coordinate in each bracket is the x coordinate.

Self-assessment questions 17.1

1. Explain the terms 'horizontal axis', 'vertical axis' and 'origin'.

2. When giving the coordinates of a point in the x–y plane, which coordinate is always given first?

3. State the coordinates of the origin.

Exercise 17.1

MyMathLab

1. Plot and label the following points in the x–y plane:
   $(-2, 0), (2, 0), (0, -2), (0, 2), (0, 0)$.

2. A set of points is plotted and then these points are joined by a straight line that is parallel to the x axis. What can you deduce about the coordinates of these points?

3. A set of points is plotted and then these points are joined by a straight line which is parallel to the y axis. What can you deduce about the coordinates of these points?

17.2 Inequalities and intervals

We often need only part of the x axis when plotting graphs. For example, we may be interested only in that part of the x axis running from $x = 1$ to $x = 3$ or from $x = -2$ to $x = 7.5$. Such parts of the x axis are called **intervals**. In order to describe concisely intervals on the x axis, we introduce some new notation.

Greater than and less than

We use the symbol > to mean 'greater than'. For example, $6 > 5$ and $-4 > -5$ are both true statements. If $x_1 > x_2$ then x_1 is to the right of x_2 on the x axis. The symbol $\geqslant$ means 'greater than or equal to', so, for example, $10 \geqslant 8$ and $8 \geqslant 8$ are both true.

The symbol < means 'less than'. We may write $4 < 5$ for example. If $x_1 < x_2$ then x_1 is to the left of x_2 on the x axis. Finally, $\leqslant$ means 'less than or equal to'. Hence $6 \leqslant 9$ and $6 \leqslant 6$ are both true statements.

Intervals

$\mathbb{R}$ is called the set of real numbers.

We often need to represent intervals on the number line. To help us do this we introduce the set $\mathbb{R}$. $\mathbb{R}$ is the symbol we use to denote all numbers from minus infinity to plus infinity. All numbers, including integers, fractions and decimals, belong to $\mathbb{R}$. To show that a number, x, is in this set, we write $x \in \mathbb{R}$ where $\in$ means 'belongs to'.

There are three different kinds of interval:

(a) *The closed interval* An interval that includes its end-points is called a **closed interval**. All the numbers from 1 to 3, including both 1 and 3, comprise a closed interval, and this is denoted using square brackets, $[1, 3]$. Any number in this closed interval must be greater than or equal to 1, and also less than or equal to 3. Thus if x is any number in the interval, then $x \geqslant 1$ and also $x \leqslant 3$. We write this compactly as $1 \leqslant x \leqslant 3$. Finally we need to show explicitly that x is any number on the x axis, and not just an integer for example. So we write $x \in \mathbb{R}$. Hence the interval $[1, 3]$ can be expressed as

$$\{x : x \in \mathbb{R},\ 1 \leqslant x \leqslant 3\}$$

This means the set contains all the numbers x with $x \geqslant 1$ and $x \leqslant 3$.

(b) *The open interval* Any interval that does not include its end-points is called an **open interval**. For example, all the numbers from 1 to 3, but excluding 1 and 3, comprise an open interval. Such an interval is denoted using round brackets, $(1, 3)$. The interval may be written using set notation as

$$\{x : x \in \mathbb{R},\ 1 < x < 3\}$$

We say that x is **strictly greater** than 1, and **strictly less** than 3, so that the values of 1 and 3 are excluded from the interval.

(c) *The semi-open or semi-closed interval* An interval may be open at one end and closed at the other. Such an interval is called **semi-open** or, as some authors say, **semi-closed**. The interval $(1, 3]$ is a semi-open interval. The square bracket next to the 3 shows that 3 is included in the interval; the round bracket next to the 1 shows that 1 is not included in the interval. Using set notation we would write

$$\{x: x \in \mathbb{R}, \ 1 < x \leqslant 3\}$$

When marking intervals on the x axis there is a notation to show whether or not the end-point is included. We use ● to show that an end-point is included (i.e. closed) whereas ○ is used to denote an end-point that is not included (i.e. open).

WORKED EXAMPLES

17.2 Describe the interval $[-3, 4]$ using set notation and illustrate it on the x axis.

Solution The interval $[-3, 4]$ is given in set notation by

$$\{x: x \in \mathbb{R}, \ -3 \leqslant x \leqslant 4\}$$

Figure 17.3 illustrates the interval.

Figure 17.3
The interval $[-3, 4]$

17.3 Describe the interval $(1, 4)$ using set notation and illustrate it on the x axis.

Solution The interval $(1, 4)$ is expressed as

$$\{x: x \in \mathbb{R}, \ 1 < x < 4\}$$

Figure 17.4 illustrates the interval on the x axis.

Figure 17.4
The interval $(1, 4)$

Self-assessment questions 17.2

1. Explain what is meant by (a) a closed interval, (b) an open interval, (c) a semi-closed interval.

2. Describe the graphical notation used to denote that an end-point is (a) included, (b) not included.

Exercise 17.2

1. Describe the following intervals using set notation. Draw the intervals on the x axis.
   (a) $[2, 6]$ (b) $(6, 8]$ (c) $(-2, 0)$ (d) $[-3, -1.5]$

2. Which of the following are true?
   (a) $8 > 3$ (b) $-3 < 8$ (c) $-3 \leqslant -3$ (d) $0.5 \geqslant 0.25$ (e) $0 > 9$
   (f) $0 \leqslant 9$ (g) $-7 \geqslant 0$ (h) $-7 < 0$

17.3 Plotting the graph of a function

The method of plotting a graph is best illustrated by example.

WORKED EXAMPLE

17.4   Plot a graph of $y = 2x - 1$ for $-3 \leqslant x \leqslant 3$.

Solution   We first calculate the value of y for several values of x. A table of x values and corresponding y values is drawn up as shown in Table 17.1.

Table 17.1
Values of x and y when
$y = 2x - 1$

x	-3	-2	-1	0	1	2	3
y	-7	-5	-3	-1	1	3	5

The independent variable x varies from -3 to 3; the dependent variable y varies from -7 to 5. In order to accommodate all these values the scale on the x axis must vary from -3 to 3, and that on the y axis from -7 to 5. Each pair of values of x and y is represented by a unique point in the x–y plane, with coordinates (x, y). Each pair of values in the table is plotted as a point and then the points are joined to form the graph (Figure 17.5). By joining the points we see that, in this example, the graph is a straight line. We were asked to plot the graph for $-3 \leqslant x \leqslant 3$. Therefore, we have indicated each end-point of the graph by a ● to show that it is included. This follows the convention used for labelling closed intervals given in §17.2.

In the previous example, although we have used x and y as independent and dependent variables, clearly other letters could be used. The points to remember are that the axis for the independent variable is horizontal and

Figure 17.5
A graph of $y = 2x - 1$

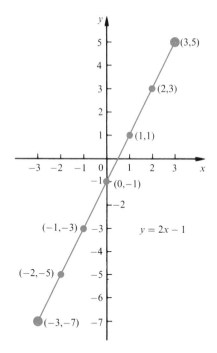

the axis for the dependent variable is vertical. When coordinates are given the value of the independent variable is always given first.

WORKED EXAMPLE

17.5   (a) Plot a graph of $y = x^2 - 2x + 2$ for $-2 \leqslant x \leqslant 3$.

(b) Use your graph to determine which of the following points lie on the graph: $(1, 2)$, $(0, 2)$, and $(1, 1)$.

Solution   (a) Table 17.2 gives values of x and the corresponding values of y. The independent variable is x and so the x axis is horizontal. The calculated points are plotted and joined. Figure 17.6 shows the graph. In this example the graph is not a straight line but rather a curve.

(b) The point $(1, 2)$ has an x coordinate of 1 and a y coordinate of 2. The point is labelled by A on Figure 17.6. Points $(0, 2)$ and $(1, 1)$ are plotted and labelled by B and C. From the figure we can see that $(0, 2)$ and $(1, 1)$ lie on the graph.

Table 17.2
Values of x and y when
$y = x^2 - 2x + 2$

x	-2	-1	0	1	2	3
y	10	5	2	1	2	5

Figure 17.6
A graph of
$y = x^2 - 2x + 2$

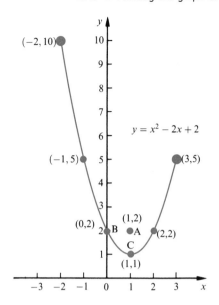

Self-assessment questions 17.3

1. Which variable is plotted vertically on a graph – the independent or dependent?

2. Suppose we wished to plot a graph of the function $g = 4t^2 + 3t - 2$. Upon which axis would the variable t be plotted?

Exercise 17.3

1. Plot graphs of the following functions:
   (a) $y = f(x) = 2x + 7$ for $-4 \leqslant x \leqslant 4$.
   (b) $y = f(x) = -2x$ for values of x between -3 and 3 inclusive.
   (c) $y = f(x) = x$ for $-5 \leqslant x \leqslant 5$.
   (d) $y = f(x) = -x$ for $-5 \leqslant x \leqslant 5$.

2. Plot a graph of the function $y = f(x) = 7$, for $-2 \leqslant x \leqslant 2$.

3. A function is given by $y = f(x) = x^3 + x$ for $-3 \leqslant x \leqslant 3$.
   (a) Plot a graph of the function.
   (b) Determine which of the following points lie on the curve: $(0, 1)$, $(1, 0)$, $(1, 2)$ and $(-1, -2)$.

4. Plot, on the same axes, graphs of the functions $f(x) = x + 2$ for $-3 \leqslant x \leqslant 3$, and $g(x) = -\frac{1}{2}x + \frac{1}{2}$ for $-3 \leqslant x \leqslant 3$. State the coordinates of the point where the graphs intersect.

5. Plot a graph of the function $f(x) = x^2 - x - 1$ for $-2 \leqslant x \leqslant 3$. State the coordinates of the points where the curve cuts (a) the horizontal axis, (b) the vertical axis.

6. (a) Plot a graph of the function $y = 1/x$ for $0 < x \leqslant 3$.
   (b) Deduce the graph of the function $y = -1/x$ for $0 < x \leqslant 3$ from your graph in part (a).

7. Plot a graph of the function defined by

$$r(t) = \begin{cases} t^2 & 0 \leqslant t \leqslant 3 \\ -2t + 15 & 3 < t \leqslant 7.5 \end{cases}$$

8. On the same axes plot a graph of $y = 2x$, $y = 2x + 1$ and $y = 2x + 3$, for $-3 \leqslant x \leqslant 3$. What observations can you make?

17.4 The domain and range of a function

The set of values that we allow the independent variable to take is called the **domain** of the function. If the domain is not actually specified in any particular example, it is taken to be the largest set possible. The set of values taken by the output is called the **range** of the function.

WORKED EXAMPLES

17.6   The function f is given by $y = f(x) = 2x$, for $1 \leqslant x \leqslant 3$.

(a) State the independent variable.

(b) State the dependent variable.

(c) State the domain of the function.

(d) Plot a graph of the function.

(e) State the range of the function.

Solution   (a) The independent variable is x.

(b) The dependent variable is y.

(c) Since we are given $1 \leqslant x \leqslant 3$ the domain of the function is the interval $[1, 3]$, that is all values from 1 to 3 inclusive.

(d) The graph of $y = f(x) = 2x$ is shown in Figure 17.7.

Figure 17.7
Graph of $y = f(x) = 2x$

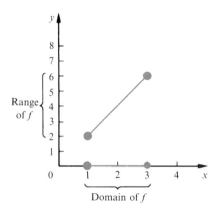

(e) The range is the set of values taken by the output, y. From the graph we see that as x varies from 1 to 3 then y varies from 2 to 6. Hence the range of the function is $[2, 6]$.

17.7 Consider the function $y = f(t) = 4t$.

(a) State which is the independent variable and which is the dependent variable.

(b) Plot a graph of the function and give its domain and range.

Solution (a) The independent variable is t and the dependent variable is y.

(b) No domain is specified so it is taken to be the largest set possible. The domain is $\mathbb{R}$, the set of all (real) numbers. However, it would be impractical to draw a graph whose domain was the whole extent of the t axis and so a selected portion is shown in Figure 17.8. Similarly only a restricted portion of the range can be drawn, although in this example the range is also $\mathbb{R}$.

Figure 17.8
A graph of $f(t) = 4t$

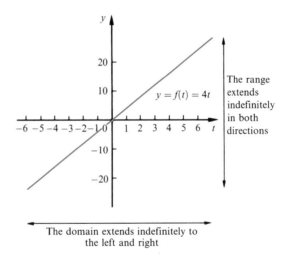

The range extends indefinitely in both directions

The domain extends indefinitely to the left and right

17.8 Consider the function $y = f(x) = x^2 - 1$: (a) state the independent variable; (b) state the dependent variable; (c) state the domain; (d) plot a graph of f and determine the range.

Solution (a) The independent variable is x.

(b) The dependent variable is y.

(c) No domain is specified so it is taken to be the largest set possible. The domain is $\mathbb{R}$.

(d) A table of values and graph of $f(x) = x^2 - 1$ are shown in Figure 17.9. From the graph we see that the smallest value of y is -1, which occurs when $x = 0$. Whether x increases or decreases from 0, the value of y increases. The range is thus all values greater than or equal to -1.

Figure 17.9
A graph of $f(x) = x^2 - 1$

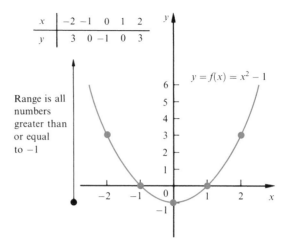

x	−2	−1	0	1	2
y	3	0	−1	0	3

Range is all numbers greater than or equal to −1

$y = f(x) = x^2 - 1$

We can write the range using the set notation introduced in §17.2 as

$$\text{range} = \{y : y \in \mathbb{R},\ y \geqslant -1\}$$

17.9   Given the function $y = 1/x$, state the domain and range.

Solution   The domain is the largest set possible. All values of x are permissible except $x = 0$ since $\frac{1}{0}$ is not defined. It is never possible to divide by 0. Thus the domain comprises all real numbers except 0. A table of values and graph are shown in Figure 17.10. As x varies, y can take on any value except 0. In this example we see that the range is thus the same as the domain. We note that the graph is split at $x = 0$. For small, positive

Figure 17.10
A graph of $y = 1/x$

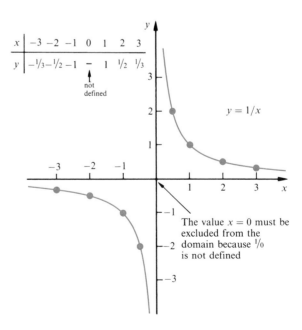

x	−3	−2	−1	0	1	2	3
y	−1/3	−1/2	−1	–	1	1/2	1/3

not defined

$y = 1/x$

The value $x = 0$ must be excluded from the domain because $1/0$ is not defined

values of x, y is large and positive. The graph approaches the y axis for very small, positive values of x. For small, negative values of x, y is large and negative. Again, the graph approaches the y axis for very small, negative values of x. We say that the y axis is an **asymptote**. In general, if the graph of a function approaches a straight line we call that line an asymptote.

WORKED EXAMPLE

17.10   Consider again the piecewise function in Worked Example 16.10:

$$y(x) = \begin{cases} x^2 + 1 & -1 \leqslant x \leqslant 2 \\ 3x & 2 < x \leqslant 6 \\ 2x + 1 & x > 6 \end{cases}$$

(a) Plot a graph of the function.

(b) State the domain of the function.

(c) State the range of the function.

Solution   (a) Figure 17.11 shows a graph of $y(x)$.

Figure 17.11
A graph of $y(x)$ for
Worked Example 17.10

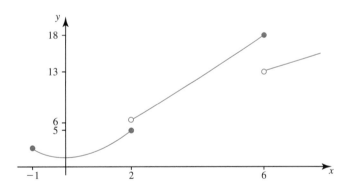

The different rules that make up the definition of y determine the three different pieces of the graph. Note the use of ● and ○ to denote the closed and open end-points.

(b) The domain is $[-1, \infty)$, that is

domain $= \{x \in \mathbb{R}, \ x \geqslant -1\}$

(c) The smallest value of y is $y = 1$, which occurs when $x = 0$. From the graph we see that the range is $[1, 5]$ and $(6, \infty)$, that is

range $= \{y \in \mathbb{R}, \ 1 \leqslant y \leqslant 5 \text{ and } y > 6\}$

Self-assessment questions 17.4

1. Explain the terms 'domain of a function' and 'range of a function'.

2. Explain why the value $x = 3$ must be excluded from any domain of the function $f(x) = 7/(x - 3)$.

3. Give an example of a function for which we must exclude the value $x = -2$ from the domain.

Exercise 17.4

1. Plot graphs of the following functions. In each case state (i) the independent variable, (ii) the dependent variable, (iii) the domain, (iv) the range.
   (a) $y = x - 1$, $6 \leqslant x \leqslant 10$ (b) $h = t^2 + 3$, $4 \leqslant t < 5$ (c) $m = 3n - 2$, $-1 < n < 1$
   (d) $y = 2x + 4$ (e) $X = y^3$ (f) $k = 4r^2 + 5$, $5 \leqslant r \leqslant 10$

2. A function is defined by

$$q(t) = \begin{cases} t^2 & 0 < t < 3 \\ 6 + t & 3 \leqslant t < 5 \\ -2t + 21 & 5 \leqslant t \leqslant 10.5 \end{cases}$$

   (a) Plot a graph of this function.   (b) State the domain of the function.
   (c) State the range of the function.   (d) Evaluate $q(1)$, $q(3)$, $q(5)$ and $q(7)$.

17.5 Solving equations using graphs

In Chapter 11 we used algebraic methods to solve equations. However, many equations cannot be solved exactly using such methods. In such cases it is often useful to use a graphical approach. The following examples illustrate the method of using graphs to solve equations.

WORKED EXAMPLES

17.11   Find graphically all solutions in the interval $[-3, 3]$ of the equation $x^2 + x - 3 = 0$.

Solution   We draw a graph of $y(x) = x^2 + x - 3$. Table 17.3 gives x and y values and Figure 17.12 shows a graph of the function. We have plotted $y = x^2 + x - 3$ and wish to solve $0 = x^2 + x - 3$. Thus we read from the graph the coordinates of the points at which $y = 0$. Such points must be on the x axis. From the graph the points are (1.3, 0) and (−2.3, 0). So the solutions of

Table 17.3
Values of x and y when
$y = x^2 + x - 3$

x	-3	-2	-1	0	1	2	3
y	3	-1	-3	-3	-1	3	9

Figure 17.12
A graph of
$y = x^2 + x - 3$

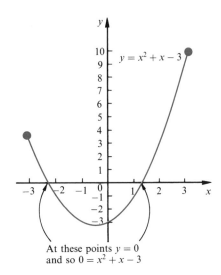

At these points $y = 0$
and so $0 = x^2 + x - 3$

$x^2 + x - 3 = 0$ are $x = 1.3$ and $x = -2.3$. Note that these are approximate solutions, being dependent upon the accuracy of the graph and the x and y scales used. Increased accuracy can be achieved by drawing an enlargement of the graph around the values $x = 1.3$ and $x = -2.3$. So, for example, we could draw $y = x^2 + x - 3$ for $1.2 \leqslant x \leqslant 1.5$ and $y = x^2 + x - 3$ for $-2.4 \leqslant x \leqslant -2.2$.

17.12 Find solutions in the interval $[-2, 2]$ of the equation $x^3 = 2x + \frac{1}{2}$ using a graphical method.

Solution The problem of solving $x^3 = 2x + \frac{1}{2}$ is identical to that of solving $x^3 - 2x - \frac{1}{2} = 0$. So we plot a graph of $y = x^3 - 2x - \frac{1}{2}$ for $-2 \leqslant x \leqslant 2$ and then locate the points where the curve cuts the x axis; that is, where the y coordinate is 0. Table 17.4 gives x and y values and Figure 17.13 shows a graph of the function. We now consider points on the graph where the y coordinate is zero. These points are marked A, B and C. Their x coordinates are -1.27, -0.26 and 1.53. Hence the solutions of $x^3 = 2x + \frac{1}{2}$ are approximately $x = -1.27$, $x = -0.26$ and $x = 1.53$.

Table 17.4
Values of x and y when
$y = x^3 - 2x - \frac{1}{2}$

x	-2	-1.5	-1	-0.5	0	0.5	1	1.5	2
y	-4.5	-0.875	0.5	0.375	-0.5	-1.375	-1.5	-0.125	3.5

Figure 17.13
A graph of
$y = x^3 - 2x - \frac{1}{2}$

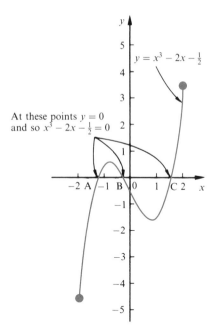

At these points $y = 0$
and so $x^3 - 2x - \frac{1}{2} = 0$

Self-assessment question 17.5

1. Explain how the graph of the function f can be used to solve the equation
   (a) $f(x) = 0$ (b) $f(x) = 1$

Exercise 17.5

1. Plot a graph of

   $$y = x^2 + \frac{3x}{2} - 2$$

   for $-3 \leqslant x \leqslant 2$. Hence solve the equation

   $$x^2 + \frac{3x}{2} - 2 = 0$$

2. Plot $y = x^2 - x - 1$ for $-3 \leqslant x \leqslant 3$.
   Hence solve $x^2 - x - 1 = 0$.

3. Plot $y = x^2 - 0.4x - 4$ for $-3 \leqslant x \leqslant 3$.
   Hence solve $x^2 - 0.4x - 4 = 0$.

4. (a) Plot $y = 3 + \frac{x}{2} - x^2$ for $-3 \leqslant x \leqslant 3$.
   (b) Use your graph to solve

   $$x^2 - \frac{x}{2} - 3 = 0$$

17.6 Solving simultaneous equations graphically

In the previous section we saw how an approximate solution to an equation could be found by drawing an appropriate graph. We now extend that technique to find an approximate solution of simultaneous equations. Recall that simultaneous equations were introduced algebraically in Chapter 11.

Given two simultaneous equations in x and y, then a solution is a pair of x and y values that satisfy both equations. For example, $x = 2$, $y = -3$ is a solution of

$$3x - y = 9$$

$$x + 2y = -4$$

because both equations are satisfied by these particular values. Worked Example 17.13 illustrates the graphical method.

WORKED EXAMPLES

17.13   Solve graphically

$$3x - y = 9 \tag{17.1}$$

$$x + 2y = -4 \tag{17.2}$$

Solution   Both equations are rearranged so that y is the subject. This gives

$$y = 3x - 9 \tag{17.3}$$

$$y = -\frac{x}{2} - 2 \tag{17.4}$$

Since Equations 17.3 and 17.4 are simple rearrangements of Equations 17.1 and 17.2 then they have the same solution. Equations 17.3 and 17.4 are drawn; Figure 17.14 illustrates this.

Figure 17.14
Points of intersection give the solution to simultaneous equations

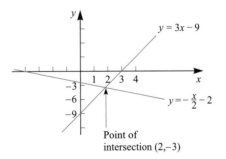

Point of
intersection (2,–3)

We seek values of x and y that fit both equations. For a point to lie on both graphs, the point must be at the intersection of the graphs. In other words, solutions to simultaneous equations are given by the points of intersection. Reading from the graph, the point of intersection is at $x = 2$,

$y = -3$. Hence $x = 2$, $y = -3$ is a solution of the given simultaneous equations.

17.14   Solve graphically

$$4x - y = 0$$

$$3x + y = 7$$

Solution   We write the equations with y as subject:

$$y = 4x$$

$$y = -3x + 7$$

These are now plotted as shown in Figure 17.15.

Figure 17.15
The graphs intersect at
$x = 1$, $y = 4$

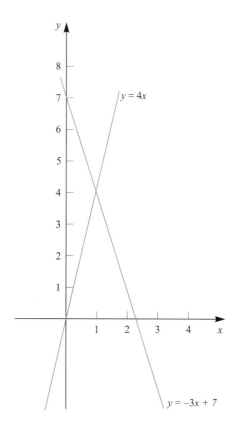

The point of intersection is $x = 1$, $y = 4$ and so this is the solution of the given simultaneous equations.

17.15   Solve graphically

$$x^2 - y = -1$$

$$2x - y = -3$$

Solution   Writing the equations with y as subject gives

$$y = x^2 + 1$$
$$y = 2x + 3$$

These are plotted as shown in Figure 17.16.

Figure 17.16
The graphs have two
points of intersection
and hence there are two
solutions

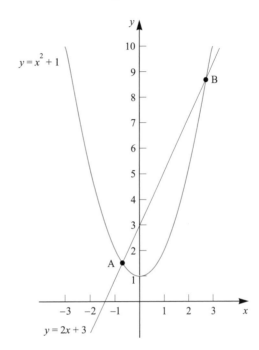

There are two points of intersection, A and B. From the graph it is difficult to extract an accurate estimate. However, by using a graphics calculator or package, greater accuracy can be achieved. The coordinates of A are $x = -0.73$, $y = 1.54$; the coordinates of B are $x = 2.73$, $y = 8.46$. Hence there are two solutions to the given equations: $x = -0.73$, $y = 1.54$ and $x = 2.73$, $y = 8.46$.

17.16   (a) On the same axes draw graphs of $y = x^3$ and $y = 2x + \frac{1}{2}$ for $-2 \leqslant x \leqslant 2$.

(b) Note the x coordinates of the points where the two graphs intersect. By referring to Worked Example 17.12 what do you conclude? Can you explain your findings?

Solution   (a) Table 17.5 gives x values and the corresponding values of x^3 and $2x + \frac{1}{2}$. Figure 17.17 shows a graph of $y = x^3$ together with a graph of $y = 2x + \frac{1}{2}$.

(b) The two graphs intersect at A, B and C. The x coordinates of these points are -1.27, -0.26 and 1.53. We note from Worked Example 17.12 that these values are the solutions of $x^3 = 2x + \frac{1}{2}$. We

Table 17.5
Values of x and y when
$y = x^3$ and $y = 2x + \frac{1}{2}$

x	-2	-1.5	-1	-0.5	0	0.5	1	1.5	2
x^3	-8	-3.375	-1	-0.125	0	0.125	1	3.375	8
$2x + \frac{1}{2}$	-3.5	-2.5	-1.5	-0.5		0.5	1.5	2.5	3.5	4.5

Figure 17.17
Graphs of $y = x^3$ and
$y = 2x + \frac{1}{2}$

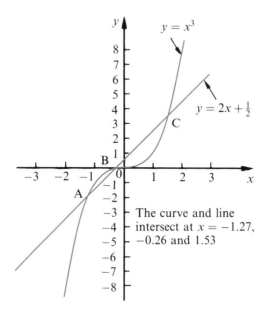

The curve and line
intersect at $x = -1.27$,
-0.26 and 1.53

explain this as follows. Where $y = x^3$ and $y = 2x + \frac{1}{2}$ intersect, their y values are identical and so at these points

$$x^3 = 2x + \frac{1}{2}$$

By reading the x coordinates at these points of intersection we are finding those x values for which $x^3 = 2x + \frac{1}{2}$.

Self-assessment questions 17.6

1. Explain the significance of points of intersection when solving simultaneous equations graphically.

2. The graphs of a pair of simultaneous equations are drawn and they intersect at three points. How many solutions do the simultaneous equations have?

3. The graphs of a pair of simultaneous equations are drawn. The graphs do not intersect. What can you conclude, if anything, about the solution of the simultaneous equations?

Exercise 17.6   *MyMathLab*

1. Solve the following pairs of simultaneous equations graphically. In each case use x values from -4 to 3.

   (a) $3x + 2y = 4$
       $x - y = 3$

   (b) $2x + y = -2$
       $-\dfrac{x}{2} + 2y = 5$

   (c) $2x + y = 4$
       $4x - 3y = -7$

   (d) $-x + 4y = 7$
       $2x - y = -7$

   (e) $x + 2y = 1$
       $\dfrac{x}{2} + 5y = -\dfrac{1}{2}$

2. Solve graphically the simultaneous equations
   $$y = x^2$$
   $$2x + y = 1$$
   Take x values from -3 to 2.

3. Solve graphically
   $$x^2 + y = 3$$
   $$4x - 3y = -3$$
   Take x values from -3 to 2.

4. Solve graphically
   $$y = \frac{x^3}{2}$$
   $$x + y = 3$$

5. Solve graphically, taking x values from -2 to 2:
   $$x^3 - y = 0$$
   $$-1.5x^2 + 5x - y = -2$$

Test and assignment exercises 17

1. Plot a graph of each of the following functions. In each case state the independent variable, the dependent variable, the domain and the range.

   (a) $y = f(x) = 3x - 7$
   (b) $y = f(x) = 3x - 7$, $-2 \leqslant x \leqslant 2$
   (c) $y = 4x^2$, $x \geqslant 0$
   (d) $y = \dfrac{1}{x^2}$, $x \neq 0$

2. A function is defined by
   $$f(x) = \begin{cases} 2x + 4 & 0 \leqslant x \leqslant 4 \\ 6 & 4 < x < 6 \\ x & 6 \leqslant x \leqslant 8 \end{cases}$$

   (a) Plot a graph of this function, stating its domain and range.
   (b) Evaluate $f(0)$, $f(4)$ and $f(6)$.

3. Plot graphs of the functions $f(x) = x + 4$ and $g(x) = 2 - x$ for values of x between -4 and 4. Hence solve the simultaneous equations $y - x = 4$, $y + x = 2$.

4. Plot a graph of the function $f(x) = 3x^2 + 10x - 8$ for values of x between -5 and 3. Hence solve the equation $3x^2 + 10x - 8 = 0$.

5. Plot a graph of the function $f(t) = t^3 + 2t^2 - t - 2$ for $-3 \leqslant t \leqslant 3$. Hence solve the equation $t^3 + 2t^2 - t - 2 = 0$.

6. Plot a graph of $h = 2 + 2x - x^2$ for $-3 \leqslant x \leqslant 3$. Hence solve the equation $2 + 2x - x^2 = 0$.

7. (a) On the same axes plot graphs of $y = x^3$ and $y = 5x^2 - 3$ for $-2 \leqslant x \leqslant 2$.
   (b) Use your graphs to solve the equation $x^3 - 5x^2 + 3 = 0$.

8. (a) Plot a graph of $y = x^3 - 2x^2 - x + 1$ for $-2 \leqslant x \leqslant 3$.
   (b) Hence find solutions of $x^3 - 2x^2 - x + 1 = 0$.
   (c) Use your graph to solve $x^3 - 2x^2 - x + 2 = 0$.

9. Solve each of the following pairs of simultaneous equations graphically. In each case take x values from -3 to 3.

   (a) $2x - y = 0$ (b) $3x + 4y = 6$ (c) $3x + y = 8$

   $x + y = -3$ $x - 2y = -8$ $x - 3y = -1.5$

   (d) $2x - 3y = 4.7$ (e) $x + 3y = 1.2$

   $\dfrac{x}{2} + y = -0.4$ $-2x + 5y = 10.8$

10. Solve graphically

    $$x^2 - y = -2$$
    $$x^2 + 2x + y = 3$$

    Take x values from -3 to 3.

11. Solve graphically

    $$x^3 - y = 0$$
    $$1.5x^2 + 6x - y = 9$$

    Take x values from -3 to 3.

The straight line

18

Objectives : This chapter:

- describes some special properties of straight line graphs
- explains the equation $y = mx + c$
- explains the terms 'vertical intercept' and 'gradient'
- shows how the equation of a line can be calculated
- explains what is meant by a tangent to a curve
- explains what is meant by the gradient of a curve
- explains how the gradient of a curve can be estimated by drawing a tangent

---

18.1 Straight line graphs

In the previous chapter we explained how to draw graphs of functions. Several of the resulting graphs were straight lines. In this chapter we focus specifically on the straight line and some of its properties.

Any equation of the form $y = mx + c$, where m and c are constants, will have a straight line graph. For example,

$$y = 3x + 7 \qquad y = -2x + \frac{1}{2} \qquad y = 3x - 1.5$$

will all result in straight line graphs. It is important to note that the variable x only occurs to the power 1. The values of m and c may be zero, so that $y = -3x$ is a straight line for which the value of c is zero, and $y = 17$ is a straight line for which the value of m is zero.

Key point   Any straight line has an equation of the form $y = mx + c$ where m and c are constants.

WORKED EXAMPLES

18.1   Which of the following equations have straight line graphs? For those that do, identify the values of m and c.

(a) $y = 7x + 5$ (b) $y = \dfrac{13x - 5}{2}$ (c) $y = 3x^2 + x$ (d) $y = -19$

Solution   (a) $y = 7x + 5$ is an equation of the form $y = mx + c$ where $m = 7$ and $c = 5$. This equation has a straight line graph.

(b) The equation

$$y = \frac{13x - 5}{2}$$

can be written as $y = \frac{13}{2}x - \frac{5}{2}$, which is in the form $y = mx + c$ with $m = \frac{13}{2}$ and $c = -\frac{5}{2}$. This has a straight line graph.

(c) $y = 3x^2 + x$ contains the term x^2. Such a term is not allowed in the equation of a straight line. The graph will not be a straight line.

(d) $y = -19$ is in the form of the equation of a straight line for which $m = 0$ and $c = -19$.

18.2   Plot each of the following graphs: $y = 2x + 3$, $y = 2x + 1$ and $y = 2x - 2$. Comment upon the resulting lines.

Solution   The three graphs are shown in Figure 18.1. Note that all three graphs have the same steepness or **slope**. However, each one cuts the vertical

Figure 18.1
Graphs of $y = 2x + 3$,
$y = 2x + 1$ and
$y = 2x - 2$

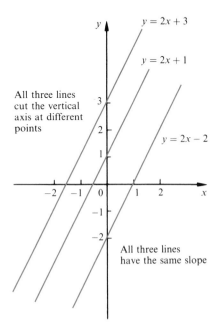

axis at a different point. This point can be obtained directly from the equation $y = 2x + c$ by looking at the value of c. For example, $y = 2x + 1$ intersects the vertical axis at $y = 1$. The graph of $y = 2x + 3$ cuts the vertical axis at $y = 3$, and the graph of $y = 2x - 2$ cuts this axis at $y = -2$.

The point where a graph cuts the vertical axis is called the **vertical intercept**. The vertical intercept can be obtained from $y = mx + c$ by looking at the value of c.

Key point

In the equation $y = mx + c$ the value of c gives the y coordinate of the point where the line cuts the vertical axis.

WORKED EXAMPLE

18.3

On the same graph plot the following: $y = x + 2$, $y = 2x + 2$, $y = 3x + 2$. Comment upon the graphs.

Solution

The three straight lines are shown in Figure 18.2. Note that the steepness of the line is determined by the coefficient of x, with $y = 3x + 2$ being steeper than $y = 2x + 2$, and this in turn being steeper than $y = x + 2$.

Figure 18.2
Graphs of $y = x + 2$, $y = 2x + 2$, and $y = 3x + 2$

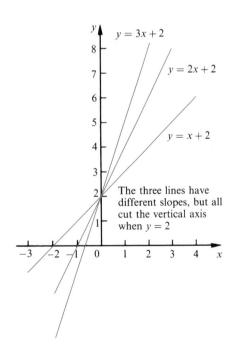

The three lines have different slopes, but all cut the vertical axis when $y = 2$

The value of m in the equation $y = mx + c$ determines the steepness of the straight line. The larger the value of m, the steeper is the line. The value m is known as the slope or **gradient** of the straight line.

Key point   In the equation $y = mx + c$ the value m is known as the gradient and is a measure of the steepness of the line.

If m is positive, the line will rise as we move from left to right. If m is negative, the line will fall. If $m = 0$ the line will be horizontal. We say it has zero gradient. These points are summarised in Figure 18.3.

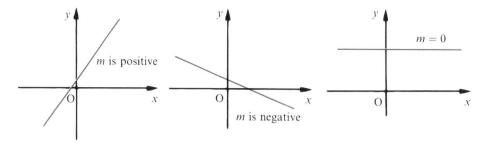

Figure 18.3
Straight line graphs with positive, negative and zero gradients

Self-assessment questions 18.1

1. Give the standard form of the equation of a straight line.

2. Explain the meaning of m and c in the equation $y = mx + c$.

Exercise 18.1

1. Identify, without plotting, which of the following functions will give straight line graphs:
    (a) $y = 3x + 9$ (b) $y = -9x + 2$ (c) $y = -6x$ (d) $y = x^2 + 3x$
    (e) $y = 17$ (f) $y = x^{-1} + 7$

2. Identify the gradient and vertical intercept of each of the following lines:
    (a) $y = 9x - 11$ (b) $y = 8x + 1.4$ (c) $y = \frac{1}{2}x - 11$ (d) $y = 17 - 2x$
    (e) $y = \frac{2x + 1}{3}$ (f) $y = \frac{4 - 2x}{5}$ (g) $y = 3(x - 1)$ (h) $y = 4$

3. Identify (i) the gradient and (ii) the vertical intercept of the following lines:
    (a) $y + x = 6$ (b) $y - 2x + 1 = 0$ (c) $2y - 4x + 3 = 0$ (d) $3x - 4y + 12 = 0$
    (e) $3x + \frac{y}{2} - 9 = 0$

18.2 Finding the equation of a straight line from its graph

If we are given the graph of a straight line it is often necessary to find its equation, $y = mx + c$. This amounts to finding the values of m and c. Finding the vertical intercept is straightforward because we can look directly for the point where the line cuts the y axis. The y coordinate of this point gives the value of c. The gradient m can be determined from knowledge of any two points on the line using the formula

Key point

$$\text{gradient} = \frac{\text{difference between the } y \text{ coordinates}}{\text{difference between the } x \text{ coordinates}}$$

WORKED EXAMPLES

18.4   A straight line graph is shown in Figure 18.4. Determine its equation.

Figure 18.4
Graph for Worked Example 18.4

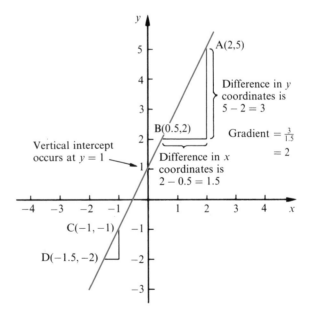

Solution   We require the equation of the line in the form $y = mx + c$. From the graph it is easy to see that the vertical intercept occurs at $y = 1$. Therefore the value of c is 1. To find the gradient m we choose any two points on the line. We have chosen the point A with coordinates (2, 5) and the point B with coordinates (0.5, 2). The difference between their y coordinates is then

$5 - 2 = 3$. The difference between their x coordinates is $2 - 0.5 = 1.5$. Then

$$\text{gradient} = \frac{\text{difference between their } y \text{ coordinates}}{\text{difference between their } x \text{ coordinates}}$$

$$= \frac{3}{1.5} = 2$$

The gradient m is equal to 2. Note that as we move from left to right the line is rising and so the value of m is positive. The equation of the line is then $y = 2x + 1$. There is nothing special about the points A and B. Any two points are sufficient to find m. For example, using the points C with coordinates $(-1, -1)$ and D with coordinates $(-1.5, -2)$ we would find

$$\text{gradient} = \frac{\text{difference between their } y \text{ coordinates}}{\text{difference between their } x \text{ coordinates}}$$

$$= \frac{-1 - (-2)}{-1 - (-1.5)}$$

$$= \frac{1}{0.5} = 2$$

as before.

18.5   A straight line graph is shown in Figure 18.5. Find its equation.

Figure 18.5
Graph for Worked
Example 18.5

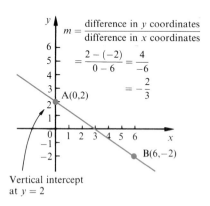

Solution   We need to find the equation in the form $y = mx + c$. From the graph we see immediately that the value of c is 2. To find the gradient we have selected any two points, A(0, 2) and B(6, -2). The difference between their y coordinates is $2 - (-2) = 4$. The difference between their x coordinates is $0 - 6 = -6$. Then

$$\text{gradient} = \frac{\text{difference between their } y \text{ coordinates}}{\text{difference between their } x \text{ coordinates}}$$

$$= \frac{4}{-6}$$

$$= -\frac{2}{3}$$

The equation of the line is therefore $y = -\frac{2}{3}x + 2$. Note in particular that, because the line is sloping downwards as we move from left to right, the gradient is negative. Note also that the coordinates of A and B both satisfy the equation of the line. That is, for A$(0, 2)$,

$$2 = -\frac{2}{3}(0) + 2$$

and for B$(6, -2)$,

$$-2 = -\frac{2}{3}(6) + 2$$

The coordinates of any other point on the line must also satisfy the equation.

The point noted at the end of Worked Example 18.5 is important:

Key point   If the point (a, b) lies on the line $y = mx + c$ then this equation is satisfied by letting $x = a$ and $y = b$.

WORKED EXAMPLE

18.6   Find the equation of the line shown in Figure 18.6.

Figure 18.6
Graph for Worked
Example 18.6

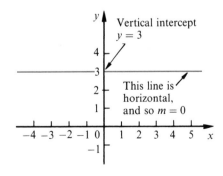

Solution   We are required to express the equation in the form $y = mx + c$. From the graph we notice that the line is horizontal. This means that its gradient is 0, that is $m = 0$. Furthermore the line cuts the vertical axis at $y = 3$ and so the equation of the line is $y = 0x + 3$ or simply $y = 3$.

It is not necessary to sketch a graph in order to find the equation. Consider the following worked examples, which illustrate an algebraic method.

WORKED EXAMPLES

18.7   A straight line passes through A(7, 1) and B(−3, 2). Find its equation.

Solution   The equation must be of the form $y = mx + c$. The gradient of the line can be found from

$$\text{gradient} = m = \frac{\text{difference between their } y \text{ coordinates}}{\text{difference between their } x \text{ coordinates}}$$

$$= \frac{1 - 2}{7 - (-3)}$$

$$= \frac{-1}{10}$$

$$= -0.1$$

Hence $y = -0.1x + c$. We can find c by noting that the line passes through (7, 1), that is the point where $x = 7$ and $y = 1$. Substituting these values into the equation $y = -0.1x + c$ gives

$$1 = -0.1(7) + c$$

so that $c = 1 + 0.7 = 1.7$. Therefore the equation of the line is $y = -0.1x + 1.7$.

18.8   Determine the equation of the line that passes through (4, −1) and has gradient −2.

Solution   Let the equation of the line be $y = mx + c$. We are told that the gradient of the line is −2, that is $m = -2$, and so we have

$$y = -2x + c$$

The point (4, −1) lies on this line: hence when $x = 4$, $y = -1$. These values are substituted into the equation of the line:

$$-1 = -2(4) + c$$

$$c = 7$$

The equation of the line is thus $y = -2x + 7$.

Self-assessment questions 18.2

1. State the formula for finding the gradient of a straight line when two points upon it are known. If the two points are (x_1, y_1) and (x_2, y_2) write down an expression for the gradient.

2. Explain how the value of c in the equation $y = mx + c$ can be found by inspecting the straight line graph.

Exercise 18.2

1. A straight line passes through the two points (1, 7) and (2, 9). Sketch a graph of the line and find its equation.

2. Find the equation of the line that passes through the two points (2, 2) and (3, 8).

3. Find the equation of the line that passes through (8, 2) and (−2, 2).

4. Find the equation of the straight line that has gradient 1 and passes through the origin.

5. Find the equation of the straight line that has gradient −1 and passes through the origin.

6. Find the equation of the straight line passing through (−1, 6) with gradient 2.

7. Which of the following points lie on the line $y = 4x - 3$?
   (a) (1, 2)   (b) (2, 5)   (c) (5, 17)
   (d) (−1, −7)   (e) (0, 2)

8. Find the equation of the straight line passing through (−3, 7) with gradient −1.

9. Determine the equation of the line passing through (−1, −6) that is parallel to the line $y = 3x + 17$.

10. Find the equation of the line with vertical intercept −2 passing through (3, 10).

18.3 Gradients of tangents to curves

Figure 18.7 shows a graph of $y = x^2$. If you study the graph you will notice that as we move from left to right, at some points the y values are decreasing, whereas at others the y values are increasing. It is intuitively obvious that the slope of the curve changes from point to point. At some points, such as A, the curve appears quite steep and falling. At points such as B the curve appears quite steep and rising. Unlike a straight line, the slope of a curve is not fixed but changes as we move from one point to another. A useful way of measuring the slope at any point is to draw a **tangent** to the curve at that point. The tangent is a straight line that just touches the curve at the point of interest. In Figure 18.7 a tangent to the curve $y = x^2$ has been drawn at the point (2, 4). If we calculate the

Figure 18.7

A graph of $y = x^2$

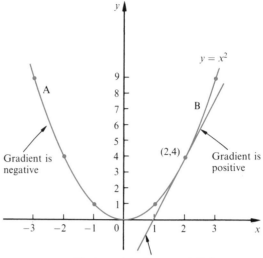

The tangent to the curve at (2,4)

gradient of this tangent, this gives the gradient of the curve at the point (2, 4).

Key point

The gradient of a curve at any point is equal to the gradient of the tangent at that point.

WORKED EXAMPLE

18.9   (a) Plot a graph of $y = x^2 - x$ for values of x between -2 and 4.

(b) Draw in tangents at the points A$(-1, 2)$ and B$(3, 6)$.

(c) By calculating the gradients of these tangents find the gradient of the curve at A and at B.

Solution   (a) A table of values and the graph are shown in Figure 18.8.

(b) We now draw tangents at A and B. At present, the best we can do is estimate these by eye.

(c) We now calculate the gradient of the tangent at A. We select any two points on the tangent and calculate the difference between their y coordinates and the difference between their x coordinates. We have chosen the points $(-3, 8)$ and $(-2, 5)$. Referring to Figure 18.8 we see that

$$\text{gradient of tangent at A} = \frac{8 - 5}{-3 - (-2)} = \frac{3}{-1}$$

$$= -3$$

Figure 18.8
A graph of $y = x^2 - x$

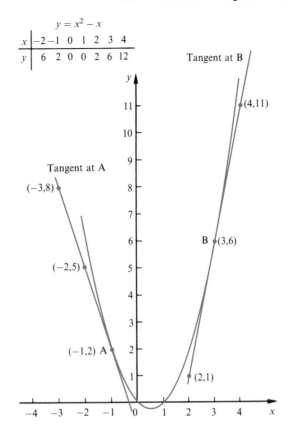

$y = x^2 - x$

x	-2	-1	0	1	2	3	4
y	6	2	0	0	2	6	12

Hence the gradient of the curve at A is -3. Similarly, to find the gradient of the tangent at B we have selected two points on this tangent, namely $(4, 11)$ and $(2, 1)$. We find

$$\text{gradient of tangent at B} = \frac{11 - 1}{4 - 2} = \frac{10}{2}$$

$$= 5$$

Hence the gradient of the tangent at B is 5. Thus the gradient of the curve at B(3, 6) is 5.

Clearly, the accuracy of our answer depends to a great extent upon how well we can draw and measure the gradient of the tangent.

WORKED EXAMPLE

18.10   (a) Sketch a graph of the curve $y = x^3$ for $-2 \leqslant x \leqslant 2$.

(b) Draw the tangent to the graph at the point where $x = 1$.

(c) Estimate the gradient of this tangent and find its equation.

Solution   (a) A graph is shown in Figure 18.9.

Figure 18.9
Graph of $y = x^3$

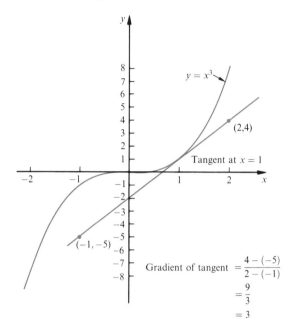

(b) The tangent has been drawn at $x = 1$.

(c) Let us write the equation of the tangent as $y = mx + c$. Two points on the tangent have been selected in order to estimate the gradient. These are $(2, 4)$ and $(-1, -5)$. From these we find

$$\text{gradient of tangent is approximately } \frac{4 - (-5)}{2 - (-1)} = \frac{9}{3} = 3$$

Therefore $m = 3$. The value of c is found by noting that the vertical intercept of the tangent is -2. The equation of the tangent is then $y = 3x - 2$.

Of course, this method will usually result in an approximation based upon how well we have drawn the graph and its tangent. A much more precise method for calculating gradients of curves is given in Chapter 33.

Self-assessment questions 18.3

1. Explain what is meant by the 'tangent' to a curve at a point.

2. Explain how a tangent is used to determine the gradient of a curve.

Exercise 18.3

1. Draw the graph of $y = 2x^2 - 1$ for values of x between -3 and 3. By drawing a tangent, estimate the gradient of the curve at A(2, 7) and B(−1, 1).

2. Draw the graph of $y = -2x^2 + 2$ for values of x between -3 and 3. Draw the tangent at the point where $x = 1$ and calculate its equation.

Test and assignment exercises 18

1. Which of the following will have straight line graphs?

   (a) $y = 2x - 11$ (b) $y = 5x + 10$ (c) $y = x^2 - 1$ (d) $y = -3 + 3x$ (e) $y = \dfrac{2x + 3}{2}$

   For each straight line, identify the gradient and vertical intercept.

2. Find the equation of the straight line that passes through the points (1, 11) and (2, 18). Show that the line also passes through (−1, −3).

3. Find the equation of the line that has gradient −2 and passes through the point (1, 1).

4. Find the equation of the line that passes through (−1, 5) and (1, 5). Does the line also pass through (2, 6)?

5. Draw a graph of $y = -x^2 + 3x$ for values of x between -3 and 3. By drawing in tangents estimate the gradient of the curve at the points (−2, −10) and (1, 2).

6. Find the equations of the lines passing through the origin with gradients (a) −2, (b) −4, (c) 4.

7. Find the equation of the line passing through (4, 10) and parallel to $y = 6x - 3$.

8. Find the equation of the line with vertical intercept 3 and passing through (−1, 9).

9. Find where the line joining (−2, 4) and (3, 10) cuts (a) the x axis, (b) the y axis.

10. A line cuts the x axis at $x = -2$ and the y axis at $y = 3$. Determine the equation of the line.

11. Determine where the line $y = 4x - 1$ cuts
    (a) the y axis   (b) the x axis   (c) the line $y = 2$

The exponential function

Objectives: This chapter:

- shows how to simplify exponential expressions
- describes the form of the exponential function
- illustrates graphs of exponential functions
- lists the properties of the exponential function
- shows how to solve equations with exponential terms using a graphical technique

19.1 Exponential expressions

An expression of the form a^x is called an **exponential expression**. The number a is called the **base** and x is the **power** or **index**. For example, 4^x and 0.5^x are both exponential expressions. The first has base 4 and the second has base 0.5. An exponential expression can be evaluated using the power button on your calculator. For example, the exponential expression with base 4 and index 0.3 is

The letter e always denotes the constant 2.71828....

$$4^{0.3} = 1.516$$

One of the most commonly used values for the base in an exponential expression is 2.71828.... This number is denoted by the letter e.

Key point

The most common exponential expression is

$$e^x$$

where e is the exponential constant, 2.71828....

The expression e^x is found to occur in the modelling of many natural phenomena, for example population growth, spread of bacteria and radioactive decay. Scientific calculators are pre-programmed to evaluate e^x for any value of x. Usually the button is marked e^x or $\exp x$. Use the next worked example to check you can use this facility on your calculator.

WORKED EXAMPLE

19.1 Use a scientific calculator to evaluate

(a) e^2 (b) e^3 (c) $e^{1.3}$ (d) $e^{-1.3}$

Solution We use the e^x button giving:

(a) $e^2 = 7.3891$

(b) $e^3 = 20.0855$

(c) $e^{1.3} = 3.6693$

(d) $e^{-1.3} = 0.2725$

Exponential expressions can be simplified using the normal rules of algebra. The laws of indices apply to exponential expressions. We note

$$e^a e^b = e^{a+b}$$

$$\frac{e^a}{e^b} = e^{a-b}$$

$$e^0 = 1$$

$$(e^a)^b = e^{ab}$$

WORKED EXAMPLES

19.2 Simplify the following exponential expressions:

(a) $e^2 e^4$ (b) $\dfrac{e^4}{e^3}$ (c) $(e^2)^{2.5}$

Solution (a) $e^2 e^4 = e^{2+4} = e^6$

(b) $\dfrac{e^4}{e^3} = e^{4-3} = e^1 = e$

(c) $(e^2)^{2.5} = e^{2 \times 2.5} = e^5$

19.3 Simplify the following exponential expressions:

(a) $e^x e^{3x}$ (b) $\dfrac{e^{4t}}{e^{3t}}$ (c) $(e^t)^4$

Solution   (a) $e^x e^{3x} = e^{x+3x} = e^{4x}$

(b) $\dfrac{e^{4t}}{e^{3t}} = e^{4t-3t} = e^t$

(c) $(e^t)^4 = e^{4t}$

19.4   Simplify

(a) $\sqrt{e^{6t}}$ (b) $e^{3y}(1+e^y) - e^{4y}$ (c) $(2e^t)^2(3e^{-t})$

Solution   (a) $\sqrt{e^{6t}} = (e^{6t})^{\frac{1}{2}} = e^{6t/2} = e^{3t}$

(b) $e^{3y}(1+e^y) - e^{4y} = e^{3y} + e^{3y}e^y - e^{4y}$

$\quad\quad = e^{3y} + e^{4y} - e^{4y}$

$\quad\quad = e^{3y}$

$\sqrt{a} = a^{\frac{1}{2}}$; that is, a square root sign is equivalent to the power $\frac{1}{2}$.

(c) $(2e^t)^2(3e^{-t}) = (2e^t)(2e^t)(3e^{-t})$

$\quad\quad = 2 \cdot 2 \cdot 3 \cdot e^t e^t e^{-t}$

$\quad\quad = 12 e^{t+t-t}$

$\quad\quad = 12 e^t$

19.5   (a) Verify that

$$(e^t + 1)^2 = e^{2t} + 2e^t + 1$$

(b) Hence simplify $\sqrt{e^{2t} + 2e^t + 1}$.

Solution   (a) $(e^t + 1)^2 = (e^t + 1)(e^t + 1)$

$\quad\quad = e^t e^t + e^t + e^t + 1$

$\quad\quad = e^{2t} + 2e^t + 1$

(b) $\sqrt{e^{2t} + 2e^t + 1} = \sqrt{(e^t + 1)^2} = e^t + 1$

Exercise 19.1

1. Use a scientific calculator to evaluate

(a) $e^{2.3}$ (b) $e^{1.9}$ (c) $e^{-0.6}$

(d) $\dfrac{1}{e^{-2} + 1}$ (e) $\dfrac{3e^2}{e^2 - 10}$

(f) $e^2 \left(e + \dfrac{1}{e} \right) - e$ (g) $e^{1.5} e^{2.7}$

(h) $\sqrt{e}\sqrt{4e}$

2. Simplify

(a) $e^2 . e^7$ (b) $\dfrac{e^7}{e^4}$ (c) $\dfrac{e^{-2}}{e^3}$

(d) $\dfrac{e^3}{e^{-1}}$ (e) $\dfrac{(4e)^2}{(2e)^3}$

3. Simplify each of the following expressions as far as possible:

(a) $e^x e^{-3x}$ (b) $e^{3x} e^{-x}$

(c) $e^{-3x} e^{-x}$ (d) $(e^{-x})^2 e^{2x}$

(e) $\dfrac{e^{3x}}{e^x}$ (f) $\dfrac{e^{-3x}}{e^{-x}}$

4. Simplify as far as possible

(a) $\dfrac{e^t}{e^3}$ (b) $\dfrac{2e^{3x}}{4e^x}$ (c) $\dfrac{(e^x)^3}{3e^x}$

(d) $e^x + 2e^{-x} + e^x$ (e) $\dfrac{e^x e^y e^z}{e^{x/2-y}}$

(f) $\dfrac{(e^{t/3})^6}{e^t + e^t}$ (g) $\dfrac{e^t}{e^{-t}}$ (h) $\dfrac{1 + e^{-t}}{e^{-t}}$

(i) $(e^z e^{-z/2})^2$ (j) $\dfrac{e^{-3+t}}{2e^t}$

(k) $\left(\dfrac{e^{-1}}{e^{-x}}\right)^{-1}$

19.2 The exponential function and its graph

The **exponential function** has the form

$$y = e^x$$

The function $y = e^x$ is found to occur in the modelling of many natural phenomena. Table 19.1 gives values of x and e^x. Using Table 19.1, a graph of $y = e^x$ can be drawn. This is shown in Figure 19.1. From the table and graph we note the following important properties of $y = e^x$:

(a) The exponential function is never negative.

(b) When $x = 0$, the function value is 1.

(c) As x increases, then e^x increases. This is known as **exponential growth**.

Table 19.1

x	-3	-2	-1	0	1	2	3
e^x	0.0498	0.1353	0.3679	1	2.7183	7.3891	20.086

Figure 19.1
A graph of $y = e^x$

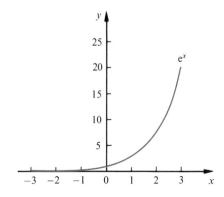

(d) By choosing x large enough, the value of e^x can be made larger than any given number. We say e^x increases without bound as x increases.

(e) As x becomes large and negative, e^x gets nearer and nearer to 0.

WORKED EXAMPLE

19.6   Sketch a graph of $y = 2e^x$ and $y = e^{2x}$ for $-3 \leqslant x \leqslant 3$.

Solution   Table 19.2 gives values of $2e^x$ and e^{2x}, and the graphs are shown in Figure 19.2.

Table 19.2

x	-3	-2	-1	0	1	2	3
$2e^x$	0.0996	0.2707	0.7358	2	5.4366	14.7781	40.1711
e^{2x}	0.0025	0.0183	0.1353	1	7.3891	54.5982	403.429

Figure 19.2
Graphs of $y = e^{2x}$ and $y = 2e^x$

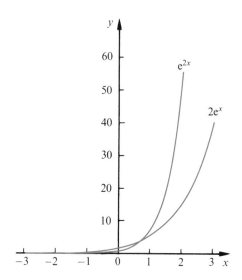

Note that when $x > 0$, the graph of $y = e^{2x}$ rises more rapidly than that of $y = 2e^x$.

We now turn attention to an associated function: $y = e^{-x}$. Values are listed in Table 19.3 and the function is illustrated in Figure 19.3. From the table and the graph we note the following properties of $y = e^{-x}$:

(a) The function is never negative.

(b) When $x = 0$, the function has a value of 1.

(c) As x increases, then e^{-x} decreases, getting nearer to 0. This is known as **exponential decay**.

Table 19.3

x	-3	-2	-1	0	1	2	3
$-x$	3	2	1	0	-1	-2	-3
e^{-x}	20.086	7.3891	2.7183	1	0.3679	0.1353	0.0498

Figure 19.3

Graph of $y = e^{-x}$

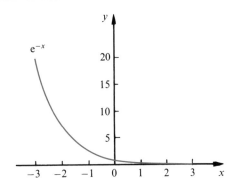

Self-assessment questions 19.2

1. List properties that are common to $y = e^x$ and $y = e^{-x}$.

2. By choosing x large enough, then e^{-x} can be made to be negative. True or false?

Exercise 19.2

1. Draw up a table of values of $y = e^{3x}$ for x between -1 and 1 at intervals of 0.2. Sketch the graph of $y = e^{3x}$. Comment upon its shape and properties.

2. A species of animal has population $P(t)$ at time t given by

   $$P = 10 - 5e^{-t}$$

   (a) Sketch a graph of P against t for $0 \leqslant t \leqslant 5$.

   (b) What is the size of the population of the species as t becomes very large?

3. The concentration, $c(t)$, of a chemical in a reaction is modelled by the equation

   $$c(t) = 6 + 3e^{-2t}$$

   (a) Sketch a graph of c against t for $0 \leqslant t \leqslant 2$.

   (b) What value does the concentration approach as t becomes large?

4. The hyperbolic functions $\sinh x$ and $\cosh x$ are defined by

   $$\sinh x = \frac{e^x - e^{-x}}{2} \qquad \cosh x = \frac{e^x + e^{-x}}{2}$$

   (a) Show $e^x = \cosh x + \sinh x$.

   (b) Show $e^{-x} = \cosh x - \sinh x$.

5. Express $3 \sinh x + 7 \cosh x$ in terms of exponential functions.

6. Express $6e^x - 9e^{-x}$ in terms of $\sinh x$ and $\cosh x$.

7. Show $(\cosh x)^2 - (\sinh x)^2 = 1$.

19.3 Solving equations involving exponential terms using a graphical method

Many equations involving exponential terms can be solved using graphs. A graphical method will yield an approximate solution, which can then be refined by choosing a smaller interval for the domain. The following example illustrates the technique.

WORKED EXAMPLE

19.7 (a) Plot $y = e^{x/2}$ and $y = 2e^{-x}$ for $-1 \leqslant x \leqslant 1$.

(b) Hence solve the equation

$$e^{x/2} - 2e^{-x} = 0$$

Solution (a) A table of values is drawn up for $e^{x/2}$ and $2e^{-x}$ so that the graphs can be drawn. Table 19.4 lists the values and Figure 19.4 shows the graphs of the functions.

Table 19.4

x	-1	-0.75	-0.5	-0.25	0	0.25	0.5	0.75	1
$e^{x/2}$	0.607	0.687	0.779	0.882	1	1.133	1.284	1.455	1.649
$2e^{-x}$	5.437	4.234	3.297	2.568	2	1.558	1.213	0.945	0.736

Figure 19.4
The graphs of $y = e^{x/2}$ and $y = 2e^{-x}$ intersect at A

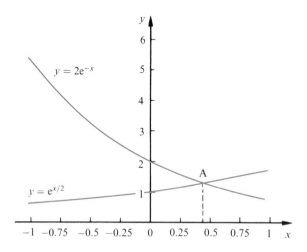

(b) We note that the graphs intersect; the point of intersection is A. The equation

$$e^{x/2} - 2e^{-x} = 0$$

is equivalent to

$$e^{x/2} = 2e^{-x}$$

The graphs of $y = e^{x/2}$ and $y = 2e^{-x}$ intersect at A, where $x = 0.44$. Hence $x = 0.44$ is an approximate solution of $e^{x/2} - 2e^{-x} = 0$. The exact answer can be shown to be 0.46 (2 d.p.).

Exercise 19.3

1. (a) Plot $y = 15 - x^2$ and $y = e^x$ for $0 \leqslant x \leqslant 3$.
   (b) Hence find an approximate solution to
   $$e^x + x^2 = 15$$

2. (a) Plot $y = 12x^2 - 1$ and $y = 3e^{-x}$ for $-1 \leqslant x \leqslant 1$.
   (b) Hence state two approximate solutions of $3e^{-x} = 12x^2 - 1$.

3. By drawing $y = e^x$ for $-3 \leqslant x \leqslant 3$ and additional appropriate lines solve the following equations:
   (a) $e^x + x = 0$

   (b) $e^x - 1.5 = 0$
   (c) $e^x - x - 5 = 0$
   (d) $\dfrac{e^x}{2} + \dfrac{x}{2} - 5 = 0$

4. (a) Draw $y = 2 + 6e^{-t}$ for $0 \leqslant t \leqslant 3$.
   (b) Use your graph to find an approximate solution to
   $$2 + 6e^{-t} = 5$$
   (c) Add the line $y = t + 4$ to your graph and hence find an approximate solution to
   $$6e^{-t} = t + 2$$

Test and assignment exercises 19

1. Use a calculator to evaluate
   (a) $e^{-3.1}$ (b) $e^{0.2}$ (c) $\dfrac{1}{e}$

2. (a) Sketch the graphs of
   $$y = e^{-x} \text{ and } y = \dfrac{x}{2}$$
   for $0 \leqslant x \leqslant 2$.
   (b) Hence solve the equation
   $$2e^{-x} - x = 0$$

3. Simplify where possible:
   (a) $e^x e^{2x} e^{-3x}$
   (b) $\dfrac{e^{2x} e^{3x}}{e^{4x}}$
   (c) $(e^{2x})^3 e^{-3x}$
   (d) $e^{2x} + e^{3x} - e^{5x}$
   (e) $\dfrac{e^{3x} + e^{5x}}{e^{2x}}$

4. Expand the brackets of

(a) $(e^x - 1)^2$ (b) $(e^{-x} - 1)^2$ (c) $e^{2x}(e^x + 1)$ (d) $2e^x(e^{-x} + 3e^x)$

5. Simplify

(a) $e^t e^{t/2}$ (b) $e^2 e^3 e$ (c) $e^{3x}(e^x - e^{2x}) - (e^{2x})^2$ (d) $e^{3+t} e^{5-2t}$

(e) $\dfrac{e^{2z+3} e^{5-z}}{e^{z+2}}$ (f) $\sqrt{e^{4z}}$ (g) $(e^{x/2} e^x)^2$ (h) $\sqrt{e^{2t} + 2e^{2t+1} + e^{2t+2}}$

(i) $\dfrac{e^t + e^{2t}}{e^{2t} + e^{3t}}$

6. Use a graphical method to solve

(a) $e^{2x} = 2x + 2$ (b) $e^{-x} = 2 - x^2$ (c) $e^x = 2 + \dfrac{x^3}{2}$

7. (a) Sketch the graphs of

   $y = 2 + e^{-x}$ and $y = e^x$

   for $0 \leqslant x \leqslant 2$.

   (b) Use your graphs to solve

   $e^x - e^{-x} = 2$

8. Remove the brackets from the following expressions:

(a) $(e^{2t})^4$ (b) $e^t(e^t - e^{-t})$ (c) $e^{2y}(e^{3y} + e^{-2y} + e^y)$ (d) $(1 + e^t)^2$

9. The height of mercury in an experiment, $H(t)$, varies according to

   $$H(t) = 10 + 3e^{-t} - 4e^{-2t}$$

   What value does H approach as t gets very large?

10. Given

   $$X(t) = \frac{3 + 2e^{2t}}{4 + e^{2t}}$$

   find the value that X approaches as t gets very large.

The logarithm function

Objectives: This chapter:

- explains the term 'base of a logarithm'
- shows how to calculate the logarithm of a number to any base
- states the laws of logarithms and uses them to simplify expressions
- shows how to solve exponential and logarithmic equations
- defines the logarithm function
- illustrates graphically the logarithm function

20.1 Introducing logarithms

Given an equation such as $125 = 5^3$, we call 5 the base and 3 the power or index. We can use **logarithms** to write the equation in another form. The logarithm form is

$$\log_5 125 = 3$$

This is read as 'logarithm to the base 5 of 125 is 3'. In general if

$$y = a^x$$

then

$$\log_a y = x$$

Key point $y = a^x$ and $\log_a y = x$ are equivalent.

'The logarithm to the base a of y is x' is equivalent to saying that 'y is a to the power x'. The word logarithm is usually shortened to just 'log'.

WORKED EXAMPLE

20.1   Write down the logarithmic form of the following:

(a) $16 = 4^2$ (b) $8 = 2^3$ (c) $25 = 5^2$

Solution   (a) $16 = 4^2$ may be written as

$$2 = \log_4 16$$

that is, 2 is the logarithm to the base 4 of 16.

(b) $8 = 2^3$ may be expressed as

$$3 = \log_2 8$$

which is read as '3 is the logarithm to the base 2 of 8'.

(c) $25 = 5^2$ may be expressed as

$$2 = \log_5 25$$

that is, 2 is the logarithm to the base 5 of 25.

Given an equation such as $16 = 4^2$ then 'taking logs' to base 4 will result in the logarithmic form $\log_4 16 = 2$.

WORKED EXAMPLE

20.2   Write the exponential form of the following:

(a) $\log_2 16 = 4$ (b) $\log_3 27 = 3$ (c) $\log_5 125 = 3$ (d) $\log_{10} 100 = 2$

Solution   (a) Here the base is 2 and so we may write $16 = 2^4$.

(b) The base is 3 and so $27 = 3^3$.

(c) The base is 5 and so $125 = 5^3$.

(d) The base is 10 and so $100 = 10^2$.

Recall that e is the constant 2.71828..., which occurs frequently in natural phenomena.

Although the base of a logarithm can be any positive number other than 1, the commonly used bases are 10 and e. Logarithms to base 10 are often denoted by 'log' or 'log$_{10}$'; logarithms to base e are denoted by 'ln' or 'log$_e$' and referred to as **natural logarithms**. Most scientific calculators possess 'log' and 'ln' buttons, which are used to evaluate logarithms to base 10 and base e.

WORKED EXAMPLES

20.3   Use a scientific calculator to evaluate the following:

(a) $\log 71$ (b) $\ln 3.7$ (c) $\log 0.4615$ (d) $\ln 0.5$ (e) $\ln 1000$

Solution   Using a scientific calculator we obtain

(a) $\log 71 = 1.8513$ (b) $\ln 3.7 = 1.3083$ (c) $\log 0.4615 = -0.3358$
(d) $\ln 0.5 = -0.6931$ (e) $\ln 1000 = 6.9078$

You should ensure that you know how to use your calculator to verify these results.

20.4   Given $10^{0.6990} = 5$ evaluate the following:

(a) $10^{1.6990}$ (b) $\log 5$ (c) $\log 500$

Solution   (a) $10^{1.6990} = 10^1.10^{0.6990} = 10(5) = 50$

(b) We are given $10^{0.6990} = 5$ and by taking logs to the base 10 we obtain $0.6990 = \log 5$.

(c) From $10^{0.6990} = 5$ we can see that

$$100(5) = 100(10^{0.6990})$$

and so

$$500 = 10^2.10^{0.6990} = 10^{2.6990}$$

Taking logs to the base 10 gives

$$\log 500 = 2.6990$$

Self-assessment question 20.1

1.   The base of a logarithm is always a positive integer. True or false?

Exercise 20.1   *MyMathLab*

1.   Use a scientific calculator to evaluate
     (a) $\log 150$ (b) $\ln 150$
     (c) $\log 0.316$ (d) $\ln 0.1$

2.   Write down the logarithmic form of the following:
     (a) $3^8 = 6561$ (b) $6^5 = 7776$
     (c) $2^{10} = 1024$ (d) $10^5 = 100000$
     (e) $4^7 = 16384$

     (f) $\left(\dfrac{1}{2}\right)^5 = 0.03125$

     (g) $12^3 = 1728$ (h) $9^4 = 6561$

3.   Write the exponential form of the following:
     (a) $\log_6 1296 = 4$ (b) $\log_{15} 225 = 2$
     (c) $\log_8 512 = 3$ (d) $\log_7 2401 = 4$
     (e) $\log_3 243 = 5$ (f) $\log_6 216 = 3$
     (g) $\log_{20} 8000 = 3$ (h) $\log_{16} 4096 = 3$
     (i) $\log_2 4096 = 12$

4.   Given $10^{0.4771} = 3$, evaluate the following:
     (a) $\log 3$ (b) $\log 300$ (c) $\log 0.03$

5.   Given $\log 7 = 0.8451$ evaluate the following:
     (a) $10^{0.8451}$ (b) $\log 700$ (c) $\log 0.07$

20.2 Calculating logarithms to any base

In §20.1 we showed how equations of the form $y = a^x$ could be written in logarithmic form as $\log_a y = x$. The number a, called the **base**, is always positive, that is $a > 0$. We also introduced the common bases 10 and e. Scientific calculators are programmed with logarithms to base 10 and to base e. Suppose that we wish to calculate logarithms to bases other than 10 or e. For example, in communication theory, logarithms to base 2 are commonly used. To calculate logarithms to base a we use one of the following two formulae:

Key point

$$\log_a X = \frac{\log_{10} X}{\log_{10} a} \qquad \log_a X = \frac{\ln X}{\ln a}$$

WORKED EXAMPLE

20.5 Evaluate (a) $\log_6 19$, (b) $\log_7 29$.

Solution (a) We use the formula

$$\log_a X = \frac{\log_{10} X}{\log_{10} a}$$

Comparing $\log_6 19$ with $\log_a X$ we see $X = 19$ and $a = 6$. So

$$\log_6 19 = \frac{\log_{10} 19}{\log_{10} 6} = \frac{1.2788}{0.7782} = 1.6433$$

We could have equally well used the formula

$$\log_a X = \frac{\ln X}{\ln a}$$

With this formula we obtain

$$\log_6 19 = \frac{\ln 19}{\ln 6} = \frac{2.9444}{1.7918} = 1.6433$$

(b) Comparing $\log_a X$ with $\log_7 29$ we see $X = 29$ and $a = 7$. Hence

$$\log_7 29 = \frac{\log_{10} 29}{\log_{10} 7} = \frac{1.4624}{0.8451} = 1.7304$$

Alternatively we use

$$\log_7 29 = \frac{\ln 29}{\ln 7} = \frac{3.3673}{1.9459} = 1.7304$$

By considering the formula

$$\log_a X = \frac{\log_{10} X}{\log_{10} a}$$

with $X = a$ we obtain

$$\log_a a = \frac{\log_{10} a}{\log_{10} a} = 1$$

Key point $\log_a a = 1$

This same result could be derived by writing the logarithm form of $a = a^1$.

Exercise 20.2   *MyMathLab*

1. Evaluate the following:
   (a) $\log_4 6$ (b) $\log_3 10$ (c) $\log_{20} 270$ (d) $\log_5 0.65$ (e) $\log_2 100$
   (f) $\log_2 0.03$ (g) $\log_{100} 10$ (h) $\log_7 7$

2. Show that
   $$2.3026 \log_{10} X = \ln X$$

3. Evaluate the following:
   (a) $\log_3 7 + \log_4 7 + \log_5 7$ (b) $\log_8 4 + \log_8 0.25$ (c) $\log_{0.7} 2$ (d) $\log_2 0.7$

20.3 Laws of logarithms

Logarithms obey several laws, which we now examine. They are introduced via examples.

WORKED EXAMPLE

20.6   Evaluate (a) $\log 7$, (b) $\log 12$, (c) $\log 84$ and $\log 7 + \log 12$. Comment on your findings.

Solution   (a) $\log 7 = 0.8451$ (b) $\log 12 = 1.0792$

(c) $\log 84 = 1.9243$, and $\log 7 + \log 12 = 0.8451 + 1.0792 = 1.9243$

We note that $\log 7 + \log 12 = \log 84$.

Worked Example 20.6 illustrates the **first law** of logarithms, which states:

Key point $\log A + \log B = \log AB$

This law holds true for any base. However, in any one calculation all bases must be the same.

WORKED EXAMPLES

20.7   Simplify to a single log term

(a) $\log 9 + \log x$
(b) $\log t + \log 4t$
(c) $\log 3x^2 + \log 2x$

Solution   (a) $\log 9 + \log x = \log 9x$

(b) $\log t + \log 4t = \log(t.4t) = \log 4t^2$

(c) $\log 3x^2 + \log 2x = \log(3x^2.2x) = \log 6x^3$

20.8   Simplify

(a) $\log 7 + \log 3 + \log 2$
(b) $\log 3x + \log x + \log 4x$

Solution   (a) We know $\log 7 + \log 3 = \log(7 \times 3) = \log 21$, and so

$$\log 7 + \log 3 + \log 2 = \log 21 + \log 2$$
$$= \log(21 \times 2) = \log 42$$

(b) We have

$$\log 3x + \log x = \log(3x.x) = \log 3x^2$$

and so

$$\log 3x + \log x + \log 4x = \log 3x^2 + \log 4x$$
$$= \log(3x^2.4x) = \log 12x^3$$

We now consider an example that introduces the second law of logarithms.

WORKED EXAMPLE

20.9   (a) Evaluate $\log 12$, $\log 4$ and $\log 3$.

(b) Compare the values of $\log 12 - \log 4$ and $\log 3$.

Solution   (a) $\log 12 = 1.0792$, $\log 4 = 0.6021$, $\log 3 = 0.4771$.

(b) From part (a),

$$\log 12 - \log 4 = 1.0792 - 0.6021 = 0.4771$$

and also

$$\log 3 = 0.4771$$

We note that $\log 12 - \log 4 = \log 3$.

This example illustrates the **second law** of logarithms, which states:

Key point

$$\log A - \log B = \log\left(\frac{A}{B}\right)$$

WORKED EXAMPLES

20.10   Use the second law of logarithms to simplify the following to a single log term:

(a) $\log 20 - \log 10$ (b) $\log 500 - \log 75$ (c) $\log 4x^3 - \log 2x$
(d) $\log 5y^3 - \log y$

Solution   (a) Using the second law of logarithms we have

$$\log 20 - \log 10 = \log\left(\frac{20}{10}\right) = \log 2$$

(b) $\log 500 - \log 75 = \log\left(\frac{500}{75}\right) = \log\left(\frac{20}{3}\right)$

(c) $\log 4x^3 - \log 2x = \log\left(\frac{4x^3}{2x}\right) = \log 2x^2$

(d) $\log 5y^3 - \log y = \log\left(\frac{5y^3}{y}\right) = \log 5y^2$

20.11   Simplify

(a) $\log 20 + \log 3 - \log 6$
(b) $\log 18 - \log 24 + \log 2$

Solution   (a) Using the first law of logarithms we see that

$$\log 20 + \log 3 = \log 60$$

and so

$$\log 20 + \log 3 - \log 6 = \log 60 - \log 6$$

Using the second law of logarithms we see that

$$\log 60 - \log 6 = \log\left(\frac{60}{6}\right) = \log 10$$

Hence

$$\log 20 + \log 3 - \log 6 = \log 10$$

(b) $\log 18 - \log 24 + \log 2 = \log\left(\dfrac{18}{24}\right) + \log 2$

$$= \log\left(\frac{3}{4}\right) + \log 2$$

$$= \log\left(\frac{3}{4} \times 2\right)$$

$$= \log 1.5$$

20.12   Simplify

(a) $\log 2 + \log 3x - \log 2x$
(b) $\log 5y^2 + \log 4y - \log 10y^2$

Solution   (a) $\log 2 + \log 3x - \log 2x = \log(2 \times 3x) - \log 2x$

$$= \log 6x - \log 2x$$

$$= \log\left(\frac{6x}{2x}\right) = \log 3$$

(b) $\log 5y^2 + \log 4y - \log 10y^2 = \log(5y^2.4y) - \log 10y^2$

$$= \log 20y^3 - \log 10y^2$$

$$= \log\left(\frac{20y^3}{10y^2}\right) = \log 2y$$

We consider a special case of the second law. Consider $\log A - \log A$. This is clearly 0. However, using the second law we may write

$$\log A - \log A = \log \left(\frac{A}{A} \right) = \log 1$$

Thus

Key point

$$\log 1 = 0$$

In any base, the logarithm of 1 equals 0.
Finally we introduce the third law of logarithms.

WORKED EXAMPLE

20.13   (a)   Evaluate $\log 16$ and $\log 2$.
(b)   Compare $\log 16$ and $4 \log 2$.

Solution   (a) $\log 16 = 1.204, \log 2 = 0.301$.

(b) $\log 16 = 1.204, 4 \log 2 = 1.204$. Hence we see that $4 \log 2 = \log 16$.

Noting that $16 = 2^4$, Worked Example 20.13 suggests the **third law** of logarithms:

Key point

$$n \log A = \log A^n$$

This law applies if n is integer, fractional, positive or negative.

WORKED EXAMPLES

20.14   Write the following as a single logarithmic expression:

(a) $3 \log 2$ (b) $2 \log 3$ (c) $4 \log 3$

Solution   (a) $3 \log 2 = \log 2^3 = \log 8$

(b) $2 \log 3 = \log 3^2 = \log 9$

(c) $4 \log 3 = \log 3^4 = \log 81$

20.15   Write as a single log term

(a) $\frac{1}{2} \log 16$ (b) $-\log 4$ (c) $-2 \log 2$ (d) $-\frac{1}{2} \log 0.5$

Solution   (a) $\frac{1}{2} \log 16 = \log 16^{\frac{1}{2}} = \log \sqrt{16} = \log 4$

(b) $-\log 4 = -1.\log 4 = \log 4^{-1} = \log\left(\dfrac{1}{4}\right) = \log 0.25$

(c) $-2\log 2 = \log 2^{-2} = \log\left(\dfrac{1}{2^2}\right) = \log\left(\dfrac{1}{4}\right) = \log 0.25$

(d) $-\dfrac{1}{2}\log 0.5 = -\dfrac{1}{2}\log\left(\dfrac{1}{2}\right) = \log\left(\dfrac{1}{2}\right)^{-\frac{1}{2}} = \log 2^{\frac{1}{2}} = \log\sqrt{2}$

20.16   Simplify

(a) $3\log x - \log x^2$
(b) $3\log t^3 - 4\log t^2$
(c) $\log Y - 3\log 2Y + 2\log 4Y$

Solution   (a) $3\log x - \log x^2 = \log x^3 - \log x^2$

$$= \log\left(\dfrac{x^3}{x^2}\right)$$

$$= \log x$$

(b) $3\log t^3 - 4\log t^2 = \log(t^3)^3 - \log(t^2)^4$

$$= \log t^9 - \log t^8$$

$$= \log\left(\dfrac{t^9}{t^8}\right)$$

$$= \log t$$

(c) $\log Y - 3\log 2Y + 2\log 4Y = \log Y - \log(2Y)^3 + \log(4Y)^2$

$$= \log Y - \log 8Y^3 + \log 16Y^2$$

$$= \log\left(\dfrac{Y.16Y^2}{8Y^3}\right)$$

$$= \log 2$$

20.17   Simplify

(a) $2\log 3x - \dfrac{1}{2}\log 16x^2$

(b) $\dfrac{3}{2}\log 4x^2 - \log\left(\dfrac{1}{x}\right)$

(c) $2\log\left(\dfrac{2}{x^2}\right) - 3\log\left(\dfrac{2}{x}\right)$

Solution   (a) $2\log 3x - \dfrac{1}{2}\log 16x^2 = \log(3x)^2 - \log(16x^2)^{\frac{1}{2}}$

$$= \log 9x^2 - \log 4x$$

$$= \log\left(\frac{9x^2}{4x}\right)$$

$$= \log\left(\frac{9x}{4}\right)$$

(b) $\dfrac{3}{2}\log 4x^2 - \log\left(\dfrac{1}{x}\right) = \log(4x^2)^{\frac{3}{2}} - \log(x^{-1})$

$$= \log 8x^3 + \log x$$

$$= \log 8x^4$$

(c) $2\log\left(\dfrac{2}{x^2}\right) - 3\log\left(\dfrac{2}{x}\right) = \log\left(\dfrac{2}{x^2}\right)^2 - \log\left(\dfrac{2}{x}\right)^3$

$$= \log\left(\frac{4}{x^4}\right) - \log\left(\frac{8}{x^3}\right)$$

$$= \log\left(\frac{4/x^4}{8/x^3}\right)$$

$$= \log\left(\frac{1}{2x}\right)$$

Self-assessment question 20.3

1. State the three laws of logarithms.

Exercise 20.3

1. Write the following as a single log term using the laws of logarithms:
    (a) $\log 5 + \log 9$ (b) $\log 9 - \log 5$
    (c) $\log 5 - \log 9$ (d) $2\log 5 + \log 1$
    (e) $2\log 4 - 3\log 2$ (f) $\log 64 - 2\log 2$
    (g) $3\log 4 + 2\log 1 + \log 27 - 3\log 12$

2. Simplify as much as possible:
    (a) $\log 3 + \log x$
    (b) $\log 4 + \log 2x$
    (c) $\log 3X - \log 2X$
    (d) $\log T^3 - \log T$
    (e) $\log 5X + \log 2X$

3. Simplify
   (a) $3 \log X - \log X^2$ (b) $\log y - 2 \log \sqrt{y}$

   (c) $5 \log x^2 + 3 \log \dfrac{1}{x}$

   (d) $4 \log X - 3 \log X^2 + \log X^3$
   (e) $3 \log y^{1.4} + 2 \log y^{0.4} - \log y^{1.2}$

4. Simplify the following as much as possible by using the laws of logarithms:
   (a) $\log 4x - \log x$ (b) $\log t^3 + \log t^4$

   (c) $\log 2t - \log \left(\dfrac{t}{4} \right)$

   (d) $\log 2 + \log \left(\dfrac{3}{x} \right) - \log \left(\dfrac{x}{2} \right)$

   (e) $\log \left(\dfrac{t^2}{3} \right) + \log \left(\dfrac{6}{t} \right) - \log \left(\dfrac{1}{t} \right)$

   (f) $2 \log y - \log y^2$

   (g) $3 \log \left(\dfrac{1}{t} \right) + \log t^2$

   (h) $4 \log \sqrt{x} + 2 \log \left(\dfrac{1}{x} \right)$

   (i) $2 \log x + 3 \log t$ (j) $\log A - \dfrac{1}{2} \log 4A$

   (k) $\dfrac{\log 9x + \log 3x^2}{3}$

   (l) $\log xy + 2 \log \left(\dfrac{x}{y} \right) + 3 \log \left(\dfrac{y}{x} \right)$

   (m) $\log \left(\dfrac{A}{B} \right) - \log \left(\dfrac{B}{A} \right)$

   (n) $\log \left(\dfrac{2t}{3} \right) + \dfrac{1}{2} \log 9t - \log \left(\dfrac{1}{t} \right)$

5. Express as a single log term:
   $$\log_{10} X + \ln X$$

6. Simplify
   (a) $\log(9x - 3) - \log(3x - 1)$
   (b) $\log(x^2 - 1) - \log(x + 1)$
   (c) $\log(x^2 + 3x) - \log(x + 3)$

20.4 Solving equations with logarithms

This section illustrates the use of logarithms in solving certain types of equations. For reference we note from §20.1 the equivalence of

$$y = a^x \quad \text{and} \quad \log_a y = x$$

and from §20.3 the laws of logarithms:

$$\log A + \log B = \log AB$$

$$\log A - \log B = \log \left(\dfrac{A}{B} \right)$$

$$n \log A = \log A^n$$

20.18 Solve the following equations:

(a) $10^x = 59$ (b) $10^x = 0.37$ (c) $e^x = 100$ (d) $e^x = 0.5$

Solution (a) $10^x = 59$
Taking logs to base 10 gives

$$x = \log 59 = 1.7709$$

(b) $10^x = 0.37$
Taking logs to base 10 gives

$$x = \log 0.37 = -0.4318$$

(c) $e^x = 100$
Taking logs to base e gives

$$x = \ln 100 = 4.6052$$

(d) $e^x = 0.5$
Taking logs to base e we have

$$x = \ln 0.5 = -0.6931$$

20.19 Solve the following equations:

(a) $\log x = 1.76$ (b) $\ln x = -0.5$ (c) $\log(3x) = 0.76$

(d) $\ln\left(\dfrac{x}{2}\right) = 2.6$ (e) $\log(2x - 4) = 1.1$ (f) $\ln(7 - 3x) = 1.75$

Solution We note that if $\log_a X = n$, then $X = a^n$.

(a) $\log x = 1.76$ and so $x = 10^{1.76}$. Using the 'x^y' button of a scientific calculator we find

$$x = 10^{1.76} = 57.5440$$

(b) $\ln x = -0.5$

$$x = e^{-0.5} = 0.6065$$

(c) $\log 3x = 0.76$

$$3x = 10^{0.76}$$

$$x = \frac{10^{0.76}}{3} = 1.9181$$

(d) $\ln\left(\dfrac{x}{2}\right) = 2.6$

$$\dfrac{x}{2} = e^{2.6}$$

$$x = 2e^{2.6} = 26.9275$$

(e) $\log(2x - 4) = 1.1$

$$2x - 4 = 10^{1.1}$$

$$2x = 10^{1.1} + 4$$

$$x = \dfrac{10^{1.1} + 4}{2} = 8.2946$$

(f) $\ln(7 - 3x) = 1.75$

$$7 - 3x = e^{1.75}$$

$$3x = 7 - e^{1.75}$$

$$x = \dfrac{7 - e^{1.75}}{3} = 0.4151$$

20.20 Solve the following:

(a) $e^{3 + x}.e^{x} = 1000$ (b) $\ln\left(\dfrac{x}{3} + 1\right) + \ln\left(\dfrac{1}{x}\right) = -1$

(c) $3e^{x - 1} = 75$ (d) $\log(x + 2) + \log(x - 2) = 1.3$

Solution   (a) Using the laws of logarithms we have

$$e^{3 + x}.e^{x} = e^{3 + 2x}$$

Hence

$$e^{3 + 2x} = 1000$$

$$3 + 2x = \ln 1000$$

$$2x = \ln(1000) - 3$$

$$x = \dfrac{\ln(1000) - 3}{2} = 1.9539$$

(b) Using the laws of logarithms we may write

$$\ln\left(\frac{x}{3}+1\right) + \ln\left(\frac{1}{x}\right) = \ln\left(\left(\frac{x}{3}+1\right)\frac{1}{x}\right) = \ln\left(\frac{1}{3}+\frac{1}{x}\right)$$

So

$$\ln\left(\frac{1}{3}+\frac{1}{x}\right) = -1$$

$$\frac{1}{3}+\frac{1}{x} = e^{-1}$$

$$\frac{1}{x} = e^{-1} - \frac{1}{3}$$

$$\frac{1}{x} = 0.0345$$

$$x = 28.947$$

(c) $3e^{x-1} = 75$

$$e^{x-1} = 25$$

$$x - 1 = \ln 25$$

$$x = \ln 25 + 1 = 4.2189$$

(d) Using the first law we may write

$$\log(x+2) + \log(x-2) = \log(x+2)(x-2) = \log(x^2-4)$$

Hence

$$\log(x^2-4) = 1.3$$

$$x^2 - 4 = 10^{1.3}$$

$$x^2 = 10^{1.3} + 4$$

$$x = \sqrt{10^{1.3}+4} = 4.8941$$

Exercise 20.4

1. Solve the following equations, giving your answer to 4 d.p.:
   (a) $\log x = 1.6000$ (b) $10^x = 75$
   (c) $\ln x = 1.2350$ (d) $e^x = 36$

2. Solve each of the following equations, giving your answer to 4 d.p.:
   (a) $\log(3t) = 1.8$ (b) $10^{2t} = 150$
   (c) $\ln(4t) = 2.8$ (d) $e^{3t} = 90$

3. Solve the following equations, giving your answer to 4 d.p.:
   (a) $\log x = 0.3940$ (b) $\ln x = 0.3940$
   (c) $10^y = 5.5$ (d) $e^z = 500$

   (e) $\log(3v) = 1.6512$ (f) $\ln\left(\dfrac{t}{6}\right) = 1$

   (g) $10^{2r+1} = 25$ (h) $e^{(2t-1)/3} = 7.6700$
   (i) $\log(4b^2) = 2.6987$

   (j) $\log\left(\dfrac{6}{2+t}\right) = 1.5$

   (k) $\ln(2r^3 + 1) = 3.0572$
   (l) $\ln(\log t) = -0.3$ (m) $10^{t^2-1} = 180$
   (n) $10^{3r^4} = 170000$
   (o) $\log(10^t) = 1.6$ (p) $\ln(e^x) = 20000$

4. Solve the following equations:
   (a) $e^{3x}.e^{2x} = 59$
   (b) $10^{3t}.10^{4-t} = 27$
   (c) $\log(5 - t) + \log(5 + t) = 1.2$
   (d) $\log x + \ln x = 4$

20.5 Properties and graph of the logarithm function

Values of $\log x$ and $\ln x$ are given in Table 20.1, and graphs of the functions $y = \log x$ and $y = \ln x$ are illustrated in Figure 20.1. The following properties are noted from the graphs:

(a) As x increases, the values of $\log x$ and $\ln x$ increase.
(b) $\log 1 = \ln 1 = 0$.

Table 20.1

x	0.01	0.1	0.5	1	2	5	10	100
$\log x$	−2	−1	−0.30	0	0.30	0.70	1	2
$\ln x$	−4.61	−2.30	−0.69	0	0.69	1.61	2.30	4.61

Figure 20.1
Graphs of $y = \log x$ and $y = \ln x$

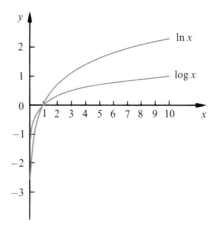

(c) As x approaches 0 the values of $\log x$ and $\ln x$ increase negatively.

(d) When $x < 1$, the values of $\log x$ and $\ln x$ are negative.

(e) $\log x$ and $\ln x$ are not defined when $x \leqslant 0$.

Self-assessment questions 20.5

1. State two properties that are common to both $y = \log x$ and $y = \ln x$.

2. It is possible to find a value of x such that the value of $\log x$ exceeds 100000000. True or false?

Exercise 20.5

1. (a) Plot a graph of $y = \log x^2$ for $0 < x \leqslant 10$.
   (b) Plot a graph of $y = \log(1/x)$ for $0 < x \leqslant 10$.

2. (a) Plot a graph of $y = \log x$ for $0 < x \leqslant 3$.
   (b) Plot on the same axes $y = 1 - \frac{x}{3}$ for $0 \leqslant x \leqslant 3$.
   (c) Use your graphs to find an approximate solution to the equation

   $$\log x = 1 - \frac{x}{3}$$

3. (a) Plot $y = \log x$ for $0 < x \leqslant 7$.
   (b) On the same axes, plot $y = 0.5 - 0.1x$.
   (c) Use your graphs to obtain an approximate solution to

   $$0.5 - 0.1x = \log x$$

Test and assignment exercises 20

1. Use a scientific calculator to evaluate
   (a) $\log 107000$ (b) $\ln 0.0371$ (c) $\log 0.1$ (d) $\ln 150$

2. Evaluate
   (a) $\log_7 20$ (b) $\log_{16} 100$ (c) $\log_2 60$ (d) $\log_8 150$ (e) $\log_6 4$

3. Simplify the following expressions to a single log term:
   (a) $\log 7 + \log t$ (b) $\log t - \log x$ (c) $\log x + 2 \log y$

   (d) $3 \log t + 4 \log r$ (e) $\frac{1}{2} \log 9v^4 - 3 \log 1$ (f) $\log(a + b) + \log(a - b)$

   (g) $\frac{2}{3} \log x + \frac{1}{3} \log xy^3$

4. Solve the following logarithmic equations, giving your answer to 4 d.p.:
    (a) $\log 7x = 2.9$ (b) $\ln 2x = 1.5$ (c) $\ln x + \ln 2x = 3.6$

    (d) $\ln\left(\dfrac{x^2}{2}\right) - \ln x = 0.7$ (e) $\log(4t - 3) = 0.9$ (f) $\log 3t + 3\log t = 2$

    (g) $3\ln\left(\dfrac{2}{x}\right) - \dfrac{1}{2}\ln x = -1$

5. (a) Plot $y = \log x$ for $0 < x \leqslant 3$.
    (b) On the same axes plot $y = \sin x$, where x is measured in radians.
    (c) Hence find an approximate solution to $\log x = \sin x$.

6. Solve the following equations, giving your answer correct to 4 d.p.:
    (a) $10^{x+1} = 70$ (b) $e^{2x+3} = 500$ (c) $3(10^{2x}) = 750$ (d) $4(e^{-x+2}) = 1000$
    (e) $\dfrac{4}{6 + e^{3x}} = 0.1500$

7. By substituting $z = e^x$ solve the equation
    $$e^{2x} - 9e^x + 14 = 0$$

8. By substituting $z = 10^x$ solve the following equation, giving your answer correct to 4 d.p.:
    $$10^{2x} - 9(10^x) + 20 = 0$$

9. Solve the following equations:
    (a) $\log(x - 1) + \log(x + 1) = 2$ (b) $\log(10 + 10^x) = 2$ (c) $\dfrac{\log 2x}{\log x} = 2$

10. Solve
    (a) $8(10^{-x}) + 10^x = 6$ (b) $10^{3x} - 4(10^x) = 0$

Measurement

Objectives: This chapter:

- provides tables of SI units and prefixes
- revises common units of length and conversions between them
- revises the concepts of areas and volume, their units of measurement and conversions between them
- explains the units 'degree' and 'radian' used to measure angles, and conversions between them
- provides a summary of important formulae needed for calculating areas of common shapes and volumes of common solids
- revises common units of mass and conversions between them
- revises common units of time
- introduces the topic of dimensional analysis

21.1 Introduction to measurement

The ability to measure quantities has been important for as long as human beings have engaged in commerce, trade and travel. Merchants have needed to count, weigh and compare in order to transact their business. As science, engineering and medicine have developed, so too has the importance of being able to measure consistently and accurately. Standard ways of doing this have been defined and there are now agreed international norms and **units of measurement** known as the Système International (SI).

For example, time can be measured using seconds, minutes, hours, days, weeks and so on. In everyday conversation we use the unit which is most appropriate. For scientific work, to enable a consistent approach, the internationally agreed unit of time is the second. We give this unit the symbol s. Table 21.1 provides a list of common physical quantities, their SI units and their symbols.

Table 21.1
SI units

Quantity	SI unit	Symbol
length	metre	m
mass	kilogram	kg
time	second	s
energy	joule	J
force	newton	N
power	watt	W

To maintain the use of SI units for very large and very small measurements we introduce a set of **prefixes**. For example, the distance between two towns may be 12000 metres but it is more convenient to measure the distance in kilometres; 12000 metres would usually be written as 12 kilometres (km). Here the prefix is 'kilo' (k), meaning 1000 (that is, 10^3), and the SI unit is still the metre. Similarly, a centimetre (cm) is $\frac{1}{100}$th part of a metre. Here the prefix is 'centi' (10^{-2}) and the SI unit is again the metre.

Table 21.2 lists common prefixes. Note that a prefix is a method of multiplying the SI unit by an appropriate power of 10 to make it larger or smaller.

Table 21.2
Common prefixes

Multiple	Prefix	Symbol
10^{12}	tera	T
10^9	giga	G
10^6	mega	M
10^3	kilo	k
10^2	hecto	h
10^1	deca	da
10^{-1}	deci	d
10^{-2}	centi	c
10^{-3}	milli	m
10^{-6}	micro	μ
10^{-9}	nano	n
10^{-12}	pico	p

21.2 Units of length

According to the SI, the standard unit of **length** is the metre, abbreviated to m. It is helpful to have an intuitive feel of this length. The work surface in a kitchen will be a little under 1 m in height, and the frame of a typical door will be about 2 m in height.

Several smaller units are commonly used as well. A metre length can be divided into 100 equal parts, and each part has length 1 centimetre (cm). So, 100 cm = 1 m. Equivalently, 1 cm is one-hundredth, or $\frac{1}{100}$ or 10^{-2}, of a metre, so

$$1 \text{ cm} = \frac{1}{100} \text{ m} = 10^{-2} \text{ m}$$

As noted earlier, the letter 'c' placed in front of the metre symbol is called a **prefix** and its use is consistent with Table 21.2 in that the standard unit of length is the metre, m, and a centimetre is 10^{-2} m.

Each centimetre can be further divided into 10 equal parts, each of length 1 millimetre (mm). So 10 mm = 1 cm. Equivalently 1 mm is one-tenth, or $\frac{1}{10}$ or 10^{-1}, of a centimetre:

$$1 \text{ mm} = \frac{1}{10} \text{ cm} = 10^{-1} \text{ cm}$$

Figure 21.1
A length of 1 centimetre can be divided into 10 equal parts of length 1 millimetre

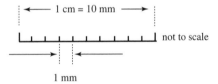

To help visualise this, see Figure 21.1 which has been enlarged to illustrate the point being made.

Because there are 10 mm in 1 cm, and 100 cm in 1 m, it follows that there are 1000 mm in 1 m, that is 1000 mm = 1 m. Equivalently,

$$1 \text{ mm} = \frac{1}{1000} \text{ m} = 10^{-3} \text{ m}$$

Note again that use of the prefix 'm' placed in front of the metre symbol m is consistent with the information in Table 21.2.

Key point

> 1 metre (m) = 100 centimetres (cm)
> = 1000 millimetres (mm)
> 1 kilometre (km) = 1000 metres (m)

There are many more prefixes which enable us to write down even larger and smaller quantities concisely. These prefixes apply not just to lengths but also to any other quantities such as mass and time. A complete set is summarised in Table 21.2.

WORKED EXAMPLES

21.1 Convert a length of 6.7 metres into centimetres.

Solution Because 1 m equals 100 cm then 6.7 metres equals $6.7 \times 100 = 670$ cm.

21.2 Convert a length of 24 mm into metres.

Solution Because 1 m equals 1000 mm, then 1 mm equals $\frac{1}{1000}$ m. Therefore 24 mm equals

$$24 \times \frac{1}{1000} = 0.024 \text{ m}$$

Exercise 21.2

1. Convert 16 cm into (a) mm, (b) m.

2. Change 356 mm into (a) cm, (b) m.

3. Change 156 m into km.

4. Convert 150 μm into (a) m, (b) cm, (c) mm.

5. Convert 0.0046 m into (a) cm, (b) mm, (c) μm.

6. Convert 0.0013 cm into μm.

21.3 Area and volume

The **area** of a shape is the number of square units it contains. One square metre, written 1 m^2, is the area of a square having sides of length 1 m. Similarly 1 square centimetre, written 1 cm^2, is the area of a square having sides of length 1 centimetre.

Converting should not cause confusion if you understand the process involved. Consider the square metre depicted in Figure 21.2. Its sides are of

Figure 21.2
A square of area 1 m^2 = 10^4 cm^2 = 10^6 mm^2

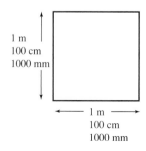

1 m
100 cm
1000 mm

1 m
100 cm
1000 mm

length 1 m = 100 cm = 1000 mm. So its area is

$$1 \times 1 = 1\,\text{m}^2 \quad \text{or} \quad 100 \times 100 = 10^4\,\text{cm}^2 \quad \text{or} \quad 1000 \times 1000 = 10^6\,\text{mm}^2$$

This means that 10^4, that is 10000, squares of side 1 cm will fit into this square metre. It means that 10^6, that is 1000000, squares of side 1 mm will fit into the same area.

Key point

Area conversions:

$$1\,\text{m}^2 = 10^4\,\text{cm}^2 = 10^6\,\text{mm}^2$$

The **volume** of a solid is the number of cubic units it contains. One cubic metre, written 1 m^3, is the volume of a cube with sides of length 1 metre. Similarly, one cubic centimetre, written 1 cm^3, is the volume of a cube having sides of length 1 centimetre. Consider the cubic metre depicted in Figure 21.3. Its sides are of length 1 m = 100 cm = 1000 mm. So its volume is

$$1 \times 1 \times 1 = 1\,\text{m}^3 \quad \text{or} \quad 100 \times 100 \times 100 = 10^6\,\text{cm}^3$$

$$\text{or} \quad 1000 \times 1000 \times 1000 = 10^9\,\text{mm}^3$$

This means that 10^6 cubes of side 1 cm will fit into this cubic metre. It means that 10^9 cubes of side 1 mm will fit into the same volume.

Key point

Volume conversions:

$$1\,\text{m}^3 = 10^6\,\text{cm}^3 = 10^9\,\text{mm}^3$$

When measuring the volume of liquids, the litre is a commonly used unit. One litre, 1 l, is that volume occupied by a cube with sides 10 cm = 100 mm as shown in Figure 21.4. Working in centimetres we have

$$1\,\text{l} = 10 \times 10 \times 10 = 1000\,\text{cm}^3 = 10^3\,\text{cm}^3$$

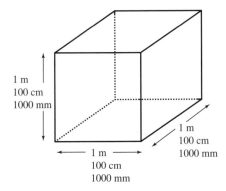

Figure 21.3
A cube of volume of 1 m^3 = 10^6 cm^3 = 10^9 mm^3

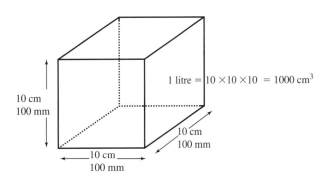

Figure 21.4
A volume of 1 litre (l) = 1000 cm^3 = 10^3 cm^3

Alternatively, working in millimetres we have

$$1\,l = 100 \times 100 \times 100 = 1000000\,mm^3 = 10^6\,mm^3$$

Key point 1 litre (l) $= 10^3\,cm^3 = 10^6\,mm^3$

Commonly, units of millilitre (ml) are used. A millilitre is one-thousandth of a litre, that is $\frac{1}{1000}$ or 10^{-3} of a litre, and so it is the same as a cubic centimetre.

Key point 1 ml $= 1\,cm^3$

WORKED EXAMPLES

21.3   Convert 256 millilitres into litres.

Solution   Because $1\,l = 1000$ ml then 1 ml is equal to $\frac{1}{1000}$ l. Therefore

$$256\ ml = 256 \times \frac{1}{1000} = 0.256\ l$$

21.4   Calculate the number of litres needed to make up a cubic metre.

Solution   Look at Figure 21.5 which depicts a cubic metre. Because there are $100 = 10 \times 10$ cm in 1 m we can think of the cubic metre as being made of smaller blocks each of side 10 cm as shown. The lowest layer of blocks will comprise 100 such blocks, and with 10 layers in the larger cube, a total of 1000 blocks will be required.

Each 10 cm cube has the capacity of 1 litre. So we see that there are 1000 litres in a cubic metre.

Figure 21.5
A cubic metre comprises 1000 cubes each of side length 10 cm, that is, 1 m³ is 1000 litres.

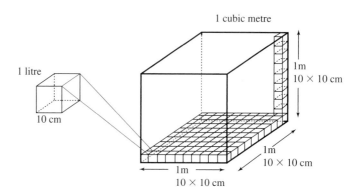

Exercise 21.3

1. Convert an area of 25 cm² to an area measured in (a) m², (b) mm².

2. Express the area 0.005 m² as an area in mm².

3. Convert 5 ml to (a) litres, (b) cm³.

4. A medicine bottle contains 200 ml. How many 5 ml teaspoons of medicine can be extracted from the bottle?

5. A medicine bottle has volume 500 cm³. How many 5 ml teaspoons of medicine can be extracted from the bottle?

6. A petrol tank has capacity 20 l. What is the capacity in cm³?

21.4 Measuring angles in degrees and radians

θ is the Greek letter 'theta' commonly used to denote angles.

We use angles to measure the amount by which a line has been turned. An angle is commonly denoted by θ. In Figure 21.6 the angle between BC and BA is θ. We write $\angle ABC = \theta$. We could also write $\angle CBA = \theta$. The two units for measuring angle are the **degree** and the **radian**. We examine each in turn.

Degrees

A full revolution is 360 degrees, denoted 360°. To put it another way, 1 degree (1°) is $\frac{1}{360}$th of a full revolution. In Figure 21.7, AB is rotated through half a revolution, that is 180°, to the position AC. CAB is a straight line and so 180° is sometimes called a **straight line angle**.

Figure 21.8 shows an angle of 90°. An angle of 90° is called a **right angle**. A triangle containing a right angle is a **right-angled triangle**. The side

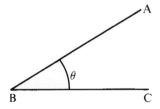

Figure 21.6
The angle ABC is θ

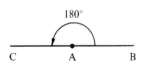

Figure 21.7
180° is a straight line angle

opposite the right angle is called the **hypotenuse**. An angle between $0°$ and $90°$ is an **acute angle**. An angle between $90°$ and $180°$ is an **obtuse angle**. An angle greater than $180°$ is a **reflex angle**.

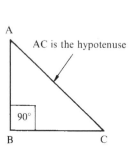

Figure 21.8
An angle of $90°$ is a right angle

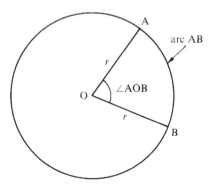

Figure 21.9
Arc AB subtends ∠AOB at the centre of the circle

Radians

The other unit of angle is the radian, and this too is based on examining a circle. Radian is short for 'radius angle'. Consider Figure 21.9, which shows a circle of radius r, centre O and an **arc** AB. We say that the arc AB **subtends** an angle AOB at the centre of the circle. Clearly if AB is short then ∠AOB will be small; if AB is long then ∠AOB will be a large angle. If arc AB has a length equal to 1 radius then we say that ∠AOB is 1 **radian**.

Key point

1 radian is the angle subtended at the centre of a circle by an arc length of 1 radius.

If the angle is in degrees we use the degree symbol, °; otherwise we assume the angle is in radians. Hence we abbreviate 1 radian to 1. From the definition we see that

an arc of length r subtends an angle of 1 radian
an arc of length $2r$ subtends an angle of 2 radians
an arc of length $3r$ subtends an angle of 3 radians

Figure 21.10
An arc of length r
subtends an angle of
1 radian

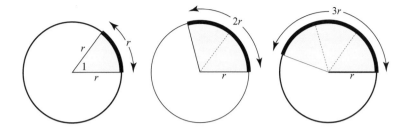

and so on, as illustrated in Figure 21.10. Study the pattern and note that

> an arc of length $2\pi r$ will subtend an angle of 2π radians

We note that $2\pi r$ is the circumference of the circle, and so the circumference subtends an angle of 2π radians. But we already know that the circumference subtends a full revolution, that is $360°$. Hence

$$2\pi = 360°$$

and so, equivalently,

Key point	$\pi = 180°$

Your scientific calculator is pre-programmed to give a value for π. Check that you can use it correctly.

Note that π is a constant whose value is approximately 3.142.

This equation allows us to convert degrees to radians and radians to degrees.

WORKED EXAMPLES

21.5   Convert to degrees

(a) $\dfrac{\pi}{2}$ (b) $\dfrac{\pi}{4}$ (c) $\dfrac{\pi}{3}$ (d) 0.7π (e) 1   (f) 3   (g) 1.3

Solution   (a) We have

$$\pi = 180°$$

and so

$$\frac{\pi}{2} = \frac{180°}{2} = 90°$$

Hence $\dfrac{\pi}{2}$ is a right angle.

(b) $\pi = 180°$

$$\frac{\pi}{4} = \frac{180°}{4}$$

$$= 45°$$

(c) $\pi = 180°$

$$\frac{\pi}{3} = \frac{180°}{3}$$

$$= 60°$$

(d) $\pi = 180°$

$$0.7\pi = 0.7 \times 180°$$

$$= 126°$$

(e) $\pi = 180°$

$$1 = \frac{180°}{\pi}$$

$$= 57.3°$$

(f) $\pi = 180°$

$$1 = \frac{180°}{\pi}$$

$$3 = 3 \times \frac{180°}{\pi}$$

$$= 171.9°$$

(g) $1 = \frac{180°}{\pi}$

$$1.3 = 1.3 \times \frac{180°}{\pi}$$

$$= 74.5°$$

21.6 Convert to radians

(a) 72° (b) 120° (c) 12° (d) 200°

Solution

(a) $180° = \pi$

$$1° = \frac{\pi}{180}$$

$$72° = 72 \times \frac{\pi}{180}$$

$$= \frac{2\pi}{5} = 1.26$$

(b) $180° = \pi$

$$1° = \frac{\pi}{180}$$

$$120° = 120 \times \frac{\pi}{180}$$

$$= \frac{2\pi}{3} = 2.09$$

(c) $180° = \pi$

$$1° = \frac{\pi}{180}$$

$$12° = 12 \times \frac{\pi}{180}$$

$$= \frac{\pi}{15} = 0.21$$

(d) $180° = \pi$

$$1° = \frac{\pi}{180}$$

$$200° = 200 \times \frac{\pi}{180}$$

$$= \frac{10\pi}{9} = 3.49$$

Note the following commonly met angles:

Key point

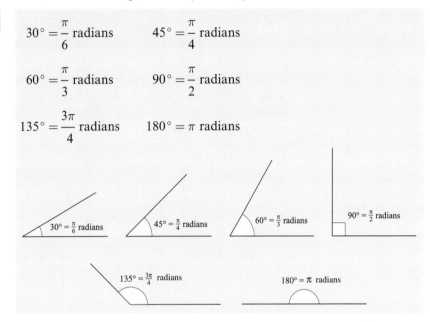

$$30° = \frac{\pi}{6} \text{ radians} \qquad 45° = \frac{\pi}{4} \text{ radians}$$

$$60° = \frac{\pi}{3} \text{ radians} \qquad 90° = \frac{\pi}{2} \text{ radians}$$

$$135° = \frac{3\pi}{4} \text{ radians} \qquad 180° = \pi \text{ radians}$$

Your calculator should be able to work with angles measured in both radians and degrees. Usually the MODE button allows you to select the appropriate measure.

Self-assessment questions 21.4

1. Define the units of angle: degree and radian.

2. How are the degree and the radian related?

Exercise 21.4

1. Convert to radians
   (a) 240° (b) 300° (c) 400° (d) 37°
   (e) 1000°

2. Convert to degrees
   (a) $\frac{\pi}{10}$ (b) $\frac{2\pi}{9}$ (c) 2 (d) 3.46
   (e) 1.75

3. Convert the following angles in radians to degrees:
   (a) 4π (b) 3.5π (c) 5 (d) 1.56

4. Convert the following angles in degrees to radians, expressing your answer as a multiple of π:
   (a) 504° (b) 216° (c) 420°
   (d) 126° (e) 324°

21.5 Areas of common shapes and volumes of common solids

The areas of common shapes and volumes of common solids can be found using formulae. Table 21.3 summarises a variety of such formulae. The calculation of an area or volume is then simply an exercise in substituting values into a formula.

Table 21.3
Formulae for areas and volumes of common shapes and solids

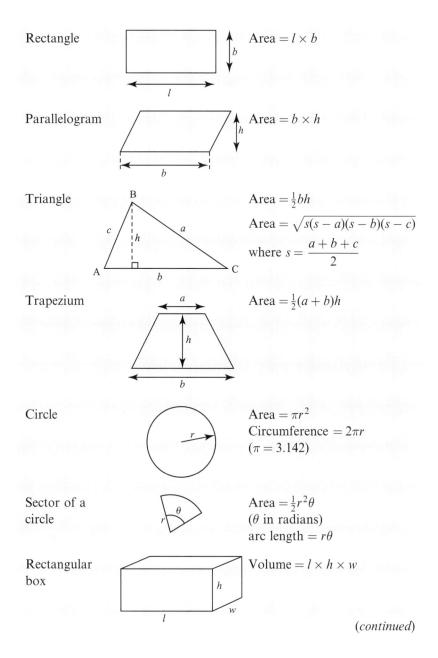

Rectangle   Area $= l \times b$

Parallelogram   Area $= b \times h$

Triangle   Area $= \frac{1}{2}bh$

Area $= \sqrt{s(s-a)(s-b)(s-c)}$

where $s = \dfrac{a+b+c}{2}$

Trapezium   Area $= \frac{1}{2}(a+b)h$

Circle   Area $= \pi r^2$
Circumference $= 2\pi r$
$(\pi = 3.142)$

Sector of a circle   Area $= \frac{1}{2}r^2\theta$
(θ in radians)
arc length $= r\theta$

Rectangular box   Volume $= l \times h \times w$

(*continued*)

Table 21.3
Continued

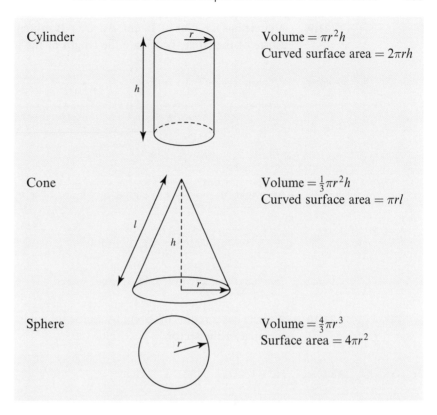

Cylinder		Volume $= \pi r^2 h$ Curved surface area $= 2\pi rh$
Cone		Volume $= \frac{1}{3}\pi r^2 h$ Curved surface area $= \pi rl$
Sphere		Volume $= \frac{4}{3}\pi r^3$ Surface area $= 4\pi r^2$

WORKED EXAMPLES

21.7   A circle has a radius of 8 cm. An angle of 1.4 is subtended at the centre by an arc. Calculate the length of the arc.

Solution   An angle of 1 is subtended by an arc of length 1 radius. So an angle of 1.4 is subtended by an arc of length $1.4 \times$ radius. Now

$$1.4 \times \text{radius} = 1.4 \times 8 = 11.2$$

The length of the arc is 11.2 cm.

21.8   A circle has radius 6 cm. An arc, AB, of the circle measures 9 cm. Calculate the angle subtended by AB at the centre of the circle.

Solution   Using arc length $= r\theta$ we have

$$9 = 6\theta$$

$$\theta = \frac{9}{6} = 1.5$$

Hence AB subtends an angle of 1.5 at the centre.

21.9   A circle has a radius of 18 cm. An arc AB subtends an angle of 142° at the centre of the circle. Calculate the length of the arc AB.

Solution   We convert 142° to radians:

$$180° = \pi$$

$$142° = \frac{142 \times \pi}{180}$$

An angle of $\dfrac{142\pi}{180}$ is subtended by an arc of length $\dfrac{142\pi}{180} \times$ radius:

$$\frac{142\pi}{180} \times \text{radius} = \frac{142\pi}{180} \times 18 = 44.61$$

The length of the arc is 44.61 cm.

21.10   A circle has a radius of 22 cm. A sector subtends an angle of 1.2. Calculate the area of the sector.

Solution   Here $r = 22$ and $\theta = 1.2$. Hence

$$\text{area of sector} = \frac{r^2\theta}{2} = \frac{(22)^2(1.2)}{2} = 290.4$$

The area of the sector is 290.4 cm².

21.11   A circle has a radius of 10 cm. The area of a sector is 170 cm². Calculate the angle at the centre.

Solution   The area of sector $= \dfrac{r^2\theta}{2}$

Substituting the values given we obtain

$$170 = \frac{(10)^2\theta}{2}$$

$$\theta = \frac{2 \times 170}{10^2}$$

$$= 3.4$$

The angle is 3.4.

Exercise 21.5

1. Find the area of a rectangle whose length is 14 cm and whose width is 7 cm.

2. A room has length 5 m and width 3.5 m. It is required to carpet this room at £4.99 per square metre. Calculate the cost of the carpet assuming there is no wastage due to cutting.

3. Find the area of a parallelogram that has a vertical height of 2 m and a base of length 8 m.

4. A triangle has sides of length 8 cm, 12 cm and 14 cm. Find its area.

5. Find the area of a trapezium whose parallel sides are 14 cm and 8 cm long and whose distance apart is 2 cm.

6. Find the area of a circle whose radius is 18 cm.

7. Find the area of a circle whose diameter is 10 cm.

8. Find the radius of the circle whose area is 26 cm^2.

9. A circular piece of cardboard of radius 70 cm has a circular piece of radius 25 cm removed from it. Calculate the area of the remaining cardboard.

10. Find the volume of a tin of beans of base radius 5 cm and height 14 cm. What is the tin's capacity in litres?

11. Find the volume of a sphere of radius 8 cm.

12. A sphere has volume 86 cm^3. Calculate its radius.

13. Calculate the area of a sector of a circle of angle 45° and radius 10 cm.

14. A garden swimming pool is designed in the form of a cylinder of base radius 3 m. Calculate how many litres of water are required to fill it to a depth of 1 m.

15. A circle has a radius of 15 cm. Calculate the length of arc that subtends an angle of
(a) 2 (b) 3 (c) 1.2 (d) 100°
(e) 217°

16. A circle has radius 12 cm. Calculate the angle subtended at the centre by an arc of length
(a) 12 cm (b) 6 cm (c) 24 cm
(d) 15 cm (e) 2 cm

17. A circle has a radius of 18 cm. Calculate the length of arc which subtends an angle of
(a) 1.5 (b) 1.1 (c) $\frac{\pi}{3}$ (d) 1.76

18. A circle has a radius of 24 cm. Calculate the angle subtended at the centre by an arc of length
(a) 18 cm (b) 30 cm (c) 20 cm
(d) 60 cm

19. A circle has a radius of 9 cm. Calculate the area of the sectors that subtend angles of
(a) 1.5 (b) 2 (c) 100° (d) 215°

20. A circle has a radius of 16 cm. Calculate the angle at the centre when a sector has an area of
(a) 100 cm^2 (b) 5 cm^2 (c) 520 cm^2

21. A circle has a radius of 18 cm. Calculate the area of a sector whose angle is
(a) 1.5 (b) 2.2 (c) 120° (d) 217°

22. A circle, centre O, has a radius of 25 cm. An arc AB has length 17 cm. Calculate the area of the sector AOB.

23. A circle, centre O, radius 12 cm, has a sector AOB of area 370 cm^2. Calculate the length of the arc AB.

24. A circle, centre O, has a sector AOB. The arc length of AB is 16 cm and the angle subtended at O is 1.2. Calculate the area of the sector AOB.

21.6 Units of mass

The standard unit of mass is the kilogram, abbreviated to kg. Sometimes smaller units, the gram (g) or milligram (mg), are used. A very sensitive balance may be able to measure in micrograms (μg). The relationship between these units is given as follows:

Key point

1 kilogram (kg) = 1000 grams (g) = 1000000 milligrams (mg)
1 gram (g) = 1000 milligrams (mg) = 1000000 micrograms (μg)
1 milligram = 1000 micrograms (μg)

Note also that 1000 kilograms equals 1 tonne (t).

WORKED EXAMPLES

21.12 Convert 8.5 kg into grams.

Solution 1 kg is equal to 1000 g and so 8.5 kg must equal $8.5 \times 1000 = 8500$ g.

21.13 Convert 50 g to kilograms.

Solution Because 1 kg equals 1000 g then 1 g equals $\frac{1}{1000}$ of a kilogram. Consequently,

$$50 \text{ g} = 50 \times \frac{1}{1000} = 0.05 \text{ kg}$$

Exercise 21.6

MyMathLab Global

1. Convert 12 kg into (a) grams, (b) milligrams.

2. Convert 168 mg into (a) grams, (b) kilograms.

3. Convert 0.005 t into kilograms.

4. Convert 875 t into kilograms.

5. Convert 3500 kg into tonnes.

21.7 Units of time

The standard unit of time is the second, abbreviated to s. Sometimes smaller units, the millisecond (ms) or microsecond (μs), are used.

21.8 Dimensional analysis

All quantities can be expressed in terms of a number of **fundamental dimensions**. Most commonly, these dimensions are length, mass and time, to which we will give the symbols L, M and T respectively. (Do not confuse this use of M and T with the symbols for prefixes in Table 21.1.) In the study of some subjects such as electrical engineering and chemistry it is necessary to consider other dimensions such as charge and temperature, but these will not concern us here.

The three fundamental dimensions have corresponding units of measurement: in the SI these are length: metre (m); mass: kilogram (kg); time: second (s).

We introduce the following square bracket [ ] notation in order to help us analyse the dimensions of a quantity:

[volume] represents the dimensions of volume
[speed] represents the dimensions of speed, and so on

Think of the volume of a rectangular box (see Table 21.3). This is given by

$$\text{volume} = \text{length} \times \text{height} \times \text{width}$$

Each of the quantities length, height and width has dimensions of length L. Thus

$$[\text{volume}] = L \times L \times L = L^3$$

WORKED EXAMPLES

21.14 Find the dimensions of speed (measured in metres per second).

Solution The unit of speed is m/s, that is a unit of length divided by a unit of time, so the dimensions are

$$[\text{speed}] = \frac{\text{length}}{\text{time}} = \frac{L}{T} = LT^{-1}$$

21.15 Find the dimensions of acceleration (measured in metres per second per second).

Solution The unit of acceleration is m/s² so the dimensions are

$$[\text{acceleration}] = \frac{\text{length}}{\text{time}^2} = \frac{L}{T^2} = LT^{-2}$$

21.16 The force, F, required to make a mass, m, accelerate with acceleration, a, is given by the formula $F = ma$. Find the dimensions of force.

Solution   The dimension of mass is M. The dimensions of acceleration are LT^{-2} as found in the previous worked example. So,

$$[\text{force}] = [\text{mass}] \times [\text{acceleration}] = MLT^{-2}$$

21.17   Pressure is defined to be force/area. Find the dimensions of pressure.

Solution $[\text{pressure}] = \dfrac{[\text{force}]}{[\text{area}]} = \dfrac{MLT^{-2}}{L^2} = ML^{-1}T^{-2}$

Dimensional analysis is a process used to check the validity of a physical formula. The dimensions of both the right-hand and left-hand side are found. If the formula is valid, the dimensions on both sides must be the same. Numbers in formulae are dimensionless. Consider the following worked example.

WORKED EXAMPLE

21.18   A commonly used formula in physics is $s = ut + \frac{1}{2}at^2$. This tells us the distance s travelled by a particle in time t when its initial speed is u and its acceleration is a. Use dimensional analysis to check whether this formula is valid.

Solution   The dimension of the left-hand side is simply the dimensions of s, distance, that is L.

On the right-hand side we consider the dimensions of each term separately. The term ut is a speed $\times$ time term and so has dimensions

$$[ut] = LT^{-1} \times T = L$$

The term $\frac{1}{2}at^2$ has dimensions of acceleration $\times$ time^2, that is

$$\left[\frac{1}{2}at^2\right] = LT^{-2} \times T^2 = L$$

So all terms on the right have dimension L. This is the same as the dimension of the term on the left and so the formula is dimensionally valid.

Exercise 21.8

1. Find the dimensions of area.

2. Energy, E, is given by the formula $F \times d$ where F is a force and d is a distance. Find the dimensions of energy.

3. The period of swing, T, of a simple pendulum is given by $T = 2\pi\sqrt{\ell/g}$ where ℓ is the length of the pendulum and g is the acceleration due to gravity. Show that the formula is dimensionally valid (the constant 2π is dimensionless).

4. The potential energy, E, of an object is given by $E = mgh$ where m is the object's mass, g is the acceleration due to gravity and h is its height. Given that energy has dimensions ML^2T^{-2} show that this formula is dimensionally valid.

5. The kinetic energy, E, of an object is given by $E = \frac{1}{2}mv^2$ where m is the object's mass and v is its speed. Given that energy has dimensions ML^2T^{-2} show that this formula is dimensionally valid.

Test and assignment exercises 21

1. Change the following to grams:
   (a) 125 mg   (b) 15 kg   (c) 0.5 mg

2. Change the following lengths to centimetres:
   (a) 65 mm   (b) 0.245 mm   (c) 5 m

3. Convert 7500 m and 125.5 m into kilometres.

4. Convert 8 l into millilitres.

5. Convert 56 ml into litres.

6. A household water tank has dimensions 1 m × 0.6 m × 0.8 m. Calculate its capacity in litres assuming it can be filled to the brim.

7. Calculate the area of the sector of a circle of radius 25 cm and angle 60°.

8. A sector of a circle of radius 10 cm has area 186 cm². Calculate the angle of the sector.

9. Calculate the volume of a cylinder whose height is 1 m and whose base radius is 10 cm. (Be careful of the units used.)

10. A sphere has diameter 16 cm. Calculate its volume and surface area.

11. A trapezium has parallel sides of length y_1 and y_2. If these sides are a distance h apart find an expression for the area of the trapezium. Calculate this area when $y_1 = 7$ cm, $y_2 = 10$ cm and $h = 1$ cm.

12. A lorry weighs 7.5 t. Express this in kilograms.

13. Convert to radians
    (a) 73° (b) 196° (c) 1000°

14. Convert to degrees
    (a) 1.75 (b) 0.004 (c) 7.9

15. A circle, centre O, has radius 22 cm. An arc AB has length 35 cm.
    (a) Calculate the angle at O subtended by AB.
    (b) Calculate the area of the sector AOB.

16. A circle, centre O, has radius 42 cm. Arc AB subtends an angle of 235° at O.
    (a) Calculate the length of the arc AB.
    (b) Calculate the area of the sector AOB.

17. A circle, centre O, has radius 20 cm. The area of sector AOB is 320 cm².
    (a) Calculate the angle at O subtended by the arc AB.
    (b) Calculate the length of the arc AB.

18. A sector AOB has an area of 250 cm² and the angle at the centre, O, is 75°.
    (a) Calculate the area of the circle.
    (b) Calculate the radius of the circle.

19. A circle, centre O, has a radius of 18 cm. An arc AB has length 22 cm. Calculate the area of sector AOB.

20. A circle, centre O, has a sector AOB. The angle subtended at O is 108°, and the arc AB has length 11 cm. Calculate the area of the sector AOB.

21. A circle, centre O, radius 14 cm, has a sector AOB of area 290 cm². Calculate the length of the arc AB.

22. Figure 21.11 shows two concentric circles of radii 5 cm and 10 cm. Arc AB has length 15 cm. Calculate the shaded area.

23. The formula for the work done W by a force F causing a displacement through a distance d is $W = F \times d$. Given that work has dimensions ML^2T^{-2} show that this formula is dimensionally valid.

Figure 21.11
Figure for Test and
assignment exercises 21,
Question 22

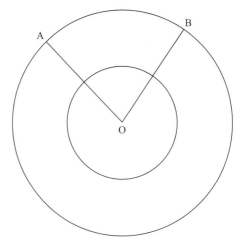

Introduction to trigonometry 22

Objectives: This chapter:

- introduces terminology associated with a right-angled triangle
- defines the trigonometrical ratios sine, cosine and tangent with reference to the lengths of the sides of a right-angled triangle
- explains how to calculate an angle given one of its trigonometrical ratios

22.1 The trigonometrical ratios

Consider the right-angled triangle ABC shown in Figure 22.1. There is a right angle at B.

The side joining points A and B is referred to as AB. The side joining points A and C is referred to as AC. The angle at A, made by the sides AB and AC, is written $\angle$BAC, $\angle A$, or simply as A. In Figure 22.1 we have labelled this angle θ.

The side opposite the right angle is always called the **hypotenuse**. So, in Figure 22.1 the hypotenuse is AC. The side **opposite** θ is BC. The remaining side, AB, is said to be **adjacent** to θ.

Figure 22.1
A right-angled triangle ABC

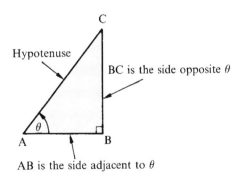

Hypotenuse

BC is the side opposite θ

A   B

AB is the side adjacent to θ

If we know the lengths BC and AC we can calculate $\frac{BC}{AC}$. This is known as the **sine** of θ, or simply $\sin\theta$. Similarly, we call $\frac{AB}{AC}$ the **cosine** of θ, written $\cos\theta$. Finally, $\frac{BC}{AB}$ is known as the **tangent** of θ, written $\tan\theta$. Sine, cosine and tangent are known as the **trigonometrical ratios**.

Key point

$$\sin\theta = \frac{\text{side opposite to }\theta}{\text{hypotenuse}} = \frac{BC}{AC}$$

$$\cos\theta = \frac{\text{side adjacent to }\theta}{\text{hypotenuse}} = \frac{AB}{AC}$$

$$\tan\theta = \frac{\text{side opposite to }\theta}{\text{side adjacent to }\theta} = \frac{BC}{AB}$$

Note that all the trigonometrical ratios are defined as the ratio of two lengths, and so they themselves have no units. Also, since the hypotenuse is always the longest side of a triangle, $\sin\theta$ and $\cos\theta$ can never be greater than 1.

WORKED EXAMPLES

22.1   Calculate $\sin\theta$, $\cos\theta$ and $\tan\theta$ for $\triangle ABC$ as shown in Figure 22.2.

Figure 22.2
$\triangle ABC$ for Worked
Example 22.1

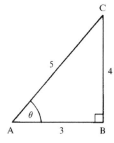

Solution

$$\sin\theta = \frac{BC}{AC} = \frac{4}{5} = 0.8$$

$$\cos\theta = \frac{AB}{AC} = \frac{3}{5} = 0.6$$

$$\tan\theta = \frac{BC}{AB} = \frac{4}{3} = 1.3333$$

Figure 22.3

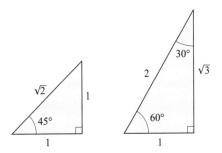

22.2   Using the triangles shown in Figure 22.3 write down expressions for

(a) sin 45°, cos 45°, tan 45°

(b) sin 60°, cos 60°, tan 60°

(c) sin 30°, cos 30°, tan 30°

Solution   (a) $\sin 45° = \frac{1}{\sqrt{2}}$, $\cos 45° = \frac{1}{\sqrt{2}}$, $\tan 45° = \frac{1}{1} = 1$

(b) $\sin 60° = \frac{\sqrt{3}}{2}$, $\cos 60° = \frac{1}{2}$, $\tan 60° = \frac{\sqrt{3}}{1} = \sqrt{3}$

(c) $\sin 30° = \frac{1}{2}$, $\cos 30° = \frac{\sqrt{3}}{2}$, $\tan 30° = \frac{1}{\sqrt{3}}$

Notice that we have left our solutions in their exact forms using square roots if necessary rather than given their decimal approximations. These, so-called, **surd forms** are frequently used.

Recall that an angle may be measured in degrees or in radians. An angle in degrees has the symbol °. Otherwise assume that the angle is measured in radians. If we know an angle, it is not necessary to draw a triangle to calculate its trigonometrical ratios. These ratios can be found directly using a scientific calculator. Your calculator must be instructed to work either in degrees or in radians. Usually the MODE button is used to select the required units.

WORKED EXAMPLES

22.3   Use a scientific calculator to evaluate

(a) sin 65°   (b) cos 17°   (c) tan 50°   (d) sin 1   (e) cos 1.5
(f) tan 0.5

Solution   For (a), (b) and (c) ensure your calculator is set to DEGREES, and not to RADIANS or GRADS.

(a) sin 65° = 0.9063   (b) cos 17° = 0.9563   (c) tan 50° = 1.1918

Now, change the MODE of your calculator in order to work in radians. Check that

(d) sin 1 = 0.8415   (e) cos 1.5 = 0.0707   (f) tan 0.5 = 0.5463

22.4   Prove

$$\frac{\sin\theta}{\cos\theta} = \tan\theta$$

Solution   From Figure 22.1, we have

$$\sin\theta = \frac{BC}{AC} \qquad \cos\theta = \frac{AB}{AC}$$

and so

$$\frac{\sin\theta}{\cos\theta} = \frac{BC/AC}{AB/AC} = \frac{BC}{AC} \times \frac{AC}{AB} = \frac{BC}{AB}$$

But

$$\tan\theta = \frac{BC}{AB}$$

and so

$$\frac{\sin\theta}{\cos\theta} = \tan\theta$$

Self-assessment questions 22.1

1. Define the trigonometrical ratios sine, cosine and tangent with reference to a right-angled triangle.

2. Explain why the trigonometrical ratios have no units.

Exercise 22.1

1. Evaluate
    (a) tan 30°   (b) sin 20°   (c) cos 75°
    (d) sin 1.2   (e) cos 0.89   (f) tan π/4

2. (a) Evaluate sin 70° and cos 20°.
    (b) Evaluate sin 25° and cos 65°.
    (c) Prove $\sin\theta = \cos(90° - \theta)$.
    (d) Prove $\cos\theta = \sin(90° - \theta)$.

3. A right-angled triangle ABC has ∠CBA = 90°, AB = 3 cm, BC = 3 cm

and AC = $\sqrt{18}$ cm. ∠CAB = θ. Without using a calculator find (a) sin θ, (b) cos θ, (c) tan θ.

4. A right-angled triangle XYZ has a right angle at Y, XY = 10 cm, YZ = 4 cm and XZ = $\sqrt{116}$ cm. If ∠ZXY = θ and ∠YZX = α, find
    (a) sin θ (b) cos θ (c) tan θ
    (d) sin α (e) cos α (f) tan α

22.2 Finding an angle given one of its trigonometrical ratios

Given an angle θ, we can use a scientific calculator to find $\sin\theta$, $\cos\theta$ and $\tan\theta$. Often we require to reverse the process: that is, given a value of $\sin\theta$, $\cos\theta$ or $\tan\theta$ we need to find the corresponding values of θ.

If

$$\sin\theta = x$$

we write

$$\theta = \sin^{-1} x$$

The superscript −1 does not denote a power. It is a notation for the inverse of the trigonometrical ratios.

and this is read as 'θ is the inverse sine of x'. This is the same as saying that θ is the angle whose sine is x. Similarly, $\cos^{-1} y$ is the angle whose cosine is y and $\tan^{-1} z$ is the angle whose tangent is z.

Most scientific calculators have inverse sine, inverse cosine and inverse tangent values programmed in. The buttons are usually denoted as $\sin^{-1}$, $\cos^{-1}$ and $\tan^{-1}$.

WORKED EXAMPLES

22.5 Use a scientific calculator to evaluate

(a) $\sin^{-1} 0.5$ (b) $\cos^{-1} 0.3$ (c) $\tan^{-1} 2$

Solution (a) Using the $\sin^{-1}$ button we have

$$\sin^{-1} 0.5 = 30°$$

This is another way of saying that $\sin 30° = 0.5$.

(b) Using a scientific calculator we see

$$\cos^{-1} 0.3 = 72.5°$$

that is,

$$\cos 72.5° = 0.3$$

(c) Using a calculator we see $\tan^{-1} 2 = 63.4°$ and hence $\tan 63.4° = 2$.

22.6 Calculate the angles θ in Figures 22.4(a), (b) and (c).

Solution (a) In Figure 22.4(a) we are given the length of the side opposite θ and the length of the adjacent side. Hence we can make use of the tangent ratio. We write

$$\tan\theta = \frac{8}{3} = 2.6667$$

Figure 22.4

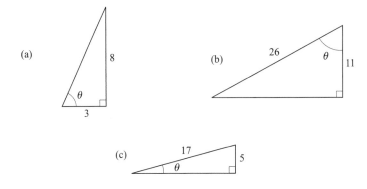

Therefore θ is the angle whose tangent is 2.6667, written

$$\theta = \tan^{-1} 2.6667$$

Using a calculator we can find the angle θ in either degrees or radians. If your calculator is set to degree mode, verify that $\theta = 69.44°$. Check also that you can obtain the equivalent answer in radians, that is 1.212.

(b) In Figure 22.4(b) we are given the length of the side adjacent to θ and the length of the hypotenuse. Hence we use the cosine ratio:

$$\cos\theta = \frac{11}{26} = 0.4231$$

Therefore

$$\theta = \cos^{-1} 0.4231$$

Using a calculator check that $\theta = 64.97°$.

(c) In Figure 22.4(c) $\sin\theta = \frac{5}{17}$. Hence

$$\theta = \sin^{-1} \frac{5}{17} = 17.10°$$

Exercise 22.2

MyMathLab Global

1. Using a calculator find the following angles in degrees:
    (a) $\sin^{-1} 0.8$ (b) $\cos^{-1} 0.2$
    (c) $\tan^{-1} 1.3$

2. Using a calculator find the following angles in radians:
    (a) $\sin^{-1} 0.63$ (b) $\cos^{-1} 0.25$
    (c) $\tan^{-1} 2.3$

3. Find the angle θ in each of the right-angled triangles shown in Figure 22.5, giving your answer in degrees.

4. Find the angle θ in each of the right-angled triangles shown in Figure 22.6, giving your answer in radians.

Figure 22.5

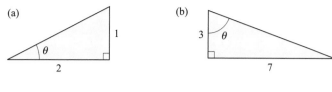

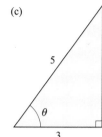

Figure 22.6

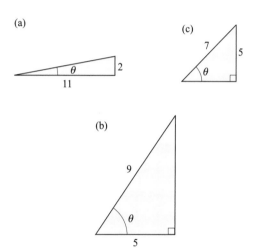

Test and assignment exercises 22

1. Use a calculator to evaluate
   (a) sin 23°   (b) cos 52°   (c) tan 77°   (d) sin 0.1   (e) cos 1.1   (f) tan 0.9
   (g) $\tan\dfrac{\pi}{6}$ (h) $\tan\dfrac{\pi}{4}$ (i) $\cos\dfrac{\pi}{3}$

2. A right-angled triangle PQR has a right angle at Q. PR = 5, PQ = 3, QR = 4, and ∠PRQ = θ. Find sin θ, cos θ and tan θ.

3. Using a calculator find the following angles in degrees:
   (a) $\sin^{-1} 0.2$ (b) $\cos^{-1} 0.5$ (c) $\tan^{-1} 2.4$

4. Using a calculator find the following angles in radians:

(a) $\sin^{-1} 0.33$ (b) $\cos^{-1} 0.67$ (c) $\tan^{-1} 0.3$

5. Find the angles α and β in Figure 22.7.

Figure 22.7

The trigonometrical functions and their graphs

23

Objectives : This chapter:

- extends the definition of trigonometrical ratios to angles greater than 90°
- introduces the trigonometrical functions and their graphs

23.1 Extended definition of the trigonometrical ratios

In §22.1 we used a right-angled triangle in order to define the three trigonometrical ratios. The angle θ is thus limited to a maximum value of 90°. To give meaning to the trigonometrical ratios of angles greater than 90° we introduce an extended definition.

Quadrants

We begin by introducing the idea of quadrants.

The x and y axes divide the plane into four quadrants as shown in Figure 23.1. The origin is O. We consider an arm OC that can rotate into any of the quadrants. In Figure 23.1 the arm is in quadrant 1. We measure the anticlockwise angle from the positive x axis to the arm and call this angle θ. When the arm is in quadrant 1, then $0° \leqslant \theta \leqslant 90°$, when in the second quadrant $90° \leqslant \theta \leqslant 180°$, when in the third quadrant $180° \leqslant \theta \leqslant 270°$ and when in the fourth quadrant $270° \leqslant \theta \leqslant 360°$.

On occasions, angles are measured in a clockwise direction from the positive x axis. In such cases these angles are conventionally taken to be negative. Figure 23.2 shows angles of $-60°$ and $-120°$. Note that for $-60°$ the arm is in the same position as for 300°. Similarly, for $-120°$ the arm is in the same position as for 240°.

Figure 23.1
The plane is divided into
four quadrants

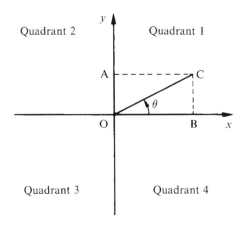

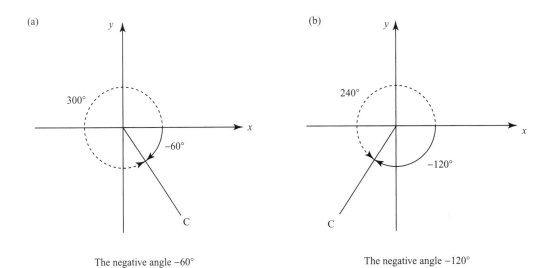

The negative angle −60°

The negative angle −120°

Figure 23.2
The negative angles of −60° and −120°

Projections

We consider the x and y projections of the arm OC. These are illustrated in Figure 23.3. We label the x projection OB and the y projection OA. One or both of these projections can be negative, depending upon the position of the arm OC. However, the length of the arm itself is always considered to be positive.

Figure 23.3
The x projection of OC
is OB; the y projection of
OC is OA

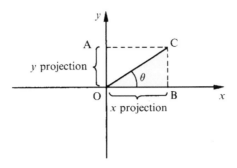

The trigonometrical ratios are now defined as

$$\sin \theta = \frac{y \text{ projection of arm OC}}{\text{OC}} = \frac{\text{OA}}{\text{OC}}$$

$$\cos \theta = \frac{x \text{ projection of arm OC}}{\text{OC}} = \frac{\text{OB}}{\text{OC}}$$

$$\tan \theta = \frac{y \text{ projection of arm OC}}{x \text{ projection of arm OC}} = \frac{\text{OA}}{\text{OB}}$$

Note that the extended definition is in terms of projections and so θ is not limited to a maximum value of $90°$.

Projections in the first quadrant

Consider the arm OC in the first quadrant, as shown in Figure 23.3. From the right-angled triangle OCB we have

$$\sin \theta = \frac{\text{BC}}{\text{OC}} \qquad \cos \theta = \frac{\text{OB}}{\text{OC}} \qquad \tan \theta = \frac{\text{BC}}{\text{OB}}$$

Alternatively, using the extended definition we could also write

$$\sin \theta = \frac{\text{OA}}{\text{OC}} \qquad \cos \theta = \frac{\text{OB}}{\text{OC}} \qquad \tan \theta = \frac{\text{OA}}{\text{OB}}$$

Noting that OA = BC, we see that the two definitions are in agreement when $0° \leqslant \theta \leqslant 90°$.

Projections in the second quadrant

We now consider the arm in the second quadrant, as shown in Figure 23.4. The x projection of OC is onto the negative part of the x axis; the y projection of OC is onto the positive part of the y axis. Hence $\sin \theta$ is positive, whereas $\cos \theta$ and $\tan \theta$ are negative.

Figure 23.4
When OC is in the
second quadrant the x
projection is negative

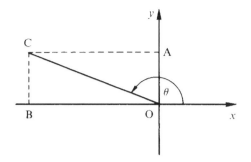

Projections in the third quadrant

When the arm is in the third quadrant, as shown in Figure 23.5, the x and
y projections are both negative. Hence for $180° < \theta < 270°$, sin θ and cos θ
are negative and tan θ is positive.

Figure 23.5
Both x and y projections
are negative

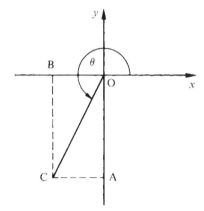

Projections in the fourth quadrant

Finally the arm OC is rotated into the fourth quadrant, as shown in
Figure 23.6. The x projection is positive, the y projection is negative, and
so sin θ and tan θ are negative and cos θ is positive.

Figure 23.6
The x projection is
positive and the y
projection is negative

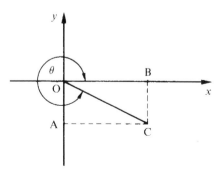

By looking at sin θ, cos θ and tan θ in the four quadrants we see that:

Key point

	Quadrant 1	Quadrant 2	Quadrant 3	Quadrant 4
sin θ	positive	positive	negative	negative
cos θ	positive	negative	negative	positive
tan θ	positive	negative	positive	negative

We have examined the trigonometrical ratios as θ varies from the first quadrant to the fourth quadrant, that is from 0° to 360°. It is possible for θ to have values outside the range 0° to 360°. Adding or subtracting 360° from an angle is equivalent to rotating the arm through a complete revolution. This will leave its position unchanged. Hence adding or subtracting 360° from an angle will leave the trigonometrical ratios unaltered. We state this mathematically as

$$\sin \theta = \sin(\theta + 360°) = \sin(\theta - 360°)$$

$$\cos \theta = \cos(\theta + 360°) = \cos(\theta - 360°)$$

$$\tan \theta = \tan(\theta + 360°) = \tan(\theta - 360°)$$

WORKED EXAMPLES

23.1   Find tan θ in Figure 23.7.

Figure 23.7

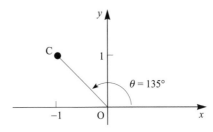

Solution   Angle θ is in the second quadrant. Hence its tangent will be negative. We can find its value from

$$\tan \theta = \frac{y \text{ projection}}{x \text{ projection}}$$

$$= \frac{1}{-1}$$

$$= -1$$

23.2   Find $\sin \theta$ in Figure 23.8.

Figure 23.8

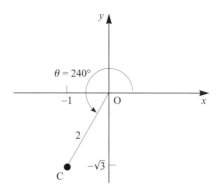

Solution   Angle θ is in the third quadrant. Hence its sine will be negative. We can find its value from

$$\sin \theta = \frac{y \text{ projection}}{\text{OC}}$$

$$= \frac{-\sqrt{3}}{2}$$

$$= -\frac{\sqrt{3}}{2}$$

23.3   Find $\cos \theta$ in Figure 23.9.

Figure 23.9

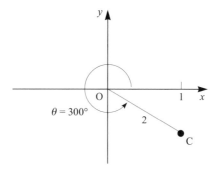

Solution   Angle θ is in the fourth quadrant. Hence its cosine will be positive. We can find its value from

$$\cos \theta = \frac{x \text{ projection}}{\text{OC}}$$

$$= \frac{1}{2}$$

Self-assessment question 23.1

1. State the sign of $\sin \theta$, $\cos \theta$ and $\tan \theta$ in each of the four quadrants.

Exercise 23.1

MyMathLab

1. An angle α is such that $\sin \alpha > 0$ and $\cos \alpha < 0$. In which quadrant does α lie?

2. An angle β is such that $\sin \beta < 0$. In which quadrants is it possible for β to lie?

3. The x projection of a rotating arm is negative. Which quadrants could the arm be in?

4. The y projection of a rotating arm is negative. Which quadrants could the arm be in?

5. Referring to Figure 23.10, state the sine, cosine and tangent of θ.

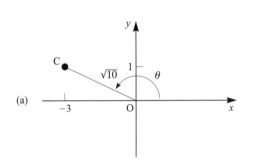

(a)

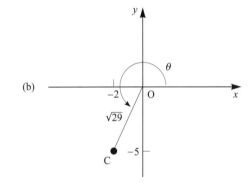

(b)

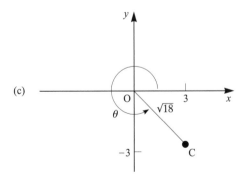

(c)

Figure 23.10

23.2 Trigonometrical functions and their graphs

Having introduced the ratios $\sin \theta$, $\cos \theta$ and $\tan \theta$, we are ready to consider the functions $y = \sin \theta$, $y = \cos \theta$ and $y = \tan \theta$. The independent variable is θ, and for every value of θ the output $\sin \theta$, $\cos \theta$ or $\tan \theta$ can be found. The graphs of these functions are illustrated in this section.

The sine function, $y = \sin \theta$

We can plot the function $y = \sin \theta$ by drawing up a table of values, as for example in Table 23.1. Plotting these values and joining them with a smooth curve produces the graph shown in Figure 23.11. It is possible to plot $y = \sin \theta$ using a graphics calculator or a graph-plotting package.

Table 23.1

θ	0°	30°	60°	90°	120°	150°	180°
$\sin \theta$	0	0.5	0.8660	1	0.8660	0.5	0

θ	210°	240°	270°	300°	330°	360°
$\sin \theta$	−0.5	−0.8660	−1	−0.8660	−0.5	0

Figure 23.11
The function $y = \sin \theta$

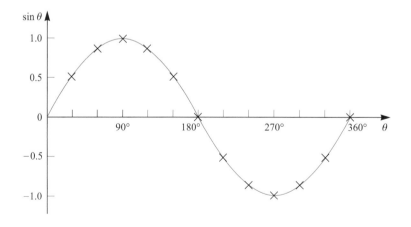

Recall from §23.1 that adding or subtracting 360° from an angle does not alter the trigonometrical ratio of that angle. Hence if we extend the values of θ below 0° and above 360°, the values of $\sin \theta$ are simply repeated. A graph of $y = \sin \theta$ for a larger domain of θ is shown in Figure 23.12. The pattern is repeated every 360°.

Figure 23.12
The values of sin θ
repeat every 360°

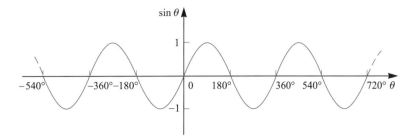

The cosine function, $y = \cos \theta$

Table 23.2 gives values of θ and $\cos \theta$. Plotting these values and joining them with a smooth curve produces the graph of $y = \cos \theta$ shown in Figure 23.13.

Table 23.2

θ	0°	30°	60°	90°	120°	150°	180°
$\cos \theta$	1	0.8660	0.5	0	−0.5	−0.8660	−1

θ	210°	240°	270°	300°	330°	360°
$\cos \theta$	−0.8660	−0.5	0	0.5	0.8660	1

Figure 23.13
The function $y = \cos \theta$

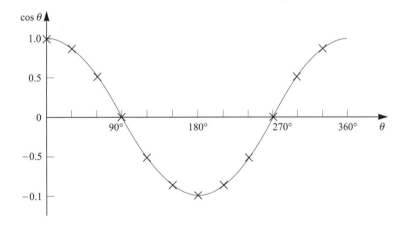

Extending the values of θ beyond 0° and 360° produces the graph shown in Figure 23.14, in which we see that the values are repeated every 360°.

Note the similarity between $y = \sin \theta$ and $y = \cos \theta$. The two graphs are identical apart from the starting point.

The tangent function, $y = \tan \theta$

By constructing a table of values and plotting points a graph of $y = \tan \theta$ may be drawn. Figure 23.15 shows a graph of $y = \tan \theta$ as θ varies from 0° to 360°. Extending the values of θ produces Figure 23.16. Note that the pattern is repeated every 180°. The values of tan θ extend from minus infinity to plus infinity.

Figure 23.14
The values of cos θ
repeat every 360°

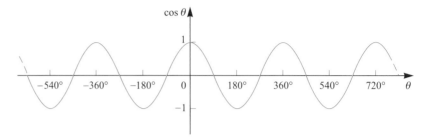

Figure 23.15
The function $y = \tan \theta$

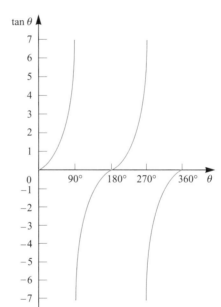

Figure 23.16
The values of tan θ
repeat every 180°

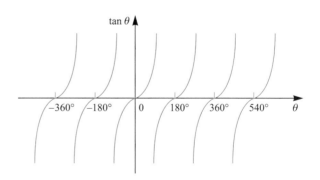

WORKED EXAMPLES

23.4   Draw $y = \sin 2\theta$ for $0° \leqslant \theta \leqslant 180°$.

Solution

Values of θ and $\sin 2\theta$ are given in Table 23.3. A graph of $y = \sin 2\theta$ is shown in Figure 23.17.

Table 23.3
Values of θ and corresponding values of $\sin 2\theta$

θ	0	15	30	45	60	75	90	105
2θ	0	30	60	90	120	150	180	210
$\sin 2\theta$	0	0.5000	0.8660	1	0.8660	0.5000	0	−0.5000

θ	120	135	150	165	180
2θ	240	270	300	330	360
$\sin 2\theta$	−0.8660	−1	−0.8660	−0.5000	0

Figure 23.17
A graph of $y = \sin 2\theta$

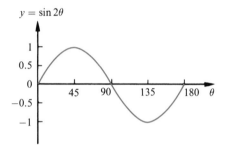

23.5 Draw a graph of $y = \cos(\theta + 30°)$ for $0° \leqslant \theta \leqslant 360°$.

Solution

A table of values is drawn up in Table 23.4. Figure 23.18 shows a graph of $y = \cos(\theta + 30°)$.

Table 23.4
Values of θ and corresponding values of $\cos(\theta + 30°)$

θ	0	30	60	90	120	150	180	210
$\theta + 30$	30	60	90	120	150	180	210	240
$\cos(\theta + 30)$	0.8660	0.5000	0	−0.5000	−0.8660	−1	−0.8660	−0.5000

θ	240	270	300	330	360
$\theta + 30$	270	300	330	360	390
$\cos(\theta + 30)$	0	0.5000	0.08660	1	0.8660

Figure 23.18
Graph of
$y = \cos(\theta + 30°)$

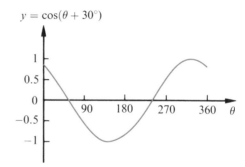

$y = \cos(\theta + 30°)$

Self-assessment questions 23.2

1. State two properties that $y = \sin\theta$ and $y = \cos\theta$ have in common.

2. State three properties that distinguish $y = \tan\theta$ from either $y = \sin\theta$ or $y = \cos\theta$.

3. State the maximum and minimum values of the functions $\sin\theta$ and $\cos\theta$.
   At what values of θ do these maximum and minimum values occur?

4. With reference to the appropriate graphs explain why $\sin(-\theta) = -\sin\theta$ and $\cos(-\theta) = \cos\theta$.

Exercise 23.2

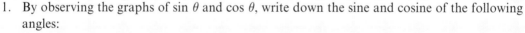

1. By observing the graphs of $\sin\theta$ and $\cos\theta$, write down the sine and cosine of the following angles:

   (a) $0°$ (b) $90°$ (c) $180°$ (d) $270°$ (e) $360°$

   These values occur frequently and should be memorised.

2. Draw the following graphs:

   (a) $y = \sin(\theta + 45°)$ for $0° \leqslant \theta \leqslant 360°$

   (b) $y = 3\cos\left(\dfrac{\theta}{2}\right)$ for $0° \leqslant \theta \leqslant 720°$

   (c) $y = 2\tan(\theta + 60°)$ for $0° \leqslant \theta \leqslant 360°$

Test and assignment exercises 23

1. State which quadrant α lies in if $\alpha =$

   (a) $30°$ (b) $-60°$ (c) $-280°$ (d) $430°$ (e) $760°$

2. State which quadrant θ lies in if $\theta =$

   (a) $\dfrac{\pi}{3}$ (b) $\dfrac{-\pi}{6}$ (c) $\dfrac{3\pi}{4}$ (d) 2.07 (e) $-\dfrac{3\pi}{4}$

3. Referring to Figure 23.19, state the sine, cosine and tangent of θ.

4. Use a graphics calculator or computer graph-plotting package to draw the following graphs:

    (a) $y = \dfrac{1}{2} \sin 2\theta$ for $0° \leqslant \theta \leqslant 360°$

    (b) $y = 3 \tan\left(\dfrac{\theta}{2}\right)$ for $0° \leqslant \theta \leqslant 720°$

    (c) $y = \cos(\theta - 90°)$ for $0° \leqslant \theta \leqslant 360°$

5. In which quadrant does the angle α lie, given
    (a) $\tan \alpha > 0$ and $\cos \alpha < 0$ (b) $\sin \alpha < 0$ and $\tan \alpha < 0$
    (c) $\sin \alpha > 0$ and $\cos \alpha < 0$ (d) $\tan \alpha < 0$ and $\cos \alpha < 0$?

Figure 23.19

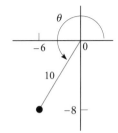

Trigonometrical identities and equations

Objectives: This chapter:

- explains what is meant by a trigonometrical identity and gives a table of important identities

- shows how identities can be used to relate the trigonometrical ratios of angles in the second, third and fourth quadrants to those in the first

- shows how complicated trigonometrical expressions can be simplified using the identities

- explains how to solve equations involving trigonometrical functions

24.1 Trigonometrical identities

At first sight, an **identity** looks like an equation. The crucial and important difference is that the left-hand side and right-hand side of an identity are equal for *all* values of the variable involved. On the other hand, an equation contains one or more unknown quantities, which must be found before the left-hand and right-hand sides are equal. We have already seen in Worked Example 22.4 that

$$\frac{\sin \theta}{\cos \theta} = \tan \theta$$

This is an identity since $\frac{\sin \theta}{\cos \theta}$ and $\tan \theta$ have the same value whatever the value of θ. Table 24.1 lists some more trigonometrical identities. In the table the symbols A and B stand for any angle we choose.

Note that $(\cos A)^2$ is usually written as $\cos^2 A$ and $(\sin A)^2$ is written as $\sin^2 A$.

There is another identity that deserves special mention:

Key point $\sin^2 A + \cos^2 A = 1$

This identity shows that sin A and cos A are closely related. Knowing sin A the identity can be used to calculate cos A and vice versa.

Table 24.1
Common trigonometrical identities

$$\frac{\sin A}{\cos A} = \tan A$$

$$\sin(A + B) = \sin A \cos B + \sin B \cos A$$

$$\sin(A - B) = \sin A \cos B - \sin B \cos A$$

$$\sin 2A = 2 \sin A \cos A$$

$$\cos(A + B) = \cos A \cos B - \sin A \sin B$$

$$\cos(A - B) = \cos A \cos B + \sin A \sin B$$

$$\cos 2A = (\cos A)^2 - (\sin A)^2 = \cos^2 A - \sin^2 A$$

$$\tan(A + B) = \frac{\tan A + \tan B}{1 - \tan A \tan B}$$

$$\tan(A - B) = \frac{\tan A - \tan B}{1 + \tan A \tan B}$$

$$\sin A + \sin B = 2 \sin\left(\frac{A + B}{2}\right) \cos\left(\frac{A - B}{2}\right)$$

$$\sin A - \sin B = 2 \sin\left(\frac{A - B}{2}\right) \cos\left(\frac{A + B}{2}\right)$$

$$\cos A + \cos B = 2 \cos\left(\frac{A + B}{2}\right) \cos\left(\frac{A - B}{2}\right)$$

$$\cos A - \cos B = -2 \sin\left(\frac{A - B}{2}\right) \sin\left(\frac{A + B}{2}\right)$$

$$\sin A = -\sin(-A)$$

$$\cos A = \cos(-A)$$

$$\tan A = -\tan(-A)$$

WORKED EXAMPLE

24.1   (a) Show that $\sin \theta = \sin(180° - \theta)$.

(b) Show that $\cos \theta = -\cos(180° - \theta)$.

(c) From parts (a) and (b) deduce that $\tan \theta = -\tan(180° - \theta)$.

Solution   (a) Consider $\sin(180° - \theta)$. Because the right-hand side is the sine of the difference of two angles we use the identity for $\sin(A - B)$:

$$\sin(A - B) = \sin A \cos B - \sin B \cos A$$

Letting $A = 180°$ and $B = \theta$ we obtain

$$\sin(180° - \theta) = \sin 180° \cos \theta - \sin \theta \cos 180°$$
$$= 0.\cos \theta - \sin \theta(-1)$$
$$= \sin \theta$$

Hence $\sin \theta = \sin(180° - \theta)$.

(b) Consider $\cos(180° - \theta)$. Here we are dealing with the cosine of the difference of two angles. We use the identity

$$\cos(A - B) = \cos A \cos B + \sin A \sin B$$

Letting $A = 180°$ and $B = \theta$ we obtain

$$\cos(180° - \theta) = \cos 180° \cos \theta + \sin 180° \sin \theta$$
$$= -1(\cos \theta) + 0(\sin \theta)$$
$$= -\cos \theta$$

Hence $\cos \theta = -\cos(180° - \theta)$.

(c) We note from Table 24.1 that $\tan \theta = \frac{\sin \theta}{\cos \theta}$ and so, using the results of parts (a) and (b),

$$\tan \theta = \frac{\sin \theta}{\cos \theta}$$

$$= \frac{\sin(180° - \theta)}{-\cos(180° - \theta)}$$

$$= -\tan(180° - \theta)$$

Because the results obtained in Worked Example 24.1 are true for any angle θ, these too are trigonometrical identities. We shall use them when solving trigonometrical equations in §24.2.

Key point $\sin \theta = \sin(180° - \theta)$

$\cos \theta = -\cos(180° - \theta)$

$\tan \theta = -\tan(180° - \theta)$

WORKED EXAMPLE

24.2   Use a calculator to evaluate the following trigonometrical ratios and hence verify the identities obtained in Worked Example 24.1.

(a) sin 30° and sin 150°

(b) cos 70° and cos 110°

(c) tan 30° and tan 150°

Solution   (a) Using a calculator verify that sin 30° = 0.5. Similarly, sin 150° = 0.5. We see that

$$\sin 30° = \sin 150° = \sin(180° - 30°)$$

This verifies the first of the identities for the case when $\theta = 30°$.

(b) cos 70° = 0.3420 and cos 110° = −0.3420. Hence

$$\cos 70° = -\cos 110° = -\cos(180° - 70°)$$

This verifies the second of the identities for the case when $\theta = 70°$.

(c) tan 30° = 0.5774, and tan 150° = −0.5774. So

$$\tan 30° = -\tan 150° = -\tan(180° - 30°)$$

This verifies the third of the identities for the case when $\theta = 30°$.

Using a similar analysis to Worked Example 24.1 it is possible to obtain the following identities:

Key point

$$\sin \theta = -\sin(\theta - 180°)$$
$$\cos \theta = -\cos(\theta - 180°)$$
$$\tan \theta = \tan(\theta - 180°)$$
$$\sin \theta = -\sin(360° - \theta)$$
$$\cos \theta = \cos(360° - \theta)$$
$$\tan \theta = -\tan(360° - \theta)$$

WORKED EXAMPLES

24.3   Evaluate the following trigonometrical ratios using a calculator and hence verify the results in the previous Key point.

(a) sin 30° and sin 210°

(b) cos 40° and cos 220°

(c) tan 50° and tan 230°

Solution (a) $\sin 30° = 0.5$ and $\sin 210° = -0.5$. Note that $\sin 210° = -\sin 30°$.

(b) $\cos 40° = 0.7660$ and $\cos 220° = -0.7660$. Note that $\cos 220° = -\cos 40°$.

(c) $\tan 50° = 1.1918$ and $\tan 230° = 1.1918$. Note that $\tan 50° = \tan 230°$.

24.4 Evaluate the following trigonometrical ratios using a calculator and hence verify the results in the previous Key point.

(a) $\sin 70°$ and $\sin 290°$

(b) $\cos 40°$ and $\cos 320°$

(c) $\tan 20°$ and $\tan 340°$

Solution (a) $\sin 70° = 0.9397$, $\sin 290° = -0.9397$ and so $\sin 290° = -\sin 70°$.

(b) $\cos 40° = 0.7660$, $\cos 320° = 0.7660$ and so $\cos 40° = \cos 320°$.

(c) $\tan 20° = 0.3640$, $\tan 340° = -0.3640$ and so $\tan 340° = -\tan 20°$.

The remaining worked examples in this section illustrate the use of a variety of identities in simplifying trigonometrical expressions. It is not always obvious which identity to use, especially for the inexperienced. The best advice is to try to experience a range of examples and work through a large number of exercises.

WORKED EXAMPLES

24.5 Simplify

$$1 - \sin A \cos A \tan A$$

Solution

$$1 - \sin A \cos A \tan A = 1 - \sin A \cos A \left(\frac{\sin A}{\cos A}\right)$$

$$= 1 - \sin^2 A$$

$$= \cos^2 A \qquad \text{since} \cos^2 A + \sin^2 A = 1$$

24.6 Show

$$\cos^4 A - \sin^4 A = \cos 2A$$

Solution We note that

$$\cos^4 A - \sin^4 A = (\cos^2 A + \sin^2 A)(\cos^2 A - \sin^2 A)$$

$$= 1(\cos^2 A - \sin^2 A) \qquad \text{using} \cos^2 A + \sin^2 A = 1$$

$$= \cos 2A \qquad \text{using Table 24.1}$$

24.7   Show

$$\frac{\sin 6\theta - \sin 4\theta}{\sin \theta} = 2 \cos 5\theta$$

Solution   We note the identity

$$\sin A - \sin B = 2 \sin \left(\frac{A - B}{2} \right) \cos \left(\frac{A + B}{2} \right)$$

Letting $A = 6\theta$, $B = 4\theta$ we obtain

$$\sin 6\theta - \sin 4\theta = 2 \sin \theta \cos 5\theta$$

and so

$$\frac{\sin 6\theta - \sin 4\theta}{\sin \theta} = 2 \cos 5\theta$$

Exercise 24.1

1. In each case state value(s) of θ in the range $0°$ to $360°$ so that the following are true:
   (a) $\sin \theta = \sin 50°$
   (b) $\cos \theta = -\cos 40°$
   (c) $\tan \theta = \tan 20°$
   (d) $\sin \theta = -\sin 70°$
   (e) $\cos \theta = \cos 10°$
   (f) $\tan \theta = -\tan 80°$
   (g) $\sin \theta = \sin 0°$
   (h) $\cos \theta = \cos 90°$

2. State values of θ in the range $0°$ to $360°$ so that
   (a) $\cos \theta = \cos 20°$
   (b) $\sin \theta = -\sin 10°$
   (c) $\tan \theta = -\tan 40°$

3. Show

$$\cos 2A = 2 \cos^2 A - 1$$

4. Show

$$\tan^2 A + 1 = \frac{1}{\cos^2 A}$$

5. Simplify

$$\frac{\cos 2\theta + \cos 8\theta}{2 \cos 3\theta}$$

6. Use the trigonometrical identities to expand and simplify if possible
   (a) $\cos(270° - \theta)$ (b) $\cos(270° + \theta)$
   (c) $\sin(270° + \theta)$ (d) $\tan(135° + \theta)$
   (e) $\sin(270° - \theta)$

7. Noting that $\tan 45° = 1$, simplify

$$\frac{1 - \tan A}{1 + \tan A}$$

8. Show that

$$\frac{1 - \cos 2\theta + \sin 2\theta}{1 + \cos 2\theta + \sin 2\theta}$$

can be simplified to $\tan \theta$.

9. Simplify

$$\frac{\sin 4\theta + \sin 2\theta}{\cos 4\theta + \cos 2\theta}$$

10. Show that

$$\frac{\tan A}{\tan^2 A + 1} = \frac{1}{2} \sin 2A$$

11. Show

(a) $\dfrac{1}{\cos A} - \cos A = \sin A \tan A$

(b) $\dfrac{1}{\sin A} - \sin A = \dfrac{\cos A}{\tan A}$

12. By using the trigonometrical identity for $\sin(A + B)$ with $A = 2\theta$ and $B = \theta$ show

$$\sin 3\theta = 3 \sin \theta - 4 \sin^3 \theta$$

24.2 Solutions of trigonometrical equations

Trigonometrical equations are equations involving the trigonometrical ratios. This section shows how to solve trigonometrical equations.

We shall need to make use of the inverse trigonometrical functions $\sin^{-1} x$, $\cos^{-1} x$ and $\tan^{-1} x$. The inverses of the trigonometrical ratios were first introduced in §22.2. Recall that since $\sin 30° = 0.5$, we write $30° = \sin^{-1} 0.5$, and say $30°$ is the angle whose sine is 0.5. Referring back to Worked Example 24.2 you will see that $\sin 150°$ is also equal to 0.5, and so it is also true that $150° = \sin^{-1} 0.5$.

Hence the inverse sine of a number can yield more than one answer. This is also true for inverse cosine and inverse tangent. Scientific calculators will simply give one value. The other values must be deduced from knowledge of the functions $y = \sin \theta$, $y = \cos \theta$ and $y = \tan \theta$.

WORKED EXAMPLES

24.8 Find all values of $\sin^{-1} 0.4$ in the range $0°$ to $360°$.

Solution Let $\theta = \sin^{-1} 0.4$, that is $\sin \theta = 0.4$. Since $\sin \theta$ is positive then one value is in the first quadrant and another value is in the second quadrant. Using a scientific calculator we see

$$\theta = \sin^{-1} 0.4 = 23.6°$$

The formula
$\sin \theta = \sin(180° - \theta)$ was
derived on page 286

This is the value in the first quadrant. The value in the second quadrant is found using $\sin \theta = \sin(180° - \theta)$. Hence the value in the second quadrant is $180° - 23.6° = 156.4°$.

24.9   Find all values of θ in the range $0°$ to $360°$ such that

(a) $\theta = \cos^{-1}(-0.5)$ (b) $\theta = \tan^{-1} 1$

Solution   (a) We have $\theta = \cos^{-1}(-0.5)$, that is $\cos \theta = -0.5$. As $\cos \theta$ is negative then θ must be in the second and third quadrants. Using a calculator we find $\cos^{-1}(-0.5) = 120°$. This is the value of θ in the second quadrant. We now seek the value of θ in the third quadrant.
We have, from the Key point on page 287,

$$\cos(\theta - 180°) = -\cos \theta$$

We are given $\cos \theta = -0.5$. So

$$\cos(\theta - 180°) = 0.5$$

Now, since θ is in the third quadrant, $\theta - 180°$ must be an acute angle whose cosine equals 0.5. That is,

$$\theta - 180° = 60°$$
$$\theta = 240°$$

This is the value of θ in the third quadrant. The required solutions are thus $\theta = 120°, 240°$.

(b) We are given $\theta = \tan^{-1} 1$, that is $\tan \theta = 1$. Since $\tan \theta$ is positive, there is a value of θ in the first and third quadrants. Using a calculator we find $\tan^{-1} 1 = 45°$. This is the value of θ in the first quadrant.
We have, from the Key point on page 287,

$$\tan(\theta - 180°) = \tan \theta$$

We are given $\tan \theta = 1$. So

$$\tan(\theta - 180°) = 1$$

Now $\theta - 180°$ must be an acute angle with tangent equal to 1. That is,

$$\theta - 180° = 45°$$
$$\theta = 225°$$

The required values of θ are $45°$ and $225°$.

24.10   Find all values of θ in the range $0°$ to $360°$ such that $\theta = \sin^{-1}(-0.5)$.

Solution   We know $\sin \theta = -0.5$ and so θ must be in the third and fourth quadrants. Using a calculator we find

$$\sin^{-1}(-0.5) = -30°$$

We require solutions in the range $0°$ to $360°$. Recall that adding $360°$ to an angle leaves the values of the trigonometrical ratios unaltered. So

$$\sin^{-1}(-0.5) = -30° + 360° = 330°$$

We have found the value of θ in the fourth quadrant.

We now seek the value in the third quadrant. We have, from the Key point on page 287,

$$\sin(\theta - 180°) = -\sin\theta$$

We are given $\sin\theta = -0.5$. Therefore

$$\sin(\theta - 180°) = 0.5$$

Now $\theta - 180°$ must be an acute angle with sine equal to 0.5. That is,

$$\theta - 180° = 30°$$
$$\theta = 210°$$

The required values of θ are $210°$ and $330°$.

24.11   Find all values of θ in the range $0°$ to $360°$ such that $\sin 2\theta = 0.5$.

Solution   We make the substitution $z = 2\theta$. As θ varies from $0°$ to $360°$ then z varies from $0°$ to $720°$. Hence the problem as given is equivalent to finding all values of z in the range $0°$ to $720°$ such that $\sin z = 0.5$.

$$\sin z = 0.5$$
$$z = \sin^{-1} 0.5$$
$$= 30°$$

Also

$$\sin(180° - 30°) = \sin 30°$$
$$\sin 150° = \sin 30° = 0.5$$
$$\sin^{-1} 0.5 = 150°$$

The values of z in the range $0°$ to $360°$ are $30°$ and $150°$. Recalling that adding $360°$ to an angle leaves its trigonometrical ratios unaltered, we see that $30° + 360° = 390°$ and $150° + 360° = 510°$ are values of z in the range $360°$ to $720°$. The solutions for z are thus

$$z = 30°, 150°, 390°, 510°$$

and hence the required values of θ are given by

$$\theta = \frac{z}{2} = 15°, 75°, 195°, 255°$$

24.12   Solve $\cos 3\theta = -0.5$ for $0° \leqslant \theta \leqslant 360°$.

Solution

We substitute $z = 3\theta$. As θ varies from $0°$ to $360°$ then z varies from $0°$ to $1080°$. Hence the problem is to find values of z in the range $0°$ to $1080°$ such that $\cos z = -0.5$.

We begin by finding values of z in the range $0°$ to $360°$. Using Worked Example 24.9(a) we see that

$$z = 120°, 240°$$

To find values of z in the range $360°$ to $720°$ we add $360°$ to each of these solutions. This gives

$$z = 120° + 360° = 480° \qquad \text{and} \qquad z = 240° + 360° = 600°$$

To find values of z in the range $720°$ to $1080°$ we add $360°$ to these solutions. This gives

$$z = 840°, 960°$$

Hence

$$z = 120°, 240°, 480°, 600°, 840°, 960°$$

Finally, using $\theta = \frac{z}{3}$ we obtain values of θ:

$$\theta = \frac{z}{3} = 40°, 80°, 160°, 200°, 280°, 320°$$

24.13

Solve

$$\tan(2\theta + 20°) = 0.3 \qquad 0° \leqslant \theta \leqslant 360°$$

Solution

We substitute $z = 2\theta + 20°$. As θ varies from $0°$ to $360°$ then z varies from $20°$ to $740°$.

First we solve

$$\tan z = 0.3 \qquad 0° \leqslant z \leqslant 360°$$

This leads to $z = 16.7°, 196.7°$. Values of z in the range $360°$ to $720°$ are

$$z = 16.7° + 360°, 196.7° + 360°$$

$$= 376.7°, 556.7°$$

By adding a further $360°$ values of z in the range $720°$ to $1080°$ are found. These are

$$z = 736.7°, 916.7°$$

Hence values of z in the range $0°$ to $1080°$ are

$$z = 16.7°, 196.7°, 376.7°, 556.7°, 736.7°, 916.7°$$

Values of z in the range $20°$ to $740°$ are thus

$$z = 196.7°, 376.7°, 556.7°, 736.7°$$

The values of θ in the range $0°$ to $360°$ are found using $\theta = (z - 20°)/2$:

$$\theta = \frac{z - 20°}{2} = 88.35°, \ 178.35°, \ 268.35°, \ 358.35°$$

Self-assessment question 24.2

1. The trigonometrical functions are $y = \sin\theta$, $y = \cos\theta$ and $y = \tan\theta$. The corresponding inverses are $\theta = \sin^{-1}y$, $\theta = \cos^{-1}y$ and $\theta = \tan^{-1}y$. Can you explain why these inverses may have several values of θ for a single value of y?

Exercise 24.2

1. Find all values in the range $0°$ to $360°$ of

   (a) $\sin^{-1} 0.9$ (b) $\cos^{-1} 0.45$ (c) $\tan^{-1} 1.3$ (d) $\sin^{-1}(-0.6)$
   (e) $\cos^{-1}(-0.75)$ (f) $\tan^{-1}(-0.3)$ (g) $\sin^{-1} 1$ (h) $\cos^{-1}(-1)$

2. Solve the following trigonometrical equations for $0° \leqslant \theta \leqslant 360°$:

   (a) $\sin(2\theta + 50°) = 0.5$ (b) $\cos(\theta + 110°) = 0.3$ (c) $\tan\left(\dfrac{\theta}{2}\right) = 1$

   (d) $\sin\left(\dfrac{\theta}{2}\right) = -0.5$ (e) $\cos(2\theta - 30°) = -0.5$ (f) $\tan(3\theta - 20°) = 0.25$

   (g) $\sin 2\theta = 2\cos 2\theta$

3. Find values of θ in the range $0°$ to $360°$ such that
   (a) $\cos(\theta - 100°) = 0.3126$ (b) $\tan(2\theta + 20°) = -1$
   (c) $\sin(\frac{2\theta}{3} + 30°) = -0.4325$

Test and assignment exercises 24

1. Solve the following trigonometrical equations for $0° \leqslant \theta \leqslant 360°$:

   (a) $13 \sin 2\theta = 5$ (b) $\cos\left(\dfrac{\theta}{3}\right) = -0.7$ (c) $\tan(\theta - 110°) = 1.5$

   (d) $\cos 3\theta = 2 \sin 3\theta$ (e) $\sin\left(\dfrac{\theta}{2} - 30°\right) = -0.65$

2. State all values in the range $0°$ to $360°$ of
   (a) $\sin^{-1} 0.85$ (b) $\cos^{-1}(-0.25)$ (c) $\tan^{-1}(-1.25)$ (d) $\sin^{-1}(-0.4)$
   (e) $\cos^{-1}\left(\dfrac{1}{3}\right)$ (f) $\tan^{-1} 1.7$

3. Show

$$\cos 2A = 1 - 2 \sin^2 A$$

4. Simplify

$$\sin^4 A + 2 \sin^2 A \cos^2 A + \cos^4 A$$

5. Simplify $\dfrac{\sin 5\theta + \sin \theta}{\cos 5\theta + \cos \theta}$

6. (a) Show

$$\cos 3A = \cos 2A \cos A - \sin 2A \sin A$$

   (b) Hence show that

$$\cos 3A = 4 \cos^3 A - 3 \cos A$$

7. Solve the following equations for $0° \leqslant \theta \leqslant 360°$:
   (a) $\sin 2\theta = -0.6$ (b) $\cos(\theta + 40°) = -0.25$ (c) $\tan(\frac{2\theta}{3}) = 1.3$

8. State all values in the range $0°$ to $360°$ for the following:
   (a) $\sin^{-1}(-0.6500)$ (b) $\cos^{-1}(-0.2500)$ (c) $\tan^{-1}(1.2500)$

9. State values of θ in the range $0°$ to $360°$ for which:
   (a) $\tan \theta = \tan 25°$ (b) $\cos \theta = -\cos 30°$ (c) $\sin \theta + \sin 80° = 0$
   (d) $\tan \theta + \tan 15° = 0$

Solution of triangles

25

Objectives: This chapter:

- explains the terms 'scalene', 'isosceles', 'equilateral' and 'right-angled' as applied to triangles

- states Pythagoras' theorem and shows how it can be used in the solution of right-angled triangles

- states the sine rule and the cosine rule and shows how they are used to solve triangles

A triangle is solved when all its angles and the lengths of all its sides are known. Before looking at the various rules used to solve triangles we define the various kinds of triangle.

25.1 Types of triangle

A **scalene triangle** is one in which all the sides are of different length. In a scalene triangle, all the angles are different too. An **isosceles triangle** is one in which two of the sides are of equal length. Figure 25.1 shows an isosceles triangle with AB = AC. In an isosceles triangle, there are also two equal angles. In Figure 25.1, ∠ABC = ∠ACB. An **equilateral triangle** has three equal sides and three equal angles, each 60°. A **right-angled triangle** is one containing a right angle, that is an angle of 90°. The side opposite the right angle is called the **hypotenuse**. Figure 25.2 shows a right-angled triangle with a right angle at C. The hypotenuse is AB.

In *any* triangle we note:

Key point   The sum of the three angles is always 180°.

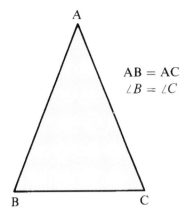

Figure 25.1
An isosceles triangle has two
equal sides and two equal angles

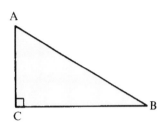

Figure 25.2
△ABC has a right angle at C;
the hypotenuse is AB

Recall from §21.4 that an acute angle is less than 90° and an obtuse angle is greater than 90° and less than 180°.

Other properties include:

(a) The longest side is opposite the largest angle.
(b) The shortest side is opposite the smallest angle.
(c) A triangle contains either three acute angles or two acute and one obtuse angle.

In △ABC, as a shorthand, we often refer to ∠ABC as ∠B, or more simply as B, ∠ACB as C and so on. The sides also have a shorthand notation. In any triangle ABC, AB is always opposite C, AC is always opposite B and BC is always opposite A. Hence we refer to AB as c, AC as b and BC as a. Figure 25.3 illustrates this. We say that A is **included** by AC and AB, B is included by AB and BC and C is included by AC and BC.

Figure 25.3
△ABC with $a = $ BC,
$b = $ AC, $c = $ AB

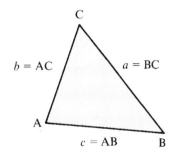

WORKED EXAMPLE

25.1   State all the angles of a right-angled isosceles triangle.

Solution   Let $\triangle ABC$ be a right-angled isosceles triangle with a right angle at B. Figure 25.4 illustrates the situation. The angles sum to $180°$ and $A = C$ because the triangle is isosceles.

$$A + B + C = 180$$

$$A + 90 + C = 180$$

$$A + C = 90$$

$$2A = 90$$

$$A = 45$$

Then $C = 45$ also. The angles are $A = 45°$, $B = 90°$, $C = 45°$.

Figure 25.4
$\triangle ABC$ is a right-angled isosceles triangle

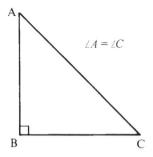

Self-assessment questions 25.1

1. Explain the terms 'scalene', 'isosceles' and 'equilateral' when applied to triangles.

2. What is the sum of the angles of a triangle?

3. Explain the meaning of the term 'hypotenuse'.

4. Can a triangle containing an obtuse angle possess a hypotenuse?

Exercise 25.1

1. In $\triangle ABC$, $A = 70°$ and $B = 42°$. Calculate C.

2. In $\triangle ABC$, $A = 42°$ and B is twice C. Calculate B and C.

3. An isosceles triangle ABC has $C = 114°$. Calculate A and B.

4. In $\triangle ABC$, $A = 50°$ and $C = 58°$. State (a) the longest side, (b) the shortest side.

5. The smallest angle of an isosceles triangle is $40°$. Calculate the angles of the triangle.

25.2 Pythagoras' theorem

One of the oldest theorems in the study of triangles is due to the ancient Greek mathematician Pythagoras. It applies to right-angled triangles. Consider any right-angled triangle ABC, as shown in Figure 25.5.

Figure 25.5
A right-angled triangle

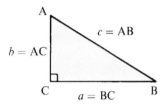

Key point

A **theorem** is an important mathematical result that, although not obvious, is possible to prove.

Pythagoras' theorem states that $c^2 = a^2 + b^2$.

In words we have: the square of the hypotenuse (c^2) equals the sum of the squares of the other two sides ($a^2 + b^2$). Note that as Pythagoras' theorem refers to a hypotenuse, it can be applied only to right-angled triangles.

WORKED EXAMPLES

25.2 In Figure 25.5, AC = 3 cm and BC = 4 cm. Calculate the length of the hypotenuse AB.

Solution We have a = BC = 4 and b = AC = 3:

$$a^2 = 4^2 = 16 \qquad b^2 = 3^2 = 9$$

We apply Pythagoras' theorem:

$$c^2 = a^2 + b^2 = 16 + 9 = 25$$

$$c = 5$$

The hypotenuse, AB, is 5 cm.

25.3 In Figure 25.5, AC = 14 cm and the hypotenuse is 22 cm. Calculate the length of BC.

Solution We have b = AC = 14 and c = AB = 22. We need to find the length of BC, that is a. We apply Pythagoras' theorem:

$$c^2 = a^2 + b^2$$

$$22^2 = a^2 + 14^2$$

$$484 = a^2 + 196$$

$$a^2 = 288$$

$$a = \sqrt{288} = 16.97$$

The length of BC is 16.97 cm.

25.4 ABC is a right-angled isosceles triangle with $A = 90°$. If BC = 12 cm, calculate the lengths of the other sides.

Solution Figure 25.6 illustrates △ABC. The triangle is isosceles and so AC = AB, that is $b = c$. We are told that the hypotenuse, BC, is 12 cm, that is $a = 12$. Applying Pythagoras' theorem gives

$$a^2 = b^2 + c^2$$

Figure 25.6
Right-angled triangle for
Worked Example 25.4

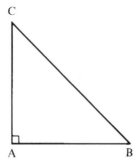

Substituting $a = 12$ and $b = c$ we get

$$12^2 = b^2 + b^2$$

$$144 = 2b^2$$

$$b^2 = 72$$

$$b = 8.49$$

The sides AC and AB are both 8.49 cm.

Self-assessment questions 25.2

1. State Pythagoras' theorem.

2. Explain what is meant by the hypotenuse of a triangle.

Exercise 25.2

MyMathLab Global

1. △ABC has $C = 90°$. If AB = 30 cm and AC = 17 cm calculate the length of BC.

2. △CDE has $D = 90°$. Given CD = 1.2 m and DE = 1.7 m calculate the length of CE.

3. △ABC has a right angle at B. Given AC : AB is 2 : 1 calculate AC : BC.

4. △XYZ has $X = 90°$. If XY = 10 cm and YZ = 2XZ calculate the lengths of XZ and YZ.

5. △LMN has LN as hypotenuse. Given LN = 30 cm and LM = 26 cm calculate the length of MN.

25.3 Solution of right-angled triangles

Recall that a triangle is solved when all angles and all lengths have been found. We can use Pythagoras' theorem and the trigonometrical ratios to solve right-angled triangles.

WORKED EXAMPLES

25.5 In △ABC, $B = 90°$, AB = 7 cm and BC = 4 cm. Solve △ABC.

Solution Figure 25.7 shows the information given. We use Pythagoras' theorem to find AC:

$$b^2 = a^2 + c^2$$
$$= 16 + 49 = 65$$
$$b = \sqrt{65} = 8.06$$

Figure 25.7
△ABC for Worked
Example 25.5

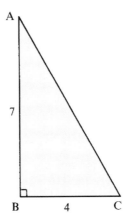

We use the trigonometrical ratios to find C:

$$\tan C = \frac{AB}{BC} = \frac{7}{4} = 1.75$$

$$C = \tan^{-1}(1.75) = 60.26°$$

The sum of the angles is 180°:

$$A + B + C = 180$$

$$A + 90 + 60.26 = 180$$

$$A = 29.74$$

The solution to △ABC is

$A = 29.74°$	$BC = 4$ cm
$B = 90°$	$AC = 8.06$ cm
$C = 60.26°$	$AB = 7$ cm

Note that the longest side is opposite the largest angle; the shortest side is opposite the smallest angle.

25.6 In △XYZ, $Y = 90°$, $Z = 35°$ and XZ = 10 cm. Solve △XYZ.

Solution Figure 25.8 illustrates the information given. We find X using the fact that the sum of the angles is 180°:

$$X + Y + Z = 180$$

$$X + 90 + 35 = 180$$

$$X = 55$$

Figure 25.8
△XYZ for Worked
Example 25.6

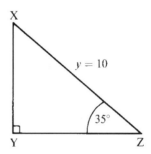

We find the length XY using the sine ratio:

$$\sin Z = \frac{XY}{XZ}$$

$$\sin 35° = \frac{XY}{10}$$

$$XY = 10 \sin 35° = 5.74$$

We find the length YZ using the cosine ratio:

$$\cos Z = \frac{\text{YZ}}{\text{XZ}}$$

$$\cos 35° = \frac{\text{YZ}}{10}$$

$$\text{YZ} = 10 \cos 35° = 8.19$$

The triangle is now completely solved:

$X = 55°$ YZ $= 8.19$ cm

$Y = 90°$ XZ $= 10$ cm

$Z = 35°$ XY $= 5.74$ cm

25.7   In $\triangle ABC$, $C = 90°$, AB $= 20$ cm and AC $= 14$ cm. Solve $\triangle ABC$.

Solution   Figure 25.9 illustrates the given information. We use Pythagoras' theorem to calculate BC:

$$c^2 = a^2 + b^2$$

$$400 = a^2 + 196$$

$$a^2 = 204$$

$$a = \sqrt{204} = 14.28$$

Figure 25.9
$\triangle ABC$ for Worked
Example 25.7

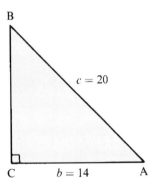

We use the cosine ratio to calculate A:

$$\cos A = \frac{\text{AC}}{\text{AB}} = \frac{14}{20} = 0.7$$

$$A = \cos^{-1}(0.7) = 45.57°$$

Since the angles sum to $180°$ then we have

$$B = 180 - C - A$$

$$= 180 - 90 - 45.57$$

$$= 44.43$$

The solution is

$A = 45.57°$	BC $= 14.28$ cm
$B = 44.43°$	AC $= 14$ cm
$C = 90°$	AB $= 20$ cm

Self-assessment question 25.3

1. Explain what is meant by 'solving a triangle'.

Exercise 25.3

1. $\triangle$ABC has $C = 90°$, $A = 37°$ and AC $=$ 36 cm. Solve $\triangle$ABC.

2. $\triangle$CDE has $E = 90°$, CE $= 14$ cm and DE $= 21$ cm. Solve $\triangle$CDE.

3. $\triangle$XYZ has $X = 90°$, $Y = 26°$ and YZ $=$ 45 mm. Solve $\triangle$XYZ.

4. $\triangle$IJK has $J = 90°$, IJ $= 15$ cm and JK $=$ 27 cm. Solve $\triangle$IJK.

5. In $\triangle$PQR, $P = 62°$, $R = 28°$ and PR $=$ 22 cm. Solve $\triangle$PQR.

6. In $\triangle$RST, $R = 70°$, $S = 20°$ and RT $=$ 12 cm. Solve $\triangle$RST.

25.4 The sine rule

In §25.3 we saw how to solve right-angled triangles. Many triangles do not contain a right angle, and in such cases we need to use other techniques. One such technique is the **sine rule**.

Consider *any* triangle ABC as shown in Figure 25.3. Recall the notation a = BC, b = AC, c = AB.

Key point

The sine rule states

$$\frac{a}{\sin A} = \frac{b}{\sin B} = \frac{c}{\sin C}$$

It must be stressed that the sine rule can be applied to any triangle. The sine rule is used when we are given either (a) two angles and one side or (b) two sides and a non-included angle.

25.8

In △ABC, $A = 30°$, $B = 84°$ and AC = 19 cm. Solve △ABC.

Solution

The information is illustrated in Figure 25.10. We are given two angles and a side and so the sine rule can be used. We can immediately find C since the angles sum to 180°:

$$C = 180° - A - B = 180° - 30° - 84° = 66°$$

Figure 25.10
△ABC for Worked Example 25.8

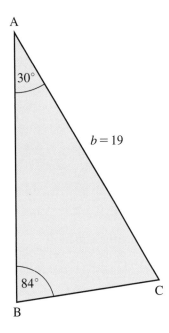

We know $A = 30°$, $B = 84°$, $C = 66°$ and $b = AC = 19$. The sine rule states

$$\frac{a}{\sin A} = \frac{b}{\sin B} = \frac{c}{\sin C}$$

Substituting in values gives

$$\frac{a}{\sin 30°} = \frac{19}{\sin 84°} = \frac{c}{\sin 66°}$$

Hence

$$a = \frac{19 \sin 30°}{\sin 84°} = 9.55 \qquad c = \frac{19 \sin 66°}{\sin 84°} = 17.45$$

The solution is

$$A = 30° \qquad a = BC = 9.55 \text{ cm}$$

$$B = 84° \qquad b = AC = 19 \text{ cm}$$

$$C = 66° \qquad c = AB = 17.45 \text{ cm}$$

25.9   In $\triangle ABC$, $B = 42°$, $AB = 12$ cm and $AC = 17$ cm. Solve $\triangle ABC$.

Solution   Figure 25.11 illustrates the situation. We know two sides and a non-included angle and so it is appropriate to use the sine rule. We are given $c = AB = 12$, $b = AC = 17$ and $B = 42°$. The sine rule states

$$\frac{a}{\sin A} = \frac{b}{\sin B} = \frac{c}{\sin C}$$

Figure 25.11
$\triangle ABC$ for Worked
Example 25.9

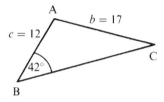

Substituting in the known values gives

$$\frac{a}{\sin A} = \frac{17}{\sin 42°} = \frac{12}{\sin C}$$

Using the equation

$$\frac{17}{\sin 42°} = \frac{12}{\sin C}$$

we see that

$$\sin C = \frac{12 \sin 42°}{17} = 0.4723$$

$$C = \sin^{-1}(0.4723) = 28.19° \qquad \text{or} \qquad 151.81°$$

As $B = 42°$, then the value $151.81°$ for C must be rejected as the sum of the angles must be $180°$. So $C = 28.19°$. Then

$$A = 180° - B - C = 180° - 42° - 28.19° = 109.81°$$

Using

$$\frac{a}{\sin A} = \frac{17}{\sin 42°}$$

we see that

$$a = \frac{17 \sin A}{\sin 42°}$$

$$= \frac{17 \sin 109.81°}{\sin 42°} = 23.90$$

The solution is

$A = 109.81°$ $a = BC = 23.90$ cm

$B = 42°$ $b = AC = 17$ cm

$C = 28.19°$ $c = AB = 12$ cm

25.10   In △ABC, AB = 23 cm, BC = 30 cm and $C = 40°$. Solve △ABC.

Solution   We are told the length of two sides and a non-included angle and so the sine rule can be applied. We have c = AB = 23, a = BC = 30 and $C = 40°$. The sine rule states

$$\frac{a}{\sin A} = \frac{b}{\sin B} = \frac{c}{\sin C}$$

Substituting in the known values gives

$$\frac{30}{\sin A} = \frac{b}{\sin B} = \frac{23}{\sin 40°}$$

Using the equation

$$\frac{30}{\sin A} = \frac{23}{\sin 40°}$$

we have

$$\sin A = \frac{30 \sin 40°}{23} = 0.8384$$

$$A = \sin^{-1}(0.8384)$$

$$= 56.97°   \text{or}   123.03°$$

There is no reason to reject either of these values and so there are two possible solutions for A. We consider each in turn.

Case 1
$A = 56.97°$ Here

$$B = 180° - 40° - 56.97° = 83.03°$$

We can now use

$$\frac{b}{\sin B} = \frac{23}{\sin 40°}$$

$$b = \frac{23 \sin B}{\sin 40°} = \frac{23 \sin 83.03°}{\sin 40°}$$

$$= 35.52$$

Case 2
$A = 123.03°$ In this case

$$B = 180° - 40° - 123.03° = 16.97°$$

We have

$$b = \frac{23 \sin B}{\sin 40°} = \frac{23 \sin 16.97°}{\sin 40°} = 10.44$$

The two solutions are

$A = 56.97°$ $a = BC = 30$ cm

$B = 83.03°$ $b = AC = 35.52$ cm

$C = 40°$ $c = AB = 23$ cm

and

$A = 123.03°$ $a = BC = 30$ cm

$B = 16.97°$ $b = AC = 10.44$ cm

$C = 40°$ $c = AB = 23$ cm

They are illustrated in Figures 25.12(a) and 25.12(b).

Figure 25.12
Solutions for Worked
Example 25.10

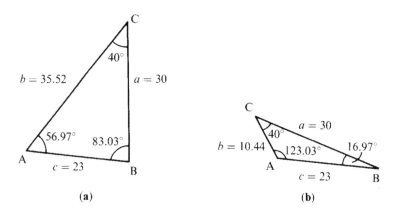

(a)

(b)

Self-assessment question 25.4

1. State the sine rule and the conditions under which it can be used to solve a triangle.

Exercise 25.4

1. Solve △ABC given
   (a) AB = 31 cm, AC = 24 cm, $C = 37°$ (b) AB = 19 cm, AC = 24 cm, $C = 37°$
   (c) BC = 17 cm, $A = 17°$, $C = 101°$ (d) $A = 53°$, AB = 9.6 cm, BC = 8.9 cm
   (e) $A = 62°$, AB = 12.2 cm, BC = 14.5 cm   (f) $B = 36°$, $C = 50°$, AC = 11 cm

25.5 The cosine rule

The cosine rule also allows us to solve any triangle. With the usual notation the cosine rule states that:

Key point

$$a^2 = b^2 + c^2 - 2bc \cos A$$

$$b^2 = a^2 + c^2 - 2ac \cos B$$

$$c^2 = a^2 + b^2 - 2ab \cos C$$

This rule is used when we are given either (a) three sides or (b) two sides and the included angle.

WORKED EXAMPLES

25.11   Solve △ABC given AB = 16 cm, AC = 23 cm and BC = 21 cm.

Solution   We are told three sides and so the cosine rule can be used. We have $a = $ BC = 21, $b = $ AC = 23, $c = $ AB = 16. Applying the cosine rule we have

$$a^2 = b^2 + c^2 - 2bc \cos A$$

$$21^2 = 23^2 + 16^2 - 2(23)(16)\cos A$$

$$736 \cos A = 344$$

$$\cos A = \frac{344}{736} = 0.4674$$

$$A = \cos^{-1}(0.4674) = 62.13°$$

We apply the cosine rule again:

$$b^2 = a^2 + c^2 - 2ac \cos B$$

$$23^2 = 21^2 + 16^2 - 2(21)(16)\cos B$$

$$672 \cos B = 168$$

$$\cos B = \frac{168}{672} = 0.25$$

$$B = \cos^{-1}(0.25) = 75.52°$$

Finally

$$C = 180° - A - B = 180° - 62.13° - 75.52° = 42.35°$$

The solution is

$$A = 62.13°   a = BC = 21 \text{ cm}$$

$$B = 75.52°   b = AC = 23 \text{ cm}$$

$$C = 42.35°   c = AB = 16 \text{ cm}$$

25.12   Solve △ABC given $A = 51°$, AC = 14 cm and AB = 24 cm.

Solution   Figure 25.13 illustrates the situation. We are given two sides and the included angle and so the cosine rule can be used. We have $A = 51°$, $c = AB = 24$ and $b = AC = 14$. We use

$$a^2 = b^2 + c^2 - 2bc \cos A$$

Figure 25.13
△ABC for Worked
Example 25.12

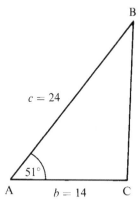

Substituting in the values given we have

$$a^2 = 14^2 + 24^2 - 2(14)(24)\cos 51°$$

$$= 196 + 576 - 672 \cos 51°$$

$$= 349.10$$

$$a = 18.68$$

We now use $b^2 = a^2 + c^2 - 2ac \cos B$ to find B:

$$14^2 = (18.68)^2 + 24^2 - 2(18.68)(24)\cos B$$

$$896.64 \cos B = 728.94$$

$$\cos B = \frac{728.94}{896.64} = 0.8130$$

$$B = \cos^{-1}(0.8130) = 35.61°$$

Finally

$$C = 180° - A - B = 180° - 51° - 35.61° = 93.39°$$

The solution is

$A = 51°$ $a = BC = 18.68$ cm

$B = 35.61°$ $b = AC = 14$ cm

$C = 93.39°$ $c = AB = 24$ cm

Self-assessment question 25.5

1. State the cosine rule and the conditions under which it can be used to solve a triangle.

Exercise 25.5

1. Solve △ABC given
   (a) AC = 119 cm, BC = 86 cm and AB = 53 cm
   (b) AB = 42 cm, AC = 30 cm and $A = 115°$
   (c) BC = 74 cm, AC = 93 cm and $C = 39°$
   (d) AB = 1.9 cm, BC = 3.6 cm and AC = 2.7 cm
   (e) AB = 29 cm, BC = 39 cm and $B = 100°$

Test and assignment exercises 25

1. Solve △ABC given
   (a) $A = 45°$, $C = 57°$, BC = 19 cm
   (b) AC = 3.9 cm, AB = 4.7 cm, $A = 64°$
   (c) AC = 41 cm, AB = 37 cm, $C = 50°$
   (d) AB = 22 cm, BC = 29 cm, $A = 37°$
   (e) AB = 123 cm, BC = 100 cm, AC = 114 cm
   (f) AB = 46 cm, $A = 33°$, $B = 76°$
   (g) BC = 64 cm, AC = 54 cm, $B = 42°$
   (h) BC = 30 cm, AC = 69 cm, $C = 97°$

2. △ABC has a right angle at B. Solve △ABC given
    (a) AB = 9 cm, BC = 14 cm   (b) BC = 10 cm, AC = 17 cm
    (c) AB = 20 cm, $A = 40°$ (d) BC = 17 cm, $A = 45°$
    (e) AC = 25 cm, $C = 27°$ (f) $A = 52°$, AC = 8.6 cm

3. Figure 25.14 illustrates △ABC. AC is 112 cm, $\angle C = 47°$ and $\angle A = 31°$. From B, a line BD is drawn that is at right angles to AC. Calculate the length of BD.

4. △ABC has $\angle A = 42°$, $\angle C = 59°$ and BC = 17 cm. Calculate the length of the longest side of △ABC.

5. A circle, centre O, has radius 10 cm. An arc AB has length 17 cm. Calculate the length of the line AB.

Figure 25.14

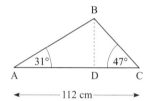

Vectors

Objectives: This chapter:

- explains what is meant by a scalar and a vector
- shows how vectors can be added and subtracted graphically
- explains how to multiply a vector by a scalar
- explains how to represent, add and subtract vectors using Cartesian coordinates
- explains how to calculate the scalar product of two vectors

26.1 Introduction to vectors and scalars

Some quantities are described by specifying a single number, together with the appropriate units. For example, we could describe a distance as 3 kilometres or a current as 10 amps. The single number is known as the **magnitude** of the quantity. Quantities that can be described by a single number in this way are known as **scalars**. Examples of scalars include temperature, length, volume and density.

Certain other quantities need not only a magnitude but also a **direction** to specify them fully. Such quantities are called **vectors**. Examples of vectors are (i) a speed of 70 kilometres an hour due north and (ii) a force of 10 newtons acting vertically down. These examples illustrate that a vector comprises both a magnitude and a direction.

Key point A vector has both magnitude and direction.

A vector is often represented graphically by a straight line with an arrowhead to show the direction. The length of the line represents the magnitude of the vector; the direction of the line shows the direction of the vector. Consider Figure 26.1.

Figure 26.1
The line AB represents
the vector **a**

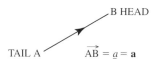

TAIL A $\overrightarrow{AB} = \underline{a} = \mathbf{a}$ B HEAD

The line joining A and B represents a vector. The length of AB represents the magnitude of the vector and the direction of the vector is given by the direction from A to B. Note the arrowhead denoting this direction. Usually the location of the line is of no consequence: it is the length (magnitude) and direction which are important.

The vector from A to B can be written $\overrightarrow{AB}$; sometimes it is abbreviated to the bold font **a** in printed work, or underlined as in $\underline{a}$ in written work. B is called the **head** of the vector; A is the **tail** of the vector.

The magnitude of $\overrightarrow{AB}$ is written as $|\overrightarrow{AB}|$ or $|\mathbf{a}|$.

Two vectors are equal if they have the same magnitude and same direction. Note that equality is needed in both aspects – direction and magnitude. A **unit vector** has a magnitude of 1. Unit vectors are often denoted by using a 'hat' (or circumflex) symbol, thus **â**.

Self-assessment questions 26.1

1. Explain what is meant by (a) a scalar, (b) a vector.

2. Explain the terms (a) magnitude of a vector, (b) unit vector, (c) head of a vector, (d) tail of a vector.

26.2 Multiplying a vector by a scalar

It is possible to multiply a vector, say **a**, by a positive number (scalar), say k. The result is a new vector whose direction is the same as **a** and whose magnitude is k times that of the original vector **a**. So, for example, if **a** has magnitude 4, then the vector 3**a** has the same direction as **a** and magnitude of 12. The vector $\frac{1}{2}$**a** has magnitude 2. Figure 26.2 illustrates this.

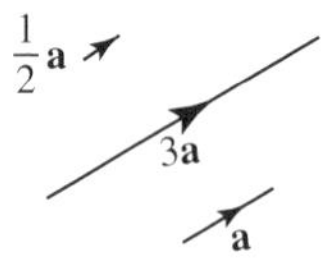

Figure 26.2
The vector 3**a** has the same direction as **a** and three times the magnitude. The vector $\frac{1}{2}$**a** has the same direction and half the magnitude

$\frac{1}{2}\mathbf{a}$

3**a**

a

A vector may also be multiplied by a negative number. Again, if **a** has magnitude 4 then $-3\mathbf{a}$ has magnitude 12. The direction of $-3\mathbf{a}$ is opposite to that of **a**. Notice that the negative sign has the effect of reversing the direction. The vector $-\mathbf{a}$ has the same magnitude as that of $\mathbf{a}$ but the opposite direction. Figure 26.3 illustrates this.

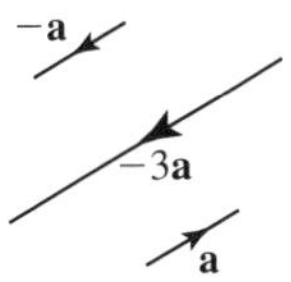

Figure 26.3
The vector $-3\mathbf{a}$ has the opposite direction to $\mathbf{a}$ and three times the magnitude. The vector $-\mathbf{a}$ has the same magnitude as $\mathbf{a}$ and the opposite direction

Recall that the magnitude of $\mathbf{a}$ is denoted by $|\mathbf{a}|$. Note that this is a scalar quantity. Then a unit vector in the direction of $\mathbf{a}$ is given by

$$\text{unit vector in direction of } \mathbf{a} = \frac{1}{|\mathbf{a}|}\mathbf{a}$$

For example, given $\mathbf{a}$ has magnitude 4, then a unit vector in the direction of $\mathbf{a}$ is $\frac{1}{4}\mathbf{a}$. If $\mathbf{b}$ has magnitude $\frac{1}{2}$ then a unit vector in the direction of $\mathbf{b}$ is $2\mathbf{b}$. Figure 26.4 illustrates some unit vectors.

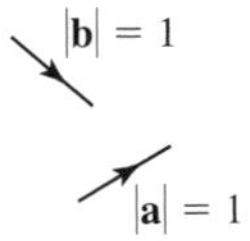

Figure 26.4
Unit vectors have a magnitude of 1

Exercise 26.2

MyMathLab Global

1. Given $\mathbf{v}$ is a vector of magnitude 2 state
   (a) a vector of magnitude 4 in the direction of $\mathbf{v}$
   (b) a vector of magnitude 0.5 in the direction of $\mathbf{v}$
   (c) a vector of magnitude 6 in the opposite direction of $\mathbf{v}$
   (d) a unit vector in the direction of $\mathbf{v}$
   (e) a unit vector in the opposite direction of $\mathbf{v}$

2. Given $|\mathbf{a}| = 5$, state (a) a unit vector in the direction of $\mathbf{a}$, (b) a unit vector in the opposite direction to $\mathbf{a}$.

3. Given $\mathbf{h}$ is a unit vector, state (a) a vector of magnitude 12 in the direction of $\mathbf{h}$, (b) a vector of magnitude 0.5 in the opposite direction to $\mathbf{h}$.

26.3 Adding and subtracting vectors

Adding vectors

If $\mathbf{a}$ and $\mathbf{b}$ are two vectors then we can form the vector $\mathbf{a} + \mathbf{b}$. We do this by moving $\mathbf{b}$, still maintaining its length and direction, so that its tail coincides with the head of **a**. Figures 26.5(a) and (b) illustrate this. Then the vector **a** + **b** goes from the tail of **a** to the head of **b**. Figure 26.5 (c) illustrates this.

The method of adding vectors illustrated in Figure 26.5 is known as the **triangle law of addition**. The sum, **a** + **b**, is the same as **b** + **a**. The sum **a** + **b** is also known as the **resultant** of **a** and **b**.

Figure 26.5
The tail of **b** is moved to join the head of **a**

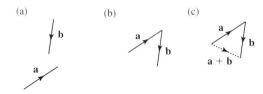

| Key point | The resultant of **a** and **b** is the sum **a** + **b**. |

WORKED EXAMPLE

26.1   Figure 26.6 illustrates a set of vectors. Use the triangle law of addition to find the resultant of (a) $\overrightarrow{AB}$ and $\overrightarrow{BC}$, (b) $\overrightarrow{BC}$ and $\overrightarrow{CD}$, (c) $\overrightarrow{AB}$ and $\overrightarrow{BD}$, (d) $\overrightarrow{AC}$ and $\overrightarrow{CD}$.

Figure 26.6
Figure for Worked Example 26.1

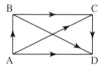

Solution   (a) $\overrightarrow{AB} + \overrightarrow{BC} = \overrightarrow{AC}$
(b) $\overrightarrow{BC} + \overrightarrow{CD} = \overrightarrow{BD}$
(c) $\overrightarrow{AB} + \overrightarrow{BD} = \overrightarrow{AD}$
(d) $\overrightarrow{AC} + \overrightarrow{CD} = \overrightarrow{AD}$

The same principle used for adding two vectors can be applied to finding the sum of three or more vectors. Consider the vectors as shown in Figure 26.7.

Figure 26.7
Addition of three vectors

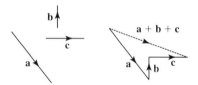

By positioning the tail of **b** at the head of **a** and the tail of **c** at the head of **b**, the sum **a** + **b** + **c** can be found. It is a vector from the tail of **a** to the head of **c** as illustrated.

Subtraction of vectors

We wish to find the difference of two vectors, that is **a** − **b**. We do this by calculating **a** + (− **b**); that is, we add the vectors **a** and −**b**. Recall that −**b** is a vector in the opposite direction to **b** and with the same magnitude.

WORKED EXAMPLE

26.2   The vectors **a** and **b** are shown in Figure 26.8. On a diagram show (a) **a** − **b**, (b) **b** − **a**.

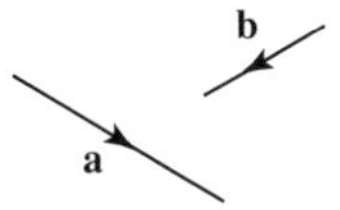

Figure 26.8
The vectors **a** and **b** for Worked Example 26.2

Solution   (a) We form the vector −**b**. This has the same magnitude (length) as **b** and the opposite direction. It is illustrated in Figure 26.9. This vector, −**b**, is now added to **a** by positioning the tail of −**b** at the head of **a** (see Figure 26.9). The sum **a** + (−**b**) is then determined using the triangle law of addition and is shown as a dotted line.

Figure 26.9
The vector **a** − **b** is found by adding **a** and (−**b**)

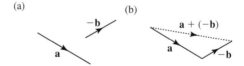

(b) To find **b** − **a** we calculate **b** + (−**a**). The vector −**a** has the same magnitude as **a** and the opposite direction. The tail of −**a** is positioned at the head of **b** and the resultant **b** + (−**a**) found using the triangle law of addition. Figure 26.10 illustrates this.

Figure 26.10
The vector **b** − **a** is found by adding **b** and (−**a**)

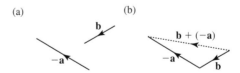

From Figures 26.8 and 26.9 we note that **b** − **a** = −(**a** − **b**).

Self-assessment questions 26.3

1. State the triangle law of addition for adding vectors **a** and **b**. Illustrate the law with a diagram.

2. Which of the following are true and which are false?

    (a) **a** + **b** is the same as **b** + **a**.

    (b) **b** − **a** is the same as **a** − **b**.

    (c) **a** + **b** + **c** is the same as **c** + **a** + **b**.

3. Explain the term 'resultant' as applied to two vectors.

Exercise 26.3

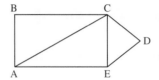

1. Using Figure 26.11 find (a) $\overrightarrow{AB} + \overrightarrow{BC}$, (b) $\overrightarrow{AC} + \overrightarrow{CD} + \overrightarrow{DE}$, (c) $\overrightarrow{AC} - \overrightarrow{EC}$,
    (d) $\overrightarrow{DE} + \overrightarrow{EA} + \overrightarrow{AC}$, (e) $\overrightarrow{AB} - \overrightarrow{CB}$, (f) $\overrightarrow{BC} - \overrightarrow{AC}$, (g) $\overrightarrow{EC} + \overrightarrow{CB} + \overrightarrow{BA}$.

Figure 26.11
Figure for Question 1

26.4 Representing vectors using Cartesian components

We can use the concept of coordinates (see Chapter 17) to describe vectors.

We begin by introducing a **unit vector** in the x direction and a unit vector in the y direction. These are denoted by **i** and **j** respectively.

Key point

The unit vectors in the x and y directions are **i** and **j** respectively.

Consider Figure 26.12 which shows the x–y plane, the point P with coordinates (2, 4) and the vector $\overrightarrow{OP}$.

Figure 26.12
The vector $\overrightarrow{OP}$ is $2\mathbf{i} + 4\mathbf{j}$

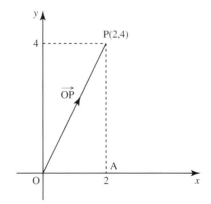

From the triangle law of addition,

$$\overrightarrow{OP} = \overrightarrow{OA} + \overrightarrow{AP}$$

Now the magnitude of $\overrightarrow{OA}$ is 2 and the direction of $\overrightarrow{OA}$ is the positive x direction, so

$$\overrightarrow{OA} = 2\mathbf{i}$$

Similarly $\overrightarrow{AP}$ is in the y direction and has magnitude 4. Hence

$$\overrightarrow{AP} = 4\mathbf{j}$$

Then finally

$$\overrightarrow{OP} = 2\mathbf{i} + 4\mathbf{j}$$

The quantities 2 and 4 are the **Cartesian components** of the vector $\overrightarrow{OP}$. Note that $2\mathbf{i} + 4\mathbf{j}$ is any vector comprising 2 units in the x direction and 4 units in the y direction. Figure 26.13 illustrates three such vectors.

Figure 26.13
Each vector is $2\mathbf{i} + 4\mathbf{j}$
regardless of position

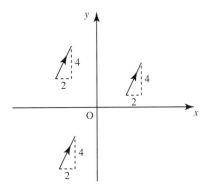

WORKED EXAMPLES

26.3 Sketch the vectors (a) $\mathbf{a} = 3\mathbf{i} + \mathbf{j}$, (b) $\mathbf{b} = -2\mathbf{i} - 3\mathbf{j}$, (c) $\mathbf{c} = -\mathbf{i} + 3\mathbf{j}$, positioning the tail of each vector at the origin.

Solution Figure 26.14 illustrates the required vectors.

Figure 26.14
Vectors $\mathbf{a} = 3\mathbf{i} + \mathbf{j}$,
$\mathbf{b} = -2\mathbf{i} - 3\mathbf{j}$, $\mathbf{c} = -\mathbf{i} + 3\mathbf{j}$

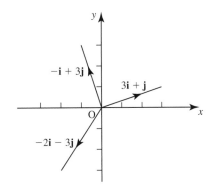

26.4   Sketch the vector $\mathbf{v} = 4\mathbf{i} - \mathbf{j}$ with the point (1, 2) as the tail of the vector.

Solution   Figure 26.15 illustrates the vector $\mathbf{v}$ with its tail at the point (1, 2).

Figure 26.15
The vector $\mathbf{v} = 4\mathbf{i} - \mathbf{j}$
starts at the point (1, 2)

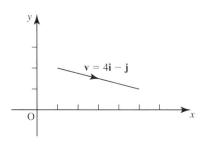

The Cartesian form of vectors is also convenient when adding or subtracting. Worked Example 26.5 illustrates this.

WORKED EXAMPLE

26.5   Given $\mathbf{a} = 7\mathbf{i} + 3\mathbf{j}$, $\mathbf{b} = \mathbf{i} - 2\mathbf{j}$ and $\mathbf{c} = -3\mathbf{i} + \mathbf{j}$ find (a) $\mathbf{a} + \mathbf{b}$, (b) $\mathbf{b} + \mathbf{c} + \mathbf{a}$, (c) $\mathbf{a} - \mathbf{b}$, (d) $\mathbf{c} - \mathbf{a}$.

Solution   We separately add or subtract the x and y components:

(a) $\mathbf{a} + \mathbf{b} = 7\mathbf{i} + 3\mathbf{j} + \mathbf{i} - 2\mathbf{j}$

$\qquad\qquad = 8\mathbf{i} + 3\mathbf{j} - 2\mathbf{j}$ combining the $\mathbf{i}$ parts

$\qquad\qquad = 8\mathbf{i} + \mathbf{j}$ combining the $\mathbf{j}$ parts

(b) $\mathbf{b} + \mathbf{c} + \mathbf{a} = \mathbf{i} - 2\mathbf{j} + (-3\mathbf{i} + \mathbf{j}) + 7\mathbf{i} + 3\mathbf{j}$

$\qquad\qquad = 5\mathbf{i} + 2\mathbf{j}$

(c) $\mathbf{a} - \mathbf{b} = 7\mathbf{i} + 3\mathbf{j} - (\mathbf{i} - 2\mathbf{j})$

$\qquad\qquad = 7\mathbf{i} + 3\mathbf{j} - \mathbf{i} + 2\mathbf{j}$

$\qquad\qquad = 6\mathbf{i} + 5\mathbf{j}$

(d) $\mathbf{c} - \mathbf{a} = -3\mathbf{i} + \mathbf{j} - (7\mathbf{i} + 3\mathbf{j})$

$\qquad\qquad = -3\mathbf{i} + \mathbf{j} - 7\mathbf{i} - 3\mathbf{j}$

$\qquad\qquad = -10\mathbf{i} - 2\mathbf{j}$

Multiplying a vector by a number (scalar) is readily performed using the Cartesian form. The number multiplies each component as shown in Worked Example 26.6.

WORKED EXAMPLE

26.6   Given $\mathbf{v} = 4\mathbf{i} + 3\mathbf{j}$ and $\mathbf{w} = -2\mathbf{i} + \mathbf{j}$ find (a) $2\mathbf{v}$, (b) $-3\mathbf{w}$, (c) $4\mathbf{v} + 5\mathbf{w}$.

Solution   (a) $2\mathbf{v} = 2(4\mathbf{i} + 3\mathbf{j})$

$\qquad\qquad = 8\mathbf{i} + 6\mathbf{j}$

(b) $-3\mathbf{w} = -3(-2\mathbf{i} + \mathbf{j})$

$\qquad\qquad = 6\mathbf{i} - 3\mathbf{j}$

(c) $4\mathbf{v} + 5\mathbf{w} = 4(4\mathbf{i} + 3\mathbf{j}) + 5(-2\mathbf{i} + \mathbf{j})$

$\qquad\qquad\quad = 16\mathbf{i} + 12\mathbf{j} - 10\mathbf{i} + 5\mathbf{j}$

$\qquad\qquad\quad = 6\mathbf{i} + 17\mathbf{j}$

The magnitude of a vector was described as the 'length' of the vector. This idea can be made exact when using Cartesian coordinates. Figure 26.16 illustrates the vector $\mathbf{r}$ where $\mathbf{r} = x\mathbf{i} + y\mathbf{j}$.

Figure 26.16
The vector $\mathbf{r} = x\mathbf{i} + y\mathbf{j}$

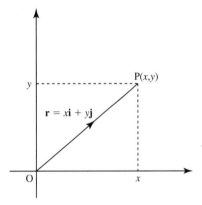

The length (magnitude) of the vector is OP, that is $|\mathbf{r}| = \mathrm{OP}$. Using Pythagoras' theorem we have

$$\mathrm{OP}^2 = x^2 + y^2$$

so

$$|\mathbf{r}| = \mathrm{OP} = \sqrt{x^2 + y^2}$$

Key point   If $\mathbf{r} = x\mathbf{i} + y\mathbf{j}$, then $|\mathbf{r}| = \sqrt{x^2 + y^2}$

WORKED EXAMPLE

26.7   Given the point A has coordinates (2, 7) and B has coordinates (4, 11) find

(a) the vector $\overrightarrow{AB}$

(b) the vector $\overrightarrow{BA}$

(c) the magnitude of $|\overrightarrow{AB}|$

Solution   Figure 26.17 illustrates the situation.

Figure 26.17
The vector $\overrightarrow{AB}$ for
Worked Example 26.7

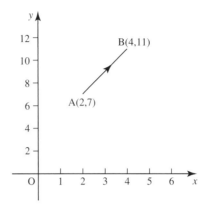

(a) Let $\overrightarrow{OA} = \mathbf{a} = 2\mathbf{i} + 7\mathbf{j}$ and $\overrightarrow{OB} = 4\mathbf{i} + 11\mathbf{j}$. From the triangle law of addition

$$\overrightarrow{OA} + \overrightarrow{AB} = \overrightarrow{OB}$$

$$\mathbf{a} + \overrightarrow{AB} = \mathbf{b}$$

$$\overrightarrow{AB} = \mathbf{b} - \mathbf{a}$$

$$= 4\mathbf{i} + 11\mathbf{j} - (2\mathbf{i} + 7\mathbf{j})$$

$$= 2\mathbf{i} + 4\mathbf{j}$$

(b) $\overrightarrow{BA} = -\overrightarrow{AB}$

$$= -(2\mathbf{i} + 4\mathbf{j})$$

$$= -2\mathbf{i} - 4\mathbf{j}$$

(c) $\overrightarrow{AB} = 2\mathbf{i} + 4\mathbf{j}$

So

$$|\overrightarrow{AB}| = \sqrt{2^2 + 4^2} = \sqrt{20}$$

Self-assessment questions 26.4

1. State a unit vector in (a) the x direction, (b) the y direction.

2. State a unit vector in (a) the negative y direction, (b) the negative x direction.

3. What is the angle between (a) $\mathbf{i}$ and $\mathbf{j}$, (b) $\mathbf{i}$ and $\mathbf{i}$, (c) $\mathbf{i}$ and $-\mathbf{i}$?

4. State an expression for the magnitude of $x\mathbf{i} + y\mathbf{j}$.

5. State a vector of magnitude 6 in the x direction.

6. State a vector of magnitude 2 in the negative y direction.

Exercise 26.4

1. Calculate the magnitude of
   (a) $\mathbf{i} + 3\mathbf{j}$ (b) $3\mathbf{i} + 4\mathbf{j}$ (c) $-\mathbf{i}$ (d) $\frac{1}{2}(\mathbf{i} + \mathbf{j})$

2. Calculate the vector from
   (a) $(0, 0)$ to $(3, 7)$ (b) $(3, 7)$ to $(5, 2)$ (c) $(5, 2)$ to $(-1, -2)$

3. Given $\mathbf{a} = 3\mathbf{i} - 2\mathbf{j}$, $\mathbf{b} = -\mathbf{i} - 3\mathbf{j}$, $\mathbf{c} = 2\mathbf{i} + 5\mathbf{j}$ find
   (a) $4\mathbf{a}$ (b) $3\mathbf{b} + 2\mathbf{c}$ (c) $2\mathbf{a} - 3\mathbf{b} - 4\mathbf{c}$ (d) $|\mathbf{a}|$ (e) $|4\mathbf{a}|$ (f) $|3\mathbf{b} + 2\mathbf{c}|$

26.5 The scalar product

We have seen how a vector may be multiplied by a number (scalar). In this section, we see one way in which a vector can be multiplied by another vector. This particular product is known as the **scalar product**. (Another method of multiplying together two vectors, known as the vector product, is not dealt with in this book. See *Mathematics for Engineers: A Modern Interactive Approach*, 3rd edition, 2008, by Croft and Davison for details.)

Key point

Given two vectors, $\mathbf{a}$ and $\mathbf{b}$, their scalar product, denoted by $\mathbf{a} \cdot \mathbf{b}$, is given by

$$\mathbf{a} \cdot \mathbf{b} = |\mathbf{a}||\mathbf{b}| \cos\theta$$

where θ is the angle between $\mathbf{a}$ and $\mathbf{b}$.

This product has applications in engineering. The scalar product is also known as the **dot product**.

WORKED EXAMPLE

26.8   Show (a) $\mathbf{i} \cdot \mathbf{i} = 1$, (b) $\mathbf{i} \cdot \mathbf{j} = 0$.

Solution   Recall that i and j are unit vectors in the x and y directions respectively.

(a) So

$$\mathbf{i} \cdot \mathbf{i} = |\mathbf{i}||\mathbf{i}| \cos 0° \quad \text{since the angle between i and itself is } 0°$$

$$= (1)(1)(1)$$

$$= 1$$

(b) $\mathbf{i} \cdot \mathbf{j} = |\mathbf{i}||\mathbf{j}| \cos 90°$ since the angle between $\mathbf{i}$ and $\mathbf{j}$ is $90°$

$\quad\quad = (1)(1)(0)$

$\quad\quad = 0$

Similarly it is easy to show that $\mathbf{j} \cdot \mathbf{j} = 1$, $\mathbf{j} \cdot \mathbf{i} = 0$.

Key point $\mathbf{i} \cdot \mathbf{i} = 1$ $\mathbf{j} \cdot \mathbf{j} = 1$ $\mathbf{i} \cdot \mathbf{j} = 0$ $\mathbf{j} \cdot \mathbf{i} = 0$

We use the above Key point to find the scalar product of two vectors from their Cartesian form. This does not require the value of the angle between the two vectors.

WORKED EXAMPLE

26.9   Given $\mathbf{a} = 3\mathbf{i} + 2\mathbf{j}$ and $\mathbf{b} = 4\mathbf{i} - \mathbf{j}$, find $\mathbf{a} \cdot \mathbf{b}$.

Solution $\mathbf{a} \cdot \mathbf{b} = (3\mathbf{i} + 2\mathbf{j}) \cdot (4\mathbf{i} - \mathbf{j})$

$\quad\quad = 3\mathbf{i} \cdot (4\mathbf{i} - \mathbf{j}) + 2\mathbf{j} \cdot (4\mathbf{i} - \mathbf{j})$

$\quad\quad = 12\mathbf{i} \cdot \mathbf{i} - 3\mathbf{i} \cdot \mathbf{j} + 8\mathbf{j} \cdot \mathbf{i} - 2\mathbf{j} \cdot \mathbf{j}$

Recalling $\mathbf{i} \cdot \mathbf{i} = \mathbf{j} \cdot \mathbf{j} = 1$, $\mathbf{i} \cdot \mathbf{j} = \mathbf{j} \cdot \mathbf{i} = 0$, we see that

$\mathbf{a} \cdot \mathbf{b} = 12(1) - 3(0) + 8(0) - 2(1) = 10$

We can use the technique of the above example to deduce the following general result:

Key point   If $\mathbf{a} = a_1\mathbf{i} + a_2\mathbf{j}$, $\mathbf{b} = b_1\mathbf{i} + b_2\mathbf{j}$ then

$$\mathbf{a} \cdot \mathbf{b} = a_1b_1 + a_2b_2$$

That is, the $\mathbf{i}$ components are multiplied together, the $\mathbf{j}$ components are multiplied together and the sum of these products is the scalar product.

WORKED EXAMPLE

26.10   Given $\mathbf{a} = 4\mathbf{i} + \mathbf{j}$ and $\mathbf{b} = 2\mathbf{i} + 3\mathbf{j}$
(a) Find the scalar product $\mathbf{a} \cdot \mathbf{b}$.
(b) Hence find the angle between $\mathbf{a}$ and $\mathbf{b}$.

Solution (a) We find the scalar product:

$$\mathbf{a} \cdot \mathbf{b} = (4\mathbf{i} + \mathbf{j}) \cdot (2\mathbf{i} + 3\mathbf{j})$$
$$= 4(2) + 1(3)$$
$$= 11$$

(b) The scalar product is also given by

$$\mathbf{a} \cdot \mathbf{b} = |\mathbf{a}||\mathbf{b}| \cos \theta$$

Now $|\mathbf{a}| = \sqrt{4^2 + 1^2} = \sqrt{17}$, $|\mathbf{b}| = \sqrt{2^2 + 3^2} = \sqrt{13}$. We have found the value of $\mathbf{a} \cdot \mathbf{b}$ in part (a). Hence using the values for $\mathbf{a} \cdot \mathbf{b}$, $|\mathbf{a}|$ and $|\mathbf{b}|$ we have

$$11 = \sqrt{17}\sqrt{13} \cos \theta$$
$$\cos \theta = \frac{11}{\sqrt{17}\sqrt{13}} = 0.7399$$

Using the inverse cosine function of a calculator we find that $\theta = 42.3°$.

Self-assessment questions 26.5

1. State two formulae for the scalar product of vectors $\mathbf{a} = a_1\mathbf{i} + a_2\mathbf{j}$ and $\mathbf{b} = b_1\mathbf{i} + b_2\mathbf{j}$.

Exercise 26.5

1. Find the scalar product of the following pairs of vectors:
   (a) $2\mathbf{i} - \mathbf{j}, 3\mathbf{i} + 2\mathbf{j}$ (b) $-\mathbf{i} + 3\mathbf{j}, 4\mathbf{i} - \mathbf{j}$ (c) $\mathbf{i} - \mathbf{j}, -2\mathbf{i} - \mathbf{j}$

2. Find the angles between the following pairs of vectors:
   (a) $3\mathbf{i} + 4\mathbf{j}, \mathbf{i} + 2\mathbf{j}$ (b) $\mathbf{i} + \mathbf{j}, 2\mathbf{i} - 3\mathbf{j}$ (c) $\mathbf{i} + \mathbf{j}, \mathbf{i} - \mathbf{j}$ (d) $\mathbf{j}, -\mathbf{j}$

3. Show that the vectors $\mathbf{a} = 2\mathbf{i} - \mathbf{j}$ and $\mathbf{b} = -\mathbf{i} - 2\mathbf{j}$ are perpendicular.

4. Points A, B and C have coordinates $(-6, 2)$, $(3, 8)$ and $(-3, -2)$ respectively.
   (a) Find $\overrightarrow{AB}$ and $\overrightarrow{BC}$.
   (b) By using the scalar product calculate the angle between $\overrightarrow{AB}$ and $\overrightarrow{BC}$.

Test and assignment exercises 26

1. Given $\mathbf{a} = 5\mathbf{i} + 7\mathbf{j}$, $\mathbf{b} = 2\mathbf{i} - \mathbf{j}$ and $\mathbf{c} = -3\mathbf{i} + 2\mathbf{j}$ find (a) $\mathbf{a} \cdot \mathbf{b}$, (b) $\mathbf{c} \cdot \mathbf{b}$, (c) $\mathbf{a} \cdot \mathbf{c}$.

2. Find the angle between the following pairs of vectors:
   (a) $5\mathbf{i} + \mathbf{j}$ and $3\mathbf{i} + 4\mathbf{j}$ (b) $4\mathbf{i} - 2\mathbf{j}$ and $-2\mathbf{i} + 7\mathbf{j}$ (c) $-\mathbf{i} + 3\mathbf{j}$ and $3\mathbf{i} - 5\mathbf{j}$
   (d) $\frac{1}{2}\mathbf{i} - \mathbf{j}$ and $\mathbf{i} + \frac{3}{4}\mathbf{j}$ (e) $0.7\mathbf{i} + 1.4\mathbf{j}$ and $-0.3\mathbf{i} + 1.9\mathbf{j}$

3. Given the points A(4, 1), B($-3, -3$), C($-4, 2$) and the origin O(0, 0) express the following vectors in Cartesian form:
   (a) $\overrightarrow{OB}$ (b) $\overrightarrow{OC}$ (c) $\overrightarrow{AB}$ (d) $\overrightarrow{CA}$ (e) $\overrightarrow{BC}$

4. Show that the following pairs of vectors are perpendicular:
   (a) $6\mathbf{i} + \mathbf{j}$ and $\mathbf{i} - 6\mathbf{j}$ (b) $-2\mathbf{i} - 3\mathbf{j}$ and $-3\mathbf{i} + 2\mathbf{j}$ (c) $\frac{\mathbf{i}+\mathbf{j}}{2}$ and $-2\mathbf{i} + 2\mathbf{j}$
   (d) $\frac{1}{4}\mathbf{i} - \frac{1}{3}\mathbf{j}$ and $4\mathbf{i} + 3\mathbf{j}$

5. Given $\mathbf{a} = 10\mathbf{i} + 3\mathbf{j}, \mathbf{b} = \mathbf{i} - 5\mathbf{j}$ and $\mathbf{c} = 2\mathbf{i} + 8\mathbf{j}$ find
   (a) $3\mathbf{b}$ (b) $2\mathbf{a} + 3\mathbf{b}$ (c) $\mathbf{a} - 2\mathbf{c}$ (d) $-2\mathbf{a} - \mathbf{b} - 4\mathbf{c}$ (e) $\frac{1}{2}\mathbf{a} - \frac{3}{2}\mathbf{b} - \mathbf{c}$

6. Find the magnitude of the vector joining:
   (a) $(2, 5)$ to $(5, 10)$ (b) $(-1, 0)$ to $(2, 1)$ (c) $(-2, -4)$ to $(4, -2)$
   (d) $(\frac{1}{2}, \frac{3}{2})$ to $(\frac{1}{4}, -\frac{3}{4})$ (e) $(6.7, -1.4)$ to $(-3.1, 2.9)$

7. Find a unit vector in the direction of $(4, 1)$ to $(3, 10)$.

8. With reference to Figure 26.18 determine
   (a) $\overrightarrow{ED} + \overrightarrow{DB}$ (b) $\overrightarrow{AD} - \overrightarrow{CD}$ (c) $\overrightarrow{EC} + \overrightarrow{CB} - \overrightarrow{DB}$
   (d) $\overrightarrow{AC} - \overrightarrow{DC} - \overrightarrow{BD}$ (e) $\overrightarrow{DE} - \overrightarrow{CE} - \overrightarrow{AC} - \overrightarrow{BA}$

9. Find the angle that the vector $2\mathbf{i} + 9\mathbf{j}$ makes with the x axis.
10. Find a vector which is perpendicular to $3\mathbf{i} - 4\mathbf{j}$.
11. Using Figure 26.19 find
    (a) $\overrightarrow{AF} + \overrightarrow{FB} + \overrightarrow{BC}$ (b) $\overrightarrow{FD} + \overrightarrow{DC} + \overrightarrow{CA}$ (c) $\overrightarrow{FE} - \overrightarrow{AE} + \overrightarrow{AC}$
    (d) $\overrightarrow{BC} - \overrightarrow{AC} - \overrightarrow{FA}$ (e) $\overrightarrow{BD} - \overrightarrow{FD} - \overrightarrow{AF}$

12. Find the vectors joining the following pairs of points:
    (a) $(7, 9)$ to $(8, 14)$ (b) $(2, -3)$ to $(0, 6)$ (c) $(-5, 1)$ to $(-2, -4)$
    (d) $(-1, 3)$ to $(2, 5)$ (e) $(-5, -2)$ to $(3, -3)$

13. Find the magnitude of the following vectors:
    (a) $4\mathbf{i} + 3\mathbf{j}$ (b) $-2\mathbf{i} + \frac{1}{2}\mathbf{j}$ (c) $-\frac{1}{2}\mathbf{i} + \mathbf{j}$ (d) $0.25\mathbf{i} - 0.75\mathbf{j}$ (e) $\frac{\mathbf{i}+\mathbf{j}}{2}$

Figure 26.18
Figure for Question 8

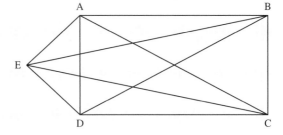

Figure 26.19
Figure for Question 11

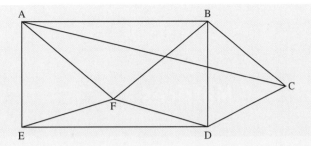

14. Given $\mathbf{u} = 5\mathbf{i} - 3\mathbf{j}$, $\mathbf{v} = -2\mathbf{i} - 7\mathbf{j}$, $\mathbf{w} = \mathbf{i} + 4\mathbf{j}$ find
    (a) $3\mathbf{w}$ (b) $-2\mathbf{v}$ (c) $\mathbf{u} + 3\mathbf{v} - 2\mathbf{w}$ (d) $2\mathbf{u} - 2\mathbf{v} + 4\mathbf{w}$ (e) $-6\mathbf{u} + 3\mathbf{v} - 2\mathbf{w}$

15. Find a unit vector in the direction from $(2, -3)$ to $(-3, 5)$.

16. Find the angle between $6\mathbf{i} - 5\mathbf{j}$ and the y axis.

17. Given $\mathbf{a} = 5\mathbf{i} - 7\mathbf{j}$, $\mathbf{b} = -\mathbf{i} + 3\mathbf{j}$, $\mathbf{c} = 2\mathbf{i} + \mathbf{j}$ find
    (a) $\mathbf{a} \cdot \mathbf{b}$ (b) $\mathbf{a} \cdot \mathbf{c}$ (c) $\mathbf{b} \cdot \mathbf{c}$ (d) $\mathbf{b} \cdot \mathbf{b}$

18. Show that for any general vector, $\mathbf{a}$,
    $\mathbf{a} \cdot \mathbf{a} = |\mathbf{a}|^2$

19. Find the angle between the following pairs of vectors:
    (a) $8\mathbf{i} + 3\mathbf{j}$ and $2\mathbf{i} + 3\mathbf{j}$
    (b) $2\mathbf{i} + 7\mathbf{j}$ and $-\mathbf{i} + 5\mathbf{j}$
    (c) $-2\mathbf{i} + 3\mathbf{j}$ and $3\mathbf{i} - 4\mathbf{j}$
    (d) $-\mathbf{i} - \mathbf{j}$ and $2\mathbf{i} + 2\mathbf{j}$
    (e) $\mathbf{i} + 2\mathbf{j}$ and $3\mathbf{i} + 6\mathbf{j}$

20. If $\mathbf{a} = 3\mathbf{i} + 4\mathbf{j}$ and $\mathbf{b} = 2\mathbf{i} + y\mathbf{j}$ are perpendicular, find y.

Matrices

Objectives: This chapter:

- explains what is meant by a matrix
- shows how matrices can be added, subtracted and multiplied
- explains what is meant by the determinant of a matrix
- shows how to find the inverse of a matrix
- shows how matrices can be used to solve simultaneous equations

---

27.1 What is a matrix?

A **matrix** is a set of numbers arranged in the form of a rectangle and enclosed in curved brackets. The plural of matrix is **matrices**. For example,

$$\begin{pmatrix} 1 & 2 & 3 \\ 3 & 4 & 6 \end{pmatrix} \qquad \begin{pmatrix} 1 \\ 2 \\ -4 \end{pmatrix} \qquad \begin{pmatrix} 1 & 1 & 2 \\ -3 & 4 & 5 \\ \frac{1}{2} & 2 & 1 \end{pmatrix}$$

are all matrices. Each number in a matrix is known as an **element**. To refer to a particular matrix we label it with a capital letter, so that we could write

$$A = \begin{pmatrix} 1 & 2 & 3 \\ 3 & 4 & 6 \end{pmatrix} \qquad B = \begin{pmatrix} 1 \\ 2 \\ -4 \end{pmatrix} \qquad C = \begin{pmatrix} 1 & 1 & 2 \\ -3 & 4 & 5 \\ \frac{1}{2} & 2 & 1 \end{pmatrix}$$

We refer to the **size** of a matrix by giving its number of rows and number of columns, in that order. So matrix A above has two rows and three columns – we say it is a 'two by three' matrix, and write this as

'2 × 3'. Similarly B has size 3 × 1 and C has size 3 × 3. Some particular types of matrix occur so frequently that they have been given special names.

A **square** matrix has the same number of rows as columns. So

$$\begin{pmatrix} 2 & -7 \\ -1 & 6 \end{pmatrix} \text{ and } \begin{pmatrix} 1 & 2 & -1 \\ 9 & 8 & 5 \\ 6 & -7 & 2 \end{pmatrix}$$

are both square matrices.

A **diagonal** matrix is a square matrix in which all the elements are 0 except those on the diagonal from the top left to the bottom right. This diagonal is called the **leading diagonal**. On the leading diagonal the elements can take any value including zero. So

$$\begin{pmatrix} 7 & 0 \\ 0 & 9 \end{pmatrix}, \quad \begin{pmatrix} 9 & 0 \\ 0 & 0 \end{pmatrix} \text{ and } \begin{pmatrix} -1 & 0 & 0 \\ 0 & 8 & 0 \\ 0 & 0 & -17 \end{pmatrix}$$

are all diagonal matrices.

An **identity** matrix is a diagonal matrix, all the diagonal entries of which are equal to 1. So

$$\begin{pmatrix} 1 & 0 \\ 0 & 1 \end{pmatrix}, \quad \begin{pmatrix} 1 & 0 & 0 \\ 0 & 1 & 0 \\ 0 & 0 & 1 \end{pmatrix} \text{ and } \begin{pmatrix} 1 & 0 & 0 & 0 \\ 0 & 1 & 0 & 0 \\ 0 & 0 & 1 & 0 \\ 0 & 0 & 0 & 1 \end{pmatrix}$$

are all identity matrices.

Self-assessment questions 27.1

1. Explain what is meant by a matrix.

2. Give an example of a 5 × 1 matrix and a 1 × 5 matrix.

3. Explain what is meant by a 'square' matrix.

4. Explain what is meant by the 'leading diagonal' of a square matrix.

5. Explain what is meant by a 'diagonal' matrix.

6. Explain what is meant by an 'identity' matrix.

7. Is it true that all identity matrices must be square?

Exercise 27.1

MyMathLab

1. Give the size of each of the following matrices:

$$A = (1 \quad 0 \quad 0 \quad 2) \qquad B = \begin{pmatrix} 1 & 2 \\ 3 & 7 \\ 9 & 8 \\ 9 & -9 \end{pmatrix} \qquad C = (1) \qquad D = \begin{pmatrix} 1 \\ -1 \\ 0 \\ 0 \\ 9 \end{pmatrix}$$

2. Classify each of the following matrices as square, diagonal or identity, as appropriate:

$$A = \begin{pmatrix} 3 & 4 \\ 2 & 9 \end{pmatrix} \qquad B = \begin{pmatrix} 1 & 0 & 0 \\ 0 & 1 & 0 \end{pmatrix} \qquad C = \begin{pmatrix} 1 & 0 & 0 \\ 0 & 1 & 0 \\ 0 & 0 & 1 \end{pmatrix} \qquad D = \begin{pmatrix} 1 & 5 & 7 \\ 0 & 1 & 0 \end{pmatrix}$$

$$E = (1 \quad 2 \quad 3 \quad 4) \qquad F = \begin{pmatrix} 9 & 0 \\ 0 & 8 \end{pmatrix} \qquad G = \begin{pmatrix} 0 & 8 \\ 9 & 0 \end{pmatrix} \qquad H = \begin{pmatrix} 0 & 1 \\ 1 & 0 \end{pmatrix}$$

3. How many elements are there in a matrix whose size is
    (a) 3×1 (b) 1×3 (c) $m \times n$ (d) $n \times n$?

27.2 Addition, subtraction and multiplication of matrices

Matrices can be added, subtracted and multiplied. In this section we shall see how to carry out these operations. However, they can never be divided.

Addition and subtraction

Two matrices having the same size can be added or subtracted by simply adding or subtracting the corresponding elements. For example,

$$\begin{pmatrix} 1 & 2 \\ 3 & 4 \end{pmatrix} + \begin{pmatrix} 5 & 2 \\ 1 & 0 \end{pmatrix} = \begin{pmatrix} 1+5 & 2+2 \\ 3+1 & 4+0 \end{pmatrix} = \begin{pmatrix} 6 & 4 \\ 4 & 4 \end{pmatrix}$$

Similarly,

$$\begin{pmatrix} 1 & 2 & 3 & 4 \\ 2 & 1 & 1 & 7 \end{pmatrix} - \begin{pmatrix} 0 & 1 & 3 & 9 \\ 7 & 0 & 0 & 1 \end{pmatrix} = \begin{pmatrix} 1-0 & 2-1 & 3-3 & 4-9 \\ 2-7 & 1-0 & 1-0 & 7-1 \end{pmatrix}$$

$$= \begin{pmatrix} 1 & 1 & 0 & -5 \\ -5 & 1 & 1 & 6 \end{pmatrix}$$

The matrices

$$\begin{pmatrix} 1 & 2 & 9 \\ -1 & \frac{1}{2} & 0 \end{pmatrix} \text{ and } \begin{pmatrix} 1 \\ 0 \\ 7 \end{pmatrix}$$

can be neither added to nor subtracted from one another because they have different sizes.

Multiplying a matrix by a number

A matrix is multiplied by a number by multiplying each element by that number. For example,

$$4 \begin{pmatrix} 1 & 2 \\ 3 & -9 \end{pmatrix} = \begin{pmatrix} 4 \times 1 & 4 \times 2 \\ 4 \times 3 & 4 \times -9 \end{pmatrix} = \begin{pmatrix} 4 & 8 \\ 12 & -36 \end{pmatrix}$$

and

$$\frac{1}{4} \begin{pmatrix} 16 \\ 8 \end{pmatrix} = \begin{pmatrix} \frac{1}{4} \times 16 \\ \frac{1}{4} \times 8 \end{pmatrix} = \begin{pmatrix} 4 \\ 2 \end{pmatrix}$$

WORKED EXAMPLE

27.1   If

$$A = \begin{pmatrix} 5 \\ 1 \\ 9 \end{pmatrix} \text{ and } B = \begin{pmatrix} 8 \\ 2 \\ -6 \end{pmatrix}$$

find (a) $A + B$ (b) $A - B$ (c) $7A$ (d) $-\frac{1}{2}B$ (e) $3A + 2B$

Solution   (a)

$$A + B = \begin{pmatrix} 5 \\ 1 \\ 9 \end{pmatrix} + \begin{pmatrix} 8 \\ 2 \\ -6 \end{pmatrix} = \begin{pmatrix} 13 \\ 3 \\ 3 \end{pmatrix}$$

(b)

$$A - B = \begin{pmatrix} 5 \\ 1 \\ 9 \end{pmatrix} - \begin{pmatrix} 8 \\ 2 \\ -6 \end{pmatrix} = \begin{pmatrix} -3 \\ -1 \\ 15 \end{pmatrix}$$

(c)

$$7A = 7 \begin{pmatrix} 5 \\ 1 \\ 9 \end{pmatrix} = \begin{pmatrix} 35 \\ 7 \\ 63 \end{pmatrix}$$

(d)

$$-\frac{1}{2}B = -\frac{1}{2}\begin{pmatrix} 8 \\ 2 \\ -6 \end{pmatrix} = \begin{pmatrix} -4 \\ -1 \\ 3 \end{pmatrix}$$

(e)

$$3A + 2B = 3\begin{pmatrix} 5 \\ 1 \\ 9 \end{pmatrix} + 2\begin{pmatrix} 8 \\ 2 \\ -6 \end{pmatrix} = \begin{pmatrix} 15 \\ 3 \\ 27 \end{pmatrix} + \begin{pmatrix} 16 \\ 4 \\ -12 \end{pmatrix} = \begin{pmatrix} 31 \\ 7 \\ 15 \end{pmatrix}$$

Multiplying two matrices together

Two matrices can only be multiplied together if the number of columns in the first is the same as the number of rows in the second. The product of two such matrices is a matrix that has the same number of rows as the first matrix and the same number of columns as the second. In symbols, this states that if A has size $p \times q$ and B has size $q \times s$, then AB has size $p \times s$. The way that the multiplication is performed may seem strange at first. You will need to work through several examples to understand how it is done.

WORKED EXAMPLES

27.2   Find AB where

$$A = \begin{pmatrix} 1 & 4 \\ 6 & 3 \end{pmatrix} \quad \text{and} \quad B = \begin{pmatrix} 2 \\ 5 \end{pmatrix}$$

Solution   The size of A is 2×2. The size of B is 2×1. Therefore the number of columns in the first matrix, A, equals the number of rows in the second matrix B. We can therefore find the product AB. The result will be a 2×1 matrix. Let us call it C. The working is as follows:

$$C = \begin{pmatrix} 1 & 4 \\ 6 & 3 \end{pmatrix}\begin{pmatrix} 2 \\ 5 \end{pmatrix} = \begin{pmatrix} 1 \times 2 + 4 \times 5 \\ 6 \times 2 + 3 \times 5 \end{pmatrix} = \begin{pmatrix} 22 \\ 27 \end{pmatrix}$$

The first row of A multiplies the first column of B to give

$$(1 \times 2) + (4 \times 5) = 2 + 20 = 22$$

This is the element in row 1, column 1 of C. The second row of A then multiplies the first column of B to give

$$(6 \times 2) + (3 \times 5) = 12 + 15 = 27$$

This is the element in row 2, column 1 of C.

27.3   Using the same A and B as the previous example, is it possible to find BA?

Solution   Recall that B has size 2×1 and A has size 2×2. When written in the order BA, the number of columns in the first matrix is 1 whereas the number of rows in the second is 2. It is not therefore possible to find BA.

27.4   Find, if possible,

$$\begin{pmatrix} 1 & 4 & 9 \\ 2 & 0 & 1 \end{pmatrix} \begin{pmatrix} 1 & 9 \\ 8 & 7 \\ -7 & 3 \end{pmatrix}$$

Solution   The number of columns in the first matrix is 3 and this is the same as the number of rows in the second. We can therefore perform the multiplication and the answer will have size 2×2.

$$\begin{pmatrix} 1 & 4 & 9 \\ 2 & 0 & 1 \end{pmatrix} \begin{pmatrix} 1 & 9 \\ 8 & 7 \\ -7 & 3 \end{pmatrix}$$

$$= \begin{pmatrix} 1 \times 1 + 4 \times 8 + 9 \times -7 & 1 \times 9 + 4 \times 7 + 9 \times 3 \\ 2 \times 1 + 0 \times 8 + 1 \times -7 & 2 \times 9 + 0 \times 7 + 1 \times 3 \end{pmatrix}$$

$$= \begin{pmatrix} -30 & 64 \\ -5 & 21 \end{pmatrix}$$

27.5   Find $(1 \quad 5)\begin{pmatrix} 2 \\ 1 \end{pmatrix}$.

Solution

$$(1 \quad 5)\begin{pmatrix} 2 \\ 1 \end{pmatrix} = (1 \times 2 + 5 \times 1) = (7)$$

Note that in this example the result is a 1×1 matrix, that is a single number.

27.6   Find $\begin{pmatrix} 3 & 2 \\ 5 & 3 \end{pmatrix}\begin{pmatrix} x \\ y \end{pmatrix}$.

Solution

$$\begin{pmatrix} 3 & 2 \\ 5 & 3 \end{pmatrix}\begin{pmatrix} x \\ y \end{pmatrix} = \begin{pmatrix} 3x + 2y \\ 5x + 3y \end{pmatrix}$$

27.7   Find IX where I is the 2×2 identity matrix

$$\begin{pmatrix} 1 & 0 \\ 0 & 1 \end{pmatrix} \quad \text{and} \quad X = \begin{pmatrix} a \\ b \end{pmatrix}$$

Solution

$$IX = \begin{pmatrix} 1 & 0 \\ 0 & 1 \end{pmatrix}\begin{pmatrix} a \\ b \end{pmatrix} = \begin{pmatrix} 1 \times a + 0 \times b \\ 0 \times a + 1 \times b \end{pmatrix} = \begin{pmatrix} a \\ b \end{pmatrix}$$

Note that the effect of multiplying the matrix X by the identity matrix is to leave X unchanged. This is a very important and useful property of identity matrices, which we shall require later. It should be remembered. This property of identity matrices should remind you of the fact that multiplying a number by 1 leaves the number unchanged: for example, $7 \times 1 = 7$, $1 \times -9 = -9$. An identity matrix plays the same role for matrices as the number 1 does when dealing with ordinary arithmetic.

Self-assessment questions 27.2

1. Under what conditions can two matrices be added or subtracted?

2. Two different types of multiplication have been described. What are these two types and how do they differ?

3. If A has size $p \times q$ and B has size $r \times s$, under what conditions will the product AB exist? If AB does exist what will be its size? Under what conditions will the product BA exist and what will be its size?

4. Suppose A and B are two matrices. The products AB and BA both exist. What can you say about the sizes of A and B?

Exercise 27.2

MyMathLab

1. If $M = \begin{pmatrix} 7 & 8 \\ 2 & -1 \end{pmatrix}$ and $N = \begin{pmatrix} 3 & 2 \\ -1 & 4 \end{pmatrix}$

    find MN and NM. Comment upon your two answers.

2. If

    $$A = \begin{pmatrix} 1 & -7 \\ -1 & 2 \end{pmatrix}$$

    $$B = \begin{pmatrix} 3 & 7 & 2 \\ -1 & 0 & -10 \end{pmatrix}$$

    $$C = (1 \quad 0 \quad -1 \quad 1)$$

    $$D = \begin{pmatrix} 7 \\ 9 \\ 8 \\ 1 \end{pmatrix} \qquad E = \begin{pmatrix} 1/2 \\ 0 \\ 0 \\ 1 \end{pmatrix}$$

    find, if possible,
    (a) $3A$ (b) $A + B$ (c) $A + C$

    (d) $5D$ (e) BC (f) CB (g) AC
    (h) CA (i) $5A - 2C$ (j) $D - E$
    (k) BA (l) AB (m) AA (that is, A times itself, usually written A^2).

3. I is the 2×2 identity matrix

    $$\begin{pmatrix} 1 & 0 \\ 0 & 1 \end{pmatrix}$$

    If

    $$A = \begin{pmatrix} 3 & 7 \\ -1 & -2 \end{pmatrix} \qquad B = \begin{pmatrix} x \\ y \end{pmatrix}$$

    and $\quad C = \begin{pmatrix} 3 & 9 & 5 \\ 4 & 2 & 8 \end{pmatrix}$

    find, where possible,

    (a) IA (b) AI (c) BI (d) IB
    (e) CI (f) IC.

    Comment upon your answers.

4. If

$$A = \begin{pmatrix} a & b \\ c & d \end{pmatrix} \quad \text{and} \quad B = \begin{pmatrix} e & f \\ g & h \end{pmatrix}$$

find
(a) $A + B$ (b) $A - B$ (c) AB
(d) BA

5. Find

$$\begin{pmatrix} -7 & 5 \\ -2 & -1 \end{pmatrix} \begin{pmatrix} x \\ y \end{pmatrix}$$

6. Find

$$\begin{pmatrix} 9 & -5 \\ 3 & 8 \end{pmatrix} \begin{pmatrix} a \\ b \end{pmatrix}$$

7. Given

$$A = \begin{pmatrix} 2 & 3 & -1 \\ 4 & 0 & 1 \end{pmatrix} \quad B = \begin{pmatrix} 1 & 2 \\ 3 & -2 \end{pmatrix}$$

find
(a) BA (b) B^2 (c) B^2A

8. Given

$$A = \begin{pmatrix} 1 & 3 \\ -1 & 4 \end{pmatrix} \quad B = \begin{pmatrix} 0 & -1 \\ 3 & 2 \end{pmatrix}$$

find
(a) AB (b) BA (c) A^2 (d) B^3

9. Given

$$A = \begin{pmatrix} 4 & 11 \\ 1 & 3 \end{pmatrix} \quad \text{and}$$

$$B = \begin{pmatrix} 3 & -11 \\ -1 & 4 \end{pmatrix}$$

verify that $AB = I$ where I is the 2×2 identity matrix.

10. Calculate

$$\begin{pmatrix} 3 & 0 \\ 0 & 2 \end{pmatrix} \begin{pmatrix} a & b \\ c & d \end{pmatrix}$$

Comment upon the effect of multiplying a matrix by

$$\begin{pmatrix} 3 & 0 \\ 0 & 2 \end{pmatrix}$$

27.3 The inverse of a 2 × 2 matrix

Consider the 2×2 matrix

$$A = \begin{pmatrix} a & b \\ c & d \end{pmatrix}$$

An important matrix that is related to A is known as the **inverse** of A and is given the symbol A^{-1}. Here, the superscript '−1' should not be read as a power, but is meant purely as a notation for the inverse matrix. A^{-1} can be found from the following formula:

Key point

If $A = \begin{pmatrix} a & b \\ c & d \end{pmatrix}$ then $A^{-1} = \dfrac{1}{ad - bc} \begin{pmatrix} d & -b \\ -c & a \end{pmatrix}$

This formula states that:

- the elements on the leading diagonal are interchanged
- the remaining elements change sign
- the resulting matrix is multiplied by $\dfrac{1}{ad - bc}$

The inverse matrix has the property that:

$$A\,A^{-1} = A^{-1}\,A = I$$

That is, when a 2×2 matrix and its inverse are multiplied together the result is the identity matrix. Consider the following example.

WORKED EXAMPLE

27.8   Find the inverse of the matrix

$$A = \begin{pmatrix} 6 & 5 \\ 2 & 2 \end{pmatrix}$$

and verify that $A\,A^{-1} = A^{-1}\,A = I$.

Solution   Using the formula for the inverse we find

$$A^{-1} = \frac{1}{(6)(2) - (5)(2)} \begin{pmatrix} 2 & -5 \\ -2 & 6 \end{pmatrix} = \frac{1}{2} \begin{pmatrix} 2 & -5 \\ -2 & 6 \end{pmatrix} = \begin{pmatrix} 1 & -\frac{5}{2} \\ -1 & 3 \end{pmatrix}$$

The inverse of $\begin{pmatrix} 6 & 5 \\ 2 & 2 \end{pmatrix}$

is therefore $\begin{pmatrix} 1 & -\frac{5}{2} \\ -1 & 3 \end{pmatrix}$

Evaluating $A\,A^{-1}$ we find

$$\begin{pmatrix} 6 & 5 \\ 2 & 2 \end{pmatrix} \begin{pmatrix} 1 & -\frac{5}{2} \\ -1 & 3 \end{pmatrix} = \begin{pmatrix} 6 \times 1 + 5 \times -1 & 6 \times (-\frac{5}{2}) + 5 \times 3 \\ 2 \times 1 + 2 \times -1 & 2 \times (-\frac{5}{2}) + 2 \times 3 \end{pmatrix}$$

$$= \begin{pmatrix} 1 & 0 \\ 0 & 1 \end{pmatrix}$$

which is the 2×2 identity matrix. Also evaluating $A^{-1}A$ we find

$$\begin{pmatrix} 1 & -\frac{5}{2} \\ -1 & 3 \end{pmatrix} \begin{pmatrix} 6 & 5 \\ 2 & 2 \end{pmatrix} = \begin{pmatrix} 1 \times 6 + (-\frac{5}{2}) \times 2 & 1 \times 5 + (-\frac{5}{2}) \times 2 \\ -1 \times 6 + 3 \times 2 & -1 \times 5 + 3 \times 2 \end{pmatrix}$$

$$= \begin{pmatrix} 1 & 0 \\ 0 & 1 \end{pmatrix}$$

which is the 2 × 2 identity matrix. We have shown that $AA^{-1} = A^{-1}A = I$.

The quantity $ad - bc$ in the formula for the inverse is known as the **determinant** of the matrix A. We often write it as $|A|$, the vertical bars indicating that we mean the determinant of A and not the matrix itself.

Key point

If $A = \begin{pmatrix} a & b \\ c & d \end{pmatrix}$ then its determinant is

$$|A| = \begin{vmatrix} a & b \\ c & d \end{vmatrix} = ad - bc$$

WORKED EXAMPLES

27.9 Find the value of the determinant

$$\begin{vmatrix} 8 & 7 \\ 9 & 2 \end{vmatrix}$$

Solution The determinant is given by $(8)(2) - (7)(9) = 16 - 63 = -47$.

27.10 Find the determinant of each of the following matrices:

(a) $C = \begin{pmatrix} 7 & 2 \\ 4 & 9 \end{pmatrix}$ (b) $D = \begin{pmatrix} 4 & -2 \\ 9 & -1 \end{pmatrix}$ (c) $I = \begin{pmatrix} 1 & 0 \\ 0 & 1 \end{pmatrix}$

(d) $E = \begin{pmatrix} 4 & 2 \\ 2 & 1 \end{pmatrix}$

Solution (a) $|C| = (7)(9) - (2)(4) = 63 - 8 = 55$. Note that the determinant is always a single number.

(b) $|D| = (4)(-1) - (-2)(9) = -4 + 18 = 14$.

(c) $|I| = (1)(1) - (0)(0) = 1 - 0 = 1$. It is always true that the determinant of an identity matrix is 1.

(d) $|E| = (4)(1) - (2)(2) = 4 - 4 = 0$. In this example the determinant of the matrix E is zero.

As we have seen in the previous example, on some occasions a matrix may be such that $ad - bc = 0$, that is its determinant is zero. Such a matrix is called **singular**. When a matrix is singular it cannot have an inverse. This is because it is impossible to evaluate the quantity $1/(ad - bc)$, which appears in the formula for the inverse; the quantity $1/0$ has no meaning in mathematics.

WORKED EXAMPLE

27.11   Show that the matrix

$$P = \begin{pmatrix} 6 & -2 \\ -24 & 8 \end{pmatrix}$$

has no inverse.

Solution   We first find the determinant of P:

$$|P| = (6)(8) - (-2)(-24) = 48 - 48 = 0$$

The determinant is zero and so the matrix P is singular. It does not have an inverse.

Self-assessment questions 27.3

1. Explain what is meant by the inverse of a 2×2 matrix. Under what conditions will this inverse exist?

2. Explain what is meant by a singular matrix.

3. A diagonal matrix has the elements on its leading diagonal equal to 7 and -9. What is the value of its determinant?

Exercise 27.3

1. Evaluate the following determinants:

   (a) $\begin{vmatrix} 9 & 7 \\ -1 & -1 \end{vmatrix}$ (b) $\begin{vmatrix} 8 & 7 \\ 2 & -1 \end{vmatrix}$

   (c) $\begin{vmatrix} 0 & 1 \\ -1 & 0 \end{vmatrix}$ (d) $\begin{vmatrix} 8 & 0 \\ 0 & 11 \end{vmatrix}$

2. Which of the following matrices are singular?

   (a) $A = \begin{pmatrix} 9 & 5 \\ -2 & 0 \end{pmatrix}$

   (b) $B = \begin{pmatrix} 6 & 45 \\ -2 & 15 \end{pmatrix}$

   (c) $C = \begin{pmatrix} 6 & 45 \\ 2 & 15 \end{pmatrix}$

   (d) $D = \begin{pmatrix} 9 & 5 \\ 0 & 1 \end{pmatrix}$

   (e) $E = \begin{pmatrix} a & b \\ a & b \end{pmatrix}$

3. By multiplying the matrices

   $$\begin{pmatrix} 4 & -9 \\ -3 & 7 \end{pmatrix} \quad \text{and} \quad \begin{pmatrix} 7 & 9 \\ 3 & 4 \end{pmatrix}$$

   together show that each one is the inverse of the other.

4. Find, if it exists, the inverse of each of the following matrices:

(a) $\begin{pmatrix} 2 & 4 \\ 6 & 10 \end{pmatrix}$ (b) $\begin{pmatrix} 1 & 0 \\ 0 & 1 \end{pmatrix}$

(c) $\begin{pmatrix} -1 & 3 \\ 2 & 2 \end{pmatrix}$ (d) $\begin{pmatrix} 8 & 4 \\ 2 & 1 \end{pmatrix}$

5. Find the inverse of each of the following matrices:

(a) $\begin{pmatrix} 9 & 2 \\ 4 & 1 \end{pmatrix}$ (b) $\begin{pmatrix} 3 & 2 \\ 5 & 4 \end{pmatrix}$

(c) $\begin{pmatrix} -2 & 1 \\ 5 & -3 \end{pmatrix}$ (d) $\begin{pmatrix} 4 & -2 \\ 9 & 5 \end{pmatrix}$

(e) $\begin{pmatrix} -2 & -3 \\ -4 & -5 \end{pmatrix}$

27.4 Application of matrices to solving simultaneous equations

Matrices can be used to solve simultaneous equations. Suppose we wish to solve the simultaneous equations

$$3x + 2y = -3$$

$$5x + 3y = -4$$

First of all we rewrite them using matrices as

$$\begin{pmatrix} 3 & 2 \\ 5 & 3 \end{pmatrix} \begin{pmatrix} x \\ y \end{pmatrix} = \begin{pmatrix} -3 \\ -4 \end{pmatrix}$$

Writing

$$A = \begin{pmatrix} 3 & 2 \\ 5 & 3 \end{pmatrix}, \quad X = \begin{pmatrix} x \\ y \end{pmatrix} \quad \text{and} \quad B = \begin{pmatrix} -3 \\ -4 \end{pmatrix}$$

we have $AX = B$. Note that we are trying to find x and y; in other words we must find X. Now, if

$$AX = B$$

then, provided A^{-1} exists, we multiply both sides by A^{-1} to obtain

$$A^{-1}AX = A^{-1}B$$

But $A^{-1}A = I$, so this becomes

$$IX = A^{-1}B$$

However, multiplying X by the identity matrix leaves X unaltered, so we can write

$$X = A^{-1}B$$

In other words if we multiply B by the inverse of A we will have X as required. Now, given

$$A = \begin{pmatrix} 3 & 2 \\ 5 & 3 \end{pmatrix}$$

then

$$A^{-1} = \frac{1}{(3)(3) - (2)(5)} \begin{pmatrix} 3 & -2 \\ -5 & 3 \end{pmatrix} = \frac{1}{-1} \begin{pmatrix} 3 & -2 \\ -5 & 3 \end{pmatrix} = \begin{pmatrix} -3 & 2 \\ 5 & -3 \end{pmatrix}$$

Finally,

$$X = A^{-1}B = \begin{pmatrix} -3 & 2 \\ 5 & -3 \end{pmatrix} \begin{pmatrix} -3 \\ -4 \end{pmatrix} = \begin{pmatrix} 1 \\ -3 \end{pmatrix}$$

Therefore the solution of the simultaneous equations is $x = 1$ and $y = -3$. We note that the crucial step is

Key point

$AX = B$

$X = A^{-1}B$ provided A^{-1} exists

WORKED EXAMPLE

27.12   Solve the simultaneous equations

$$x + 2y = 13$$

$$2x - 5y = 8$$

Solution   First of all we rewrite the equations using matrices as

$$\begin{pmatrix} 1 & 2 \\ 2 & -5 \end{pmatrix} \begin{pmatrix} x \\ y \end{pmatrix} = \begin{pmatrix} 13 \\ 8 \end{pmatrix}$$

Writing

$$A = \begin{pmatrix} 1 & 2 \\ 2 & -5 \end{pmatrix}, \quad X = \begin{pmatrix} x \\ y \end{pmatrix} \quad \text{and} \quad B = \begin{pmatrix} 13 \\ 8 \end{pmatrix}$$

we have $AX = B$. We must now find the matrix X. If $AX = B$ then multiplying both sides by A^{-1} gives $A^{-1}AX = A^{-1}B$. But $A^{-1}A = I$, so this becomes $IX = A^{-1}B$. It follows that

$$X = A^{-1}B$$

We must multiply B by the inverse of A. The inverse of A is

$$A^{-1} = \frac{1}{(1)(-5) - (2)(2)} \begin{pmatrix} -5 & -2 \\ -2 & 1 \end{pmatrix} = \frac{1}{-9} \begin{pmatrix} -5 & -2 \\ -2 & 1 \end{pmatrix}$$

Finally,

$$X = A^{-1}B = \frac{1}{-9}\begin{pmatrix} -5 & -2 \\ -2 & 1 \end{pmatrix}\begin{pmatrix} 13 \\ 8 \end{pmatrix} = -\frac{1}{9}\begin{pmatrix} -81 \\ -18 \end{pmatrix} = \begin{pmatrix} 9 \\ 2 \end{pmatrix}$$

Therefore the solution of the simultaneous equations is $x = 9$ and $y = 2$.

Self-assessment question 27.4

1. Explain how the simultaneous equations $ax + by = c$, $dx + ey = f$, where a, b, c, d, e and f are given numbers and x and y are the unknowns, can be written using matrices.

Exercise 27.4

1. Use matrices to solve the following simultaneous equations:
   (a) $x + y = 17$, $2x - y = 10$ (b) $2x - y = 10$, $x + 3y = -2$
   (c) $x + 2y = 15$, $3x - y = 10$

2. Try to solve the simultaneous equations $2x - 3y = 5$, $6x - 9y = 15$ using the matrix method. What do you find? Can you explain this?

3. Use matrices to solve each of the following simultaneous equations:

   (a) $3x - y = 3$ (b) $2x + 3y = 10$ (c) $x + \dfrac{y}{2} = 2$ (d) $\dfrac{x}{2} + 3y = 5$

   $\quad\ x + y = 5$ $\quad\ x - 2y = -9$ $\quad 4x - 3y = 18$ $\quad 2x + 5y = 6$

Test and assignment exercises 27

1. Express

$$\frac{1}{2}\begin{pmatrix} 6 & 6 \\ -18 & 4 \end{pmatrix} + 7\begin{pmatrix} 8 & 2 \\ 0 & 2 \end{pmatrix}$$

as a single matrix.

2. If

$$A = \begin{pmatrix} 7 & 7 & 1 \\ 2 & 0 & 9 \end{pmatrix} \quad \text{and} \quad B = \begin{pmatrix} 9 \\ 3 \\ 6 \end{pmatrix}$$

find, if possible, AB and BA.

3. Find the value of the following determinants:

   (a) $\begin{vmatrix} 9 & -7 \\ 8 & 8 \end{vmatrix}$ (b) $\begin{vmatrix} 9 & 0 \\ 0 & 7 \end{vmatrix}$ (c) $\begin{vmatrix} 0 & 11 \\ -7 & 0 \end{vmatrix}$ (d) $\begin{vmatrix} \alpha & \beta \\ \gamma & \delta \end{vmatrix}$

4. Use matrices to solve the following simultaneous equations:

   (a) $x + 2y = 15$, $x - 2y = -5$ (b) $7x + 8y = 56$, $2x - 2y = 16$
   (c) $2x - 3y = 1$, $x + 2y = -3$

5. Find the inverse of each of the following matrices, if it exists:

   (a) $\begin{pmatrix} -1 & 0 \\ 0 & 1 \end{pmatrix}$ (b) $\begin{pmatrix} 10 & 7 \\ 6 & 4 \end{pmatrix}$ (c) $\begin{pmatrix} \frac{1}{2} & 1 \\ 3 & 7 \end{pmatrix}$

6. Show that each of the following matrices has no inverse:

   (a) $\begin{pmatrix} 16 & 8 \\ 4 & 2 \end{pmatrix}$ (b) $\begin{pmatrix} a & b \\ a^2 & ab \end{pmatrix}$

7. If

   $$A = \begin{pmatrix} 7 & 5 \\ 2 & -3 \end{pmatrix} \quad \text{and} \quad B = \begin{pmatrix} -2 & 2 \\ 3 & 7 \end{pmatrix}$$

   find AB and BA.

8. Solve the simultaneous equations $2x + 2y = 4$, $7x + 5y = 8$ using the matrix method.

9. Find AB and BA when

   $$A = \begin{pmatrix} 1 & 2 & 1 \\ 3 & 4 & 0 \\ 0 & 1 & -1 \end{pmatrix} \quad \text{and} \quad B = \begin{pmatrix} 2 & 4 & 1 \\ 0 & 3 & 1 \\ 0 & 0 & 4 \end{pmatrix}$$

10. Given

    $$A = \begin{pmatrix} 1 & 2 & -1 \\ 0 & 3 & 1 \\ 4 & 2 & -2 \end{pmatrix} \quad B = \begin{pmatrix} 3 & -1 \\ 4 & 1 \\ 0 & 2 \end{pmatrix}$$

    find

    (a) AB (b) A^2

Tables and charts

Objectives: This chapter:

■ explains the distinction between discrete and continuous data

■ shows how raw data can be organised using a tally chart

■ explains what is meant by a frequency distribution and a relative frequency distribution

■ shows how data can be represented in the form of bar charts, pie charts, pictograms and histograms

---

28.1 Introduction to data

In the modern world, information from a wide range of sources is gathered, presented and interpreted. Most newspapers, television news programmes and documentaries contain vast numbers of facts and figures concerning almost every aspect of life, including environmental issues, the lengths of hospital waiting lists, crime statistics and economic performance indicators. Information such as this, which is gathered by carrying out surveys and doing research, is known as **data**.

Sometimes data must take on a value from a specific set of numbers, and no other values are possible. For example, the number of children in a family must be 0, 1, 2, 3, 4 and so on, and no intermediate values are allowed. It is impossible to have 3.32 children, say. Usually one allowable value differs from the next by a fixed amount. Such data is said to be **discrete**. Other examples of discrete data include:

■ the number of car thefts in a city in one week – this must be 0, 1, 2, 3 and so on. You cannot have three-and-a-half thefts.

■ shoe sizes – these can be ..., $3\frac{1}{2}$, 4, $4\frac{1}{2}$, 5, $5\frac{1}{2}$, 6, $6\frac{1}{2}$, 7 and so on. You cannot have a shoe size of 9.82.

Sometimes data can take on *any* value within a specified range. Such data is called **continuous**. The volume of liquid in a 1 litre jug can take any value from 0 to 1 litre. The lifespan of an electric light bulb could be any non-negative value.

Frequently data is presented in the form of tables and charts, the intention being to make the information readily understandable. When data is first collected, and before it is processed in any way, it is known as **raw data**. For example, in a test the mathematics marks out of 10 of a group of 30 students are

7	5	5	8	9	10	9	10	7	8	6	3	5	9	6
10	8	8	7	8	6	7	8	8	10	9	4	5	9	8

This is raw data. To try to make sense of this data it is helpful to find out how many students scored each particular mark. This can be done by means of a **tally chart**. To produce this we write down in a column all the possible marks, in order. We then go through the raw data and indicate the occurrence of each mark with a vertical line or tally like /. Every time a fifth tally is recorded this is shown by striking through the previous four, as in $\cancel{||||}$. This makes counting up the tallies particularly easy. The tallies for all the marks are shown in Table 28.1. The number of occurrences of each mark is called its **frequency**, and the frequencies can be found from the number of tallies. We see that the tally chart is a useful way of organising the raw data into a form that will enable us to answer questions and obtain useful information about it.

Table 28.1
Tally chart for
mathematics marks

Mark	Tally	Frequency
0		0
1		0
2		0
3	/	1
4	/	1
5	////	4
6	///	3
7	////	4
8	$\cancel{||||}$ ///	8
9	$\cancel{||||}$	5
10	////	4
Total		30

Nowadays much of the tedium of producing tally charts for very large amounts of data can be avoided using computer programs available specifically for this purpose.

Self-assessment questions 28.1

1. Explain the distinction between discrete data and continuous data. Give a new example of each.

2. Explain the purpose of a tally chart.

Exercise 28.1

1. State whether each of the following is an example of discrete data or continuous data:
   (a) the number of matches found in a matchbox
   (b) the percentage of sulphur dioxide found in an air sample taken above a city
   (c) the number of peas in a packet of frozen peas
   (d) the radius of a ball bearing
   (e) the weight of a baby
   (f) the number of students in a particular type of accommodation
   (g) the weight of a soil sample
   (h) the temperature of a patient in hospital
   (i) the number of telephone calls received by an answering machine during one day

2. Twenty-five people were asked to state their year of birth. The information given was

   1975   1975   1974   1974   1975   1976   1976   1974   1972   1973   1973   1970   1975
   1975   1973   1970   1975   1974   1975   1974   1970   1973   1974   1974   1975

   Produce a tally chart to organise this data and state the frequency of occurrence of each year of birth.

28.2 Frequency tables and distributions

Data is often presented in a **table**. For example, in a recent survey, 1000 students were asked to state the type of accommodation in which they lived during term time. The data gathered is given in Table 28.2. As we saw from the work in the previous section, the number of occurrences of each entry in the table is known as its **frequency**. So, for example, the frequency of students in private rented accommodation is 250. The table summarises the various frequencies and is known as a **frequency distribution**.

A **relative frequency distribution** is found by expressing each frequency as a proportion of the total frequency.

Table 28.2
Frequency distribution

Type of accommodation	Number of students
Hall of residence	675
Parental home	50
Private rented	250
Other	25
Total	1000

Key point

The relative frequency is found by dividing a frequency by the total frequency.

In this example, because the total frequency is 1000, the relative frequency is easy to calculate. The number of students living at their parental home is 50 out of a total of 1000. Therefore the relative frequency of this group of students is $\frac{50}{1000}$ or 0.050. Table 28.3 shows the relative frequency distribution of student accommodation. Note that the relative frequencies always sum to 1.

Table 28.3
Relative frequency
distribution

Type of accommodation	Number of students	Relative frequency
Hall of residence	675	0.675
Parental home	50	0.050
Private rented	250	0.250
Other	25	0.025
Total	1000	1.000

It is sometimes helpful to express the frequencies as percentages. This is done by multiplying each relative frequency by 100, which gives the results shown in Table 28.4. We see from this that 5% of students questioned live at their parental home.

Table 28.4
Frequencies expressed as
percentages

Type of accommodation	Number of students	Relative frequency $\times$ 100
Hall of residence	675	67.5%
Parental home	50	5.0%
Private rented	250	25.0%
Other	25	2.5%
Total	1000	100%

| Key point | A frequency can be expressed as a percentage by multiplying its relative frequency by 100. |

WORKED EXAMPLE

28.1   In 1988 there were 4320 areas throughout the world that were designated as national parks. Table 28.5 is a frequency distribution showing how these parks are distributed in different parts of the world.

(a)   Express this data as a relative frequency distribution and also in terms of percentages.
(b)   What percentage of the world's national parks are in Africa?

Table 28.5
The number of national parks throughout the world

Region	Number of national parks
Africa	486
N. America	587
S. America	315
Former Soviet Union	168
Asia	960
Europe	1032
Oceania	767
Antarctica	5
Total	4320

Solution

Table 28.6
The number of national parks throughout the world

Region	Number	Relative frequency	Rel. freq. × 100
Africa	486	0.1125	11.25%
N. America	587	0.1359	13.59%
S. America	315	0.0729	7.29%
Former Soviet Union	168	0.0389	3.89%
Asia	960	0.2222	22.22%
Europe	1032	0.2389	23.89%
Oceania	767	0.1775	17.75%
Antarctica	5	0.0012	0.12%
Total	4320	1.000	100%

(a) The relative frequencies are found by expressing each frequency as a proportion of the total. For example, Europe has 1032 national parks out of a total of 4320, and so its relative frequency is $\frac{1032}{4320} = 0.2389$. Figures for the other regions have been calculated and are shown in Table 28.6. To express a relative frequency as a percentage it is multiplied by 100. Thus we find that Europe has 23.89% of the world's national parks. The percentages for the other regions are also shown in Table 28.6.

(b) We note from Table 28.6 that 11.25% of the world's national parks are in Africa.

When dealing with large amounts of data it is usual to group the data into classes. For example, in an environmental experiment the weights in grams of 30 soil samples were measured using a balance. We could list all 30 weights here but such a long list of data is cumbersome. Instead, we have already grouped the data into several weight ranges or **classes**. Table 28.7 shows the number of samples in each class. Such a table is called a **grouped frequency distribution**.

Table 28.7
Grouped frequency
distribution

Weight range (grams)	Number of samples in that range
100–109	5
110–119	8
120–129	11
130–139	5
140–149	1
Total	30

A disadvantage of grouping the data in this way is that information about individual weights is lost. However, this disadvantage is usually outweighed by having a more compact set of data that is easier to work with. There are five samples that lie in the first class, eight samples in the second and so on. For the first class, the numbers 100 and 109 are called **class limits**, the smaller number being the **lower class limit** and the larger being the **upper class limit**. Theoretically, the class 100–109 includes all weights from 99.5 g up to but not including 109.5 g. These numbers are called the **class boundaries**. In practice the class boundaries are found by adding the upper class limit of one class to the lower limit of the next class and dividing by 2. The **class width** is the difference between the larger class boundary and the smaller. The class 100–109 has width $109.5 - 99.5 = 10$. We do not know the actual values of the weights in each class. Should we require an estimate, the best we can do is use the value in the middle of the class, known as the class **midpoint**. The midpoint of the first class is 104.5, the midpoint of the second class is 114.5 and so on. The midpoint can be found by adding half the class width to the lower class boundary.

WORKED EXAMPLE

28.2   The systolic blood pressure in millimetres of mercury of 20 workers was recorded to the nearest millimetre. The data collected was as follows:

121   123   124   129   130   119   129   124   119   121   122   124
124   128   129   136   120   119   121   136

(a) Group this data into classes 115–119, 120–124, and so on.
(b) State the class limits of the first three classes.
(c) State the class boundaries of the first three classes.
(d) What is the class width of the third class?
(e) What is the midpoint of the third class?

Solution   (a)   To group the data a tally chart is used as shown in Table 28.8.

Table 28.8
Tally chart for blood pressures

Pressure (mm mercury)	Tally	Frequency
115–119	///	3
120–124	ℍℍ ℍℍ	10
125–129	////	4
130–134	/	1
135–139	//	2
Total		20

(b)   The class limits of the first class are 115 and 119; those of the second class are 120 and 124; those of the third are 125 and 129.

(c)   Theoretically, the class 115–119 will contain all blood pressures between 114.5 and 119.5 and so the class boundaries of the first class are 114.5 and 119.5. The class boundaries of the second class are 119.5 and 124.5, and those of the third class are 124.5 and 129.5.

(d)   The upper class boundary of the third class is 129.5. The lower class boundary is 124.5. The difference between these values gives the class width, that is $129.5 - 124.5 = 5$.

(e)   The midpoint of the third class is found by adding half the class width to the lower class boundary, that is $2.5 + 124.5 = 127$.

Self-assessment questions 28.2

1. Why is it often useful to present data in the form of a *grouped* frequency distribution?

2. Explain the distinction between class boundaries and class limits.

3. How is the class width calculated?

4. How is the class midpoint calculated?

Exercise 28.2

MyMathLab

1. The age of each patient over the age of 14 visiting a doctor's practice in one day was recorded as follows:

    18 18 76 15 72 45 48 62 21
    27 45 43 28 19 17 37 35 34
    23 25 46 56 32 18 23 34 32
    56 29 43

    (a) Use a tally chart to produce a grouped frequency distribution with age groupings 15–19, 20–24, 25–29, and so on.

    (b) What is the relative frequency of patients in the age group 70–74?

    (c) What is the relative frequency of patients aged 70 and over?

    (d) Express the number of patients in the age group 25–29 as a percentage.

2. Consider the following table, which shows the lifetimes, to the nearest hour, of 100 energy saving light bulbs.

    (a) If a light bulb has a life of 5499.4 hours into which class will it be put?

    (b) If a light bulb has a life of 5499.8 hours into which class will it be put?

    (c) State the class boundaries of each class.

    (d) Find the class width.

    (e) State the class midpoint of each class.

Lifetime (hours)	Frequency
5000–5499	6
5500–5999	32
6000–6499	58
6500–6999	4
Total	100

3. The percentages of 60 students in an Information Technology test are given as follows:

    45 92 81 76 51 46 82 65 61
    19 62 58 72 65 66 97 61 57
    63 93 61 46 47 61 56 45 39
    47 55 58 81 71 52 38 53 59
    82 92 87 86 51 29 19 79 55
    18 53 29 87 87 77 85 67 89
    17 29 86 57 59 57

    Form a frequency distribution by drawing up a tally chart using class intervals 0–9, 10–19 and so on.

4. The data in the table opposite gives the radius of 20 ball bearings. State the class boundaries of the second class and find the class width.

Radius (mm)	Frequency
20.56–20.58	3
20.59–20.61	6
20.62–20.64	8
20.65–20.67	3

28.3 Bar charts, pie charts, pictograms and histograms

In a **bar chart** information is represented by rectangles or bars. The bars may be drawn horizontally or vertically. The length of each bar corresponds to a frequency.

The data given earlier in Table 28.2 concerning student accommodation is presented in the form of a horizontal bar chart in Figure 28.1. Note that the scale on the horizontal axis must be uniform, that is it must be evenly spaced, and that the type of accommodation is clearly identified on each bar. The bar chart must be given a title to explain the information that is being presented.

Figure 28.1
Horizontal bar chart showing student accommodation

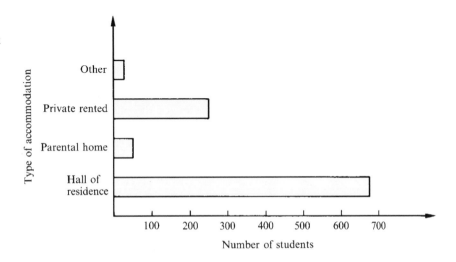

28.3   A reproduction antique furniture manufacturer has been producing pine wardrobes for the past six years. The number of wardrobes sold each year is given in Table 28.9. Produce a vertical bar chart to illustrate this information.

Table 28.9
Wardrobes sold

Year	Number sold
2003	15
2004	16
2005	20
2006	20
2007	24
2008	30

Solution   The information is presented in the form of a vertical bar chart in Figure 28.2.

Figure 28.2
Vertical bar chart showing the number of wardrobes sold each year

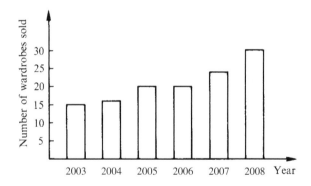

In a **pie chart**, a circular 'pie' is divided into a number of portions, with each portion, or **sector**, of the pie representing a different category. The whole pie represents all categories together. The size of a particular portion must represent the number in its category, and this is done by dividing the circle proportionately. Usually a protractor will be required.

28.4   A sample of 360 people were asked to state their favourite flavour of potato crisp: 75 preferred 'plain' crisps, 120 preferred 'cheese and onion',

90 preferred 'salt and vinegar', and the remaining 75 preferred 'beef'. Draw a pie chart to present this data.

Solution   All categories together, that is all 360 people asked, make up the whole pie. Because there are 360° in a circle this makes the pie chart particularly easy to draw. A sector of angle 75° will represent 'plain crisps', a sector of 120° will represent 'cheese and onion' and so on. The pie chart is shown in Figure 28.3.

Figure 28.3
Pie chart showing preferred flavour of crisps of 360 people

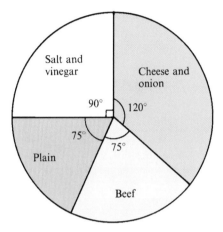

The pie chart in the previous worked example was simple to draw because the total number of all categories was 360, the same as the number of degrees in a circle. Let us now see how to deal with a situation where this is not the case. First each relative frequency is found and then this is multiplied by 360. The result is the angle, in degrees, of the corresponding sector.

Key point   The angle, in degrees, of each sector in a pie chart is found by multiplying the corresponding relative frequency by 360. This gives the angle of each sector.

WORKED EXAMPLE

28.5   Five hundred students were asked to state how they usually travelled to and from their place of study. The results are given in Table 28.10. Present this information in a pie chart.

Solution   To find the angle of each sector of the pie chart we first find the corresponding relative frequencies. Recall that this is done by dividing each frequency by the total, 500. Then each relative frequency is multiplied by 360. The results are shown in Table 28.11. The table gives the required

Table 28.10
Means of travel

Means of travel	Frequency f
Bus	50
Walk	180
Cycle	200
Car	40
Other	30
Total	500

Table 28.11
Means of travel

Means of travel	Frequency f	Rel. freq. $f/500$	Rel. freq. $\times 360$
Bus	50	$\frac{50}{500}$	$\frac{50}{500} \times 360 = 36.0°$
Walk	180	$\frac{180}{500}$	$\frac{180}{500} \times 360 = 129.6°$
Cycle	200	$\frac{200}{500}$	$\frac{200}{500} \times 360 = 144.0°$
Car	40	$\frac{40}{500}$	$\frac{40}{500} \times 360 = 28.8°$
Other	30	$\frac{30}{500}$	$\frac{30}{500} \times 360 = 21.6°$
Total	500		$360°$

Figure 28.4
Pie chart showing how
students travel to work

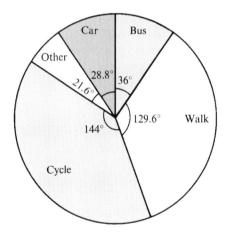

angles. Using a protractor the pie chart can then be drawn. This is shown in Figure 28.4.

A **pictogram** uses a picture to represent data in an eye-catching way. There are many ways to produce pictograms but generally the number of objects drawn represents the number of items in a particular category. This is a common form of representation popular on television when the viewer sees the information for only a few seconds.

WORKED EXAMPLE

28.6   The number of hours of sunshine in the month of August 1975 in four popular holiday resorts is given in Table 28.12. Show this information using a pictogram. Use one 'sun' to represent 100 hours of sunshine.

Table 28.12
Hours of sunshine in August 1975 in four resorts

Resort	Number of hours
Torbay	300
Scarborough	275
Blackpool	350
Brighton	325

Solution   The pictogram is shown in Figure 28.5.

Figure 28.5
Pictogram showing hours of sunshine in August 1975

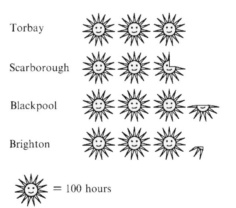

A **histogram** is often used to illustrate data that is continuous or has been grouped. It is similar to a vertical bar chart in that it is drawn by constructing vertical rectangles. However, in a histogram it is the *area* of each rectangle, and not its length, that is proportional to the frequency of each class. When all class widths are the same this point is not important and you can assume that the length of a rectangle represents the frequency. Class boundaries and not class limits must be used on the horizontal axis to distinguish one class from the next. Consider the following example.

28.7   The heights, to the nearest centimetre, of 100 students are given in Table 28.13.

Table 28.13
Height of 100 students

Height (cm)	Frequency
164–165	4
166–167	8
168–169	10
170–171	27
172–173	30
174–175	10
176–177	6
178–179	5
Total	100

(a) State the class limits of the second class.

(b) Identify the class boundaries of the first class and the last class.

(c) In which class would a student whose actual height is 167.4 cm be placed?

(d) In which class would a student whose actual height is 167.5 cm be placed?

(e) Draw a histogram to depict this data.

Solution   (a) The lower class limit of the second class is 166 cm. The upper class limit is 167 cm.

(b) The class boundaries of the first class are 163.5 cm and 165.5 cm. The class boundaries of the last class are 177.5 cm and 179.5 cm.

(c) A student with height 167.4 cm will have his height recorded to the nearest centimetre as 167. He will be placed in the class 166–167.

(d) A student with height 167.5 cm will have her height recorded to the nearest centimetre as 168. She will be placed in the class 168–169.

(e) The histogram is shown in Figure 28.6. Note that the class boundaries are used on the horizontal axis to distinguish one class from the next. The area of each rectangle is proportional to the frequency of each class.

Figure 28.6
Histogram showing the
distribution of heights of
100 students

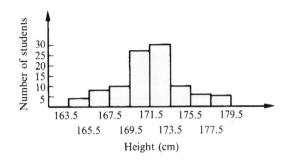

Self-assessment questions 28.3

1. There is an important distinction between a histogram and a vertical bar chart. Can you explain this?

2. Give one advantage and one disadvantage of using a pictogram to represent data.

3. On a histogram the class boundaries are used on the horizontal axis to distinguish one class from the next. Why do you think class boundaries as opposed to class limits are used?

Exercise 28.3

1. A local council wants to produce a leaflet explaining how the council tax is spent. For every one pound spent, 40p went to Education, 15p went to the Police, 15p went to Cleansing and Refuse, 20p went to Highways and the remaining 10p went to a variety of other services. Show this information (a) using a pie chart, and (b) using a pictogram.

2. The composition of the atmosphere of the Earth is 78% nitrogen, 21% oxygen and 1% other gases. Draw a pie chart to show this information.

3. An insurance company keeps records on the distribution of the amount of small claims for theft it receives. Amounts are given to the nearest pound. This information is given in the following table.

Amount (£)	Frequency
0–99	5
100–199	8
200–299	16
300–399	24
400–499	26

(a) State the class boundaries.
(b) In which class would a claim for £199.75 be placed?
(c) Draw a histogram to show this information.

1. State whether the following are examples of discrete or continuous data:

   (a) the temperature of a furnace
   (b) the number of goals scored in a football match
   (c) the amount of money in a worker's wage packet at the end of the week

2. A group of students were interviewed to obtain information on their termly expenditure. On average, for every pound spent, 60p was spent on accommodation, fuel etc., 25p was spent on food, 10p was spent on entertainment and the remainder on a variety of other items. Draw a pie chart to show this expenditure.

3. Forty lecturers were asked to state which newspaper they bought most regularly. The results were: *Guardian*, 15; *Independent*, 8; *The Times*, 2; *Daily Express*, 7; *Daily Mail*, 2; None, 6. Produce a pie chart to show this data.

4. The percentage of total body weight formed by various parts of the body of a newborn baby is as follows: muscles, 26%; skeleton, 17%; skin, 20%; heart, lungs etc., 11%; remainder, 26%. Depict this information (a) on a pie chart, (b) using a bar chart.

5. The following marks were obtained by 27 students in a test. Draw up a frequency distribution by means of a tally chart.

   5 4 6 2 9 1 10 10 5 5 6 5 5 9 8 7 9 6 7 6 8 7 6 9 8 8 7

6. The number of cars sold by a major manufacturer of luxury cars in the last 10 years is given in the table below. Show this data on a horizontal bar chart.

Year	Number
1999	2000
2000	2320
2001	2380
2002	2600
2003	2650
2004	2750
2005	2750
2006	2800
2007	1500
2008	1425

7. The table below shows the frequency distribution of the lifetimes in hours of 100 light bulbs tested.

   (a) State the class limits of the first and second classes.

(b) State the class boundaries of all classes.
(c) Determine each class width and midpoint.
(d) Draw a histogram for this frequency distribution.

Lifetime (hours)	Frequency
600–699	13
700–799	25
800–899	32
900–999	10
1000–1099	15
1100–1199	5

8. In each year from 1988 to 1993 the numbers of visitors to a museum were respectively 12000, 13000, 14500, 18000, 18000 and 17500. Show this information in a pictogram.

Statistics

29

Objectives: This chapter:

- introduces the three common averages – mean, median and mode – and shows how to calculate them
- explains what is meant by the variance and standard deviation and shows how to calculate them

29.1 Introduction

In the previous chapter we outlined how data gathered from a range of sources can be represented. In order to get the most out of such data and be able to interpret it sensibly and reliably, techniques have been developed for its analysis. **Statistics** is the name given to this science. One important statistical quantity is an 'average', the purpose of which is to represent a whole set of data by a single number that in some way represents the set. This chapter explains three types of average and shows how to calculate them. Finally the 'variance' and 'standard deviation' are introduced. These are numbers that describe how widely the data is spread.

29.2 Averages: the mean, median and mode

When presented with a large amount of data it is often useful to ask 'is there a single number that typifies the data?' For example, following an examination taken by a large number of candidates, examiners may be dealing with hundreds of examination scripts. The marks on these will vary from those with very poor marks to those with very high or even top marks. In order to judge whether the group of students as a whole found the examination difficult, the examiner may be asked to provide a single value that gives a measure of how well the students have performed. Such a value is called an **average**. In statistics there are three important averages: the arithmetic mean, the median and the mode.

The arithmetic mean

The **arithmetic mean**, or simply the **mean**, of a set of values is found by adding up all the values and dividing the result by the total number of values in the set. It is given by the following formula:

Key point	$$\text{mean} = \frac{\text{sum of the values}}{\text{total number of values}}$$

WORKED EXAMPLES

29.1 Eight students sit a mathematics test and their marks out of 10 are 4, 6, 6, 7, 7, 7, 8 and 10. Find the mean mark.

Solution The sum of the marks is $4 + 6 + 6 + 7 + 7 + 7 + 8 + 10 = 55$. The total number of values is 8. Therefore,

$$\text{mean} = \frac{\text{sum of the values}}{\text{total number of values}} = \frac{55}{8} = 6.875$$

The examiner can quote 6.875 out of 10 as the 'average mark' of the group of students.

29.2 In a hospital a patient's body temperature is recorded every hour for six hours. Find the mean temperature over the six-hour period if the six temperatures, in °C, were

36.5   36.8   36.9   36.9   36.9   37.0

Solution To find the mean temperature the sum of all six values is found and the result is divided by 6. That is,

$$\text{mean} = \frac{36.5 + 36.8 + 36.9 + 36.9 + 36.9 + 37.0}{6}$$

$$= \frac{221.0}{6} = 36.83 \text{ °C}$$

In more advanced work we make use of a formula for the mean that requires knowledge of some special notation. Suppose we have n values and we call these x_1, x_2, x_3 and so on up to x_n. The mean of these values is given the symbol $\bar{x}$, pronounced 'x bar'. To calculate the mean we must add up these values and divide by n, that is

$$\text{mean} = \bar{x} = \frac{x_1 + x_2 + x_3 + \cdots + x_n}{n}$$

A notation is often used to shorten this formula. In mathematics, the Greek letter sigma, written $\sum$, stands for a sum. The sum $x_1 + x_2$ is written

$$\sum_{i=1}^{2} x_i$$

and the sum $x_1 + x_2 + x_3 + x_4$ is written

$$\sum_{i=1}^{4} x_i$$

Note that i runs through all integer values from 1 to n.

where the values above and below the sigma sign give the first and last values in the sum. Similarly $x_1 + x_2 + x_3 + \cdots + x_n$ is written $\sum_{i=1}^{n} x_i$. Using this notation, the formula for the mean can be written in the following way:

Key point

$$\text{mean} = \bar{x} = \frac{\sum_{i=1}^{n} x_i}{n}$$

There is a further treatment of sigma notation in Chapter 12, §12.5.

WORKED EXAMPLES

29.3   Express the following in sigma notation:

(a) $x_1 + x_2 + x_3 + \cdots + x_8 + x_9$ (b) $x_{10} + x_{11} + \cdots + x_{100}$

Solution   (a) $x_1 + x_2 + x_3 + \cdots + x_8 + x_9 = \sum_{i=1}^{9} x_i$

(b) $x_{10} + x_{11} + \cdots + x_{100} = \sum_{i=10}^{100} x_i$

29.4   Find the mean of the values $x_1 = 5$, $x_2 = 7$, $x_3 = 13$, $x_4 = 21$ and $x_5 = 29$.

Solution   The number of values equals 5, so we let $n = 5$. The sum of the values is

$$\sum_{i=1}^{5} x_i = x_1 + x_2 + x_3 + x_4 + x_5 = 5 + 7 + 13 + 21 + 29 = 75$$

The mean is

$$\text{mean} = \bar{x} = \frac{\sum_{i=1}^{5} x_i}{n} = \frac{75}{5} = 15$$

When the data is presented in the form of a frequency distribution the mean is found by first multiplying each data value by its frequency. The results are added. This is equivalent to adding up all the data values. The mean is found by dividing this sum by the sum of all the frequencies. Note that the sum of the frequencies is equal to the total number of values. Consider the following example.

WORKED EXAMPLE

29.5   Thirty-eight students sit a mathematics test and their marks out of 10 are shown in Table 29.1. Find the mean mark.

Solution   Each data value, in this case the mark, is multiplied by its frequency, and the results are added. This is equivalent to adding up all the 38 individual marks. This is shown in Table 29.2.

Table 29.1
Marks of 38 students in a test

Table 29.2
Marks of 38 students multiplied by frequency

Mark, m	Frequency, f	Mark, m	Frequency, f	$m \times f$
0	0	0	0	0
1	0	1	0	0
2	1	2	1	2
3	0	3	0	0
4	1	4	1	4
5	7	5	7	35
6	16	6	16	96
7	8	7	8	56
8	3	8	3	24
9	1	9	1	9
10	1	10	1	10
Total	38	Totals	38	236

Note that the sum of all the frequencies is equal to the number of students taking the test. The number 236 is equal to the sum of all the individual marks. The mean is found by dividing this sum by the sum of all the frequencies:

$$\text{mean} = \frac{236}{38} = 6.21$$

The mean mark is 6.21 out of 10.

Using the sigma notation the formula for the mean mark of a frequency distribution with N classes, where the frequency of value x_i is f_i, becomes

Key point

$$\text{mean} = \bar{x} = \frac{\sum_{i=1}^{N} f_i \times x_i}{\sum_{i=1}^{N} f_i}$$

Note that $\sum_{i=1}^{N} f_i = n$; that is, the sum of all the frequencies equals the total number of values.

When the data is in the form of a grouped distribution the class midpoint is used to calculate the mean. Consider the following example.

WORKED EXAMPLE

29.6 The heights, to the nearest centimetre, of 100 students are given in Table 29.3. Find the mean height.

Solution Because the actual heights of students in each class are not known we use the midpoint of the class as an estimate. The midpoint of the class 164–165 is 164.5. Other midpoints and the calculation of the mean are shown in Table 29.4.

Then

$$\text{mean} = \bar{x} = \frac{\sum_{i=1}^{N} f_i \times x_i}{\sum_{i=1}^{N} f_i} = \frac{17150}{100} = 171.5$$

The mean height is 171.5 cm.

Table 29.3
Heights of 100 students

Height (cm)	Frequency
164–165	4
166–167	8
168–169	10
170–171	27
172–173	30
174–175	10
176–177	6
178–179	5
Total	100

Table 29.4
Heights of 100 students with midpoints multiplied by frequency

Height (cm)	Frequency f_i	Midpoint x_i	$f_i \times x_i$
164–165	4	164.5	658.0
166–167	8	166.5	1332.0
168–169	10	168.5	1685.0
170–171	27	170.5	4603.5
172–173	30	172.5	5175.0
174–175	10	174.5	1745.0
176–177	6	176.5	1059.0
178–179	5	178.5	892.5
Total	100		17150.0

The median

A second average that also typifies a set of data is the **median**.

The **median** of a set of numbers is found by listing the numbers in ascending order and then selecting the value that lies halfway along the list.

WORKED EXAMPLE

29.7   Find the median of the numbers

1   2   6   7   9   11   11   11   14

Solution   The set of numbers is already given in order. The number halfway along the list is 9, because there are four numbers before it and four numbers after it in the list. Hence the median is 9.

When there is an even number of values, the median is found by taking the mean of the two middle values.

WORKED EXAMPLE

29.8   Find the median of the following salaries: £24,000, £12,000, £16,000, £22,000, £10,000 and £25,000.

Solution   The numbers are first arranged in order as £10,000, £12,000, £16,000, £22,000, £24,000 and £25,000. Because there is an even number of values there are two middle figures: £16,000 and £22,000. The mean of these is

$$\frac{16,000 + 22,000}{2} = 19,000$$

The median salary is therefore £19,000.

The mode

A third average is the **mode**.

The **mode** of a set of values is that value that occurs most often.

WORKED EXAMPLE

29.9   Find the mode of the set of numbers

1   1   4   4   5   6   8   8   8   9

Solution   The number that occurs most often is 8, which occurs three times. Therefore 8 is the mode. Usually a mode is quoted when we want to represent the most popular value in a set.

Sometimes a set of data may have more than one mode.

WORKED EXAMPLE

29.10   Find the mode of the set of numbers

20   20   21   21   21   48   48   49   49   49

Solution   In this example there is no single value that occurs most frequently. The number 21 occurs three times, but so does the number 49. This set has two modes. The data is said to be **bimodal**.

Self-assessment questions 29.2

1. State the three different types of average commonly used in statistical calculations.

2. In an annual report, an employer of a small firm claims that the median salary for the workforce is £18,500. However, over discussions in the canteen it is apparent that no worker earns this amount. Explain how this might have arisen.

Exercise 29.2

MyMathLab

1. Find the mean, median and mode of the following set of values: 2, 3, 5, 5, 5, 5, 8, 8, 9.

2. Find the mean of the set of numbers 1, 1, 1, 1, 1, 1, 256. Explain why the mean does not represent the data adequately. Which average might it have been more appropriate to use?

3. The marks of seven students in a test were 45%, 83%, 99%, 65%, 68%, 72% and 66%. Find the mean mark and the median mark.

4. Write out fully each of the following expressions:

   (a) $\sum_{i=1}^{4} x_i$ (b) $\sum_{i=1}^{7} x_i$

   (c) $\sum_{i=1}^{3} (x_i - 3)^2$ (d) $\sum_{i=1}^{4} (2 - x_i)^2$

5. Write the following concisely using sigma notation:

   (a) $x_3 + x_4 + x_5 + x_6$
   (b) $(x_1 - 1) + (x_2 - 1) + (x_3 - 1)$

6. Calculate the mean, median and mode of the following numbers: 1.00, 1.15, 1.25, 1.38, 1.39 and 1.40.

7. The prices of the eight executive homes advertised by a local estate agent are

£290,000 £375,000 £325,000 £299,950
£319,950 £327,500 £299,500 £329,500

Find the mean price of these houses.

8. The data in the table gives the radius of 20 ball bearings. Find the class midpoints and hence calculate the average radius.

Radius (mm)	Frequency
20.56–20.58	3
20.59–20.61	6
20.62–20.64	8
20.65–20.67	3

29.3 The variance and standard deviation

Suppose we consider the test marks of two groups of three students. Suppose that the marks out of 10 for the first group are

4   7   and   10

while those of the second group are

7   7   and   7

It is easy to calculate the mean mark of each group: the first group has mean mark

$$\frac{4+7+10}{3} = \frac{21}{3} = 7$$

The second group has mean mark

$$\frac{7+7+7}{3} = \frac{21}{3} = 7$$

We see that both groups have the same mean mark even though the marks in the first group are widely spread, whereas the marks in the second group are all the same. If the teacher quotes just the mean mark of each group this gives no information about how widely the marks are spread. The **variance** and **standard deviation** are important and widely used statistical quantities that contain this information. Most calculators are pre-programmed to calculate these quantities. Once you understand the processes involved, check to see if your calculator can be used to find the variance and standard deviation of a set of data.

Suppose we have n values x_1, x_2, x_3 up to x_n. Their mean is $\bar{x}$ given by

$$\bar{x} = \frac{\sum_{i=1}^{n} x_i}{n}$$

The variance is found from the following formula:

Key point

$$\text{variance} = \frac{\sum_{i=1}^{n} (x_i - \bar{x})^2}{n}$$

If you study this carefully you will see that to calculate the variance we must:

- calculate the mean value $\bar{x}$
- subtract the mean from each value in turn, that is find $x_i - \bar{x}$
- square each answer to get $(x_i - \bar{x})^2$
- add up all these squared quantities to get $\sum_{i=1}^{n} (x_i - \bar{x})^2$
- divide the result by n to get

$$\frac{\sum_{i=1}^{n} (x_i - \bar{x})^2}{n}$$

which is the variance

The standard deviation is found by taking the square root of the variance:

Key point

$$\text{standard deviation} = \sqrt{\frac{\sum_{i=1}^{n} (x_i - \bar{x})^2}{n}}$$

Let us calculate the variance and standard deviation of each of the two sets of marks 4, 7, 10 and 7, 7, 7. We have already noted that the mean of each set is 7. Each stage of the calculation of the variance of the set 4, 7, 10 is shown in Table 29.5. The mean is subtracted from each number, and the results are squared and then added. Note that when a negative number is squared the result is positive. The squares are added to give 18. In this

Table 29.5

x_i	$x_i - \bar{x}$	$(x_i - \bar{x})^2$
4	$4 - 7 = -3$	$(-3)^2 = 9$
7	$7 - 7 = 0$	$0^2 = 0$
10	$10 - 7 = 3$	$3^2 = 9$
Total		18

example, the number of values equals 3. So, taking $n = 3$,

$$\text{variance} = \frac{\sum_{i=1}^{n} (x_i - \bar{x})^2}{n} = \frac{18}{3} = 6$$

The standard deviation is the square root of the variance, that is $\sqrt{6} = 2.449$. Similarly, calculation of the variance and standard deviation of the set 7, 7, 7 is shown in Table 29.6. Again, n equals 3. So

$$\text{variance} = \frac{\sum_{i=1}^{n} (x_i - \bar{x})^2}{n} = \frac{0}{3} = 0$$

Table 29.6

x_i	$x_i - \bar{x}$	$(x_i - \bar{x})^2$
7	$7 - 7 = 0$	$0^2 = 0$
7	$7 - 7 = 0$	$0^2 = 0$
7	$7 - 7 = 0$	$0^2 = 0$
Total		0

The standard deviation is the square root of the variance and so it also equals zero. The fact that the standard deviation is zero reflects that there is no spread of values. By comparison it is easy to check that the standard deviation of the set 6, 7 and 8, which also has mean 7, is equal to 0.816. The fact that the set 4, 7 and 10 has standard deviation 2.449 shows that the values are more widely spread than in the set 6, 7 and 8.

WORKED EXAMPLE

29.11 Find the variance and standard deviation of 10, 15.8, 19.2 and 8.7.

Solution First the mean is found:

$$\bar{x} = \frac{10 + 15.8 + 19.2 + 8.7}{4} = \frac{53.7}{4} = 13.425$$

The calculation to find the variance is given in Table 29.7.

$$\text{variance} = \frac{\sum_{i=1}^{n} (x_i - \bar{x})^2}{n} = \frac{73.049}{4} = 18.262$$

The standard deviation is the square root of the variance:

$$\sqrt{18.262} = 4.273$$

Table 29.7

x_i	$x_i - \bar{x}$	$(x_i - \bar{x})^2$
10	$10 - 13.425 = -3.425$	$(-3.425)^2 = 11.731$
15.8	$15.8 - 13.425 = 2.375$	$2.375^2 = 5.641$
19.2	$19.2 - 13.425 = 5.775$	$5.775^2 = 33.351$
8.7	$8.7 - 13.425 = -4.725$	$(-4.725)^2 = 22.326$
Total		73.049

When dealing with a grouped frequency distribution with N classes the formula for the variance becomes:

Key point

$$\text{variance} = \frac{\sum_{i=1}^{N} f_i (x_i - \bar{x})^2}{\sum_{i=1}^{N} f_i}$$

As before, the standard deviation is the square root of the variance.

WORKED EXAMPLE

29.12 In a period of 30 consecutive days in July the temperature in °C was recorded as follows:

```
18   19   20   23   24   24   21   18   17   16
16   17   17   17   18   19   20   20   22   23
24   24   25   23   21   21   20   19   19   18
```

(a) Produce a grouped frequency distribution showing data grouped from 16 to 17 °C, 18 to 19 °C and so on.
(b) Find the mean temperature of the grouped data.
(c) Find the standard deviation of the grouped data.

Solution (a) The data is grouped using a tally chart as in Table 29.8.

Table 29.8

Temperature range (°C)	Tally	Frequency
16–17	ЖҺ I	6
18–19	ЖҺ III	8
20–21	ЖҺ II	7
22–23	IIII	4
24–25	ЖҺ	5
Total		30

(b) When calculating the mean temperature from the grouped data we do not know the actual temperatures. We do know the frequency of each class. The best we can do is use the midpoint of each class as an estimate of the values in that class. The class midpoints and the calculation to determine the mean are shown in Table 29.9. The mean is then

$$\bar{x} = \frac{\sum_{i=1}^{N} f_i \times x_i}{\sum_{i=1}^{N} f_i} = \frac{603}{30} = 20.1 \ °C$$

Table 29.9

Temperature range (°C)	Frequency f_i	Class midpoint x_i	$f_i \times x_i$
16–17	6	16.5	99.0
18–19	8	18.5	148.0
20–21	7	20.5	143.5
22–23	4	22.5	90.0
24–25	5	24.5	122.5
Total	30		603

(c) To find the variance and hence the standard deviation we must subtract the mean, 20.1, from each value, square and then add the results. Finally this sum is divided by 30. The complete calculation is shown in Table 29.10.

Table 29.10

Temperature range (°C)	Frequency f_i	Class midpoint x_i	$x_i - \bar{x}$	$(x_i - \bar{x})^2$	$f_i(x_i - \bar{x})^2$
16–17	6	16.5	−3.60	12.96	77.76
18–19	8	18.5	−1.60	2.56	20.48
20–21	7	20.5	0.40	0.16	1.12
22–23	4	22.5	2.40	5.76	23.04
24–25	5	24.5	4.40	19.36	96.80
Total	30				219.20

Finally the variance is given by

$$\text{variance} = \frac{\sum_{i=1}^{N} f_i(x_i - \bar{x})^2}{\sum_{i=1}^{N} f_i} = \frac{219.20}{30} = 7.307$$

The standard deviation is the square root of the variance, that is $\sqrt{7.307} = 2.703$.

Self-assessment questions 29.3

1. Why is an average often an insufficient way of describing a set of values?

2. Describe the stages involved in calculating the variance and standard deviation of a set of values $x_1, x_2, \ldots, x_n$.

Exercise 29.3

1. Find the standard deviation of each of the following sets of numbers and comment upon your answers:
   (a) 20, 20, 20, 20   (b) 16, 17, 23, 24
   (c) 0, 20, 20, 40

2. Calculate the variance and standard deviation of the following set of numbers: 1, 2, 3, 3, 3, 4, 7, 8, 9, 10.

3. The examination results of two students, Jane and Tony, are shown in the table. Find the mean and standard deviation of each of them. Comment upon the results.

	Jane	Tony
Maths	50	29
Physics	42	41
Chemistry	69	60
French	34	48
Spanish	62	80

4. The marks of 50 students in a mathematics examination are given as

```
45  50  62  62  68  47  45  44  48
73  62  63  62  67  80  45  41  40
23  55  21  83  67  49  48  48  62
63  79  58  71  37  32  58  54  50
62  66  68  62  81  92  62  45  49
71  72  70  49  51
```

   (a) Produce a grouped frequency distribution using classes 25–29, 30–34, 35–39 and so on.
   (b) The modal class is the class with the highest frequency. State the modal class.
   (c) Calculate the mean, variance and standard deviation of the grouped frequency distribution.

Test and assignment exercises 29

1. Calculate the mean, median and mode of the following numbers: 2.7, 2.8, 3.1, 3.1, 3.1, 3.4, 3.8, 4.1.

2. Calculate the mean of
   (a) the first 10 whole numbers,

   $$1, 2, \ldots, 10$$

(b) the first 12 whole numbers,

$$1, 2, \ldots, 12$$

(c) the first n whole numbers,

$$1, 2, \ldots, n$$

3. A manufacturer of breakfast cereals uses two machines that pack 250 g packets of cereal automatically. In order to check the average weight, sample packets are taken and weighed to the nearest gram. The results of checking eight packets from each of the two machines are given in the table.

Machine

| 1 | 250 | 250 | 251 | 257 | 253 | 250 | 268 | 259 |
| 2 | 249 | 251 | 250 | 251 | 252 | 258 | 254 | 250 |

(a) Find the mean weight of each sample of eight packets.
(b) Find the standard deviation of each sample.
(c) Comment upon the performance of the two machines.

4. What is the median of the five numbers 18, 28, 39, 42, 43? If 50 is added as the sixth number what would the median become?

5. Express the following sums in sigma notation:

(a) $y_1 + y_2 + y_3 + \cdots + y_7 + y_8$
(b) $y_1^2 + y_2^2 + y_3^2 + \cdots + y_7^2 + y_8^2$
(c) $(y_1 - \bar{y})^2 + (y_2 - \bar{y})^2 + \cdots + (y_7 - \bar{y})^2 + (y_8 - \bar{y})^2$

Probability [03]

Objectives: This chapter:

- introduces theoretical and experimental probability and how to calculate them
- explains the meaning of the term 'complementary events'
- explains the meaning of the term 'independent events'

30.1 Introduction

Some events in life are impossible; other events are quite certain to happen. For example, it is impossible for a human being to live for 1000 years. We say that the probability of a human being living for 1000 years is 0. On the other hand it is quite certain that all of us will die someday. We say that the probability that all of us will die is 1. Using the letter P to stand for probability we can write

P(a human will live for 1000 years) $= 0$

and

P(all of us will die someday) $= 1$

Many other events in life are neither impossible nor certain. These have varying degrees of likelihood, or chance. For example, it is quite unlikely but not impossible that the British weather throughout a particular summer will be dominated by snowfall. It is quite likely but not certain that the life expectancy of the UK population will continue to rise in the foreseeable future. Events such as these can be assigned probabilities varying from 0 up to 1. Those having probabilities close to 1 are quite likely to happen. Those having probabilities close to 0 are almost impossible. Several events and their probabilities are shown on the

Figure 30.1
The probability scale

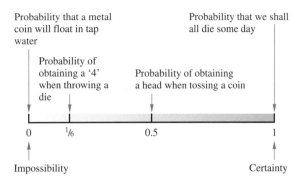

probability scale in Figure 30.1. It is important to note that no probability can lie outside the range 0 to 1.

All probabilities lie in the range [0, 1].

Complementary events

Consider the following situation. A light bulb is tested. Clearly it either works or it does not work. Here we have two events: the first is that the light bulb works and the second is that the light bulb does not work. When the bulb is tested, one or other of these events must occur. Furthermore, each event excludes the other. In such a situation we say the two events are **complementary**. In general, two events are complementary if one of them must happen and, when it does, the other event cannot. The sum of the probabilities of the two complementary events must always equal 1. This is known as **total probability**. We shall see how this result can be used to calculate probabilities shortly.

Self-assessment questions 30.1

1. All probabilities must lie in a certain range. What is this range?

2. Explain the meaning of the term 'complementary events'.

30.2 Calculating theoretical probabilities

Sometimes we have sufficient information about a set of circumstances to calculate the probability of an event occurring. For example, suppose we roll a die and ask what the probability is of obtaining a score of '5'. If the die is fair, or **unbiased**, the chance of getting a '5' is the same as the chance of getting any other score. You would expect to get a '5' one time in every six. That is,

the probability of throwing a '5' is $\frac{1}{6}$ or 0.167

In other words there is a one in six chance of scoring a '5'. The fact that the probability is closer to 0 than to 1 means that it is quite unlikely that a '5' will be thrown, although it is not impossible. Such a probability is known as a **theoretical probability** and, when all events are equally likely, it is calculated from the following formula:

Key point

When all events are equally likely

$$P\left(\begin{array}{c} \text{obtaining our} \\ \text{chosen event} \end{array}\right) = \frac{\text{number of ways the chosen event can occur}}{\text{total number of possibilities}}$$

For example, suppose we ask what the probability is of obtaining a score more than 4. A score more than 4 can occur in two ways, by scoring a '5' or a '6'. If the die is fair, all possible events are equally likely. The total number of possibilities is 6, therefore,

$$P(\text{score more than 4}) = \frac{2}{6} = \frac{1}{3}$$

WORKED EXAMPLES

30.1   A fair die is thrown. What is the probability of obtaining an even score?

Solution   The chosen event is throwing an even score, that is a '2', '4' or '6'. There are therefore three ways that this chosen event can occur out of a total of six, equally likely, ways. So

$$P(\text{even score}) = \frac{3}{6} = \frac{1}{2}$$

30.2   A fair coin is tossed. What is the probability that it will land with its head uppermost?

Solution   There are two equally likely ways the coin can land: head uppermost or tail uppermost. The chosen event, that the coin lands with its head uppermost, is just one of these ways. The chance of getting a head is the same as the chance of getting a tail. Therefore,

$$P(\text{head}) = \frac{1}{2}$$

Note from the previous example that $P(\text{tail}) = \frac{1}{2}$ and that also

$$P(\text{tail}) + P(\text{head}) = 1$$

Note that the two events, getting a head and getting a tail, are complementary because one of them must happen and either one excludes the other. Therefore the sum of the two probabilities, the total probability, equals 1.

WORKED EXAMPLE

30.3   Two coins are tossed.

(a) Write down all the possible outcomes.

(b) What is the probability of obtaining two tails?

Solution   (a) Letting H stand for head and T for tail, the possible outcomes are

$$H, H \qquad H, T \qquad T, H \qquad T, T$$

There are four possible outcomes, each one equally likely to occur.

(b) Obtaining two tails is just one of the four possible outcomes. Therefore, the probability of obtaining two tails is $\frac{1}{4}$.

Exercise 30.2

MyMathLab Global

1. A die is thrown. Find
    (a) the probability of obtaining a score less than 6
    (b) the probability of obtaining a score more than 6
    (c) the probability of obtaining an even score less than 5
    (d) the probability of obtaining an even score less than 2

2. There are four Aces in a pack of 52 playing cards. What is the probability that a card selected at random is not an Ace?

3. A drawer contains six red socks, six black socks and eight blue socks. Find the probability that a sock selected at random from the drawer is
    (a) black   (b) red   (c) red or blue

4. Two dice are thrown together and their scores are added together. By considering all the possible outcomes, find the probability that the total score will be
    (a) 12   (b) 0   (c) 1   (d) 2   (e) more than 5

5. A basket contains 87 good apples and three bad ones. What is the probability that an apple chosen at random is bad?

6. A box contains 16 red blocks, 20 blue blocks, 24 orange blocks and 10 black blocks. A block is picked at random. Calculate the probability that the block is
    (a) black   (b) orange   (c) blue
    (d) red or blue   (e) red or blue or orange
    (f) not orange

7. Three coins are tossed. By considering all possible outcomes calculate the probability of obtaining
    (a) two heads and one tail
    (b) at least two heads
    (c) no heads

30.3 Calculating experimental probabilities

In some circumstances we do not have sufficient information to calculate a theoretical probability. We know that if a coin is unbiased the probability of obtaining a head is $\frac{1}{2}$. But suppose the coin is biased so that it is more likely to land with its tail uppermost. We can experiment by tossing the coin a large number of times and counting the number of tails obtained. Suppose we toss the coin 100 times and obtain 65 tails. We can then estimate the probability of obtaining a tail as $\frac{65}{100}$ or 0.65. Such a probability is known as an **experimental probability** and is accurate only if a very large number of experiments are performed. Generally, we can calculate an experimental probability from the following formula:

Key point

$$P\left(\begin{array}{c}\text{chosen event}\\ \text{occurs}\end{array}\right) = \frac{\text{number of ways the chosen event occurs}}{\text{total number of times the experiment is repeated}}$$

WORKED EXAMPLES

30.4 A biased die is thrown 1000 times and a score of '6' is obtained on 200 occasions. If the die is now thrown again what is the probability of obtaining a score of '6'?

Solution Using the formula for the experimental probability we find

$$P(\text{throwing a '6'}) = \frac{200}{1000} = 0.2$$

If the die were unbiased the theoretical probability of throwing a '6' would be $\frac{1}{6} = 0.167$, so the die has been biased in favour of throwing a '6'.

30.5 A manufacturer produces microwave ovens. It is known from experience that the probability that a microwave oven is of an acceptable standard is 0.92. Find the probability that an oven selected at random is not of an acceptable standard.

Solution When an oven is tested either it is of an acceptable standard or it is not. An oven cannot be both acceptable and unacceptable. The two events, that the oven is acceptable or that the oven is unacceptable, are therefore complementary. Recall that the sum of the probabilities of complementary events is 1 and so

$$P(\text{oven is not acceptable}) = 1 - 0.92 = 0.08$$

Self-assessment questions 30.3

1. In what circumstances is it appropriate to use an experimental probability?

2. A series of experiments is performed three times. In the first of the series an experiment is carried out 100 times. In the second it is carried out 1000 times and in the third 10000 times. Which of the series of experiments is likely to lead to the best estimate of probability? Why?

Exercise 30.3

1. A new component is fitted to a washing machine. In a sample of 150 machines tested, 7 failed to function correctly. Calculate the probability that a machine fitted with the new component (a) works correctly, (b) does not work correctly.

2. In a sample containing 5000 nails manufactured by a company, 5% are too short or too long. A nail is picked at random from the production line. Estimate the probability that it is of the right length.

3. The probability that the Post Office delivers first-class mail on the following working day after posting is 0.96. If 93500 first-class letters are posted on Wednesday, how many are likely to be delivered on Thursday?

4. The probability that a car rescue service will reach a car in less than one hour is 0.87. If the rescue service is called out 17300 times in one day, calculate the number of cars reached in less than one hour.

5. Out of 50000 components tested, 48700 were found to be working well. A batch of 3000 components is delivered to a depot. How many are likely to be not working well?

30.4 Independent events

Two events are **independent** if the occurrence of either one in no way affects the occurrence of the other. For example, if an unbiased die is thrown twice the score on the second throw is in no way affected by the score on the first. The two scores are independent. The **multiplication law** for independent events states the following:

Key point

If events A and B are independent, then the probability of obtaining A and B is given by

$$P(A \text{ and } B) = P(A) \times P(B)$$

WORKED EXAMPLE

30.6   A die is thrown and a coin is tossed. What is the probability of obtaining a '6' and a head?

Solution   These events are independent since the score on the die in no way affects the result of tossing the coin, and vice versa. Therefore

$$P(\text{throwing a '6' and tossing a head}) = P(\text{throwing a '6'})$$
$$\times P(\text{tossing a head})$$

$$= \frac{1}{6} \times \frac{1}{2}$$

$$= \frac{1}{12}$$

When several events are independent of each other the multiplication law becomes

$$P(A \text{ and } B \text{ and } C \text{ and } D \dots) = P(A) \times P(B) \times P(C) \times P(D) \dots$$

WORKED EXAMPLE

30.7   A coin is tossed three times. What is the probability of obtaining three heads?

Solution   The three tosses are all independent events since the result of any one has no effect on the others. Therefore

$$P(3 \text{ heads}) = \frac{1}{2} \times \frac{1}{2} \times \frac{1}{2} = \frac{1}{8}$$

Self-assessment questions 30.4

1. Explain what is meant by saying two events are independent.

2. Suppose we have two packs each of 52 playing cards. A card is selected from each pack. Event A is that the card from the first pack is the Ace of Spades. Event B is that the card from the second pack is a Spade. Are these two events dependent or independent?

3. From a single pack of cards, two are removed. The first is examined. Suppose event A is that the first card is the Ace of Spades. The second card is examined. Event B is that the second card is a Spade. Are the two events dependent or independent?

Exercise 30.4

1. A die is thrown and a coin is tossed. What is the probability of getting an even score on the die and a tail?

2. Suppose you have two packs each of 52 playing cards. A card is drawn from the first and a card is drawn from the second. What is the probability that both cards are the Ace of Spades?

3. A coin is tossed eight times. What is the probability of obtaining eight tails?

4. A die is thrown four times. What is the probability of obtaining four '1's?

5. Suppose that there is an equal chance of a mother giving birth to a boy or a girl.
   (a) If one child is born find the probability that it is a boy.

   (b) If two children are born find the probability that they are both boys, assuming that the sex of neither one can influence the sex of the other.

6. The probability that a component is working well is 0.96. If four components are picked at random calculate the probability that
   (a) they all work well
   (b) none of them work well

7. The probability that a student passes a module is 0.91. If three modules are studied calculate the probability that the student passes
   (a) all three modules   (b) two modules
   (b) one module   (d) no modules

Test and assignment exercises 30

1. The following numbers are probabilities of a certain event happening. One of them is an error. Which one?
   (a) 0.5   (b) $\dfrac{3}{4}$ (c) 0.001   (d) $\dfrac{13}{4}$

2. Which of the following cannot be a probability?
   (a) 0.125   (b) −0.2   (c) 1   (d) 0

3. Events A, B and C are defined as follows. In each case state the complementary event.
   A: the lifespan of a light bulb is greater than 1000 hours
   B: it will rain on Christmas Day 2020
   C: your car will be stolen sometime in the next 100 days

4. The probability of a component manufactured in a factory being defective is 0.01. If three components are selected at random what is the probability that they all work as required?

5. A parcel delivery company guarantees 'next-day' delivery for 99% of its parcels. Out of a sample of 1800 deliveries, 15 were delivered later than the following day. Is the company living up to its promises?

6. At a certain bus stop the probability of a bus arriving late is $\frac{3}{20}$. The probability of its arriving on time is $\frac{4}{5}$. Find
   (a) the probability that the bus arrives early
   (b) the probability that the bus does not arrive late

7. Suppose you have two packs each of 52 playing cards. A card is drawn from the first and a card is drawn from the second. What is the probability that
   (a) both cards are diamonds?
   (b) both cards are black?
   (c) both cards are Kings?

8. Find the probability of obtaining an odd number when throwing a fair die once.

9. A bag contains eight red beads, four white beads and five blue beads. A bead is drawn at random. Determine the probability that it is
   (a) red   (b) white   (c) black   (d) blue   (e) red or white   (f) not blue

10. Two cards are selected from a pack of 52. Find the probability that they are both Aces if the first card is
    (a) replaced and   (b) not replaced

11. A pack contains 20 cards, numbered 1 to 20. A card is picked at random. Calculate the probability that the number on the card is
    (a) even   (b) 16 or more   (c) divisible by 3

12. Out of 42300 components, 846 were defective. How many defective components would you expect in a batch of 1500?

13. A biased die has the following probabilities:
    $$P(1) = 0.1, \ P(2) = 0.15, \ P(3) = 0.1,$$
    $$P(4) = 0.2, \ P(5) = 0.2, \ P(6) = 0.25$$

    The die is thrown twice. Calculate the probability that
    (a) both scores are '6's
    (b) both scores are '1's
    (c) the first score is odd and the second is even
    (d) the total score is 10

Correlation

31

Objectives: This chapter:

- explains how to explore the association between two variables using a scatter diagram.
- explains what is meant by positive, negative and zero correlation
- introduces the product-moment correlation coefficient
- introduces Spearman's coefficient of rank correlation

31.1 Introduction

In the sciences, business, health studies, geography, and many other fields, we are often interested in exploring relationships that may exist between two variables. For example, might it be true that, in general, taller people have greater weight than shorter people? Do cars with larger engine sizes travel fewer miles using a gallon of petrol than those with smaller engine sizes? Do business studies students who are good at statistics perform well in another aspect of their course?

The answers to some of these questions can be obtained using the statistics of **correlation**. As a first step, a graphical representation of the data is prepared using a **scatter diagram**. This is a simple method which helps us quickly draw some conclusions about the data. When a more quantitative, or rigorous, approach is needed a quantity known as the **correlation coefficient** can be calculated. This associates a number with the relationship between the variables and enables us to describe the strength of the relationship, or correlation, as perfect, strong, weak, or non-existent. Under certain conditions we may be able to predict the value of one variable from knowledge of the other.

This chapter explains how to draw and interpret a scatter diagram, and how to calculate two types of correlation coefficient: the product-moment correlation coefficient, and Spearman's coefficient of rank correlation. Chapter 32 explains how we can use a technique known as **regression** to predict values of one variable when we know values of the other.

31.2 Scatter diagrams

A simple and practical way of exploring relationships between two variables is the scatter diagram. Consider the following worked example.

WORKED EXAMPLE

31.1   In a study of food production in a developing country the total crop of wheat (in units of 10000 tonnes) was measured over a period of five summers. The summer rainfall (in cm) was also recorded. The data is provided in Table 31.1. Use a scatter diagram to explore the relationship between rainfall and wheat crop.

Table 31.1

Year	1	2	3	4	5
Wheat crop (10000 tonnes)	42	38	48	51	45
Rainfall (cm)	20	19	30	34	24

To draw a scatter diagram we think of each data pair (that is, each pair of values of rainfall and wheat crop, such as 20 and 42) as the coordinates of a point.

We choose one of the variables to be plotted on the horizontal axis and one on the vertical axis. At this stage, when we are simply exploring whether a relationship may exist, it does not matter which. In Figure 31.1

Figure 31.1
Scatter diagram showing the association between rainfall and wheat crop

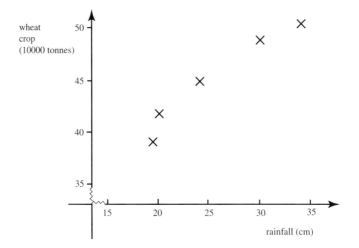

rainfall has been plotted on the horizontal axis and wheat crop on the vertical. Each point is represented by a ×.

The scatter diagram in Figure 31.1 shows that the points appear to cluster around an imaginary line which slopes upwards from left to right. There appears to be a **linear relationship** between the two variables.

The line around which the points cluster has been drawn in by eye in Figure 31.2, although this line is not part of the scatter diagram. This line, which is closest, in some sense, to all of the data points, is called **the line of best fit**. We shall see in Chapter 32 how the equation of this line can be calculated exactly. Note that this line has a positive slope, or gradient. (You may find it useful to refresh your knowledge by re-reading Chapter 18.) When the points cluster closely around a line with a positive gradient we are observing **strong positive correlation** between the two variables.

Figure 31.2
The scatter diagram of Worked Example 31.1 showing the line of best fit

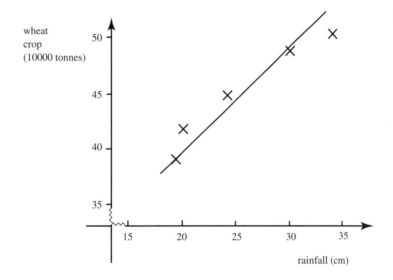

The scatter diagram suggests that there is indeed a strong (linear) relationship between the two variables with low wheat crop being associated with low rainfall and high wheat crop associated with high rainfall.

Scatter diagrams can take different forms. For example, the one in Figure 31.3(a) shows that the points cluster around a line with positive gradient but not so closely as in Figure 31.2. This is an example of weak, but positive correlation. There appears to be a relationship between the two variables but this is not as strong as in Figure 31.2.

Figure 31.3(b) shows that the points cluster closely around a line with a negative gradient. The variables here exhibit strong negative correlation. This means that high values of one variable are associated with low values of the other, and vice versa. In Figure 31.3(c) there is weak negative correlation. Finally, in Figure 31.3(d) the points do not lie on or close to a straight line at all. These points exhibit no correlation whatsoever.

When all points lie exactly on a straight line, whether with positive or negative gradient, the correlation is said to be perfect.

Key point

If X and Y are positively correlated then, on the whole, as X increases, Y increases.

If X and Y are negatively correlated then, on the whole, as X increases, Y decreases.

Figure 31.3
Scatter diagrams can illustrate different forms of association

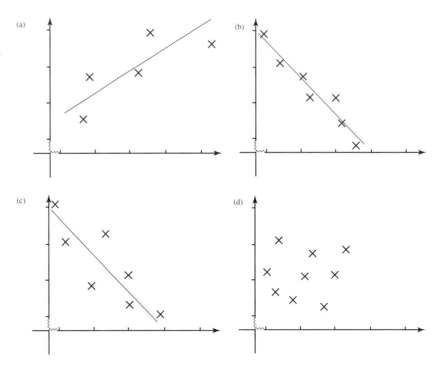

It is also possible that a relationship exists between two variables but that this relationship is not of a linear nature. For example, the scatter diagram in Figure 31.4 seems to indicate that the points lie on a well defined curve, but not on a straight line. Statistical techniques to explore non-linear relationships such as this are available but are beyond the scope of this book.

Figure 31.4
A non-linear association
between two variables

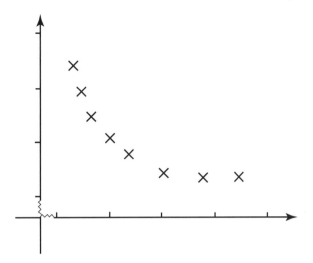

Self-assessment questions 31.2

1. Explain the purpose of a scatter diagram.

2. Explain what is meant by positive correlation, negative correlation and zero correlation.

3. Give an example of your own of two variables which might be positively correlated.

4. Give an example of your own of two variables which might be negatively correlated.

Exercise 31.2   MyMathLab Global

1. For each of the following sets of data plot a scatter diagram. If possible, draw conclusions about the nature of a possible relationship between the two variables.
   (a)

X	2	8	5	4	10
Y	9	6	8	7	4

   (b)

X	50	52	54	56	58	60
Y	42	48	36	50	37	46

2. A Human Resources department in a company gathers data on the age (in years) and salary (in £1000's) of a particular group of employees. The data is given in the table on the next page. Draw a scatter diagram for this data and comment upon the nature of the relationship between age and salary.

| age | 18 | 19 | 19 | 28 | 35 | 36 | 49 | 50 | 50 |
| salary | 12 | 13 | 14 | 18 | 20 | 21 | 30 | 38 | 21 |

31.3 Correlation coefficient

As we have seen, a scatter diagram can give us useful graphical information about the relationship between two variables. But when we require a more precise measure of a relationship, a useful quantity is the **correlation coefficient**, denoted by r. This is a number which measures the strength of the correlation. It lies between -1 and $+1$. When r takes the value 1 there is an exact straight line relationship with a positive gradient, and thus perfect positive correlation. When it takes the value -1 there is also a straight line relationship, but this time with a negative gradient. This is perfect negative correlation. When the value of r is 0 there is no correlation whatsoever. When r is close to 1 we have strong positive correlation. When r is close to -1 there is strong negative correlation.

Suppose we have gathered data in the form of n pairs of values of quantities X and Y. Suppose we denote the values of X by x_i and the values of Y by y_i where i takes values $1 \ldots n$. We can write the n pairs of values in the form (x_i, y_i). The **product-moment correlation coefficient**, r, is given by

$$r = \frac{\sum_{i=1}^{n}(x_i - \bar{x})(y_i - \bar{y})}{\sqrt{\sum_{i=1}^{n}(x_i - \bar{x})^2 \sum_{i=1}^{n}(y_i - \bar{y})^2}}$$

where $\bar{x}$ is the mean of the x values, and $\bar{y}$ is the mean of the y values. In practice, when calculating the correlation coefficient by hand some regard it as easier to work with an alternative, but equivalent, version of the formula. This version avoids prior calculation of the mean values of x and y and also the differences from the mean values. It states

$$r = \frac{n\sum_{i=1}^{n} x_i y_i - \sum_{i=1}^{n} x_i \sum_{i=1}^{n} y_i}{\sqrt{n\sum_{i=1}^{n} x_i^2 - \left(\sum_{i=1}^{n} x_i\right)^2} \sqrt{n\sum_{i=1}^{n} y_i^2 - \left(\sum_{i=1}^{n} y_i\right)^2}}$$

We shall provide examples illustrating both versions.

Key point

When we have n pairs of values (x_i, y_i) of two variables X and Y, the product-moment correlation coefficient, r, is given either by

$$r = \frac{\displaystyle\sum_{i=1}^{n}(x_i - \bar{x})(y_i - \bar{y})}{\sqrt{\displaystyle\sum_{i=1}^{n}(x_i - \bar{x})^2 \sum_{i=1}^{n}(y_i - \bar{y})^2}}$$

or, equivalently, by

$$r = \frac{\displaystyle n\sum_{i=1}^{n}x_i y_i - \sum_{i=1}^{n}x_i \sum_{i=1}^{n}y_i}{\sqrt{n\displaystyle\sum_{i=1}^{n}x_i^2 - \left(\sum_{i=1}^{n}x_i\right)^2}\sqrt{n\displaystyle\sum_{i=1}^{n}y_i^2 - \left(\sum_{i=1}^{n}y_i\right)^2}}$$

If $r = 0$ there is no correlation.
If $0 < r \leq 1$ there is positive correlation.
If $-1 \leq r < 0$ there is negative correlation.
If r is close to 1 or -1 the correlation is strong.
If r is close to 0 the correlation is weak.

These formulae seem daunting, especially when first met. Work through the following example to see how to use them to calculate a correlation coefficient. In practice, once you understand what is going on you will be able to use a computer package such as Excel which has built-in commands for performing such calculations.

WORKED EXAMPLES

31.2 Four pairs of observations (x_i, y_i) of two quantities, X and Y, are listed in Table 31.2. To explore a possible relationship between the values of X and Y calculate the correlation coefficient, r, using the first of the given formulae.

Table 31.2

x_i	y_i
9	20
21	42
33	61
45	79

Solution   To calculate r it is first necessary to find the mean of the four x values, $\bar{x}$, and the mean of the four y values, $\bar{y}$. These calculations are shown in Table 31.3.

Table 31.3

x_i	y_i
9	20
21	42
33	61
45	79
108	202

$$\bar{x} = \frac{108}{4} = 27 \quad \bar{y} = \frac{202}{4} = 50.5$$

Then $\bar{x}$ is subtracted from each x value and $\bar{y}$ is subtracted from each y value, before various products are found. This information is best presented in a table, such as that in Table 31.4.

Table 31.4

x_i	y_i	$x_i - \bar{x}$	$y_i - \bar{y}$	$(x_i - \bar{x})(y_i - \bar{y})$	$(x_i - \bar{x})^2$	$(y_i - \bar{y})^2$
9	20	-18	-30.5	549	324	930.25
21	42	-6	-8.5	51	36	72.25
33	61	6	10.5	63	36	110.25
45	79	18	28.5	513	324	812.25
$\bar{x} = 27$	$\bar{y} = 50.5$	0	0	1176	720	1925

With the table complete all the quantities required in the formula for r are available:

$$r = \frac{\sum_{i=1}^{n}(x_i - \bar{x})(y_i - \bar{y})}{\sqrt{\sum_{i=1}^{n}(x_i - \bar{x})^2 \sum_{i=1}^{n}(y_i - \bar{y})^2}} = \frac{1176}{\sqrt{(720)(1925)}} = 0.9989$$

With a value of r so close to $+1$ we conclude that there is very strong positive correlation between the variables x and y.

31.3   For the data in Worked Example 31.2 recalculate r using the second of the formulae to confirm that the result is the same.

Solution   As before, we draw up a table showing the quantities required (Table 31.5):

Table 31.5

x_i	y_i	x_i^2	y_i^2	$x_i y_i$
9	20	81	400	180
21	42	441	1764	882
33	61	1089	3721	2013
45	79	2025	6241	3555
108	202	3636	12126	6630

This data can be substituted into the formula:

$$r = \frac{n\sum_{i=1}^{n} x_i y_i - \sum_{i=1}^{n} x_i \sum_{i=1}^{n} y_i}{\sqrt{n\sum_{i=1}^{n} x_i^2 - \left(\sum_{i=1}^{n} x_i\right)^2} \sqrt{n\sum_{i=1}^{n} y_i^2 - \left(\sum_{i=1}^{n} y_i\right)^2}}$$

$$= \frac{4(6630) - (108)(202)}{\sqrt{4(3636) - 108^2}\sqrt{4(12126) - 202^2}}$$

$$= \frac{4704}{\sqrt{2880}\sqrt{7700}}$$

$$= 0.9989 \quad \text{to 4 d.p.}$$

This confirms the same result is obtained using either version of the formulae.

31.4   In a recent survey of new cars data was collected on the engine size (in cubic centimetres (cc)) and the fuel economy measured in miles per gallon (mpg). The fuel economy is calculated by simulating driving in both urban and extra-urban conditions. The data is presented in Table 31.6. Calculate the correlation coefficient for this data and comment upon the answer.

Table 31.6

x = engine size (cc)	y = economy (mpg)
999	57.6
1498	45.6
1596	37.7
1998	38.2
2295	28.8
6750	13.7

Solution   First we calculate the mean of the x and y values as shown in Table 31.7.

Table 31.7

x = engine size (cc)	y = economy (mpg)
999	57.6
1498	45.6
1596	37.7
1998	38.2
2295	28.8
6750	13.7

$$\bar{x} = \frac{15136}{6} = 2522.7 \quad \bar{y} = \frac{221.6}{6} = 36.9$$

Then $\bar{x}$ is subtracted from each x value and $\bar{y}$ is subtracted from each y value, before various products are found. This information is best presented in a table, such as that in Table 31.8.

Table 31.8

x	y	$x_i - \bar{x}$	$y_i - \bar{y}$	$(x_i - \bar{x})(y_i - \bar{y})$	$(x_i - \bar{x})^2$	$(y_i - \bar{y})^2$
999	57.6	−1523.7	20.7	−31540.59	2321661.69	428.49
1498	45.6	−1024.7	8.7	−8914.89	1050010.09	75.69
1596	37.7	−926.7	0.8	−741.36	858772.89	0.64
1998	38.2	−524.7	1.3	−682.11	275310.09	1.69
2295	28.8	−227.7	−8.1	1844.37	51847.29	65.61
6750	13.7	4227.3	−23.2	−98073.36	17870065.29	538.24

$\bar{x} = 2522.7 \quad \bar{y} = 36.9$ −138107.94   22427667.34   1110.36

With the table complete all the quantities required in the formula for r can be found:

$$r = \frac{\displaystyle\sum_{i=1}^{n}(x_i - \bar{x})(y_i - \bar{y})}{\sqrt{\displaystyle\sum_{i=1}^{n}(x_i - \bar{x})^2 \sum_{i=1}^{n}(y_i - \bar{y})^2}} = \frac{-138107.94}{\sqrt{(22427667.34)(1110.36)}} = -0.8752 \ (4\,\text{d.p.})$$

With a value of r so close to -1 we conclude that there is very strong negative correlation between the variables x and y. This means that, on the basis of the data supplied here, there is strong negative correlation between engine size and fuel economy. In general, smaller-sized engines in our sample are more economical than the larger engines.

You should rework this example using the second of the formulae for r.

Self-assessment question 31.3

1. Describe in words the extent of association between two variables when (a) $r=1$, (b) $r=-0.82$, (c) $r=-0.33$, (d) $r=0$.

Exercise 31.3

MyMathLab Global

1. Calculate the correlation coefficient r for the following data. Interpret your result.

x_i	y_i
1	11
3	51
4	96
5	115

2. Calculate the correlation coefficient r for the following data. Interpret your result.

x_i	y_i
0	2
1	5
2	8
3	11

3. Investigate the availability of computer software for the calculation of the correlation coefficient. Calculate the correlation coefficient for the data of Exercise 31.2, Question 2.

31.4 Spearman's coefficient of rank correlation

Sometimes specific values of two variables, X and Y say, are not available, but we may have data on their order, or **rank**. For example, suppose we have data on seven students, A, B, ..., F, G, which tells us who came first, second etc., in each of two athletic events – swimming and running. The data would take the form given in Table 31.9. So, for example, student A came first in both events, whereas student G came last in the swimming event and second from last in the running event.

Table 31.9

student	swimming rank	running rank
A	1	1
B	3	7
C	6	5
D	2	3
E	4	2
F	5	4
G	7	6

Suppose we want to know whether there is correlation between the two. Is there an association between good performance in the running event and good performance when swimming?

Spearman's formula can be used to calculate a **coefficient of rank correlation**, which is a number lying between -1 and 1 which provides information on the strength of the relationship exhibited by the data.

To apply the formula we need to find, for each pair of rank values, the **difference** between the ranks. So for student A, the difference between the ranks is $1 - 1 = 0$. For student G the difference is $7 - 6 = 1$. We do this for all pairs of values. Then each difference is squared. The detail is shown in Table 31.10.

Table 31.10

student	swimming rank	running rank	difference D	D^2
A	1	1	0	0
B	3	7	-4	16
C	6	5	1	1
D	2	3	-1	1
E	4	2	2	4
F	5	4	1	1
G	7	6	1	1

24

If there are n pairs of values, the coefficient of rank correlation is given by

$$r = 1 - \frac{6\sum_{i=1}^{n} D^2}{n(n^2 - 1)}$$

In this case $n = 7$ and so

$$r = 1 - \frac{6 \times 24}{7(7^2 - 1)} = 1 - \frac{144}{(7)(48)} = 0.57 \quad (2 \text{ d.p.})$$

On the basis of this data we can conclude there is a moderate positive correlation between performance in the two athletic events.

Key point

Spearman's coefficient of rank correlation: Given n pairs of values which are the rank orders of two variables X and Y, the **coefficient of rank correlation** is given by

$$r = 1 - \frac{6\sum_{i=1}^{n} D^2}{n(n^2 - 1)}$$

where D is the difference between the ranks of the corresponding values of X and Y.

Self assessment question 31.4

1. In what circumstances might you use the coefficient of rank correlation rather than the product-moment correlation coefficient?

Exercise 31.4

1. The table shows the rank order of six students in their mathematics and their science tests:

student	maths rank	science rank
A	1	1
B	3	6
C	6	5
D	2	3
E	4	2
F	5	4

Find the coefficient of rank correlation and comment upon the strength of any relationship.

2. The table shows the rank order of 10 countries in terms of the amount per capita spent upon education and the wealth of that country:

country	education spend per capita	wealth
A	4	6
B	3	4
C	6	5
D	5	7
E	7	8
F	8	9
G	10	10
H	9	1
I	1	3
J	2	2

Find the coefficient of rank correlation and comment upon the strength of any relationship.

3. Suppose you are given the rank orders of two variables and suppose that the ranks agree exactly: that is, when the rank of X is 1, the rank of Y is 1, and so on. Show that Spearman's coefficient of rank correlation, r, is 1. Further, show that when one ranking is the exact reverse of the other, the value of r is -1.

Test and assignment exercises 31

1. Calculate the correlation coefficient r for the following data. Interpret your result.

x_i	y_i
0	11
1	7
2	3

2. Figure 31.5 shows several scatter diagrams. Comment upon the nature of possible relationships between the two variables.

Figure 31.5

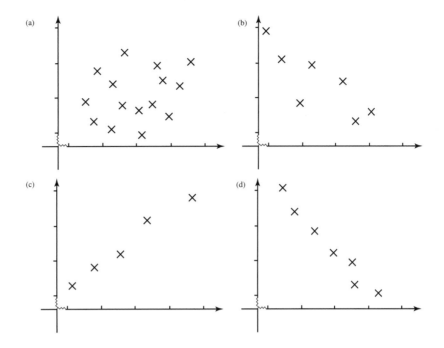

3. The table gives the mean summer temperature (in °C) at a holiday resort during the past six years together with volume (in litres) of ice cream purchased from local shops and vendors:

Year	1	2	3	4	5	6
Mean temperature (°C)	10	14	18	11	12	10
Volume (1000 litres)	20	32	33	24	23	18

Calculate the correlation coefficient for this data and comment upon any possible relationship.

4. The table provides data on seven families. It shows annual income and the amount spent in 2004 on family holidays. Rank the data and calculate Spearman's coefficient of rank correlation. Comment upon the calculated value.

family	annual income	holiday spend
A	24,000	875
B	56,000	5400
C	36,000	2100
D	29,500	400
E	100,000	1250
F	58,000	4250
G	38,000	2259

Regression

Objectives: This chapter:

- explains the concept of simple linear regression

- shows how to calculate the regression line of Y on X

- shows how to use the regression equation to predict values of the dependent variable, Y, given values of the independent variable, X

32.1 Introduction

In Chapter 31 we introduced **correlation** in order to explore the relationship between two variables, X and Y say, measured on some object of interest. We drew scatter diagrams to explore possible relationships visually and calculated the correlation coefficient, which gave us an indication of the strength of the relationship between the two variables.

In correlation, the two variables, X and Y, have equal 'status' and when a scatter diagram is drawn it does not matter which variable is plotted on the horizontal axis and which is plotted on the vertical axis. Whilst the diagrams will differ if we switch axes, the correlation coefficient will be the same either way. This is because it measures the degree of association between the two variables but does not attribute any dependency, cause or effect.

We now move on to study **regression**. The purpose of regression is to predict the value of one of the variables given a value of the other variable. When we use the data to predict the value of Y corresponding to a value of X that we choose, we refer to this as 'regression of Y on X'. So regression is subtly different from correlation. The status of the two variables is not the same since we are regarding Y as being a variable whose value depends upon the value we choose for X. In such cases we refer to X as the **independent variable**, and Y as the **dependent variable**. Statistical textbooks often refer to X as the **predictor** and Y as the **response variable** because from a value of X we can *predict* the corresponding value of Y.

For example, we might be interested in trying to predict the blood pressure of a member of a group of workers given his or her age. To do this we could select several members of the group, chosen by age, X, and measure their blood pressure, Y. We then use this data to calculate the regression of blood pressure on age. This produces a regression equation which can then predict the blood pressure (without actually measuring it) of another worker given only his or her age.

In any regression analysis, values of the independent variable, X, should be preselected. For example, we may decide to select workers aged 20, 25, 30, 35, 40, ..., 60 and measure the blood pressures, Y, of these chosen individuals. A scatter diagram is then drawn. Provided the diagram indicates that the points are scattered around an imaginary straight line then it is appropriate to calculate a **regression equation** for this line. This is the equation of the line which best fits the data and which can then be used to predict unknown values of Y.

32.2 The regression equation

Suppose we have gathered a set of data by selecting values of the variable, X, and measuring values of the corresponding variable, Y, upon each of n objects. Let the values of X be $x_1, x_2, \ldots, x_n$ and the corresponding values of Y be $y_1, y_2, \ldots, y_n$.

The regression equation is the equation of the straight line which most closely fits all of the given data. It is given in the standard form of an equation of a straight line which we shall write as

$$y = a + bx$$

(You may find it helpful to re-read Chapter 18.) So here, a is the vertical intercept of the line with the y axis, and b is the gradient of the line. The value of b is calculated first using the formula

$$b = \frac{n \sum_{i=1}^{n} x_i y_i - \left(\sum_{i=1}^{n} x_i \right)\left(\sum_{i=1}^{n} y_i \right)}{n \sum_{i=1}^{n} x_i^2 - \left(\sum_{i=1}^{n} x_i \right)^2}$$

The vertical intercept, a, is then given by

$$a = \frac{\sum_{i=1}^{n} y_i - b \sum_{i=1}^{n} x_i}{n}$$

The derivation of these formulae, which is beyond the scope of this book, relies on a technique which, loosely, makes the square of the vertical difference between each point on the scatter diagram and the line of best fit as small as possible. The proof can be found in most statistical textbooks. The formulae look very complicated when met for the first time, but in essence they require us simply to multiply some values together and add them up, as we shall see.

Key point

Given values of variables X and Y measured on n objects of interest, the regression line of Y on X is given by $y = a + bx$ where

$$b = \frac{n\sum_{i=1}^{n} x_i y_i - \left(\sum_{i=1}^{n} x_i\right)\left(\sum_{i=1}^{n} y_i\right)}{n\sum_{i=1}^{n} x_i^2 - \left(\sum_{i=1}^{n} x_i\right)^2} \qquad a = \frac{\sum_{i=1}^{n} y_i - b\sum_{i=1}^{n} x_i}{n}$$

WORKED EXAMPLE

32.1   Table 32.1 shows values of X and Y measured on five individuals.

Table 32.1

individual	x_i	y_i
A	10	28
B	20	32
C	30	38
D	40	39
E	50	48

(a)   Use a scatter diagram to show that there appears to be a linear relationship between X and Y.
(b)   Find the equation which represents the regression of Y on X.
(c)   Plot the regression line on the scatter diagram.
(d)   Use the equation to predict the value of Y when $X = 37$.

Solution   (a)   The given data has been used to draw the scatter diagram in Figure 32.1.
    Inspection of the scatter diagram gives us some confidence that the points are scattered around an imaginary line and so it is appropriate to try to find this line using the regression equation.

Figure 32.1
Scatter diagram for
Worked Example 32.1

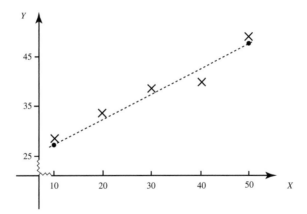

(b) Note from the formulae that we need to add up various quantities, for
     example

$$\sum_{i=1}^{n} x_i \quad \sum_{i=1}^{n} x_i y_i$$

In this example, in which there are five individuals, $n = 5$. It is easier to
keep track of the calculations if the intermediate quantities are recorded
in a table of values such as that shown in Table 32.2.

Table 32.2

individual	x_i	y_i	$x_i y_i$	x_i^2
A	10	28	280	100
B	20	32	640	400
C	30	38	1140	900
D	40	39	1560	1600
E	50	48	2400	2500
	$\sum_{i=1}^{5} x_i = 150$	$\sum_{i=1}^{5} y_i = 185$	$\sum_{i=1}^{5} x_i y_i = 6020$	$\sum_{i=1}^{5} x_i^2 = 5500$

We now have all the quantities we need to calculate b and then a:

$$b = \frac{n \sum_{i=1}^{n} x_i y_i - \left(\sum_{i=1}^{n} x_i\right)\left(\sum_{i=1}^{n} y_i\right)}{n \sum_{i=1}^{n} x_i^2 - \left(\sum_{i=1}^{n} x_i\right)^2}$$

$$= \frac{5(6020) - (150)(185)}{5(5500) - (150)^2}$$

$$= 0.47$$

Then

$$a = \frac{\sum_{i=1}^{n} y_i - b \sum_{i=1}^{n} x_i}{n}$$

$$= \frac{185 - (0.47)(150)}{5}$$

$$= 22.9$$

So the regression equation, $y = a + bx$, is $y = 22.9 + 0.47x$. We can use this equation to predict values of y for any values of x we choose, although it is not advisable to use it to predict values outside the interval of given values of X (that is, 10 to 50).

(c) The easiest way to draw the regression line, $y = 22.9 + 0.47x$, is to find two points which lie on the line. Note that when $x = 10$, $y = 27.6$, so the point (10, 27.6) lies on the line. When $x = 50$, $y = 22.9 + 0.47(50) = 46.4$, so the point (50, 46.4) also lies on the line. The points have been plotted and the line has been drawn in Figure 32.1.

(d) When the value of X is $x = 37$ we find $y = 22.9 + 0.47(37) = 40.29$. You could also use the straight line graph to read off the value of Y corresponding to $X = 37$ although this is not likely to be as accurate as using the equation.

It is relevant to note that computer software packages such as Excel have in-built functions which will calculate regression equations. You should seek advice about the availability of such software within your own institution because with data sets of a realistic size it is pointless trying to perform these calculations by hand.

It is also important to realise that even though we have calculated the regression equation and used it to predict a value of Y, we have no idea how good a prediction this might be. Advanced statistical techniques make it possible to state how confident we can be in our predicted value but these techniques are beyond the scope of this book.

Self-assessment questions 32.2

1. When drawing a scatter diagram for a regression problem, which variable is plotted on the horizontal axis and which on the vertical axis?

2. What is meant by the phrase 'the regression of Y on X'?

3. Explain the meaning of the terms 'predictor' and 'response variable'.

Exercise 32.2

1. Table 32.3 shows values of X and Y measured on six students.

Table 32.3

student	x_i	y_i
A	5	3
B	10	18
C	15	42
D	20	67
E	25	80
F	30	81

   (a) Use a scatter diagram to show that there appears to be a linear relationship between X and Y.

   (b) Find the equation which represents the regression of Y on X.

   (c) Plot the regression line on the scatter diagram.

   (d) Use the equation to predict the value of Y when $X = 12$.

2. It is believed that a certain drug can reduce a patient's pulse rate. In an experiment it was found that the larger the dose of the drug, the more the pulse rate reduced. A doctor wishes to find a regression equation which could be used to predict the pulse rate achieved for specific dosages. An experiment is carried out on four healthy individuals, varying the dose, X (in μg), and measuring the corresponding pulse rate, Y (in beats per minute). The data is shown in Table 32.4.

Table 32.4

individual	dose x_i	pulse rate y_i
A	2	72
B	2.5	71
C	3	65
D	3.5	60

   (a) Use a scatter diagram to show that there appears to be a linear relationship between X and Y.

   (b) Find the equation which represents the regression of Y on X.

   (c) Plot the regression line on the scatter diagram.

   (d) Use the equation to predict the pulse rate expected by administering a dose of 3.2 μg.

Test and assignment exercises 32

1. Calculate the Y on X regression line for the data in Table 32.5. Use the equation to find the value of the response variable when the predictor is 2.5.

Table 32.5

item	x_i	y_i
A	1	2.8
B	2	7.9
C	3	13.4
D	4	18.0

2. Table 32.6 shows values of Statistics Module marks, (X), and Econometrics Module marks, (Y), measured on six students.

Table 32.6

student	x_i	y_i
A	35	22
B	40	38
C	45	52
D	50	68
E	55	65
F	60	73

(a) Use a scatter diagram to show that there appears to be a linear relationship between X and Y.

(b) Find the equation which represents the regression of Y on X.

(c) Plot the regression line on the scatter diagram.

(d) Another student, G, scored 48 in her Statistics Module. What mark in Econometrics would the regression equation predict?

Objectives: This chapter:

- introduces a technique called differentiation for calculating the gradient of a curve at any point
- illustrates how to find the gradient function of several common functions using a table
- introduces some simple rules for finding gradient functions
- explains what is meant by the terms 'first derivative' and 'second derivative'
- explains the terms 'maximum' and 'minimum' when applied to functions
- applies the technique of differentiation to locating maximum and minimum values of a function

33.1 The gradient function

y' stands for the gradient function of $y = f(x)$. We can also denote the gradient function by $f'(x)$.

Suppose we have a function, $y = f(x)$, and are interested in its slope, or gradient, at several points. For example, Figure 33.1 shows a graph of the function $y = 2x^2 + 3x$. Imagine tracing the graph from the left to the right. At point A the graph is falling rapidly. At point B the graph is falling but less rapidly than at A. Point C lies at the bottom of the dip. At point D the graph is rising. At E it is rising but more quickly than at D. The important point is that the gradient of the curve changes from point to point. The following section describes a mathematical technique for measuring the gradient at different points. We introduce another function called the **gradient function**, which we write as $\frac{dy}{dx}$. This is read as 'dee y by dee x'. We sometimes simplify this notation to y', read as 'y dash'. For almost all of the functions you will meet this gradient function can be found by using a formula, applying a rule or checking in a table. Knowing the gradient function we can find the gradient of the curve at any point. The following sections show how to find the gradient function of a number of common functions.

Figure 33.1
Graph of $y = 2x^2 + 3x$

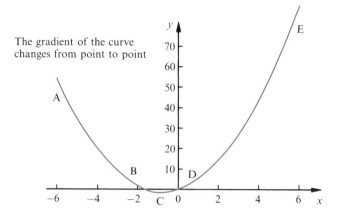

The gradient of the curve
changes from point to point

Given a function $y = f(x)$ we denote its gradient function by $\frac{dy}{dx}$ or simply by y'.

33.2 Gradient function of $y = x^n$

For any function of the form $y = x^n$ the gradient function is found from the following formula:

If $y = x^n$ then $y' = nx^{n-1}$.

WORKED EXAMPLES

33.1   Find the gradient function of (a) $y = x^3$, (b) $y = x^4$.

Solution   (a) Comparing $y = x^3$ with $y = x^n$ we see that $n = 3$. Then $y' = 3x^{3-1} = 3x^2$.

(b) Applying the formula with $n = 4$ we find that if $y = x^4$ then $y' = 4x^{4-1} = 4x^3$.

33.2   Find the gradient function of (a) $y = x^2$, (b) $y = x$.

Solution   (a) Applying the formula with $n = 2$ we find that if $y = x^2$ then $y' = 2x^{2-1} = 2x^1$. Because x^1 is simply x we find that the gradient function is $y' = 2x$.

(b) Applying the formula with $n = 1$ we find that if $y = x^1$ then $y' = 1x^{1-1} = 1x^0$. Because x^0 is simply 1 we find that the gradient function is $y' = 1$.

Finding the gradient of a graph is now a simple matter. Once the gradient function has been found, the gradient at any value of x is found by substituting that value into the gradient function. If, after carrying out this substitution, the result is negative, then the curve is falling. If the result is positive, the curve is rising. The size of the gradient function is a measure of how rapidly this fall or rise is taking place. We write $y'(x = 2)$ or simply $y'(2)$ to denote the value of the gradient function when $x = 2$.

WORKED EXAMPLES

33.3   Find the gradient of $y = x^2$ at the points where

(a) $x = -1$ (b) $x = 0$ (c) $x = 2$ (d) $x = 3$

Solution   From Worked Example 33.2, or from the formula, we know that the gradient function of $y = x^2$ is given by $y' = 2x$.

(a)   When $x = -1$ the gradient of the graph is then $y'(-1) = 2(-1) = -2$. The fact that the gradient is negative means that the curve is falling at the point.

(b)   When $x = 0$ the gradient is $y'(0) = 2(0) = 0$. The gradient of the curve is zero at this point. This means that the curve is neither falling nor rising.

(c)   When $x = 2$ the gradient is $y'(2) = 2(2) = 4$. The fact that the gradient is positive means that the curve is rising.

(d)   When $x = 3$ the gradient is $y'(3) = 2(3) = 6$ and so the curve is rising here. Comparing this answer with that of part (c) we conclude that the curve is rising more rapidly at $x = 3$ than at $x = 2$, where the gradient was found to be 4.

      The graph of $y = x^2$ is shown in Figure 33.2. If we compare our results with the graph we see that the curve is indeed falling when

Figure 33.2
Graph of $y = x^2$

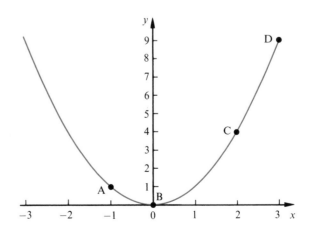

$x = -1$ (point A) and is rising when $x = 2$ (point C), and $x = 3$ (point D). At the point where $x = 0$ (point B) the curve is neither rising nor falling.

33.4   Find the gradient function of $y = x^{-3}$. Hence find the gradient of $y = x^{-3}$ when $x = 4$.

Solution   Using the formula with $n = -3$ we find that if $y = x^{-3}$ then $y' = -3x^{-3-1} = -3x^{-4}$. Because x^{-4} can also be written as $1/x^4$ we could write $y' = -3(1/x^4)$ or $-3/x^4$. When $x = 4$ we find $y'(4) = -3/4^4 = -\frac{3}{256}$. This number is very small and negative, which means that when $x = 4$ the curve is falling, but only slowly.

33.5   Find the gradient function of $y = 1$.

Solution   Before we use the formula to calculate the gradient function let us think about the graph of $y = 1$. Whatever the value of x, this function takes the value 1. Its graph must then be a horizontal line – it neither rises nor falls. We conclude that the gradient function must be zero, that is $y' = 0$. To obtain the same result using the formula we must rewrite 1 as x^0. Then, using the formula with $n = 0$, we find that if $y = x^0$ then $y' = 0x^{0-1} = 0$.

The previous worked example illustrates an important result. Any constant function has a gradient function equal to zero because its graph is a horizontal line, and is therefore neither rising nor falling.

The technique introduced in this section has a very long history and as a consequence a wide variety of alternative names for the gradient function

Table 33.1
The gradient function of some common functions

$y = f(x)$	$y' = f'(x)$	Notes
constant	0	
x	1	
x^2	$2x$	
x^n	nx^{n-1}	
e^x	e^x	
e^{kx}	ke^{kx}	k is a constant
$\sin x$	$\cos x$	
$\cos x$	$-\sin x$	
$\sin kx$	$k \cos kx$	k is a constant
$\cos kx$	$-k \sin kx$	k is a constant
$\ln kx$	$1/x$	k is a constant

have emerged. For example, the gradient function is also called the **first derivative**, or simply the **derivative**. The process of obtaining this is also known as **differentiation.** Being asked to **differentiate** $y = x^5$, say, is equivalent to being asked to find its gradient function, y'. Furthermore, because the gradient function measures how rapidly a graph is changing, it is also referred to as the **rate of change** of y. The area of study that is concerned with differentiation is known as *differential calculus*. When you meet these words you should realise that they are simply technical terms associated with finding the gradients of graphs.

It is possible to find the gradient function of a wide range of functions by simply referring to standard results. Some of these are given in Table 33.1. Note that, for technical reasons, when finding the gradient functions of trigonometrical functions the angle x must always be measured in radians.

WORKED EXAMPLES

33.6   Use Table 33.1 to find the gradient function y' when y is

(a) $\sin x$ (b) $\sin 2x$ (c) $\cos 3x$ (d) e^x

Solution   (a) Directly from the table we see that if $y = \sin x$ then its gradient function is given by $y' = \cos x$. This result occurs frequently and is worth remembering.

(b) Using the table and taking $k = 2$ we find that if $y = \sin 2x$ then $y' = 2 \cos 2x$.

(c) Using the table and taking $k = 3$ we find that if $y = \cos 3x$ then $y' = -3 \sin 3x$.

(d) Using the table we see directly that if $y = e^x$ then $y' = e^x$. Note that the exponential function e^x is very special because its gradient function y' is the same as y.

33.7   Find the gradient function of $y = e^{-x}$. Hence find the gradient of the graph of y at the point where $x = 1$.

Solution   Noting that $e^{-x} = e^{-1x}$ and using Table 33.1 with $k = -1$ we find that if $y = e^{-x}$ then $y' = -1e^{-x} = -e^{-x}$. Using a calculator to evaluate this when $x = 1$ we find $y'(1) = -e^{-1} = -0.368$.

33.8   Find the gradient function of $y = \sin 4x$ where $x = 0.3$.

Solution   Using Table 33.1 with $k = 4$ gives $y' = 4 \cos 4x$. Remembering to measure x in radians we evaluate this when $x = 0.3$:

$$y'(0.3) = 4 \cos 4(0.3) = 4 \cos 1.2 = 1.4494$$

Self-assessment questions 33.2

1. What is the purpose of the gradient function?

2. State the formula for finding the gradient function when $y = x^n$.

3. State three alternative names for the gradient function.

4. State an alternative notation for y'.

5. What is the derivative of any constant? What is the graphical explanation of this answer?

6. When evaluating the derivative of the sine and cosine functions, in which units must the angle x be measured – radians or degrees?

7. Which function is equal to its derivative?

Exercise 33.2

MyMathLab

1. Find y' when y is given by
   (a) x^8 (b) x^7 (c) x^{-1} (d) x^{-5}
   (e) x^{13} (f) x^5 (g) x^{-2}

2. Find y' when y is given by
   (a) $x^{3/2}$ (b) $x^{5/2}$ (c) $x^{-1/2}$
   (d) $x^{1/2}$ (e) $\sqrt{x}$ (f) $x^{0.2}$

3. Sketch a graph of $y = 8$. State the gradient function, y'.

4. Find the first derivative of each of the following functions:
   (a) $y = x^{16}$ (b) $y = x^{0.5}$ (c) $y = x^{-3.5}$
   (d) $y = \dfrac{1}{x^{1/2}}$ (e) $y = \dfrac{1}{\sqrt{x}}$

5. Find the gradient of the graph of the function $y = x^4$ when
   (a) $x = -4$ (b) $x = -1$ (c) $x = 0$
   (d) $x = 1$ (e) $x = 4$

   Can you infer anything about the shape of the graph from this information?

6. Find the gradient of the graphs of each of the following functions at the points given:
   (a) $y = x^3$ at $x = -3$, at $x = 0$ and at $x = 3$
   (b) $y = x^4$ at $x = -2$ and at $x = 2$

   (c) $y = x^{1/2}$ at $x = 1$ and $x = 2$
   (d) $y = x^{-2}$ at $x = -2$ and $x = 2$

7. Sometimes a function is given in terms of variables other than x. This should pose no problems. Find the gradient functions of
   (a) $y = t^2$ (b) $y = t^{-3}$ (c) $y = t^{1/2}$
   (d) $y = t^{-1.5}$

8. Use Table 33.1 to find the gradient function y' when y is equal to
   (a) $\cos 7x$ (b) $\sin \frac{1}{2}x$ (c) $\cos \frac{x}{2}$
   (d) e^{4x} (e) e^{-3x}

9. Find the derivative of each of the following functions:
   (a) $y = \ln 3x$ (b) $y = \sin \frac{1}{3}x$
   (c) $y = e^{x/2}$

10. Find the gradient of the graph of the function $y = \sin x$ at the points where
    (a) $x = 0$ (b) $x = \pi/2$ (c) $x = \pi$

11. Table 33.1, although written in terms of the variable x, can still be applied when other variables are involved. Find y' when
    (a) $y = t^7$ (b) $y = \sin 3t$ (c) $y = e^{2t}$
    (d) $y = \ln 4t$ (e) $y = \cos \frac{t}{3}$

33.3 Some rules for finding gradient functions

Up to now the formulae we have used for finding gradient functions have applied to single terms. It is possible to apply these much more widely with the addition of three more rules. The first applies to the sum of two functions, the second to their difference and the third to a constant multiple of a function.

Key point   *Rule 1:* If $y = f(x) + g(x)$ then $y' = f'(x) + g'(x)$.

In words this says that to find the gradient function of a sum of two functions we simply find the two gradient functions separately and add these together.

WORKED EXAMPLE

33.9   Find the gradient function of $y = x^2 + x^4$.

Solution   The gradient function of x^2 is $2x$. The gradient function of x^4 is $4x^3$. Therefore

$$\text{if } y = x^2 + x^4 \quad \text{then} \quad y' = 2x + 4x^3$$

The second rule is really an extension of the first, but we state it here for clarity.

Key point   *Rule 2:* If $y = f(x) - g(x)$ then $y' = f'(x) - g'(x)$.

WORKED EXAMPLE

33.10   Find the gradient function of $y = x^5 - x^7$.

Solution   We find the gradient function of each term separately and subtract them. That is, $y' = 5x^4 - 7x^6$.

Key point   *Rule 3:* If $y = k f(x)$, where k is a number, then $y' = k f'(x)$.

WORKED EXAMPLE

33.11   Find the gradient function of $y = 3x^2$.

Solution   This function is 3 times x^2. The gradient function of x^2 is $2x$. Therefore, using Rule 3, we find that

$$\text{if } y = 3x^2 \quad \text{then } y' = 3(2x) = 6x$$

The rules can be applied at the same time. Consider the following examples.

WORKED EXAMPLES

33.12   Find the derivative of $y = 4x^2 + 3x^{-3}$.

Solution   The derivative of $4x^2$ is $4(2x) = 8x$. The derivative of $3x^{-3}$ is $3(-3x^{-4}) = -9x^{-4}$. Therefore, if $y = 4x^2 + 3x^{-3}$ then $y' = 8x - 9x^{-4}$.

33.13   Find the derivative of

(a) $y = 4 \sin t - 3 \cos 2t$ (b) $y = \dfrac{e^{2t}}{3} + 6 + \dfrac{\ln(2t)}{5}$

Solution   (a)   We differentiate each quantity in turn using Table 33.1:

$$y' = 4 \cos t - 3(-2 \sin 2t) = 4 \cos t + 6 \sin 2t$$

(b)   Writing y as $\frac{1}{3}e^{2t} + 6 + \frac{1}{5}\ln(2t)$ we find

$$y' = \frac{2}{3} e^{2t} + 0 + \frac{1}{5} \left(\frac{1}{t} \right) = \frac{2e^{2t}}{3} + \frac{1}{5t}$$

Self-assessment question 33.3

1.   State the rules for finding the derivative of (a) $f(x) + g(x)$, (b) $f(x) - g(x)$, (c) $kf(x)$.

Exercise 33.3

MyMathLab

1.   Find the gradient function of each of the following:
(a) $y = x^6 - x^4$ (b) $y = 6x^2$
(c) $y = 9x^{-2}$ (d) $y = \dfrac{1}{2} x$
(e) $y = 2x^3 - 3x^2$

2.   Find the gradient function of each of the following:
(a) $y = 2 \sin x$ (b) $y = 3 \sin 4x$
(c) $y = 7 \cos 9x + 3 \sin 4x$
(d) $y = e^{3x} - 5e^{-2x}$

3.   Find the gradient function of $y = 3x^3 - 9x + 2$. Find the value of the gradient function at $x = 1$, $x = 0$ and $x = -1$. At which of these points is the curve steepest?

4.   Write down a function which when differentiated gives $2x$. Are there any other functions you can differentiate to give $2x$?

5. Find the gradient of
   $y = 3 \sin 2t + 4 \cos 2t$ when $t = 2$.
   Remember to use radians.

6. Find the gradient of $y = 2x - x^3 + e^{2x}$
   when $x = 1$.

7. Find the gradient of $y = (2x - 1)^2$ when
   $x = 0$.

8. Show that the derivative of $f(x) = x^2 - 2x$ is 0 when $x = 1$.

9. Find the values of x such that the
   gradient of $y = \frac{x^3}{3} - x + 7$ is 0.

10. The function $y(x)$ is given by

   $$y(x) = x + \cos x \qquad 0 \leqslant x \leqslant 2\pi$$

   Find the value(s) of x where the gradient
   of y is 0.

33.4 Higher derivatives

In some applications, and in more advanced work, it is necessary to find the derivative of the gradient function itself. This is termed the **second derivative** and is written y'' and read 'y double dash'. Some books write

$$y'' \text{ as } \frac{d^2y}{dx^2}$$

Key point

y'' or $\dfrac{d^2y}{dx^2}$ is found by differentiating y'.

It is a simple matter to find the second derivative by differentiating the first derivative. No new techniques or tables are required.

WORKED EXAMPLES

33.14 Find the first and second derivatives of $y = x^4$.

Solution From Table 33.1, if $y = x^4$ then $y' = 4x^3$. The second derivative is found by differentiating the first derivative. Therefore

if $y' = 4x^3$ then $y'' = 4(3x^2) = 12x^2$

33.15 Find the first and second derivatives of $y = 3x^2 - 7x + 2$.

Solution Using Table 33.1 we find $y' = 6x - 7$. The derivative of the constant 2 equals 0. Differentiating again we find $y'' = 6$, because the derivative of the constant -7 equals 0.

33.16 If $y = \sin x$, find

(a) $\dfrac{dy}{dx}$ (b) $\dfrac{d^2y}{dx^2}$

Solution   (a) Recall that $\frac{dy}{dx}$ is the first derivative of y. From Table 33.1 this is given by

$$\frac{dy}{dx} = \cos x$$

(b) $\frac{d^2y}{dx^2}$ is the second derivative of y. This is found by differentiating the first derivative. From Table 33.1 we find that the derivative of $\cos x$ is $-\sin x$, and so

$$\frac{d^2y}{dx^2} = -\sin x$$

Self-assessment questions 33.4

1. Explain how the second derivative of y is found.

2. Give two alternative notations for the second derivative of y.

Exercise 33.4

1. Find the first and second derivatives of the following functions:
   (a) $y = x^2$ (b) $y = 3x^8$ (c) $y = 9x^5$
   (d) $y = 3x^{1/2}$ (e) $y = 3x^4 + 5x^2$
   (f) $y = 9x^3 - 14x + 11$
   (g) $y = x^{1/2} + 4x^{-1/2}$

2. Find the first and second derivatives of
   (a) $y = \sin 3x + \cos 2x$
   (b) $y = e^x + e^{-x}$

3. Find the second derivative of $y = \sin 4x$. Show that the sum of $16y$ and y'' is zero, that is $y'' + 16y = 0$.

4. If $y = x^4 - 2x^3 - 36x^2$, find the values of x for which $y'' = 0$.

5. Determine y'' given y is
   (a) $2 \sin 3x + 4 \cos 2x$
   (b) $\sin kt$, k constant
   (c) $\cos kt$, k constant
   (d) $A \sin kt + B \cos kt$
       A, B, k constants

6. Find y'' when y is given by
   (a) $6 + e^{3x}$ (b) $2 + 3x + 2e^{4x}$
   (c) $e^{2x} - e^{-2x}$ (d) $\frac{1}{e^x}$ (e) $e^x(e^x + 3)$

7. Given $y = 1000 + 200x + 6x^2 - x^3$, find the value(s) of x for which $y'' = 0$.

8. Given $y = 2e^x + \sin 2x$, evaluate y'' when $x = 1$.

9. Given $y = 2 \sin 4t - 5 \cos 2t$, evaluate y'' when $t = 1.2$.

33.5 Finding maximum and minimum points of a curve

Consider the graph sketched in Figure 33.3. There are a number of important points marked on this graph, all of which have one thing in

Figure 33.3
The curve has a
maximum at A, a
minimum at C and
points of inflexion at B
and D

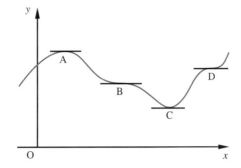

common. At each point A, B, C and D the gradient of the graph is zero. If
you were travelling from the left to the right then at these points the graph
would appear to be flat. Points where the gradient is zero are known as
stationary points. To the left of point A the curve is rising; to its right the
curve is falling. A point such as A is called a **maximum turning point** or
simply a **maximum**. To the left of point C the curve is falling; to its right
the curve is rising. A point such as C is called a **minimum turning point**, or
simply a **minimum**. Points B and D are known as **points of inflexion**. At
these points the slope of the curve is momentarily zero but then the curve
continues rising or falling as before.

Because at all these points the gradient is zero, they can be located by
looking for values of x that make the gradient function zero.

Key point

Stationary points are located by setting the gradient function equal to
zero, that is $y' = 0$.

WORKED EXAMPLE

33.17   Find the stationary points of $y = 3x^2 - 6x + 8$.

Solution   We first determine the gradient function y' by differentiating y. This is
found to be $y' = 6x - 6$. Stationary points occur when the gradient is zero,
that is when $y' = 6x - 6 = 0$. From this

$$6x - 6 = 0$$
$$6x = 6$$
$$x = 1$$

When $x = 1$ the gradient is zero. When $x = 1$, $y = 5$. Therefore $(1, 5)$ is
a stationary point. At this stage we cannot tell whether to expect a
maximum, minimum or point of inflexion; all we know is that one of these
occurs when $x = 1$. However, a sketch of the graph of $y = 3x^2 - 6x + 8$ is
shown in Figure 33.4, which reveals that the point is a minimum.

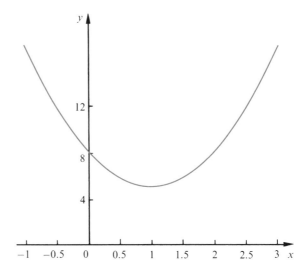

There are a number of ways to determine the nature of a stationary point once its location has been found. One way, as we have seen, is to sketch a graph of the function. Two alternative methods are now described.

Finding the nature of a stationary point by looking at the gradient on either side

At a stationary point we know that the gradient is zero. If this stationary point is a minimum, then we see from Figure 33.5 that, as we move from left to right, the gradient changes from negative to zero to positive. Therefore a little to the left of a minimum the gradient function will be negative; a little to the right it will be positive. Conversely, if the stationary point is a maximum, as we move from left to right, the gradient changes from positive to zero to negative. A little to the left of a maximum the gradient function will be positive; a little to the right it will be negative.

Minimum

Maximum

Point of inflexion

The behaviour of the gradient function close to a point of inflexion is also shown. The sign of the gradient function is the same on both sides of a point of inflexion. With this information we can determine the nature of a stationary point without first plotting a graph.

WORKED EXAMPLES

33.18   Find the location and nature of the stationary points of
$y = 2x^3 - 6x^2 - 18x$.

Solution   In order to find the location of the stationary points we must calculate the gradient function. This is $y' = 6x^2 - 12x - 18$. At a stationary point $y' = 0$ and so the equation that must be solved is $6x^2 - 12x - 18 = 0$. This quadratic equation is solved as follows:

$$6x^2 - 12x - 18 = 0$$

Factorising:

$$6(x^2 - 2x - 3) = 0$$

$$6(x + 1)(x - 3) = 0$$

$$\text{so that } x = -1 \text{ and } 3$$

The stationary points are located at $x = -1$ and $x = 3$. At these values, $y = 10$ and $y = -54$ respectively. We now find their nature:

When $x = -1$ A little to the left of $x = -1$, say at $x = -2$, we calculate the gradient of the graph using the gradient function. That is, $y'(-2) = 6(-2)^2 - 12(-2) - 18 = 24 + 24 - 18 = 30$. Therefore the graph is rising when $x = -2$. A little to the right of $x = -1$, say at $x = 0$, we calculate the gradient of the graph using the gradient function. That is, $y'(0) = 6(0)^2 - 12(0) - 18 = -18$. Therefore the graph is falling when $x = 0$. From this information we conclude that the turning point at $x = -1$ must be a maximum.

When $x = 3$ A little to the left of $x = 3$, say at $x = 2$, we calculate the gradient of the graph using the gradient function. That is, $y'(2) = 6(2)^2 - 12(2) - 18 = 24 - 24 - 18 = -18$. Therefore the graph is falling when $x = 2$. A little to the right of $x = 3$, say at $x = 4$, we calculate the gradient of the graph using the gradient function. That is, $y'(4) = 6(4)^2 - 12(4) - 18 = 96 - 48 - 18 = 30$. Therefore the graph is rising when $x = 4$. From this information we conclude that the turning point at $x = 3$ must be a minimum.

To show this behaviour a graph of $y = 2x^3 - 6x^2 - 18x$ is shown in Figure 33.6 where these turning points can be clearly seen.

Figure 33.6
Graph of
$y = 2x^3 - 6x^2 - 18x$

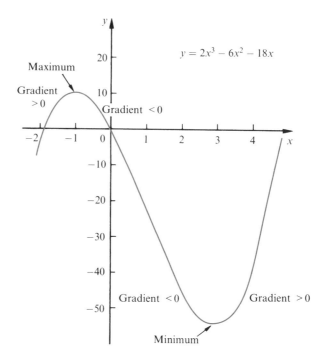

33.19   Find the location and nature of the stationary points of
$y = x^3 - 3x^2 + 3x - 1$.

Solution   First we find the gradient function: $y' = 3x^2 - 6x + 3$. At a stationary
point $y' = 0$ and so

$$3x^2 - 6x + 3 = 0$$

$$3(x^2 - 2x + 1) = 0$$

$$3(x - 1)(x - 1) = 0$$

$$\text{and so } x = 1$$

When $x = 1$, $y = 0$.

We conclude that there is only one stationary point, and this is at $(1, 0)$.
To determine its nature we look at the gradient function on either side of

Figure 33.7
Graph of
$y = x^3 - 3x^2 + 3x - 1$
showing the point of
inflexion

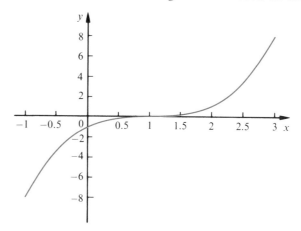

this point. At $x = 0$, say, $y'(0) = 3$. At $x = 2$, say, $y'(2) = 3(2^2) - 6(2) + 3 = 3$. The gradient function changes from positive to zero to positive as we move through the stationary point. We conclude that the stationary point is a point of inflexion. For completeness a graph is sketched in Figure 33.7.

The second-derivative test for maximum and minimum points

A simple test exists that uses the second derivative for determining the nature of a stationary point once it has been located.

Key point

If y'' is positive at the stationary point, the point is a minimum.
If y'' is negative at the stationary point, the point is a maximum.
If y'' is equal to zero, this test does not tell us anything and the previous method should be used.

WORKED EXAMPLE

33.20   Locate the stationary points of

$$y = \frac{x^3}{3} + \frac{x^2}{2} - 12x + 5$$

and determine their nature using the second-derivative test.

Solution   We first find the gradient function: $y' = x^2 + x - 12$. Setting this equal to zero to locate the stationary points we find

$$x^2 + x - 12 = 0$$

$$(x + 4)(x - 3) = 0$$

$$x = -4 \text{ and } 3$$

When $x = -4$, $y = \frac{119}{3}$. When $x = 3$, $y = -\frac{35}{2}$.

There are two stationary points: at $(-4, \frac{119}{3})$ and $(3, -\frac{35}{2})$. To apply the second-derivative test we need to find y''. This is found by differentiating y' to give $y'' = 2x + 1$. We now evaluate this at each stationary point in turn:

$x = -4$ When $x = -4$, $y''(-4) = 2(-4) + 1 = -7$. This is negative and we conclude from the second-derivative test that the point is a maximum point.

$x = 3$ When $x = 3$, $y''(3) = 2(3) + 1 = 7$. This is positive and we conclude from the second-derivative test that the point is a minimum point. Remember, if in doubt, that these results could be confirmed by sketching a graph of the function.

Self-assessment questions 33.5

1. State the three types of stationary point.

2. State the second-derivative test for determining the nature of a stationary point.

3. Under what condition will the second-derivative test fail?

Exercise 33.5

1. Determine the location and nature of any stationary point of the following functions:
   (a) $y = x^2 + 1$ (b) $y = -x^2$
   (c) $y = 2x^3 + 9x^2$
   (d) $y = -2x^3 + 27x^2$
   (e) $y = x^3 - 3x^2 + 3x + 1$

2. Determine the location of the maximum and minimum points of $y = \sin x$, $0 \leqslant x \leqslant 2\pi$.

3. Find the maximum and minimum points of $y = e^x - x$.

4. Find the maximum and minimum points of $y = 2x^3 - 3x^2 - 12x + 1$.

5. Determine the location and nature of the stationary points of $y = -\frac{x^3}{3} + x$.

6. Determine the location and nature of the stationary points of $y = x^5 - 5x + 1$.

Test and assignment exercises 33

1. Find y' when y equals
   (a) $7x$ (b) x^{15} (c) $x^{3/2}$ (d) $3x^{-1}$ (e) $5x^4$ (f) $16x^{-3}$

2. Find y' when y equals
   (a) $\dfrac{x+1}{x}$ (b) $3x^2 - 7x + 14$ (c) $\sin 3x$ (d) $e^{5x} + e^{4x}$ (e) $3e^{3x}$

3. Find the first and second derivatives of
   (a) $f(x) = e^{-3x}$ (b) $f(x) = \ln 2x$ (c) $f(x) = x + \dfrac{x^2}{2} + \dfrac{x^3}{3}$ (d) $f(x) = -15$
   (e) $f(x) = 3x$

4. Find y' when
   (a) $y = 3t - 7$ (b) $y = 16t^2 + 7t + 9$

5. Find the first derivative of each of the following functions:
   (a) $f(x) = x(x + 1)$ (b) $f(x) = x(x - 2)$ (c) $f(x) = x^2(x + 1)$
   (d) $f(x) = (x - 2)(x + 3)$ (e) $f(t) = 3t^2 + 7t + 11$ (f) $f(t) = 7(t - 1)(t + 1)$

6. Find the first and second derivatives of each of the following:
   (a) $2x^2$ (b) $4x^3$ (c) $8x^7 - 5x^2$ (d) $\dfrac{1}{x}$

7. Find the location and nature of the maxima and minima of
$y = 2x^3 + 15x^2 - 84x$

8. Find a function which when differentiated gives $\frac{1}{2}x^3$.

9. Find the gradient of y at the specified points:
(a) $y = 3x + x^2$ at $x = -1$ (b) $y = 4\cos 3x + \dfrac{\sin 4x}{2}$ at $x = 0.5$
(c) $y = 3e^{2x} - 2e^{-3x}$ at $x = 0.1$

10. Evaluate the second derivative of y at the specified point:
(a) $y = \frac{1}{2}\sin 4x$ $x = 1$ (b) $y = e^{-2x} + x^2$ $x = 0$
(c) $y = 3\ln 5x$ $x = 1$

11. Find the value(s) of t for which y' is zero:
(a) $y = t^3 - 12t^2 + 1$ (b) $y = \ln(2t) - 2t^2$

12. Determine the location and nature of the stationary points of
(a) $y = x^4 - 108x$ (b) $y = \ln 2x + \frac{1}{x}$

Techniques of differentiation

34

34.1 Introduction

In this chapter we study three rules of differentiation: the product, quotient and chain rules. These rules extend the range of functions that we are able to differentiate. They build upon the list of functions and their derivatives as given in Table 33.1 on page 408. This chapter assumes that you are familiar with this table; if you are not then please go back and study Chapter 33.

Using the product rule we are able to differentiate products of functions. Products of functions occur when one function is multiplied by another. For example, the product of the functions x^3 and $\sin 2x$ is $x^3 \sin 2x$. Using the product rule enables us to differentiate the product $y = x^3 \sin 2x$.

Using the quotient rule we can differentiate quotients of functions. A quotient is formed when one function is divided by another. For example, when x^2 is divided by $x^3 + 1$ we obtain the quotient $\frac{x^2}{x^3+1}$. The quotient rule enables us to differentiate functions of the form $y = \frac{x^2}{x^3+1}$.

Sometimes we will come across a function $y(x)$, say, where the variable x is itself a function of another variable, t say. So we have $y = y(x)$ and $x = x(t)$. For example, suppose $y(x) = x^3$ and $x(t) = \sin t$. We say that y is a **function of a function** and we can write

$$y = x^3 = (\sin t)^3$$

The chain rule enables us to differentiate a function of a function.

We now study each rule in turn.

34.2 The product rule

Suppose we are able to differentiate individually expressions $u(x)$ and $v(x)$. For example, we may have $u(x) = x^3$ and $v(x) = \sin 2x$. Using Table 33.1 we are able to find the derivatives $\frac{du}{dx}$ and $\frac{dv}{dx}$. The product rule then allows us to differentiate the product of u and v; that is, we can use the product rule to differentiate $y = u \times v$, in this case $y = x^3 \sin 2x$.

Key point

The product rule: if $y = uv$ then

$$\frac{dy}{dx} = \frac{du}{dx} v + u \frac{dv}{dx}$$

WORKED EXAMPLES

34.1   Use the product rule to differentiate $y = x^3 \sin 2x$.

Solution   We let $u(x) = x^3$ and $v(x) = \sin 2x$.
Then clearly

$$y = uv$$

and using Table 33.1 we have

$$\frac{du}{dx} = 3x^2 \qquad \frac{dv}{dx} = 2\cos 2x$$

Using the product rule we have

$$\frac{dy}{dx} = \frac{du}{dx} v + u \frac{dv}{dx}$$

$$= 3x^2 \sin 2x + x^3(2\cos 2x)$$

$$= x^2(3\sin 2x + 2x\cos 2x)$$

34.2   Use the product rule to differentiate $y = e^{2x}\cos x$.

Solution   We let

$$u(x) = e^{2x} \qquad v(x) = \cos x$$

so that

$$y = e^{2x}\cos x = uv$$

Using Table 33.1 we have

$$\frac{du}{dx} = 2e^{2x} \qquad \frac{dv}{dx} = -\sin x$$

Using the product rule we have

$$\frac{dy}{dx} = \frac{du}{dx}v + u\frac{dv}{dx}$$

$$= 2e^{2x}\cos x + e^{2x}(-\sin x)$$

$$= e^{2x}(2\cos x - \sin x)$$

34.3   Use the product rule to differentiate $y = x \ln x$.

Solution   Let $u = x$, $v = \ln x$. Then

$$y = x \quad \ln x = uv$$

and

$$\frac{du}{dx} = 1 \qquad \frac{dv}{dx} = \frac{1}{x}$$

So using the product rule

$$\frac{dy}{dx} = \frac{du}{dx}v + u\frac{dv}{dx}$$

$$= 1 \ln x + x\left(\frac{1}{x}\right)$$

$$= \ln x + 1$$

34.4   Find the gradient of $y = 2x^2 e^{-x}$ when $x = 1$.

Solution   Let $u = 2x^2$, $v = e^{-x}$ so that

$$y = 2x^2 e^{-x} = uv$$

and

$$\frac{du}{dx} = 4x \qquad \frac{dv}{dx} = -e^{-x}$$

Using the product rule we see

$$\frac{dy}{dx} = \frac{du}{dx}v + u\frac{dv}{dx}$$

$$= 4xe^{-x} + 2x^2(-e^{-x})$$

$$= 2xe^{-x}(2 - x)$$

Recall that the derivative, $\frac{dy}{dx}$, is the gradient of y. When $x = 1$

$$\frac{dy}{dx} = 2e^{-1}(2 - 1) = 0.7358$$

so that the gradient of $y = 2x^2 e^{-x}$ when $x = 1$ is 0.7358.

Self-assessment question 34.2

1. State the product rule for differentiating $y(x) = u(x)v(x)$.

Exercise 34.2

1. Use the product rule to differentiate
   (a) $x \sin x$ (b) $x^2 \cos x$
   (c) $\sin x \cos x$ (d) xe^x (e) xe^{-x}

2. Find the derivative of the following expressions:
   (a) $2x \ln(3x)$ (b) $e^{3x} \sin x$
   (c) $e^{-\frac{x}{2}} \cos 4x$ (d) $\sin 2x \cos 3x$
   (e) $e^x \ln(2x)$

3. By writing $\sin x/e^x$ in the form $e^{-x} \sin x$ find the gradient of
   $$y = \frac{\sin x}{e^x}$$
   when $x = 1$.

4. Find the derivative of
   $$y = \frac{\cos 2x}{x^2}$$
   (Hint: See the technique of Question 3.)

5. Differentiate
   $$y = (\sin x + x)^2$$

6. Differentiate
   (a) $e^{-x}x^2$ (b) $e^{-x}x^2 \sin x$

7. Given that
   $$y = x^3 e^x$$
   calculate the value(s) of x for which $\frac{dy}{dx} = 0$.

8. If u, v and w are functions of x and
   $$y = uvw$$
   show, using the product rule twice, that
   $$\frac{dy}{dx} = \frac{du}{dx}vw + u\frac{dv}{dx}w + uv\frac{dw}{dx}$$

34.3 The quotient rule

The quotient rule is the partner of the product rule. Whereas the product rule allows us to differentiate a product, that is uv, the quotient rule allows us to differentiate a quotient, that is u/v.

Suppose we are able to differentiate individually the expressions $u(x)$ and $v(x)$. The quotient rule then allows us to differentiate the quotient of u and v: that is, the quotient rule allows us to differentiate $y = u/v$.

Key point   The quotient rule: if

$$y = \frac{u}{v}$$

then

$$\frac{dy}{dx} = \frac{v\,\dfrac{du}{dx} - u\,\dfrac{dv}{dx}}{v^2}$$

WORKED EXAMPLES

34.5   Use the quotient rule to differentiate

$$y = \frac{\sin x}{x^2}$$

Solution   We let $u(x) = \sin x$ and $v(x) = x^2$. Then

$$y = \frac{u}{v} \qquad \text{and} \qquad \frac{du}{dx} = \cos x \qquad \frac{dv}{dx} = 2x$$

Using the quotient rule we have

$$\frac{dy}{dx} = \frac{v\,\dfrac{du}{dx} - u\,\dfrac{dv}{dx}}{v^2}$$

$$= \frac{x^2 \cos x - \sin x(2x)}{(x^2)^2}$$

$$= \frac{x(x \cos x - 2 \sin x)}{x^4}$$

$$= \frac{x \cos x - 2 \sin x}{x^3}$$

34.6   Use the quotient rule to differentiate $y = \dfrac{x+1}{x^2+1}$.

Solution   We let

$$u(x) = x + 1 \qquad v(x) = x^2 + 1$$

and so clearly

$$y = \frac{u}{v}$$

Now

$$\frac{du}{dx} = 1 \qquad \frac{dv}{dx} = 2x$$

Applying the quotient rule we have

$$\frac{dy}{dx} = \frac{v\dfrac{du}{dx} - u\dfrac{dv}{dx}}{v^2}$$

$$= \frac{(x^2+1)1 - (x+1)2x}{(x^2+1)^2}$$

$$= \frac{x^2 + 1 - 2x^2 - 2x}{(x^2+1)^2}$$

$$= \frac{1 - 2x - x^2}{(x^2+1)^2}$$

34.7 Use the quotient rule to differentiate $y = \dfrac{e^x + x}{\sin x}$.

Solution We let

$$u(x) = e^x + x \qquad v(x) = \sin x$$

and so

$$\frac{du}{dx} = e^x + 1 \qquad \frac{dv}{dx} = \cos x$$

Clearly

$$y = \frac{u}{v}$$

and so

$$\frac{dy}{dx} = \frac{v\dfrac{du}{dx} - u\dfrac{dv}{dx}}{v^2}$$

$$= \frac{\sin x(e^x + 1) - (e^x + x)\cos x}{(\sin x)^2}$$

Self-assessment question 34.3

1. State the quotient rule for differentiating $y(x) = \dfrac{u(x)}{v(x)}$.

Exercise 34.3

1. Use the quotient rule to differentiate the following functions:

   (a) $y = \dfrac{e^x}{x+1}$ (b) $y = \dfrac{\ln x}{x}$

   (c) $y = \dfrac{x}{\ln x}$ (d) $y = \dfrac{\cos t}{\sin t}$

   (e) $y = \dfrac{t^2 + t}{2t + 1}$

2. Evaluate the derivative of the following functions when $x = 2$:

   (a) $y = \dfrac{x+1}{x+2}$ (b) $y = \dfrac{2 \ln(3x)}{x+1}$

   (c) $y = \dfrac{\sin 2x}{\cos 3x}$ (d) $y = \dfrac{e^{2x}+1}{e^x + 1}$

   (e) $y = \dfrac{x^3 + 1}{x^2 + 1}$

3. Noting that

   $$\tan \theta = \frac{\sin \theta}{\cos \theta}$$

   find the derivative of $y = \tan \theta$.

4. Noting that

   $$\tan(k\theta) = \frac{\sin(k\theta)}{\cos(k\theta)} \qquad k \text{ constant}$$

   find the derivative of $y = \tan(k\theta)$.

34.4 The chain rule

Suppose we are given a function $y(x)$ where the variable x is itself a function of another variable, t say. We say that y is a **function of a function**. For example, suppose

$$y(x) = \cos x \qquad \text{and} \qquad x(t) = t^2$$

Then we can write

$$y = \cos(t^2)$$

There will be occasions when it is necessary to calculate $\frac{dy}{dt}$. This can be done by first finding $\frac{dy}{dx}$ and $\frac{dx}{dt}$ and then using the chain rule.

Key point

The **chain rule** states that if $y = y(x)$ and $x = x(t)$, then

$$\frac{dy}{dt} = \frac{dy}{dx} \times \frac{dx}{dt}$$

WORKED EXAMPLE

34.8   Use the chain rule to find $\frac{dy}{dt}$ when $y = \cos x$ and $x = t^2$.

Solution   When $y = \cos x$ our previous knowledge of differentiation (or use of Table 33.1) tells us that $\frac{dy}{dx} = -\sin x$. Also, when $x = t^2$, $\frac{dx}{dt} = 2t$. Using the chain rule,

$$\frac{dy}{dt} = \frac{dy}{dx} \times \frac{dx}{dt}$$

$$= (-\sin x) \times 2t$$

Using the fact that $x = t^2$ enables us to write this as

$$\frac{dy}{dt} = -2t \sin t^2$$

The chain rule is more useful when we are given a complicated function of a function and we are able to break it down into two simpler functions. This is usually done by making a suitable substitution. Consider carefully the following examples.

WORKED EXAMPLES

34.9   If $y = (7t + 3)^4$ find $\dfrac{dy}{dt}$.

Solution   Note that if we introduce a new variable x and make the substitution $7t + 3 = x$ then y takes the much simpler form $y = x^4$. Then, from $y = x^4$,

$$\frac{dy}{dx} = 4x^3$$

and from $x = 7t + 3$,

$$\frac{dx}{dt} = 7$$

Using the chain rule

$$\frac{dy}{dt} = \frac{dy}{dx} \times \frac{dx}{dt}$$

$$= 4x^3 \times 7$$

$$= 28x^3$$

$$= 28(7t + 3)^3$$

34.10   If $y = (3t^2 + 4)^{1/2}$, find the derivative $\dfrac{dy}{dt}$.

Solution   If we let $3t^2 + 4 = x$, then y looks much simpler: $y = x^{1/2}$. Then

$$\frac{dy}{dx} = \frac{1}{2}x^{-1/2} \qquad \text{and} \qquad \frac{dx}{dt} = 6t$$

Using the chain rule,

$$\frac{dy}{dt} = \frac{dy}{dx} \times \frac{dx}{dt}$$

$$= \frac{1}{2}x^{-1/2} \times 6t$$

$$= 3t(3t^2 + 4)^{-1/2}$$

34.11   Find the gradient of the function $y = e^{(t^2)}$ when $t = 0.5$.

Solution   Note that by writing $t^2 = x$ then $y = e^x$. Using the chain rule,

$$\frac{dy}{dt} = \frac{dy}{dx} \times \frac{dx}{dt}$$

$$= e^x \times 2t$$

$$= 2te^{(t^2)}$$

This is the gradient function of $y = e^{(t^2)}$. So, when $t = 0.5$ the value of the gradient is $(2)(0.5)e^{(0.5^2)} = e^{0.25} = 1.284$.

Self-assessment questions 34.4

1. State the chain rule for differentiating $y(t)$ when $y = y(x)$ and $x = x(t)$.

2. How would the chain rule be written if we wanted to differentiate $y(x)$ when y is given as $y = y(z)$ and $z = z(x)$?

3. Explain what is meant by a function of a function. Give an example.

4. Decide which rule – product, quotient or chain – it might be appropriate to use in order to differentiate the following functions:

    (a) $\dfrac{1}{x}\cos x$ (b) $x\sin x$ (c) $\sin\left(\dfrac{1}{x}\right)$

Exercise 34.4

1. Use the chain rule to find $\frac{dy}{dt}$ when

   (a) $y = \sin(x)$ and $x = t^2$

   (b) $y = x^{1/2}$ and $x = (t + 1)$

   (c) $y = e^x$ and $x = 2t^2$

   (d) $y = 3x^{5/2}$ and $x = (t^3 + 2t)$

2. Use the chain rule to differentiate

   (a) $y = \cos 3t^2$

   (b) $y = \sin(2t + 1)$

   (c) $y = \ln(3t - 2)$

   (d) $y = \sin(t^2 + 3t + 2)$

3. Use the chain rule to differentiate

   (a) $y = \sin(3x + 1)$

   (b) $y = e^{4x-3}$

   (c) $y = \cos(2x - 4)$

   (d) $y = (x^2 + 5x - 4)^{1/2}$

   (e) $y = \tan\frac{1}{x}$ (Hint: See Exercise 34.3, Question 4.)

4. In the special case of a function of a function when $y = \ln f(x)$, it can be shown from the chain rule that

   $$\frac{dy}{dx} = \frac{f'(x)}{f(x)}$$

   Verify this result using the chain rule in each of the following cases:

   (a) $y = \ln(4x - 3)$

   (b) $y = \ln(3x^2)$

   (c) $y = \ln(\sin x)$

   (d) $y = \ln(x^2 + 3x)$

Test and assignment exercises 34

1. Use the product rule to differentiate the following:

   (a) $e^x \sin 3x$ (b) $x^3 \ln x$ (c) $2x^2 \cos 3x$ (d) $(e^{2x} + 1)(x^2 + 3)$ (e) $\ln 2x \ln 3x$

2. Use the quotient rule to differentiate the following:

   (a) $\dfrac{x^2 + 3}{x^2 - 3}$ (b) $\dfrac{1}{x + 1}$ (c) $\dfrac{\cos x}{\sin x}$ (d) $\dfrac{e^x + 1}{e^x + 2}$ (e) $\dfrac{\ln 2x}{2x}$

3. Evaluate the derivative, $\frac{dy}{dx}$, when $x = 1$ given

   (a) $y = x^2 \sin 3x$ (b) $y = \dfrac{\sin 3x}{x^2}$ (c) $y = \dfrac{x^2}{\sin 3x}$

4. Differentiate

   (a) $x \ln x$ (b) $e^{2x} x \ln x$ (c) $\dfrac{e^{2x}}{x \ln x}$

5. (a) Differentiate $y = e^{-3x} x$ using the product rule.

   (b) Differentiate $y = x/e^{3x}$ using the quotient rule.

   (c) What do you notice about your answers? Can you explain this?

6. Calculate the value(s) of x for which the function

$$y = (x^2 - 3)e^{-x}$$

has a zero gradient.

7. Use the chain rule to find $\frac{dy}{dt}$ when

(a) $y = (t^2 + 1)^{1/3}$

(b) $y = \sin(3t^2 - 4t + 9)$

(c) $y = e^{-2t+3}$

(d) $y = (5t + 3)^4$

8. Evaluate the derivative of $P = (4t + 3)^5$ when $t = 1$.

9. Find the gradient of $y = 2\ln(x^2 + 1)$ when $x = 1$

10. Differentiate, with respect to x,

(a) $y = x^2 \sin x$

(b) $y = \dfrac{\sin x}{x}$

(c) $y = x\sin(x^2)$

(d) $y = \dfrac{\sin(x^2)}{x}$

Integration and areas under curves

35

Objectives: This chapter:

- introduces the reverse process of differentiation, which is called 'integration'
- explains what is meant by an 'indefinite integral' and a 'definite integral'
- shows how integration can be used to find the area under a curve

35.1 Introduction

In some applications it is necessary to find the area under a curve and above the x axis between two values of x, say a and b. Such an area is shown shaded in Figure 35.1. In this chapter we shall show how this area can be found using a technique known as **integration**. This area of study is known as **integral calculus**. However, before we do this it is necessary to consider two types of integration: **indefinite** and **definite** integration.

Figure 35.1
The shaded area is found by integration

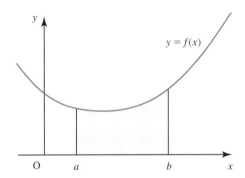

35.2 Indefinite integration: the reverse of differentiation

Suppose we ask 'what function can we differentiate to give $2x$?' In fact, there are lots of correct answers to this question. We can use our knowledge of differentiation to find these. The function $y = x^2$ has derivative $\frac{dy}{dx} = 2x$, so clearly x^2 is one answer. Furthermore, $y = x^2 + 13$ also has derivative $2x$ as do $y = x^2 + 5$ and $y = x^2 - 7$, because differentiation of any constant equals zero. Indeed any function of the form $y = x^2 + c$ where c is a constant has derivative equal to $2x$. The answer to the original question 'what function can we differentiate to give $2x$?' is then any function of the form $y = x^2 + c$. Here we have started with the derivative of a function and then found the function itself. This is differentiation in reverse. We call this process **integration** or more precisely **indefinite integration**. We say that $2x$ has been **integrated** with respect to x to give $x^2 + c$. The quantity $2x$ is called the **integrand**. The quantity $x^2 + c$ is called the **indefinite integral** of $2x$. Both differentiation and indefinite integration are shown schematically in Figure 35.2.

Figure 35.2
The indefinite integral of $2x$ is $x^2 + c$

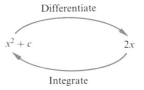

We now introduce a notation used to state this compactly. We write

$$\int 2x \, dx = x^2 + c$$

The $\int$ sign indicates that the reverse process to differentiation is to be performed. Following the function to be integrated is always a term of the form dx, which indicates the independent variable, in this case x. When finding indefinite integrals there will always be a constant known as the **constant of integration**. Every time we integrate, the constant of integration must be added.

As a further illustration, if $y = \sin x$ then we know from a table of derivatives that $y' = \cos x$. We can write this as

$$\int \cos x \, dx = \sin x + c$$

To find integrals it is often possible to use a table of derivatives such as that on page 408, and to read the table from right to left. However, it is more helpful to have a table of integrals such as that given in Table 35.1. As with the table of derivatives given in Chapter 33, when dealing with trigonometrical functions the angle x must always be measured in radians. The rule

$$\int x^n \, dx = \frac{1}{n+1} x^{n+1} + c$$

in the table applies for all values of n except $n = -1$.

Table 35.1
Table of indefinite integrals

$f(x)$	$\int f(x) \, dx$, all '$+c$'	Notes
$2x$	x^2	
x	$\frac{1}{2}x^2$	
k, constant	kx	
x^n	$\dfrac{1}{n+1} x^{n+1}$	$n \neq -1$
$\dfrac{1}{x}$	$\ln x$	
e^x	e^x	
e^{kx}	$\dfrac{e^{kx}}{k}$	k is a constant
$\sin x$	$-\cos x$	
$\cos x$	$\sin x$	
$\sin kx$	$-\dfrac{\cos kx}{k}$	k is a constant
$\cos kx$	$\dfrac{\sin kx}{k}$	k is a constant

WORKED EXAMPLES

35.1 Use the table to find the following indefinite integrals:

(a) $\displaystyle\int x^4 \, dx$ (b) $\displaystyle\int x^{-2} \, dx$

Solution (a) From the table we note that

$$\int x^n \, dx = \frac{1}{n+1} x^{n+1} + c$$

Therefore, using this formula with $n = 4$, we find

$$\int x^4 \, dx = \frac{1}{4+1} x^{4+1} + c = \frac{1}{5} x^5 + c$$

(b) Similarly, using the formula with $n = -2$ we find

$$\int x^{-2} \, dx = \frac{1}{-2+1} x^{-2+1} + c = \frac{1}{-1} x^{-1} + c = -x^{-1} + c$$

Note that because integration is the reverse of differentiation, in both cases the answer can be checked by differentiating.

35.2   Use Table 35.1 to find

(a) $\int e^x \, dx$ (b) $\int x^{-1} \, dx$

Solution   (a) Directly from the table we see that $\int e^x \, dx = e^x + c$.

(b) Noting that $x^{-1} = 1/x$, Table 35.1 gives $\int x^{-1} \, dx = \ln x + c$. It is important to realise that the formula for integrating x^n cannot be used here; the formula specifically excludes use of the value $n = -1$.

35.3   Use Table 35.1 to find the following indefinite integrals:

(a) $\int e^{2x} \, dx$ (b) $\int \cos 3x \, dx$

Solution   (a) Using Table 35.1 we note that

$$\int e^{kx} \, dx = \frac{e^{kx}}{k} + c$$

Using this result with $k = 2$ we find

$$\int e^{2x} \, dx = \frac{e^{2x}}{2} + c$$

(b) Using Table 35.1 we note that

$$\int \cos kx \, dx = \frac{\sin kx}{k} + c$$

Using this result with $k = 3$ we find

$$\int \cos 3x \, dx = \frac{\sin 3x}{3} + c$$

35.4   Find $\int \sqrt{x} \, dx$.

Solution Recall that $\sqrt{x}$ is the same as $x^{1/2}$. Therefore

$$\int \sqrt{x}\, dx = \int x^{1/2}\, dx = \frac{x^{3/2}}{3/2} + c = \frac{2}{3} x^{3/2} + c$$

Self-assessment questions 35.2

1. Explain what you understand by an indefinite integral.

2. Why is it always necessary to add a constant of integration when finding an indefinite integral?

Exercise 35.2

1. Use Table 35.1 to find the following indefinite integrals:

   (a) $\int x\, dx$ (b) $\int 2x\, dx$ (c) $\int x^5\, dx$

   (d) $\int x^7\, dx$ (e) $\int x^{-5}\, dx$

   (f) $\int x^{-1/2}\, dx$ (g) $\int x^0\, dx$

   (h) $\int x^{-1/4}\, dx$

2. Find

   (a) $\int 9\, dx$ (b) $\int \frac{1}{2}\, dx$ (c) $\int -7\, dx$

   (d) $\int 0.5\, dx$

3. Use Table 35.1 to find the following integrals:

   (a) $\int \sin 5x\, dx$ (b) $\int \sin \frac{1}{2} x\, dx$

   (c) $\int \cos \frac{x}{2}\, dx$ (d) $\int e^{-3x}\, dx$

   (e) $\int \sin 0.25x\, dx$ (f) $\int e^{-0.5x}\, dx$

   (g) $\int e^{x/2}\, dx$

4. Find the following integrals:

   (a) $\int \cos 5x\, dx$ (b) $\int \sin 4x\, dx$

   (c) $\int e^{6t}\, dt$ (d) $\int \cos\left(\frac{t}{3}\right)\, dt$

5. Find the following integrals:

   (a) $\int \cos 3y\, dy$ (b) $\int \frac{1}{\sqrt{x}}\, dx$

   (c) $\int \sin\left(\frac{3t}{2}\right)\, dt$ (d) $\int \sin(-x)\, dx$

35.3 Some rules for finding other indefinite integrals

Up to now the formulae we have used for finding indefinite integrals have applied to single terms. More complicated functions can be integrated if we make use of three simple rules. The first applies to the sum of two functions, the second to their difference and the third to a constant multiple of a function.

Rule 1:

$$\int [f(x) + g(x)] \, dx = \int f(x) \, dx + \int g(x) \, dx$$

In words, this says that to find the integral of a sum of two functions we simply find the integrals of the two separate parts and add these together.

WORKED EXAMPLE

35.5   Find the integral $\int (x^2 + x^3) \, dx$.

Solution   From Table 35.1 the integral of x^2 is $\frac{x^3}{3} + c_1$. The integral of x^3 is $\frac{x^4}{4} + c_2$. Note that we have called the constants of integration c_1 and c_2. The integral of $x^2 + x^3$ is simply the sum of these separate integrals, that is

$$\int (x^2 + x^3) \, dx = \frac{x^3}{3} + \frac{x^4}{4} + c_1 + c_2$$

The two constants can be combined into a single constant c, say, so that we would usually write

$$\int (x^2 + x^3) \, dx = \frac{x^3}{3} + \frac{x^4}{4} + c$$

The second rule is really an extension of the first, but we state it here for clarity.

Rule 2:

$$\int [f(x) - g(x)] \, dx = \int f(x) \, dx - \int g(x) \, dx$$

In words, this says that to find the integral of the difference of two functions we simply find the integrals of the two separate parts and find their difference.

WORKED EXAMPLES

35.6   Find the integral $\int (x^3 - x^6) \, dx$.

Solution

From Table 35.1 the integral of x^3 is $\frac{x^4}{4} + c_1$. The integral of x^6 is $\frac{x^7}{7} + c_2$. Note once again that we have called the constants of integration c_1 and c_2. The integral of $x^3 - x^6$ is simply the difference of these separate integrals, that is

$$\int (x^3 - x^6)\, dx = \frac{x^4}{4} - \frac{x^7}{7} + c_1 - c_2$$

As before the two constants can be combined into a single constant c, say, so that we would usually write

$$\int (x^3 - x^6)\, dx = \frac{x^4}{4} - \frac{x^7}{7} + c$$

35.7

Find $\int (\sin 2x - \cos 2x)\, dx$.

Solution

Using the second rule we can find the integral of each term separately. From Table 35.1 $\int \sin 2x\, dx = -(\cos 2x)/2 + c_1$ and $\int \cos 2x\, dx = (\sin 2x)/2 + c_2$. Therefore

$$\int (\sin 2x - \cos 2x)\, dx = \left(-\frac{\cos 2x}{2} + c_1 \right) - \left(\frac{\sin 2x}{2} + c_2 \right)$$

$$= -\frac{\cos 2x}{2} - \frac{\sin 2x}{2} + c$$

where we have combined the two constants into a single constant c.

Key point

Rule 3: If k is a constant, then

$$\int [k \times f(x)]\, dx = k \times \int f(x)\, dx$$

In words, the integral of k times a function is simply k times the integral of the function. This means that a constant factor can be taken through the integral sign and written at the front, as in the following example.

WORKED EXAMPLE

35.8

Find $\int 3x^4\, dx$.

Solution

Using the third rule we can write

$$\int 3x^4\, dx = 3 \times \int x^4\, dx$$

taking the constant factor out. Now $\int x^4 \, dx = \frac{x^5}{5} + c$ so that

$$\int 3x^4 \, dx = 3 \times \int x^4 \, dx = 3 \times \left(\frac{x^5}{5} + c \right)$$

$$= \frac{3x^5}{5} + 3c$$

It is common practice to write the constant $3c$ as simply another constant K, say, to give $\int 3x^4 \, dx = \frac{3x^5}{5} + K$.

Self-assessment question 35.3

1. State the rules for finding the indefinite integrals (a) $\int [f(x) + g(x)] \, dx$, (b) $\int [f(x) - g(x)] \, dx$ and (c) $\int [k \times f(x)] \, dx$, where k is a constant.

Exercise 35.3

1. Using Table 35.1 and the rules of integration given in this section, find

    (a) $\int (x^2 + x) \, dx$

    (b) $\int (x^2 + x + 1) \, dx$

    (c) $\int 4x^6 \, dx$ (d) $\int (5x + 7) \, dx$

    (e) $\int \frac{1}{x} \, dx$ (f) $\int \frac{1}{x^2} \, dx$

    (g) $\int \left(\frac{3}{x} + \frac{7}{x^2} \right) \, dx$ (h) $\int x^{1/2} \, dx$

    (i) $\int 4x^{-1/2} \, dx$ (j) $\int \frac{1}{\sqrt{x}} \, dx$

    (k) $\int (3x^3 + 7x^2) \, dx$

    (l) $\int (2x^4 - 3x^2) \, dx$

    (m) $\int (x^5 - 7) \, dx$

2. Find the following integrals:

    (a) $\int 3e^x \, dx$ (b) $\int \frac{1}{2} e^x \, dx$

    (c) $\int (3e^x - 2e^{-x}) \, dx$

    (d) $\int 5e^{2x} \, dx$

    (e) $\int (\sin x - \cos 3x) \, dx$

3. Integrate each of the following functions with respect to x:

    (a) $3x^2 - 2e^x$ (b) $\frac{x}{2} + 1$

    (c) $2(\sin 3x - 4 \cos 3x)$

    (d) $3 \sin 2x + 5 \cos 3x$ (e) $\frac{5}{x}$

4. Find $\int \left(\frac{5}{t} - t^2 \right) \, dt$

5. Find the following integrals:

(a) $\displaystyle\int (3 \sin 2t + 4 \cos 2t)\, dt$

(b) $\displaystyle\int (-\sin 3x - 2 \cos 4x)\, dx$

(c) $\displaystyle\int \left[1 + 2 \sin\left(\frac{x}{2}\right)\right] dx$

(d) $\displaystyle\int \left[\frac{1}{2} - \frac{1}{3} \cos\left(\frac{x}{3}\right)\right] dx$

6. Find the following integrals:

(a) $\displaystyle\int [e^x(1 + e^x)]\, dx$

(b) $\displaystyle\int [(x + 2)(x + 3)]\, dx$

(c) $\displaystyle\int (5 \cos 2t - 2 \sin 5t)\, dt$

(d) $\displaystyle\int \left[\frac{2}{x}(x + 3)\right] dx$

35.4 Definite integrals

There is a second form of integral known as a definite integral. This is written as

$$\int_a^b f(x)\, dx$$

and is called the **definite integral** of $f(x)$ between the **limits** $x = a$ and $x = b$. This definite integral looks just like the indefinite integral $\int f(x)\, dx$ except values a and b are placed below and above the integral sign. The value $x = a$ is called the **lower limit** and $x = b$ is called the **upper limit**. Unlike the indefinite integral, which is a function depending upon x, the definite integral has a specific numerical value. To find this value we first find the indefinite integral of $f(x)$, and then calculate

its value at the upper limit − its value at the lower limit

The resulting number is the value of the definite integral. To see how this value is obtained consider the following examples.

WORKED EXAMPLES

35.9 Evaluate the definite integral $\int_0^1 x^3\, dx$.

Solution We first find the indefinite integral $\int x^3\, dx$. This is

$$\int x^3\, dx = \frac{x^4}{4} + c$$

This is evaluated at the upper limit, where $x = 1$, giving $\frac{1}{4} + c$, and at the lower limit, where $x = 0$, giving $0 + c$, or simply c. The difference is found:

value at the upper limit $-$ value at the lower limit

$$= \left(\frac{1}{4} + c \right) - (c) = \frac{1}{4}$$

We note that the constant c has cancelled out; this will always happen and so for definite integrals it is usually omitted altogether. The calculation is written more compactly as

When evaluating definite integrals, the constant of integration may be omitted.

$$\int_0^1 x^3 \, dx = \left[\frac{x^4}{4} \right]_0^1 \quad \text{the indefinite integral of } x^3 \text{ is written in square brackets}$$

$$= \frac{1}{4} - 0$$

$$= \frac{1}{4}$$

Note that the definite integral $\int_0^1 x^3 \, dx$ has a definite numerical value, $\frac{1}{4}$.

35.10   Find the definite integral $\int_{-1}^1 (3x^2 + 7) \, dx$.

Solution

$$\int_{-1}^1 (3x^2 + 7) \, dx = [x^3 + 7x]_{-1}^1 \quad \text{(the indefinite integral of } 3x^2 + 7 \text{ is written in square brackets)}$$

$$= (\text{value of } x^3 + 7x \text{ when } x = 1)$$

$$- (\text{value of } x^3 + 7x \text{ when } x = -1)$$

$$= (1^3 + 7) - \{(-1)^3 - 7\}$$

$$= (8) - (-8)$$

$$= 16$$

35.11   Evaluate $\int_0^1 (e^x + 1) \, dx$.

Solution

$$\int_0^1 (e^x + 1) \, dx = [e^x + x]_0^1$$

$$= (e^1 + 1) - (e^0 + 0)$$

$$= e^1 + 1 - 1$$

$$= e^1$$

$$= 2.718$$

35.12   Find $\int_{\pi/4}^{\pi/2} \cos 2x \, dx$.

Solution   Note that the calculation must be carried out using radians and not degrees.

$$\int_{\pi/4}^{\pi/2} \cos 2x \, dx = \left[\frac{\sin 2x}{2} \right]_{\pi/4}^{\pi/2}$$

$$= \left(\frac{\sin \pi}{2} \right) - \left(\frac{\sin \pi/2}{2} \right)$$

$$= 0 - \frac{1}{2}$$

$$= -\frac{1}{2}$$

Self-assessment question 35.4

1. Explain the important way in which a definite integral differs from an indefinite integral.

Exercise 35.4

1. Evaluate the following definite integrals:

(a) $\int_{-1}^{1} 7x \, dx$ (b) $\int_{0}^{3} 7x \, dx$

(c) $\int_{-1}^{1} x^2 \, dx$ (d) $\int_{-1}^{1} (-x^2) \, dx$

(e) $\int_{0}^{2} x^2 \, dx$ (f) $\int_{0}^{3} (x^2 + 4x) \, dx$

(g) $\int_{0}^{\pi} \cos x \, dx$ (h) $\int_{1}^{2} \frac{1}{x} \, dx$

(i) $\int_{0}^{\pi} -\sin x \, dx$ (j) $\int_{1}^{4} 13 \, dx$

(k) $\int_{0}^{\pi} (x^2 + \sin x) \, dx$ (l) $\int_{1}^{3} 5x^2 \, dx$

(m) $\int_{0}^{1} (-x) \, dx$ (n) $\int_{1}^{3} x^{-2} \, dx$

(o) $\int_{0}^{1} \sin 3t \, dt$

2. Evaluate the following definite integrals:

(a) $\int_{0}^{2} (3e^x + 1) \, dx$

(b) $\int_{0}^{1} (2e^{2x} + \sin x) \, dx$

(c) $\int_{1}^{2} (2 \cos 2t - 3 \sin 2t) \, dt$

(d) $\int_{1}^{3} \left(\frac{2}{x} + \frac{x}{2} \right) \, dx$

3. Evaluate the following definite integrals:

(a) $\int_{0}^{2} [x(2x + 3)] \, dx$

(b) $\int_{-1}^{1} \left(3e^{2x} - \frac{3}{e^{2x}} \right) \, dx$

(c) $\displaystyle\int_0^{\pi} \left[2 \sin\left(\frac{t}{2}\right) - 4 \cos 2t \right] dt$

(b) $\displaystyle\int_0^1 [e^{-2x}(e^x - 2e^{-x})]\, dx$

(d) $\displaystyle\int_1^2 \left(\sin 3x - \frac{3}{x} \right) dx$

(c) $\displaystyle\int_0^{0.5} (\sin \pi t + \cos \pi t)\, dt$

(d) $\displaystyle\int_0^1 (2x - 3)^2\, dx$

4. Evaluate the following:

(a) $\displaystyle\int_1^2 \left[\frac{2}{x}\left(3 + \frac{1}{x} \right) \right] dx$

35.5 Areas under curves

As we stated at the beginning of this chapter, integration can be used to find the area underneath a curve and above the x axis. Consider the shaded area in Figure 35.3 which lies under the curve $y = f(x)$, above the x axis, between the values $x = a$ and $x = b$. Provided all of the required area lies above the x axis, this area is given by the following formula:

Key point

$$\text{area} = \int_a^b f(x)\, dx$$

Figure 35.3
The shaded area is given by $\int_a^b f(x)\, dx$

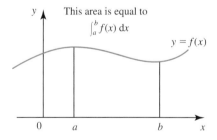

WORKED EXAMPLES

35.13   Find the area under $y = 7x + 3$ between $x = 0$ and $x = 2$.

Solution This area is shown in Figure 35.4. To find this area we must evaluate the definite integral $\int_0^2 (7x + 3)\, \mathrm{d}x$:

$$\text{area} = \int_0^2 (7x + 3)\, \mathrm{d}x = \left[\frac{7x^2}{2} + 3x \right]_0^2$$

$$= \left\{ \frac{7(2)^2}{2} + 3(2) \right\} - (0)$$

$$= 20$$

Figure 35.4
The shaded area is given by $\int_0^2 (7x + 3)\, \mathrm{d}x$

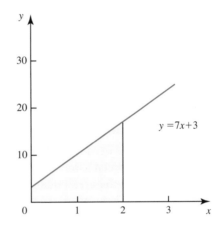

The required area is 20 square units. Convince yourself that this is correct by calculating the area of the shaded trapezium using the formula given in Table 21.3.

35.14 Find the area under the curve $y = x^2$ and above the x axis, between $x = 2$ and $x = 5$.

Solution The required area is shown in Figure 35.5.

$$\text{area} = \int_2^5 x^2\, \mathrm{d}x = \left[\frac{x^3}{3} \right]_2^5$$

$$= \left(\frac{5^3}{3} \right) - \left(\frac{2^3}{3} \right)$$

$$= \left(\frac{125}{3} \right) - \left(\frac{8}{3} \right)$$

Figure 35.5
The shaded area is given
by $\int_2^5 x^2 \, dx$

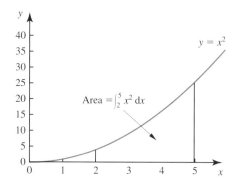

$$= \frac{117}{3}$$

$$= 39$$

The required area is 39 square units.

Note that the definite integral $\int_a^b f(x) \, dx$ represents the area under $y = f(x)$ between $x = a$ and $x = b$ only when all this area lies above the x axis. If between a and b the curve dips below the x axis, the result obtained will not be the area bounded by the curve. We shall see how to overcome this now.

Finding areas when some or all lie below the x axis

Consider the graph of $y = x^3$ shown in Figure 35.6. Suppose we want to know the area enclosed by the graph and the x axis between $x = -1$ and

Figure 35.6
Graph of $y = x^3$

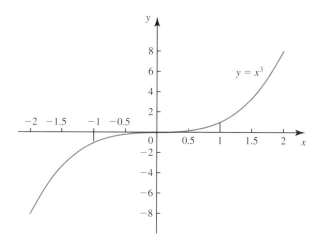

$x = 1$. Evaluating $\int_{-1}^{1} x^3 \, \mathrm{d}x$ we find

$$\int_{-1}^{1} x^3 \, \mathrm{d}x = \left[\frac{x^4}{4} \right]_{-1}^{1}$$

$$= \left\{ \frac{(1)^4}{4} \right\} - \left\{ \frac{(-1)^4}{4} \right\}$$

$$= \frac{1}{4} - \frac{1}{4}$$

$$= 0$$

The result of finding the definite integral is 0, even though there is clearly a region enclosed by the curve and the x axis. The reason for this is that part of the curve lies below the x axis. Let us evaluate two separate integrals, one representing the area below the x axis and the other representing the area above the x axis. The area below the x axis is found from

$$\int_{-1}^{0} x^3 \, \mathrm{d}x = \left[\frac{x^4}{4} \right]_{-1}^{0}$$

$$= \left\{ \frac{(0)^4}{4} \right\} - \left\{ \frac{(-1)^4}{4} \right\}$$

$$= 0 - \left(\frac{1}{4} \right)$$

$$= -\frac{1}{4}$$

The result is negative, that is $-\frac{1}{4}$. This means that the area bounded by the curve and the x axis between $x = -1$ and $x = 0$ is equal to $\frac{1}{4}$, the negative sign indicating that this area is below the x axis. The area above the x axis is found from

$$\int_{0}^{1} x^3 \, \mathrm{d}x = \left[\frac{x^4}{4} \right]_{0}^{1}$$

$$= \left\{ \frac{(1)^4}{4} \right\} - \left\{ \frac{(0)^4}{4} \right\}$$

$$= \frac{1}{4}$$

So the area between $x = 0$ and $x = 1$ is $\frac{1}{4}$. Note that we could have deduced this from the symmetry in the graph. The total area between $x = -1$ and $x = 1$ is therefore equal to $\frac{1}{4} + \frac{1}{4} = \frac{1}{2}$. The important point to note is that when some or all of the area lies below the x axis it is necessary to find the total area by finding the area of each part. It is therefore essential to draw a graph of the function to be integrated first.

Self-assessment question 35.5

1. Explain how the area bounded by a curve and the x axis can be found using a definite integral. What precautions must be taken when doing this?

Exercise 35.5

1. Find the area bounded by the curve $y = x^3 + 3$, the x axis and the lines $x = 1$ and $x = 3$.

2. Find the area bounded by the curve $y = 4 - x^2$ and the x axis between $x = -1$ and $x = 4$.

3. Find the area under $y = e^{2t}$ from $t = 1$ to $t = 4$.

4. Find the area under $y = \frac{1}{t}$ for $1 \leqslant t \leqslant 5$.

5. Find the area under $y = \sqrt{x}$ between $x = 1$ and $x = 2$.

6. Find the area under $y = \frac{1}{x}$ from $x = 1$ to $x = 4$.

7. Find the area under $y = \cos 2t$ from $t = 0$ to $t = 0.5$.

Test and assignment exercises 35

1. Find the following indefinite integrals:

(a) $\displaystyle\int 3x \, dx$

(b) $\displaystyle\int (2x + 4) \, dx$

(c) $\displaystyle\int (x^2 + 7) \, dx$

(d) $\displaystyle\int (3x^2 + 7x - 7) \, dx$

(e) $\displaystyle\int e^{1.5x} \, dx$

(f) $\displaystyle\int (\sin 5x - \cos 4x) \, dx$

(g) $\displaystyle\int 5e^{-x} \, dx$

2. Evaluate the following definite integrals:

(a) $\displaystyle\int_0^4 x^2 \, dx$

(b) $\displaystyle\int_0^4 x^4 \, dx$

(c) $\displaystyle\int_{-1}^2 3x^3 \, dx$

(d) $\displaystyle\int_0^1 (x^2 + 3x + 4) \, dx$

(e) $\displaystyle\int_{-2}^2 (x^3 + 2) \, dx$

(f) $\displaystyle\int_{-3}^2 (2x - 13) \, dx$

(g) $\displaystyle\int_3^5 (x + 2x + 3x^2) \, dx$

(h) $\displaystyle\int_2^7 5 \, dx$

(i) $\displaystyle\int_1^2 11x^4 \, dx$

3. Calculate the area bounded by the curve $y = x^2 - 4x + 3$ and the x axis from $x = 1$ to $x = 3$.

4. Sketch a graph of $y = \sin x$ from $x = 0$ to $x = \pi$. Calculate the area bounded by your curve and the x axis.

5. Find $\int 4e^{2t}\, dt$.

6. Find

$$\int \left(5 + x + x^2 + \frac{5}{x} \right) dx$$

7. Find

$$\int_1^2 \frac{1}{x^2}\, dx$$

8. Calculate the area bounded by $y = e^{-x}$ and the x axis between

(a) $x = 0$ and $x = 1$ (b) $x = 1$ and $x = 2$ (c) $x = 0$ and $x = 2$

Deduce that

$$\int_0^2 e^{-x}\, dx = \int_0^1 e^{-x}\, dx + \int_1^2 e^{-x}\, dx$$

9. Find the area under $y = 2e^{-t} + t$ from $t = 0$ to $t = 2$.

10. Evaluate

$$\int_1^2 \frac{3}{x^3}\, dx$$

11. Calculate the area above the x axis and below $y = 4 - x^2$.

12. Calculate the area above the positive x axis and below $y = e^{-x}$.

13. Evaluate the following:

(a) $\displaystyle\int_0^\pi (\sin 3t + 2 \cos 3t)\, dt$

(b) $\displaystyle\int_0^\pi \left[\cos \left(\frac{t}{2} \right) - \sin \left(\frac{t}{3} \right) \right] dt$

(c) $\displaystyle\int_0^1 \left(\frac{1}{2} \sin 2t - \cos 3t \right) dt$

14. Calculate $\int_{-1}^2 [x^2(3 + x)]\, dx$.

15. By using a suitable trigonometrical identity, find $\int 2 \sin t \cos t\, dt$.

16. By using appropriate trigonometrical identities, evaluate the following definite integrals:

(a) $\displaystyle\int_0^1 (\sin^2 t + \cos^2 t)\, dt$

(b) $\displaystyle\int_0^1 \cos^2 t\, dt$

(c) $\displaystyle\int_0^1 \sin^2 t\, dt$

Techniques of integration

36

Objectives: This chapter:

- explains what is meant by the product of two functions
- explains how to integrate products of functions using a method called 'integration by parts'
- explains how to use the method when finding definite integrals
- explains how to integrate using a substitution
- explains the use of partial fractions in integration

36.1 Products of functions

When asked to find the product of two functions we must multiply them together. So, the product of the functions $3x^2$ and e^x is $3x^2e^x$. Similarly the product of the functions x and $\ln x$ is $x \ln x$.

Key point

Given two functions $f(x)$ and $g(x)$, say, their product is the function $f(x)g(x)$.

When given a product of functions it is often necessary to identify the individual functions making up the product.

WORKED EXAMPLES

36.1 Identify suitable functions, $f(x)$ and $g(x)$, which make up the product function $f(x)g(x) = (3x^2 + 4x + 7)\sin x$.

Solution Here we can choose $f(x) = 3x^2 + 4x + 7$ and $g(x) = \sin x$ as the individual functions which make up the product $(3x^2 + 4x + 7)\sin x$.

36.2   Identify suitable functions $f(x)$ and $g(x)$ which make up the product function $f(x)g(x) = \dfrac{\ln x}{x}$.

Solution   At first sight the function $\dfrac{\ln x}{x}$ may not appear to be a product. But if we rewrite it as

$$\frac{\ln x}{x} = \frac{1}{x}\ln x$$

we can see the individual functions $f(x) = \dfrac{1}{x}$ and $g(x) = \ln x$.

36.3   Identify suitable functions, $f(x)$ and $g(x)$, which make up the product function $x^3\,e^x$.

Solution   Here we can choose $f(x) = x^3$ and $g(x) = e^x$. Note, however, that we could have $f(x) = x^2$ and $g(x) = xe^x$, or even $f(x) = x$ and $g(x) = x^2e^x$. So, often we have a choice over the functions $f(x)$ and $g(x)$ which can result in the same product $f(x)g(x)$. Sometimes writing a product in an alternative way will make it easier to integrate.

Even when a function is not strictly a product of functions it is sometimes convenient to think of it as a product with the constant function $f(x) = 1$. For example, we can think of $\ln x$ as $1 \times \ln x$, and so $\ln x$ is a product of the functions $f(x) = 1$ and $g(x) = \ln x$. This trick is sometimes useful in integration, as we shall see.

36.2 Integrating products of functions

When it is necessary to integrate a product of two functions we make use of a rule, known as **integration by parts**. Consider an integral of the form

$$\int u\frac{dv}{dx}\,dx$$

and notice that it is the integral of the product of a term u and a term $\dfrac{dv}{dx}$. The rule tells us how to integrate quantities of this form. It states:

Key point

Integration by parts:

$$\int u\frac{dv}{dx}\,dx = uv - \int v\frac{du}{dx}\,dx$$

To apply this formula to integrate a product we must let one of the functions in the product equal u. It must be possible to differentiate this function in order to calculate $\dfrac{du}{dx}$ which you will notice also appears in the formula. We let the other term in the product equal $\dfrac{dv}{dx}$. It must be possible to find the integral of this function because when we integrate $\dfrac{dv}{dx}$ we obtain v.

Note that v appears on the right-hand side of the formula, so we will need to calculate it.

The formula for integration by parts replaces the given integral by another which, it is hoped, is easier to evaluate. Consider the following worked examples.

WORKED EXAMPLES

36.4   Find $\displaystyle\int x \sin x \, dx$.

Solution   We must first match the given problem with the left-hand side of the formula for integration by parts; that is, we must let

$$u\frac{dv}{dx} = x \sin x$$

We choose $u = x$, and $\dfrac{dv}{dx} = \sin x$. Looking ahead to the right-hand side of the formula we see it is necessary to calculate $\dfrac{du}{dx}$ and v. Clearly,

$$\text{if } u = x \quad \text{then} \quad \frac{du}{dx} = 1$$

We find v by integration:

$$\text{if } \frac{dv}{dx} = \sin x \quad \text{then} \quad v = \int \sin x \, dx = -\cos x$$

When finding v there is no need to worry about a constant of integration. (When you have had a lot of practice, think about why this is the case.)

We can now apply the formula

$$\int u \frac{dv}{dx} \, dx = uv - \int v \frac{du}{dx} \, dx$$

$$\int x \sin x \, dx = x(-\cos x) - \int -\cos x \, (1) \, dx$$

$$= -x \cos x + \int \cos x \, dx$$

$$= -x \cos x + \sin x + c$$

36.5   Find $\displaystyle\int 5x e^{2x} \, dx$.

Solution   We must first match the given problem with the left-hand side of the formula for integration by parts; that is, we must let

$$u\frac{dv}{dx} = 5x e^{2x}$$

We choose $u = 5x$, and $\dfrac{dv}{dx} = e^{2x}$. To apply the formula it is necessary to calculate $\dfrac{du}{dx}$ and v. Clearly,

$$\text{if } u = 5x \quad \text{then} \quad \frac{du}{dx} = 5$$

We find v by integration:

$$\text{if } \frac{dv}{dx} = e^{2x} \quad \text{then} \quad v = \int e^{2x}\,dx = \frac{1}{2}e^{2x}$$

We can now apply the formula:

$$\int u \frac{dv}{dx}\,dx = uv - \int v \frac{du}{dx}\,dx$$

$$\int 5xe^{2x}\,dx = 5x\left(\frac{1}{2}e^{2x}\right) - \int \left(\frac{1}{2}e^{2x}\right)(5)\,dx$$

$$= \frac{5xe^{2x}}{2} - \frac{5}{2}\int e^{2x}\,dx$$

$$= \frac{5xe^{2x}}{2} - \frac{5}{2}\left(\frac{1}{2}e^{2x}\right) + c$$

$$= \frac{5xe^{2x}}{2} - \frac{5}{4}e^{2x} + c$$

Sometimes it is necessary to apply the formula more than once as in the following worked example.

36.6 Find $\int x^2 \cos x\,dx$.

Solution We must first match the given problem with the left-hand side of the formula for integration by parts; that is, we must let

$$u \frac{dv}{dx} = x^2 \cos x$$

We choose $u = x^2$, and $\dfrac{dv}{dx} = \cos x$. We need to calculate $\dfrac{du}{dx}$ and v. Clearly,

$$\text{if } u = x^2 \quad \text{then} \quad \frac{du}{dx} = 2x$$

We find v by integration:

$$\text{if } \frac{dv}{dx} = \cos x \quad \text{then} \quad v = \int \cos x\,dx = \sin x$$

We can now apply the formula

$$\int u \frac{dv}{dx} \, dx = uv - \int v \frac{du}{dx} \, dx$$

$$\int x^2 \cos x \, dx = x^2 \sin x - \int \sin x \, (2x) \, dx$$

The new integral, on the right-hand side, can be written $2 \int x \sin x \, dx$ and this has already been found in Worked Example 36.4. Using that result we find

$$\int x^2 \cos x \, dx = x^2 \sin x - 2(-x \cos x + \sin x) + c$$

$$= x^2 \sin x + 2x \cos x - 2 \sin x + c$$

36.7   Find $\int \ln x \, dx$.

Solution   In this worked example there appears to be no product of functions. However, in § 36.1 we noted that we can think of any function as a product with the constant function 1, so we write $\ln x = (\ln x) \cdot 1$.

The integral of $\ln x$ is neither a standard nor commonly known integral. However, the derivative of $\ln x$ is known to be $\frac{1}{x}$. Hence to apply the formula for integration by parts we choose

$$u = \ln x \qquad \text{and} \qquad \frac{dv}{dx} = 1$$

so that

$$\frac{du}{dx} = \frac{1}{x} \qquad v = \int 1 \, dx = x$$

Then

$$\int u \frac{dv}{dx} \, dx = uv - \int v \frac{du}{dx} \, dx$$

$$\int (\ln x) \cdot 1 \, dx = (\ln x)x - \int x \frac{1}{x} \, dx$$

$$= x \ln x - \int 1 \, dx$$

$$= x \ln x - x + c$$

Self-assessment questions 36.2

1. What is meant by the product of two functions?

2. State the formula for integration by parts.

Exercise 36.2

1. Find the following integrals:

   (a) $\int x \cos x \, dx$ (b) $\int 3te^{2t} \, dt$ (c) $\int xe^{-3x} \, dx$ (d) $\int 4z \ln z \, dz$

2. Use integration by parts twice to find $\int x^2 \sin x \, dx$.

36.3 Definite integrals

When it is necessary to evaluate a definite integral involving products of functions we still use the formula for integration by parts. It is modified to take account of the limits on the integral. The formula states:

Key point

Integration by parts formula for definite integrals:

$$\int_a^b u \frac{dv}{dx} \, dx = [uv]_a^b - \int_a^b v \frac{du}{dx} \, dx$$

WORKED EXAMPLES

36.8   Find $\int_0^2 xe^{-x} \, dx$.

Solution   Let $u \dfrac{dv}{dx} = xe^{-x}$ and choose $u = x$ and $\dfrac{dv}{dx} = e^{-x}$ so that

$$\frac{du}{dx} = 1 \quad \text{and} \quad v = \int e^{-x} \, dx = -e^{-x}$$

We can now apply the formula with limits of integration $a = 0$ and $b = 2$:

$$\int_a^b u \frac{dv}{dx} \, dx = [uv]_a^b - \int_a^b v \frac{du}{dx} \, dx$$

$$\int_0^2 xe^{-x} \, dx = [x(-e^{-x})]_0^2 - \int_0^2 -e^{-x}(1) \, dx$$

$$= [-xe^{-x}]_0^2 + \int_0^2 e^{-x} \, dx$$

$$= [-xe^{-x}]_0^2 + [-e^{-x}]_0^2$$

$$= \{-2e^{-2} - (0)\} + \{-e^{-2} - (-1)\}$$

$$= 1 - 3e^{-2}$$

36.9   Find $\int_{\pi/4}^{\pi/2} 7x \sin 3x \, dx$.

Solution   Remember that when doing calculus with trigonometrical functions any calculations must be carried out using radians and not degrees. You may need to alter the mode of your calculator.

Let $u\dfrac{dv}{dx} = 7x \sin 3x$ and choose $u = 7x$ and $\dfrac{dv}{dx} = \sin 3x$ so that

$$\frac{du}{dx} = 7 \quad \text{and} \quad v = \int \sin 3x\,dx = -\frac{1}{3}\cos 3x$$

We can now apply the formula with $a = \frac{\pi}{4}$ and $b = \frac{\pi}{2}$:

$$\int_a^b u\frac{dv}{dx}\,dx = [uv]_a^b - \int_a^b v\frac{du}{dx}\,dx$$

$$\int_{\pi/4}^{\pi/2} 7x \sin 3x\,dx = \left[7x\left(-\frac{1}{3}\cos 3x\right)\right]_{\pi/4}^{\pi/2} - \int_{\pi/4}^{\pi/2}\left(-\frac{1}{3}\cos 3x\right)(7)\,dx$$

$$= \left[7x\left(-\frac{1}{3}\cos 3x\right)\right]_{\pi/4}^{\pi/2} + \frac{7}{3}\int_{\pi/4}^{\pi/2}\cos 3x\,dx$$

$$= \left[7x\left(-\frac{1}{3}\cos 3x\right)\right]_{\pi/4}^{\pi/2} + \left[\frac{7}{9}\sin 3x\right]_{\pi/4}^{\pi/2}$$

$$= \left\{\frac{7\pi}{2}\left(-\frac{1}{3}\cos\frac{3\pi}{2}\right)\right\} - \left\{\frac{7\pi}{4}\left(-\frac{1}{3}\cos\frac{3\pi}{4}\right)\right\}$$

$$+ \left(\frac{7}{9}\sin\frac{3\pi}{2}\right) - \left(\frac{7}{9}\sin\frac{3\pi}{4}\right)$$

Now

$$\cos\frac{3\pi}{2} = 0 \quad \cos\frac{3\pi}{4} = -\frac{1}{\sqrt{2}} \quad \sin\frac{3\pi}{2} = -1 \quad \sin\frac{3\pi}{4} = \frac{1}{\sqrt{2}}$$

and so this simplifies to

$$\int_{\pi/4}^{\pi/2} 7x \sin 3x\,dx = -\frac{7\pi}{12\sqrt{2}} - \frac{7}{9} - \frac{7}{9\sqrt{2}} = -2.6236 \text{ to 4 d.p.}$$

Exercise 36.3

1. Evaluate the following integrals:

   (a) $\displaystyle\int_0^1 xe^x\,dx$ (b) $\displaystyle\int_1^2 xe^x\,dx$ (c) $\displaystyle\int_1^2 xe^{-x}\,dx$ (d) $\displaystyle\int_{-1}^1 xe^{-x}\,dx$

   (e) $\displaystyle\int_1^2 \ln x\,dx$ (f) $\displaystyle\int_0^\pi x\cos 2x\,dx$ (g) $\displaystyle\int_0^{\pi/2} x\cos 2x\,dx$ (h) $\displaystyle\int_{-\pi/2}^{\pi/2} x\sin x\,dx$

36.4 Integration by substitution

Integration by substitution allows us to extend considerably the range of functions that we can integrate. Suppose we wish to integrate $f(x)$, that is find $\int f(x)\,\mathrm{d}x$. In some cases it is more useful to introduce another function, $y = g(x)$, and write the required integral in terms of y, rather than in terms of x. With this substitution we write $\int f(x)\,\mathrm{d}x$ in terms of y. The following examples illustrate the technique.

It is not always obvious that a particular integral can be found using a substitution. Even if it is clear that a substitution will help, the actual substitution to be used may not be obvious. As your experience grows you will be able to judge better when a substitution is needed and what that substitution needs to be. Although the following examples illustrate the process, it is essential that you try the exercises to gain experience in this area.

WORKED EXAMPLES

36.10   Find $\int (3x + 1)^4 \,\mathrm{d}x$.

Solution   Although we could expand $(3x + 1)^4$ and then integrate term by term, we will use integration by substitution. We make the substitution $y = 3x + 1$. The integrand, $(3x + 1)^4$, is then simply y^4. Now we need to write 'dx' in terms of y also. Since $y = 3x + 1$ then

$$\frac{\mathrm{d}y}{\mathrm{d}x} = 3$$

and so $\mathrm{d}x = \frac{1}{3}\mathrm{d}y$. Hence we have

$$\int (3x + 1)^4 \,\mathrm{d}x = \int y^4 \frac{1}{3}\,\mathrm{d}y$$

$$= \frac{1}{3} \int y^4 \,\mathrm{d}y$$

$$= \frac{1}{3}\frac{y^5}{5} + c$$

$$= \frac{y^5}{15} + c$$

$$= \frac{(3x + 1)^5}{15} + c$$

36.11   Find $\int x \sin(2x^2)\,\mathrm{d}x$.

Solution   We introduce the substitution $y = 2x^2$. Then

$$\frac{dy}{dx} = 4x$$

and so $dx = \dfrac{1}{4x}\,dy$. We are now able to write the integral in terms of y:

$$\int x \sin(2x^2)\,dx = \int x \sin y \frac{1}{4x}\,dy$$

$$= \int \frac{1}{4} \sin y \,dy$$

$$= -\frac{\cos y}{4} + c$$

$$= -\frac{\cos(2x^2)}{4} + c$$

36.12   Evaluate $\displaystyle\int_1^5 t\sqrt{t^2 + 1}\,dt$.

Solution   We use the substitution $y = t^2 + 1$. Then

$$\frac{dy}{dt} = 2t \quad\text{from which}\quad dt = \frac{1}{2t}\,dy$$

Turning attention to the limits we recall that in the original integral the limits are limits on t, that is the limits are $t = 1$ to $t = 5$. When we write the integral in terms of y we must remember that the limits must be updated also.
   When $t = 1$, $y = 1^2 + 1 = 2$; when $t = 5$, $y = 5^2 + 1 = 26$. We are now ready to write the integral in terms of y:

$$\int_1^5 t\sqrt{t^2 + 1}\,dt = \int_2^{26} t\sqrt{y}\frac{1}{2t}\,dy \qquad \text{(note the limits on } y)$$

$$= \frac{1}{2}\int_2^{26} \sqrt{y}\,dy$$

$$= \frac{1}{2}\left[\frac{2y^{3/2}}{3}\right]_2^{26}$$

$$= \frac{1}{3}\left[y^{3/2}\right]_2^{26}$$

$$= \frac{1}{3}\left(26^{3/2} - 2^{3/2}\right)$$

$$= 43.25$$

36.13   Use integration by substitution to evaluate $\displaystyle\int_1^2 \frac{3}{(2t+1)^2}\,dt$.

Solution   We introduce $y = 2t + 1$. The integrand, $\dfrac{3}{(2t+1)^2}$, may then be written as $\dfrac{3}{y^2}$. Since $y = 2t + 1$, then

$$\frac{dy}{dt} = 2 \quad\text{from which}\quad dt = \frac{1}{2}\,dy$$

The limits on t are 1 and 2. The corresponding limits on y need to be found. When $t = 1$, $y = 3$; when $t = 2$, $y = 5$. Hence

$$\int_1^2 \frac{3}{(2t+1)^2} \, dt = \int_3^5 \frac{3}{y^2} \frac{1}{2} \, dy$$

$$= \frac{3}{2} \int_3^5 \frac{1}{y^2} \, dy$$

$$= \frac{3}{2} \left[-\frac{1}{y} \right]_3^5$$

$$= \frac{3}{2} \left(-\frac{1}{5} + \frac{1}{3} \right) = 0.2$$

36.14 By using a suitable substitution find $\int \frac{1}{ax+b} \, dx$ where a and b are constants.

Solution Let $y = ax + b$ so that $\dfrac{dy}{dx} = a$, $\dfrac{1}{a} dy = dx$. Hence

$$\int \frac{1}{ax+b} \, dx = \int \frac{1}{y} \left(\frac{1}{a} \right) \, dy$$

$$= \frac{1}{a} \ln y + c$$

$$= \frac{1}{a} \ln(ax+b) + c$$

Key point

$$\int \frac{1}{ax+b} \, dx = \frac{1}{a} \ln(ax+b) + c \quad \text{where } a \text{ and } b \text{ are constants}$$

WORKED EXAMPLES

36.15 Find

(a) $\displaystyle\int \frac{1}{x+1} \, dx$ (b) $\displaystyle\int \frac{1}{x+3} \, dx$ (c) $\displaystyle\int \frac{1}{4x+7} \, dx$ (d) $\displaystyle\int \frac{1}{3-2x} \, dx$

Solution Using the above Key point we have

(a) $\displaystyle\int \frac{1}{x+1} \, dx = \ln(x+1) + c$ $\qquad a = 1, b = 1$

(b) $\displaystyle\int \frac{1}{x+3} \, dx = \ln(x+3) + c$ $\qquad a = 1, b = 3$

(c) $\displaystyle\int \frac{1}{4x+7} \, dx = \frac{1}{4} \ln(4x+7) + c$ $\qquad a = 4, b = 7$

(d) $\displaystyle\int \frac{1}{3-2x} \, dx = -\frac{1}{2} \ln(3-2x) + c$ $\qquad a = -2, b = 3$

Exercise 36.4

1. Use integration by substitution to find

   (a) $\displaystyle\int (4x+3)^3 \, \mathrm{d}x$ (b) $\displaystyle\int \sqrt{3x-1} \, \mathrm{d}x$ (c) $\displaystyle\int \frac{2}{1-x} \, \mathrm{d}x$ (d) $\displaystyle\int \sin(6x+5) \, \mathrm{d}x$

   (e) $\displaystyle\int \cos\left(\frac{x}{2}+3\right) \mathrm{d}x$

2. By using the substitution $y = x^2$ find

   $$\int x e^{(x^2)} \, \mathrm{d}x$$

3. By using the substitution $y = \frac{1}{x}$, find

   $$\int \frac{e^{\frac{1}{x}}}{x^2} \, \mathrm{d}x$$

4. Find

   (a) $\displaystyle\int \left(\frac{\theta}{3}-1\right)^5 \mathrm{d}\theta$ (b) $\displaystyle\int t^2 \sqrt{t^3-1} \, \mathrm{d}t$ (c) $\displaystyle\int \cos\left(\frac{y+1}{2}\right) \mathrm{d}y$ (d) $\displaystyle\int \sin(10-x) \, \mathrm{d}x$

   (e) $\displaystyle\int \frac{1}{e^{2x-1}} \, \mathrm{d}x$

5. Evaluate

   (a) $\displaystyle\int_{2}^{6} \left(\frac{x}{2}-1\right)^4 \mathrm{d}x$ (b) $\displaystyle\int_{-0.5}^{1} \cos(4\theta+2) \, \mathrm{d}\theta$ (c) $\displaystyle\int_{\frac{\pi}{2}}^{\pi} 2\sin(\pi-2\theta) \, \mathrm{d}\theta$

   (d) $\displaystyle\int_{0}^{1} x e^{(-x^2)} \, \mathrm{d}x$ (e) $\displaystyle\int_{0}^{5} \sqrt{10-2x} \, \mathrm{d}x$

36.5   Integration by partial fractions

We have seen in §9.5 how to express a single fraction as the sum of its partial fractions. By expressing a fraction as its partial fractions, integration may then be possible. The following examples illustrate this.

WORKED EXAMPLES

36.16   Express $\dfrac{3x+5}{x^2+4x+3}$ as its partial fractions. Hence find $\displaystyle\int \frac{3x+5}{x^2+4x+3} \, \mathrm{d}x$.

Solution
$$\frac{3x+5}{x^2+4x+3} = \frac{3x+5}{(x+1)(x+3)} = \frac{A}{x+1} + \frac{B}{x+3}$$

from which

$$3x + 5 = A(x + 3) + B(x + 1)$$

Using the above equation, putting $x = -1$ results in $A = 1$; putting $x = -3$ results in $B = 2$. So

$$\frac{3x + 5}{x^2 + 4x + 3} = \frac{1}{x + 1} + \frac{2}{x + 3}$$

Hence

$$\int \frac{3x + 5}{x^2 + 4x + 3} \, dx = \int \frac{1}{x + 1} + \frac{2}{x + 3} \, dx$$

$$= \ln(x + 1) + 2\ln(x + 3) + c$$

(see Worked Example 36.15).

36.17 Express $\dfrac{2x + 3}{4x^2 + 4x + 1}$ as its partial fractions and hence find

$$\int \frac{2x + 3}{4x^2 + 4x + 1} \, dx.$$

Solution

$$\frac{2x + 3}{4x^2 + 4x + 1} = \frac{2x + 3}{(2x + 1)^2} = \frac{A}{2x + 1} + \frac{B}{(2x + 1)^2}$$

from which

$$2x + 3 = A(2x + 1) + B$$

With $x = -0.5$, we see that $B = 2$. With $x = 0$ we have $3 = A(1) + B$ from which $A = 1$. So

$$\frac{2x + 3}{4x^2 + 4x + 1} = \frac{1}{2x + 1} + \frac{2}{(2x + 1)^2}$$

Hence

$$\int \frac{2x + 3}{4x^2 + 4x + 1} \, dx = \int \frac{1}{2x + 1} \, dx + \frac{2}{(2x + 1)^2} \, dx$$

$$= \frac{\ln(2x + 1)}{2} - \frac{1}{2x + 1} + c$$

by using the substitution $y = 2x + 1$.

36.18 Evaluate $\displaystyle\int_0^1 \frac{5x + 8}{2x^2 + 7x + 5} \, dx.$

Solution Expressing $\dfrac{5x + 8}{2x^2 + 7x + 5}$ as its partial fractions yields

$$\frac{3}{2x + 5} + \frac{1}{x + 1}$$

So

$$\int_0^1 \frac{5x+8}{2x^2+7x+5}\,dx = \int_0^1 \frac{3}{2x+5} + \frac{1}{x+1}\,dx$$

$$= \left[\frac{3}{2}\ln(2x+5) + \ln(x+1) \right]_0^1$$

$$= \frac{3}{2}\ln 7 + \ln 2 - \frac{3}{2}\ln 5 - \ln 1$$

$$= 1.1979$$

Exercise 36.5

1. Express $\dfrac{x+6}{x^2+6x+8}$ as its partial fractions and hence find $\displaystyle\int \frac{x+6}{x^2+6x+8}\,dx$.

2. Find $\displaystyle\int \frac{10x-1}{4x^2-1}\,dx$.

3. Find $\displaystyle\int \frac{3x-2}{9x^2-6x+1}\,dx$ by first expressing the integrand as its partial fractions.

4. Find $\displaystyle\int_1^2 \frac{3x^2+6x+2}{x^3+3x^2+2x}\,dx$.

Test and assignment exercises 36

1. Find the following integrals:

    (a) $\displaystyle\int 4x\sin x\,dx$ (b) $\displaystyle\int xe^{4x}\,dx$ (c) $\displaystyle\int xe^{\frac{1}{2}x}\,dx$ (d) $\displaystyle\int x\ln 4x\,dx$

2. Evaluate the following integrals:

    (a) $\displaystyle\int_{-1}^2 xe^x\,dx$ (b) $\displaystyle\int_1^4 xe^{2x}\,dx$ (c) $\displaystyle\int_1^3 4\ln x\,dx$ (d) $\displaystyle\int_0^\pi x\sin 5x\,dx$

    (e) $\displaystyle\int_{-\pi/3}^{\pi/3} x\cos x\,dx$

3. Use integration by parts twice to find $\displaystyle\int_0^\pi x^2\sin x\,dx$.

4. Find $\displaystyle\int_1^e \frac{\ln x}{x^3}\,dx$.

5. (Harder) Let $I = \displaystyle\int e^x\sin x\,dx$. Using integration by parts twice show that

    $$I = \frac{1}{2}e^x(\sin x - \cos x)$$

6. By means of the given substitution find the following integrals:

(a) $\int (3x - 7)^6 \, dx \quad y = 3x - 7$

(b) $\int \sin(2\theta + \pi) \, d\theta \quad y = 2\theta + \pi$

(c) $\int \sqrt{9 - t} \, dt \quad y = 9 - t$

(d) $\int 3\cos(\pi - \theta) \, d\theta \quad y = \pi - \theta$

(e) $\int e^{4x+1} \, dx \quad y = 4x + 1$

7. Find

(a) $\int 2\sin\left(\dfrac{\theta}{2} + 1\right) d\theta$ (b) $\int 3x\cos(x^2) \, dx$ (c) $\int xe^{3-x^2} \, dx$

(d) $\int \dfrac{3}{(2y + 3)^2} \, dy$ (e) $\int \sin\left(\dfrac{\theta + \pi}{2}\right) d\theta$

8. By means of a suitable substitution show

$$\int \sin(k\theta + \alpha) \, d\theta = \frac{-\cos(k\theta + \alpha)}{k} + c \quad k, \alpha \text{ constants}$$

9. By means of a suitable substitution show

$$\int \cos(k\theta + \alpha) \, d\theta = \frac{\sin(k\theta + \alpha)}{k} + c \quad k, \alpha \text{ constants}$$

10. Evaluate

(a) $\displaystyle\int_2^3 (7 - 2x)^8 \, dx$ (b) $\displaystyle\int_1^2 2\sin(3\theta + 2) \, d\theta$ (c) $\displaystyle\int_0^\pi 4\cos(2\pi - 3\theta) \, d\theta$

(d) $\displaystyle\int_0^1 (x + 1)e^{x^2+2x} \, dx$ (e) $\displaystyle\int_0^1 \dfrac{1}{2 + y} \, dy$

11. By use of partial fractions find the following integrals:

(a) $\int \dfrac{5x + 14}{x^2 + 6x + 8} \, dx$ (b) $\int \dfrac{x - 5}{x^2 - 7x + 12} \, dx$ (c) $\int \dfrac{3x + 14}{x^2 + 10x + 25} \, dx$

(d) $\int \dfrac{x - 2}{x^2 - 8x + 16} \, dx$ (e) $\int \dfrac{3x^2 - 1}{x^3 - x} \, dx$

12. Evaluate the following:

(a) $\displaystyle\int_8^9 \dfrac{5x - 8}{x^2 - 5x - 14} \, dx$ (b) $\displaystyle\int_6^8 \dfrac{x + 11}{x^2 - 6x + 5} \, dx$ (c) $\displaystyle\int_0^2 \dfrac{2x + 3}{4x^2 + 4x + 1} \, dx$

Functions of more than one variable and partial differentiation

Objectives: This chapter:

- introduces functions that have two (or more) inputs
- shows how functions with two inputs can be represented graphically
- introduces partial differentiation

37.1 Functions of two independent variables

In Chapter 16 a function was defined as a mathematical rule that operates upon an input to produce a single output. We saw that the input is referred to as the independent variable because we are free, within reason, to choose its value. The output is called the dependent variable because its value *depends* upon the value of the input. Commonly we use the letter x to represent the input, y the output and f the function, in which case we write $y = f(x)$. Examples include $y = 5x^2$, $y = 1/x$ and so on. In such cases, there is a single independent variable, x.

We now move on to consider examples in which there are two independent variables. This means that there will be two inputs to a function, each of which can be chosen independently. Once these values have been chosen, the function rule will be used to process them in order to produce a single output – the dependent variable. Notice that although there is now more than one input, there is still a single output.

WORKED EXAMPLES

37.1 A function f is defined by $f(x, y) = 3x + 4y$. Calculate the output when the input values are $x = 5$ and $y = 6$.

Solution   In this example, the two independent variables are labelled x and y. The function is labelled f. When the input x takes the value 5, and the input y takes the value 6, the output is $3 \times 5 + 4 \times 6 = 15 + 24 = 39$. We could write this as $f(5, 6) = 39$.

37.2   A function f is defined by $f(x, y) = 11x^2 - 7y + 2$. Calculate the output when the inputs are $x = -2$ and $y = 3$.

Solution   In this example we are required to find $f(-2, 3)$. Substitution into the function rule produces

$$f(-2, 3) = 11 \times (-2)^2 - 7 \times (3) + 2 = 44 - 21 + 2 = 25$$

37.3   A function is defined by $f(x, y) = x/y$. Calculate the output when the inputs are $x = 9$ and $y = 3$.

Solution $f(9, 3) = 9/3 = 3$

37.4   A function is defined by $f(x, y) = e^{x+y}$.

(a) Calculate the output when the inputs are $x = -1$ and $y = 2$.
(b) Show that this function can be written in the equivalent form $f(x, y) = e^x e^y$.

Solution   (a) $f(-1, 2) = e^{-1+2} = e^1 = e = 2.718$ (3 d.p.)

(b) Using the first law of indices we can write e^{x+y} as $e^x e^y$.

Note that in the above worked examples both x and y are independent variables. We can choose a value for y quite independently of the value we have chosen for x. This is quite different from the case of a function of a single variable when, for example, if we write $y = 5x^2$, choosing x automatically determines y.

It is common to introduce another symbol to stand for the output. So, in Worked Example 37.1 we may write $z = f(x, y) = 3x + 4y$ or simply $z = 3x + 4y$. In Worked Example 37.2 we can write simply $z = 11x^2 - 7y + 2$. Here the value of z depends upon the values chosen for x and y, and so z is referred to as the dependent variable.

Key point   A function of two variables is a rule that produces a single output when values of two independent variables are chosen.

Functions of two variables are introduced because they arise naturally in applications in business, economics, physical sciences, engineering etc. Consider the following examples, which illustrate this.

WORKED EXAMPLES

37.5   The volume V of the cylinder shown in Figure 37.1 is given by the formula $V = \pi r^2 h$, where r is the radius and h is the height of the cylinder. Suppose

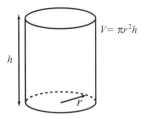

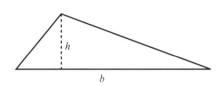

Figure 37.1
The volume V depends upon the two independent variables r and h, and is given by $V = \pi r^2 h$

Figure 37.2
The area A depends upon the two independent variables b and h, and is given by $A = \frac{1}{2}bh$

we choose a value for the radius and a value for the height. Note that we can choose these values independently. We can then use the formula to determine the volume of the cylinder. We can regard V as depending upon the two independent variables r and h and write $V = f(r, h) = \pi r^2 h$. Here, the dependent variable is V since the value of the volume depends upon the values chosen for the radius and the height.

37.6   The area A of the triangle shown in Figure 37.2 is given by the formula $A = \frac{1}{2} bh$, where b is the base length and h is the height. We can choose values for b and h independently, and we can use the formula to find the area. We see that the area depends upon the two independent variables b and h, and so $A = f(b, h)$. For example, if we choose $b = 7$ cm and $h = 3$ cm then the function rule tells us $A = f(7, 3) = \frac{1}{2} \times 7 \times 3 = 10.5$. The area of the triangle is 10.5 cm^2.

37.7   When a sum of money is invested for one year the simple interest earned, I, depends upon the two independent variables P, the amount invested, and i, the interest rate per year expressed as a decimal, that is $I = f(P, i)$. The function rule is $I = Pi$. Calculate the interest earned when £500 is invested for one year at an interest rate of 0.06 (that is, 6%).

Solution   We seek $f(500, 0.06)$, which equals $500 \times 0.06 = 30$. That is, the simple interest earned is £30.

Self-assessment questions 37.1

1. Explain what is meant by a function of two variables.

2. Explain the different roles played by the variable y when (a) f is a function of two variables such that $z = f(x, y)$, and (b) f is a function of a single variable such that $y = f(x)$.

3. Suppose $P = f(V, T) = RT/V$ where R is a constant. State which variables are dependent and which are independent.

4. Suppose $y = f(x, t)$. State which variables are dependent and which are independent.

Exercise 37.1

1. Given $z = f(x, y) = 7x + 2y$ find the output when $x = 8$ and $y = 2$.

2. If $z = f(x, y) = -11x + y$ find (a) $f(2, 3)$, (b) $f(11, 1)$.

3. If $z = f(x, y) = 3e^x - 2e^y + x^2y^3$ find $z(1, 1)$.

4. If $w = g(x, y) = 7 - xy$ find the value of the dependent variable w when $x = -3$ and $y = -9$.

5. If $z = f(x, y) = \sin(x + y)$ find $f(20°, 30°)$ where the inputs are angles measured in degrees.

6. If $f(x, t) = e^{2xt}$ find $f(0.5, 3)$.

37.2 Representing a function of two independent variables graphically

By now you will be familiar with the way in which a function of one variable is represented graphically. For example, you have seen in Chapter 17 how the graph of the function $y = f(x)$ is drawn by plotting the independent variable x on the horizontal axis, and the dependent variable y on the vertical axis. Given a value of x, the function rule enables us to calculate a value for y, and the point with coordinates (x, y) is then plotted. Joining all such points produces a graph of the function.

When two independent variables are involved, as in $z = f(x, y)$, the plotting of the graph is rather more difficult because we now need two axes for the independent variables and a third axis for the dependent variable. This means that the graph is drawn in three dimensions – a task that is particularly difficult to do in the two-dimensional plane of the paper. Easily usable computer software is readily available for plotting graphs in three dimensions, so rather than attempt this manually we shall be content with understanding the process involved in producing the graph, and illustrating this with several examples.

Three axes are drawn at right angles, and these are labelled x, y and z as shown in Figure 37.3. There is more than one way of doing this, but the usual convention is shown here. When labelled in this way the axes are said to form a **right-handed set**. The axes intersect at the origin O.

Figure 37.3
Three perpendicular axes, labelled x, y and z

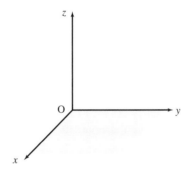

WORKED EXAMPLE

37.8   Given the function of two variables $z = f(x, y) = 3x + 4y$ calculate the value of the function when x and y take the following values. Represent each point graphically.

(a) $x = 3$, $y = 2$ (b) $x = 5$, $y = 0$ (c) $x = 0$, $y = 5$

Solution   In each case we use the function rule $z = f(x, y) = 3x + 4y$ to calculate the corresponding value of z:

(a) $z = 3(3) + 4(2) = 17$ (b) $z = 3(5) + 4(0) = 15$
(c) $z = 3(0) + 4(5) = 20$

It is conventional to write the coordinates of the points in the form (x, y, z) so that the three points in this example are $(3, 2, 17)$, $(5, 0, 15)$ and $(0, 5, 20)$.

Each of the points can then be drawn as shown in Figure 37.4, where they have been labelled A, B and C respectively. Notice that the z coordinate of point A, which is 17, gives the height of A above the point $(3, 2)$ in the x–y plane. Similar comments apply to points B and C.

Figure 37.4
Three points in three-dimensional space with coordinates A$(3, 2, 17)$, B$(5, 0, 15)$ and C$(0, 5, 20)$

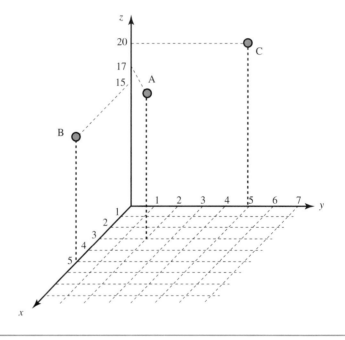

If we were to continue selecting more points and plotting them we would find that all the points lie in a plane. This plane is shown in Figure 37.5. Notice that in each case the z coordinate is the height of the plane above the point (x, y) in the x–y plane. For example, the point C is 20 units above the point $(0, 5)$ in the x–y plane.

Figure 37.5
The plane $z = 3x + 4y$

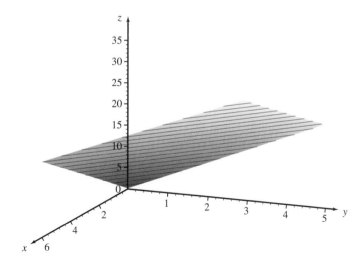

In the previous worked example we saw that when we drew a graph of the function of two variables $z = 3x + 4y$ we obtained a plane. In more general cases the graph of $z = f(x, y)$ will be a curved surface, and the z coordinate is the height of the surface above the point (x, y). Some more examples of functions of two variables and their graphs are shown in Figures 37.6 and 37.7. It would be a useful exercise for you to try to reproduce these graphs for yourself using computer software available in your college or university.

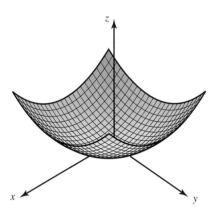

Figure 37.6
The function $z = x^2 + y^2$

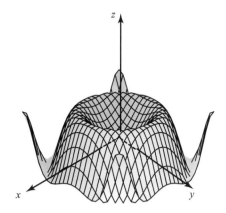

Figure 37.7
The function $z = \sin(x^2 + y^2)$

Self-assessment question 37.2

1. Draw three perpendicular axes and label the axes in the correct order according to the convention stated in this section.

1. Given the function $z = 3x - 6y$, find the z coordinate corresponding to each of the following points, which lie in the $x - y$ plane:
   (a) $(4, 4)$ (b) $(0, 3)$ (c) $(5, 0)$
   Plot the points on a three-dimensional graph.

37.3   Partial differentiation

Before attempting this section it is essential that you have a thorough understanding of differentiation of functions of one variable, and you should revise Chapters 33 and 34 if necessary. In particular, you may need to refer frequently to Table 33.1 on page 408, which gives the derivatives of some common functions.

In Chapter 33 we introduced differentiation of a function of a single variable and showed how the derivative of such a function can be calculated. Recall that this gives useful information about the gradient of the graph of the function at different points. In this section we explain how functions of two variables are differentiated.

In Chapter 33 we differentiated y, which was a function of x, to obtain the derivative $\frac{dy}{dx}$. Note that the dependent variable (in this case y) is differentiated with respect to the independent variable (in this case x).

Consider now z, which depends upon two variables x and y. Recall that z is then the dependent variable, and x and y are the independent variables. Hence z can be differentiated with respect to x to produce a derivative, and it can also be differentiated with respect to y to produce another, different derivative. So for functions of two variables there are two derivatives: we can no longer talk about *the* derivative of z. This is a fundamental difference between functions of one variable and functions of two variables.

When differentiating functions of two variables we refer to this process as **partial differentiation**, and instead of using a normal letter d, as in $\frac{dy}{dx}$, we use a curly d instead and write ∂. Do not be put off by this notation – you will soon get used to it.

When we differentiate z with respect to (w.r.t.) x we denote the derivative produced by $\frac{\partial z}{\partial x}$. When we differentiate z w.r.t. y we denote the derivative produced by $\frac{\partial z}{\partial y}$.

Key point

If $z = f(x, y)$, then the derivatives $\frac{\partial z}{\partial x}$ and $\frac{\partial z}{\partial y}$ are called the (first) partial derivatives of z.

We now explain how these derivatives are calculated.

Partial differentiation with respect to *x*

Suppose we have a function $z = 5x + 11$. You will recall that its derivative $\frac{dz}{dx} = 5$. Note that the derivative of the constant 11 is zero. Similarly if $z = 6x + 12$, $\frac{dz}{dx} = 6$. Here the derivative of the constant 12 is zero. We are now ready to introduce partial differentiation with respect to *x*.

Consider the function $z = 5x + y$. When we differentiate with respect to *x* we treat any occurrence of the variable *y* as *though it were a constant*. Hence in this case the derivative of *y* is zero, and we write

$$\text{if } z = 5x + y \quad \text{then} \quad \frac{\partial z}{\partial x} = 5 + 0 = 5$$

Similarly,

$$\text{if } z = 7x^2 - y \quad \text{then} \quad \frac{\partial z}{\partial x} = 14x - 0 = 14x$$

If *y* is treated as a constant, then so too will be multiples of *y*, such as $7y$ and $-3y$. Furthermore, any functions of *y*, such as y^2 and e^y, will also be regarded as constants.

WORKED EXAMPLES

37.9 Calculate $\frac{\partial z}{\partial x}$ when $z = 9x + 2$.

Solution We must differentiate $z = 9x + 2$ with respect to *x*. This function is particularly simple because *y* does not appear at all. The derivative of $9x$ is 9, and the derivative of the constant 2 is zero. Hence $\frac{\partial z}{\partial x} = 9$.

37.10 Calculate $\frac{\partial z}{\partial x}$ when $z = 9x + y$.

Solution To find $\frac{\partial z}{\partial x}$ we treat *y* as though it were a constant. Imagine it were just a number like the '2' in the previous example. Then $\frac{\partial z}{\partial x} = 9$.

37.11 Calculate $\frac{\partial z}{\partial x}$ when $z = y - 19x$.

Solution
$$\frac{\partial z}{\partial x} = -19$$

37.12 Calculate $\frac{\partial z}{\partial x}$ when $z = 3x + 4y + 11$.

Solution
$$\frac{\partial z}{\partial x} = 3$$

In this case 11 is a constant and we treat *y*, and hence $4y$, as a constant.

37.13 Calculate $\frac{\partial z}{\partial x}$ when $z = 4x^2 - 3y$.

Solution
$$\frac{\partial z}{\partial x} = 8x - 0 = 8x$$

The quantity $-3y$ is treated as a constant.

37.14   Calculate $\frac{\partial z}{\partial x}$ when $z = x^2 - y^2$.

Solution y, and hence y^2, is treated as a constant. When $-y^2$ is differentiated with respect to x the result will be zero. Hence if $z = x^2 - y^2$ then $\frac{\partial z}{\partial x} = 2x$.

37.15   Calculate $\frac{\partial z}{\partial x}$ when $z = 5y^3 - x^4$.

Solution y, and hence $5y^3$, is treated as a constant. If $z = 5y^3 - x^4$, then

$$\frac{\partial z}{\partial x} = -4x^3$$

37.16   Calculate $\frac{\partial z}{\partial x}$ when $z = 3x - 4y^3$.

Solution   If $z = 3x - 4y^3$, then $\frac{\partial z}{\partial x} = 3$.

Recall that

if $4x$ is differentiated with respect to x the result is 4
if $5x$ is differentiated with respect to x the result is 5

Extending this to the following function of two variables,

if $z = yx$ is differentiated with respect to x the result is y

We see that because y is treated as if it were a constant then $\frac{\partial z}{\partial x} = y$.

WORKED EXAMPLES

37.17   Calculate $\frac{\partial z}{\partial x}$ when $z = yx^2$.

Solution   To find $\frac{\partial z}{\partial x}$ imagine that you were trying to differentiate $3x^2$, say. The result would be $3(2x) = 6x$. Hence if $z = yx^2$ then $\frac{\partial z}{\partial x} = y(2x)$ or simply $2xy$.

37.18   Calculate $\frac{\partial z}{\partial x}$ when $z = 3yx^2$.

Solution   Treating y as a constant we see that this function is of the form constant $\times x^2$, and so $\frac{\partial z}{\partial x} = (3y)(2x) = 6xy$.

37.19   Calculate $\frac{\partial z}{\partial x}$ when $z = 3x^2 + 4xy + 11$.

Solution $$\frac{\partial z}{\partial x} = 6x + 4y$$

37.20   Calculate $\frac{\partial z}{\partial x}$ when

(a) $z = x^2 e^y$ (b) $z = x^2 + e^y$ (c) $z = 3x^2 \cos y$
(d) $z = 3x^2 + \cos y$ (e) $z = 3x^2 + 4x \sin y$

Solutions   (a) y and hence e^y are treated as constants. Then $\frac{\partial z}{\partial x} = 2x e^y$.

(b) $\frac{\partial z}{\partial x} = 2x$.

(c) y and hence $\cos y$ are treated as constants. Then $\frac{\partial z}{\partial x} = 6x \cos y$.

(d) $\frac{\partial z}{\partial x} = 6x$.

(e) $\frac{\partial z}{\partial x} = 6x + 4\sin y$.

37.21 Calculate $\frac{\partial z}{\partial x}$ when (a) $z = \sin 3x$, (b) $z = \sin yx$.

Solution

(a) If $z = \sin 3x$ then $\frac{\partial z}{\partial x} = 3\cos 3x$.

(b) If $z = \sin yx$ then $\frac{\partial z}{\partial x} = y\cos yx$ because y is treated as a constant.

Key point

The partial derivative with respect to x of a function $z = f(x, y)$ is denoted by $\frac{\partial z}{\partial x}$ and is calculated by differentiating the function with respect to x and treating y as though it were a constant.

Partial differentiation with respect to y

When we differentiate a function $f(x, y)$ with respect to y we treat any occurrence of the variable x *as though it were a constant*. The partial derivative with respect to y of a function $z = f(x, y)$ is denoted by $\frac{\partial z}{\partial y}$. Consider the following examples.

WORKED EXAMPLES

37.22 Find $\frac{\partial z}{\partial y}$ when $z = 3y^4 + 4x^2 + 8$.

Solution

When calculating $\frac{\partial z}{\partial y}$ we treat any occurrence of x as if it were a constant. So the term $4x^2$ is treated as a constant, and its partial derivative with respect to y is zero. That is,

$$\text{if } z = 3y^4 + 4x^2 + 8 \quad \text{then} \quad \frac{\partial z}{\partial y} = 12y^3 + 0 + 0 = 12y^3$$

37.23 Find $\frac{\partial z}{\partial y}$ when $z = 3x^2y$.

Solution

Because x is treated as a constant, we are dealing with a function of the form $z = \text{constant} \times y$. The derivative with respect to y will be simply the constant factor. That is,

$$\text{if } z = 3x^2y \quad \text{then} \quad \frac{\partial z}{\partial y} = 3x^2$$

37.24 Find $\frac{\partial z}{\partial y}$ when $z = 4xy^3$.

Solution

Because x is treated as a constant, we are dealing with a function of the form $z = \text{constant} \times y^3$. The derivative with respect to y will be the constant $\times 3y^2$, that is

$$\text{if } z = 4xy^3 \quad \text{then} \quad \frac{\partial z}{\partial y} = (4x)(3y^2) = 12xy^2$$

Key point

The partial derivative with respect to y of a function $z = f(x, y)$ is denoted by $\frac{\partial z}{\partial y}$ and is calculated by differentiating the function with respect to y and treating x as though it were a constant.

It will be necessary to work with symbols other than z, x and y. Consider the following worked example.

WORKED EXAMPLE

37.25   Consider the function $w = f(p, t)$. Find $\frac{\partial w}{\partial p}$ and $\frac{\partial w}{\partial t}$ when $w = 3t^7 + 4pt + p^2$.

Solution

$$\frac{\partial w}{\partial p} = 4t + 2p \qquad \frac{\partial w}{\partial t} = 21t^6 + 4p$$

Self-assessment questions 37.3

1. State the derivative of (a) 7, (b) 11, (c) k, a constant.

2. Suppose $z = f(x, y)$. When y is differentiated partially with respect to x, state the result.

3. What is the symbol for the partial derivative of z with respect to x?

4. What is the symbol for the partial derivative of z with respect to y?

5. If y is a function of x and t, that is $y = f(x, t)$, write down the symbol for the partial derivative of y with respect to x and the symbol for the partial derivative of y with respect to t.

Exercise 37.3

1. In each case, given $z = f(x, y)$, find $\frac{\partial z}{\partial x}$ and $\frac{\partial z}{\partial y}$.

   (a) $z = 5x + 11y$ (b) $z = -7y - 14x$
   (c) $z = 8x$ (d) $z = -5y$
   (e) $z = 3x + 8y - 2$
   (f) $z = 17 - 3x + 2y$ (g) $z = 8$
   (h) $z = 8 - 3y$ (i) $z = 2x^2 - 7y$
   (j) $z = 9 - 3y^3 + 7x$
   (k) $z = 9 - 9(x - y)$
   (l) $z = 9(x + y + 3)$

2. In each case, given $z = f(x, y)$, find $\frac{\partial z}{\partial x}$ and $\frac{\partial z}{\partial y}$.

   (a) $z = xy$ (b) $z = 3xy$

   (c) $z = -9yx$ (d) $z = x^2y$
   (e) $z = 9x^2y$ (f) $z = 8xy^2$

3. If $z = 9x + y^2$ evaluate $\frac{\partial z}{\partial x}$ and $\frac{\partial z}{\partial y}$ at the point $(4, -2)$.

4. Find $\frac{\partial z}{\partial x}$ and $\frac{\partial z}{\partial y}$ when

   (a) $z = e^{2x}$ (b) $z = e^{5y}$
   (c) $z = e^{xy}$ (d) $z = 4e^{2y}$

5. If $y = x \sin t$ find $\frac{\partial y}{\partial x}$ and $\frac{\partial y}{\partial t}$.

37.4 Partial derivatives requiring the product or quotient rules

Consider the following more demanding worked examples, which use the rules developed in Chapter 34.

WORKED EXAMPLES

37.26   Find $\frac{\partial z}{\partial x}$ and $\frac{\partial z}{\partial y}$ when $z = yxe^{2x}$.

Solution   To find $\frac{\partial z}{\partial x}$ we treat y as constant. We are dealing with a function of the form constant $\times xe^{2x}$. Note that this function itself contains a product of the functions x and e^{2x}, and so we must use the product rule for differentiation. The derivative of xe^{2x} is $e^{2x}(1) + x(2e^{2x}) = e^{2x}(1 + 2x)$. Hence

$$\text{if } z = yxe^{2x} \quad \text{then} \quad \frac{\partial z}{\partial x} = y(e^{2x}(1 + 2x)) = ye^{2x}(1 + 2x)$$

To find $\frac{\partial z}{\partial y}$ we treat x as constant. In turn, this means that xe^{2x} is constant too. This time the calculation is much simpler because we are dealing with a function of the form $z = \text{constant} \times y$. So,

$$\text{if } z = yxe^{2x} \quad \text{then} \quad \frac{\partial z}{\partial y} = xe^{2x}$$

37.27   Find $\frac{\partial z}{\partial x}$ and $\frac{\partial z}{\partial y}$ when $z = ye^x/x$.

Solution   To find $\frac{\partial z}{\partial x}$ we treat y as constant. We are dealing with a function of the form constant $\times e^x/x$. Note that this function itself contains a quotient of the functions e^x and x and so we must use the quotient rule for differentiation. The derivative of e^x/x is $(e^x x - e^x(1))/x^2 = e^x(x - 1)/x^2$. Hence

$$\text{if } z = \frac{ye^x}{x} \quad \text{then} \quad \frac{\partial z}{\partial x} = \frac{ye^x(x - 1)}{x^2}$$

To find $\frac{\partial z}{\partial y}$ we treat x as constant. In turn, this means that e^x/x is constant too. This calculation is simple because we are dealing with a function of the form $z = \text{constant} \times y$. So,

$$\text{if } z = \frac{ye^x}{x} \quad \text{then} \quad \frac{\partial z}{\partial y} = \frac{e^x}{x}$$

Exercise 37.4

1. Find $\frac{\partial z}{\partial x}$ and $\frac{\partial z}{\partial y}$ when

   (a) $z = yxe^x$ (b) $z = xye^y$ (c) $z = xe^{xy}$ (d) $z = ye^{xy}$ (e) $z = x^2 \sin(xy)$

   (f) $z = y\cos(xy)$ (g) $z = x\ln(xy)$ (h) $z = 3xy^3e^x$

37.5 Higher-order derivatives

Just as functions of one variable have second derivatives found by differentiating the first derivative, so too do functions of two variables. If $z = f(x, y)$ the first partial derivatives are $\frac{\partial z}{\partial x}$ and $\frac{\partial z}{\partial y}$. The second partial derivatives are found by differentiating the first partial derivatives. We can differentiate either first partial derivative with respect to x or with respect to y to obtain various second partial derivatives as summarised below:

Key point

$$\text{differentiating } \frac{\partial z}{\partial x} \text{ w.r.t. } x \text{ produces } \frac{\partial}{\partial x}\left(\frac{\partial z}{\partial x}\right) = \frac{\partial^2 z}{\partial x^2}$$

$$\text{differentiating } \frac{\partial z}{\partial x} \text{ w.r.t. } y \text{ produces } \frac{\partial}{\partial y}\left(\frac{\partial z}{\partial x}\right) = \frac{\partial^2 z}{\partial y \partial x}$$

$$\text{differentiating } \frac{\partial z}{\partial y} \text{ w.r.t. } x \text{ produces } \frac{\partial}{\partial x}\left(\frac{\partial z}{\partial y}\right) = \frac{\partial^2 z}{\partial x \partial y}$$

$$\text{differentiating } \frac{\partial z}{\partial y} \text{ w.r.t. } y \text{ produces } \frac{\partial}{\partial y}\left(\frac{\partial z}{\partial y}\right) = \frac{\partial^2 z}{\partial y^2}$$

The second partial derivatives of z are

$$\frac{\partial^2 z}{\partial x^2} \qquad \frac{\partial^2 z}{\partial y \partial x} \qquad \frac{\partial^2 z}{\partial x \partial y} \qquad \frac{\partial^2 z}{\partial y^2}$$

WORKED EXAMPLES

37.28   Given $z = 3xy^3 - 2xy$ find all the second partial derivatives of z.

Solution   First of all the first partial derivatives must be found:

$$\frac{\partial z}{\partial x} = 3y^3 - 2y \qquad \frac{\partial z}{\partial y} = 9xy^2 - 2x$$

Then each of these is differentiated with respect to x:

$$\frac{\partial^2 z}{\partial x^2} = \frac{\partial}{\partial x}\left(\frac{\partial z}{\partial x}\right) = 0$$

$$\frac{\partial^2 z}{\partial x \partial y} = \frac{\partial}{\partial x}\left(\frac{\partial z}{\partial y}\right) = 9y^2 - 2$$

Now, each of the first partial derivatives must be differentiated with respect to y:

$$\frac{\partial^2 z}{\partial y \partial x} = \frac{\partial}{\partial y}\left(\frac{\partial z}{\partial x}\right) = 9y^2 - 2$$

$$\frac{\partial^2 z}{\partial y^2} = \frac{\partial}{\partial y}\left(\frac{\partial z}{\partial y}\right) = 18xy$$

Note that

$$\frac{\partial^2 z}{\partial y \partial x} = \frac{\partial^2 z}{\partial x \partial y}$$

It is usually the case that the result is the same either way.

37.29   Find all second partial derivatives of $z = \sin(xy)$.

Solution   First of all the first partial derivatives are found:

$$\frac{\partial z}{\partial x} = y\,\cos(xy) \qquad \frac{\partial z}{\partial y} = x\,\cos(xy)$$

Then each of these is differentiated with respect to x:

$$\frac{\partial^2 z}{\partial x^2} = \frac{\partial}{\partial x}\left(\frac{\partial z}{\partial x}\right) = -y^2\,\sin(xy)$$

$$\frac{\partial^2 z}{\partial x \partial y} = \frac{\partial}{\partial x}\left(\frac{\partial z}{\partial y}\right) = -xy\,\sin(xy) + \cos(xy)$$

Note here the need to use the product rule to differentiate $x\,\cos(xy)$ with respect to x.

Now each of the first partial derivatives must be differentiated with respect to y:

$$\frac{\partial^2 z}{\partial y \partial x} = \frac{\partial}{\partial y}\left(\frac{\partial z}{\partial x}\right) = -xy\,\sin(xy) + \cos(xy)$$

$$\frac{\partial^2 z}{\partial y^2} = \frac{\partial}{\partial y}\left(\frac{\partial z}{\partial y}\right) = -x^2\,\sin(xy)$$

Self-assessment question 37.5

1. Given $z = f(x, y)$, explain what is meant by

$$\frac{\partial^2 z}{\partial y \partial x} \quad \text{and} \quad \frac{\partial^2 z}{\partial x \partial y}$$

Exercise 37.5

MyMathLab

1. Find all the second partial derivatives in each of the following cases:
   (a) $z = xy$ (b) $z = 7xy$
   (c) $z = 8x + 9y + 10$
   (d) $z = 8y^2 x + 11$ (e) $z = -2y^3 x^2$
   (f) $z = x + y$

2. Find all the second partial derivatives in each of the following cases:
   (a) $z = \frac{1}{x}$ (b) $z = \frac{y}{x}$ (c) $z = \frac{x}{y}$
   (d) $z = \frac{1}{x} + \frac{1}{y}$

3. Find all the second partial derivatives in each of the following cases:
   (a) $z = x \sin y$ (b) $z = y \cos x$
   (c) $z = ye^{2x}$ (d) $z = ye^{-x}$

4. Find all the second partial derivatives in each of the following cases:
   (a) $z = 8e^{xy}$ (b) $z = -3e^x \sin y$
   (c) $z = 4e^y \cos x$

5. Find all the second partial derivatives in each of the following cases:
   (a) $z = \ln x$ (b) $z = \ln y$ (c) $z = \ln xy$
   (d) $z = x \ln y$ (e) $z = y \ln x$

37.6 Functions of several variables

In more advanced applications, functions of several variables arise. For example, suppose w is a function of three independent variables x, y and z, that is $w = f(x, y, z)$. It is not possible to represent such a function graphically in the usual manner because this would require four dimensions. Nevertheless, partial derivatives can be calculated in exactly the same way as we have done for two independent variables.

WORKED EXAMPLES

37.30   If $w = 3x + 5y - 4z$ find $\frac{\partial w}{\partial x}$, $\frac{\partial w}{\partial y}$ and $\frac{\partial w}{\partial z}$.

Solution   When finding $\frac{\partial w}{\partial x}$ the variables y and z are treated as constants. So $\frac{\partial w}{\partial x} = 3$.

When finding $\frac{\partial w}{\partial y}$ the variables x and z are treated as constants. So $\frac{\partial w}{\partial y} = 5$.

When finding $\frac{\partial w}{\partial z}$ the variables x and y are treated as constants. So $\frac{\partial w}{\partial z} = -4$.

37.31   Find all second derivatives of $w = f(x, y, z) = x + y + z + xy + yz + xz$.

Solution   The first partial derivatives must be found first:

$$\frac{\partial w}{\partial x} = 1 + y + z \qquad \frac{\partial w}{\partial y} = 1 + x + z \qquad \frac{\partial w}{\partial z} = 1 + y + x$$

The second partial derivatives are found by differentiating the first partial derivatives w.r.t. x, w.r.t. y and w.r.t. z. This gives

$$\frac{\partial}{\partial x}\left(\frac{\partial w}{\partial x}\right) = 0 \qquad \frac{\partial}{\partial x}\left(\frac{\partial w}{\partial y}\right) = 1 \qquad \frac{\partial}{\partial x}\left(\frac{\partial w}{\partial z}\right) = 1$$

Similarly,

$$\frac{\partial}{\partial y}\left(\frac{\partial w}{\partial x}\right) = 1 \qquad \frac{\partial}{\partial y}\left(\frac{\partial w}{\partial y}\right) = 0 \qquad \frac{\partial}{\partial y}\left(\frac{\partial w}{\partial z}\right) = 1$$

And finally,

$$\frac{\partial}{\partial z}\left(\frac{\partial w}{\partial x}\right) = 1 \qquad \frac{\partial}{\partial z}\left(\frac{\partial w}{\partial y}\right) = 1 \qquad \frac{\partial}{\partial z}\left(\frac{\partial w}{\partial z}\right) = 0$$

Self-assessment question 37.6

1. If $w = f(x, y, z)$ explain how

$$\frac{\partial^3 w}{\partial x \partial y \partial z}$$

is calculated.

Exercise 37.6

1. If $w = f(x, y, t)$ such that $w = 8x - 3y - 4t$, find $\frac{\partial w}{\partial x}$, $\frac{\partial w}{\partial y}$ and $\frac{\partial w}{\partial t}$.

2. If $w = f(x, y, t)$ such that $w = x^2 + y^3 - 5t^2$, find $\frac{\partial w}{\partial x}$, $\frac{\partial w}{\partial y}$ and $\frac{\partial w}{\partial t}$.

3. Find all the first partial derivatives of $w = xyz$.

Test and assignment exercises 37

1. If $z = 14x - 13y$ state $\frac{\partial z}{\partial x}$ and $\frac{\partial z}{\partial y}$.

2. If $w = 5y - 2x$ state $\frac{\partial^2 w}{\partial x^2}$ and $\frac{\partial^2 w}{\partial y^2}$.

3. If $z = 3x^2 + 7xy - y^2$ find $\frac{\partial^2 z}{\partial y \partial x}$ and $\frac{\partial^2 z}{\partial x \partial y}$.

4. If $z = 14 - 4xy$ evaluate $\frac{\partial z}{\partial x}$ and $\frac{\partial z}{\partial y}$ at the point $(1, 2)$.

5. If $z = 4e^{5xy}$ find $\frac{\partial z}{\partial x}$ and $\frac{\partial z}{\partial y}$.

6. If $y = x \cos t$ find $\frac{\partial y}{\partial x}$ and $\frac{\partial y}{\partial t}$.

7. If $w = 3xy^2 + 2yz^2$ find all first partial derivatives of w at the point with coordinates $(1, 2, 3)$.

8. If $V = D^{1/4} T^{-5/6}$ find $\frac{\partial V}{\partial T}$ and $\frac{\partial V}{\partial D}$.

9. If $p = RT/V$ where R is a constant, find $\frac{\partial p}{\partial V}$ and $\frac{\partial p}{\partial T}$.

10. Find $\frac{\partial f}{\partial x}$, $\frac{\partial f}{\partial y}$, $\frac{\partial^2 f}{\partial x^2}$, $\frac{\partial^2 f}{\partial y^2}$ and $\frac{\partial^2 f}{\partial x \partial y}$ if $f = (x - y)^2$.

Solutions

Chapter 1

Self-assessment questions 1.1

1. An integer is a whole number, that is a number from the set

   $$... -4, -3, -2, -1, 0, 1, 2, 3, 4 ...$$

   The positive integers are 1, 2, 3, 4 .... The negative integers are $...-4, -3, -2, -1$.
2. The sum is the result of adding numbers. The difference is found by subtracting one number from another. The product of numbers is found by multiplying the numbers. The quotient of two numbers is found by dividing one number by the other.
3. (a) positive (b) negative (c) negative (d) positive

Exercise 1.1

1. (a) 3 (b) 9 (c) 11 (d) 21 (e) 30
   (f) 56 (g) -13 (h) -19 (i) -19
   (j) -13 (k) 29 (l) -75 (m) -75
   (n) 29
2. (a) -24 (b) -32 (c) -30 (d) 16
   (e) -42
3. (a) -5 (b) 3 (c) -3 (d) 3 (e) -3
   (f) -6 (g) 6 (h) -6
4. (a) Sum $= 3 + 6 = 9$
      Product $= 3 \times 6 = 18$

(b) Sum $= 17$, Product $= 70$
(c) Sum $= 2 + 3 + 6 = 11$
    Product $= 2 \times 3 \times 6 = 36$
5. (a) Difference $= 18 - 9 = 9$
      Quotient $= \frac{18}{9} = 2$
   (b) Difference $= 20 - 5 = 15$
      Quotient $= \frac{20}{5} = 4$
   (c) Difference $= 100 - 20 = 80$
      Quotient $= \frac{100}{20} = 5$

Self-assessment questions 1.2

1. BODMAS is a priority rule used when evaluating expressions: Brackets (do first), Of, Division, Multiplication (do secondly), Addition, Subtraction (do thirdly).
2. False. For example, $(12 - 4) - 3$ is not the same as $12 - (4 - 3)$. The former is equal to 5 whereas the latter is equal to 11. The position of the brackets is clearly important.

Exercise 1.2

1. (a) $6 - 2 \times 2 = 6 - 4 = 2$
   (b) $(6 - 2) \times 2 = 4 \times 2 = 8$
   (c) $6 \div 2 - 2 = 3 - 2 = 1$
   (d) $(6 \div 2) - 2 = 3 - 2 = 1$
   (e) $6 - 2 + 3 \times 2 = 6 - 2 + 6 = 10$
   (f) $6 - (2 + 3) \times 2 = 6 - 5 \times 2$
       $= 6 - 10 = -4$

(g) $(6-2)+3\times2=4+3\times2=4+6$
$\quad=10$

(h) $\frac{16}{-2}=-8$ (i) $\frac{-24}{-3}=8$

(j) $(-6)\times(-2)=12$

(k) $(-2)(-3)(-4)=-24$

2. (a) $6\times(12-3)+1=6\times9+1$
$\quad\quad=54+1=55$

(b) $6\times12-(3+1)=72-4=68$

(c) $6\times(12-3+1)=6\times10=60$

(d) $5\times(4-3)+2=5\times1+2=5+2=7$

(e) $5\times4-(3+2)=20-5=15$
$\quad$ or $5\times(4-3+2)=5\times3=15$

(f) $5\times(4-(3+2))=5\times(4-5)$
$\quad\quad=5\times(-1)=-5$

Self-assessment questions 1.3

1. A prime number is a positive integer larger than 1 that cannot be expressed as the product of two smaller positive integers.

2. 2, 3, 5, 7, 11, 13, 17, 19, 23, 29

3. All even numbers have 2 as a factor and so can be expressed as the product of two smaller numbers. The exception to this is 2 itself, which can only be expressed as 1×2, and since these numbers are not both smaller than 2, then 2 is prime.

Exercise 1.3

1. 13, 2 and 29 are prime.

2. (a) 2×13 (b) $2\times2\times5\times5$

(c) $3\times3\times3$ (d) 71

(e) $2\times2\times2\times2\times2\times2$ (f) 3×29

(g) 19×23 (h) 29×31

3. $30=2\times3\times5$
$42=2\times3\times7$
2 and 3 are common prime factors.

Self-assessment questions 1.4

1. H.c.f. stands for 'highest common factor'. The h.c.f. is the largest number that is a factor of each of the numbers in the original set.

2. L.c.m. stands for 'lowest common multiple'. It is the smallest number that can be divided exactly by each of the numbers in the set.

Exercise 1.4

1. (a) $12=2\times2\times3$ $15=3\times5$
$21=3\times7$
Hence h.c.f. $=3$

(b) $16=2\times2\times2\times2$
$24=2\times2\times2\times3$
$40=2\times2\times2\times5$
So h.c.f. $=2\times2\times2=8$

(c) $28=2\times2\times7$ $70=2\times5\times7$
$120=2\times2\times2\times3\times5$
$160=2\times2\times2\times2\times2\times5$
So h.c.f. $=2$

(d) $35=5\times7$ $38=2\times19$
$42=2\times3\times7$
So h.c.f. $=1$

(e) $96=2\times2\times2\times2\times2\times3$
$120=2\times2\times2\times3\times5$
$144=2\times2\times2\times2\times3\times3$
So h.c.f. $=2\times2\times2\times3=24$

2. (a) 5 $6=2\times3$ $8=2\times2\times2$
So l.c.m. $=2\times2\times2\times3\times5$
$\quad\quad=120$

(b) $20=2\times2\times5$ $30=2\times3\times5$
So l.c.m. $=2\times2\times3\times5=60$

(c) 7 $9=3\times3$ $12=2\times2\times3$
So l.c.m. $=2\times2\times3\times3\times7$
$\quad\quad=252$

(d) $100=2\times2\times5\times5$
$150=2\times3\times5\times5$
$235=5\times47$
So l.c.m. $=$
$2\times2\times3\times5\times5\times47=14100$

(e) $96=2\times2\times2\times2\times2\times3$
$120=2\times2\times2\times3\times5$
$144=2\times2\times2\times2\times3\times3$
So l.c.m. $=$
$2\times2\times2\times2\times2\times3\times3\times5$
$\quad\quad=1440$

Chapter 2

Self-assessment questions 2.1

1. (a) A fraction is formed by dividing a whole number by another whole number, for example $\frac{11}{3}$.
   (b) If the top number (the numerator) of a fraction is greater than or equal to the bottom number (the denominator) then the fraction is improper. For example, $\frac{101}{100}$ and $\frac{9}{9}$ are both improper fractions.
   (c) If the top number is less than the bottom number then the fraction is proper. For example, $\frac{99}{100}$ is a proper fraction.
2. (a) The numerator is the 'top number' of a fraction.
   (b) The denominator is the 'bottom number' of a fraction. For example, in $\frac{17}{14}$, the numerator is 17 and the denominator is 14.

Exercise 2.1

1. (a) Proper (b) Proper (c) Improper
   (d) Proper (e) Improper

Self-assessment questions 2.2

1. True. For example $7 = \frac{7}{1}$.
2. The numerator and denominator can both be divided by their h.c.f. This simplifies the fraction.
3. $\frac{3}{4}, \frac{9}{12}, \frac{75}{100}$.

Exercise 2.2

1. (a) $\dfrac{18}{27} = \dfrac{2}{3}$ (b) $\dfrac{12}{20} = \dfrac{3}{5}$ (c) $\dfrac{15}{45} = \dfrac{1}{3}$
   (d) $\dfrac{25}{80} = \dfrac{5}{16}$ (e) $\dfrac{15}{60} = \dfrac{1}{4}$ (f) $\dfrac{90}{200} = \dfrac{9}{20}$
   (g) $\dfrac{15}{20} = \dfrac{3}{4}$ (h) $\dfrac{2}{18} = \dfrac{1}{9}$ (i) $\dfrac{16}{24} = \dfrac{2}{3}$

(j) $\dfrac{30}{65} = \dfrac{6}{13}$ (k) $\dfrac{12}{21} = \dfrac{4}{7}$ (l) $\dfrac{100}{45} = \dfrac{20}{9}$
(m) $\dfrac{6}{9} = \dfrac{2}{3}$ (n) $\dfrac{12}{16} = \dfrac{3}{4}$ (o) $\dfrac{13}{42}$
(p) $\dfrac{13}{39} = \dfrac{1}{3}$ (q) $\dfrac{11}{33} = \dfrac{1}{3}$ (r) $\dfrac{14}{30} = \dfrac{7}{15}$
(s) $-\dfrac{12}{16} = -\dfrac{3}{4}$ (t) $\dfrac{11}{-33} = -\dfrac{1}{3}$
(u) $\dfrac{-14}{-30} = \dfrac{7}{15}$

2. $\dfrac{3}{4} = \dfrac{21}{28}$

3. $4 = \dfrac{20}{5}$

4. $\dfrac{5}{12} = \dfrac{15}{36}$

5. $2 = \dfrac{8}{4}$

6. $6 = \dfrac{18}{3}$

7. $\dfrac{2}{3} = \dfrac{8}{12}, \dfrac{5}{4} = \dfrac{15}{12}, \dfrac{5}{6} = \dfrac{10}{12}$

8. $\dfrac{4}{9} = \dfrac{8}{18}, \dfrac{1}{2} = \dfrac{9}{18}, \dfrac{5}{6} = \dfrac{15}{18}$

9. (a) $\dfrac{1}{2} = \dfrac{6}{12}$ (b) $\dfrac{3}{4} = \dfrac{9}{12}$ (c) $\dfrac{5}{2} = \dfrac{30}{12}$
   (d) $5 = \dfrac{60}{12}$ (e) $4 = \dfrac{48}{12}$ (f) $12 = \dfrac{144}{12}$

Self-assessment question 2.3

1. The l.c.m. of the denominators is found. Each fraction is expressed in an equivalent form with this l.c.m. as denominator. Addition and subtraction can then take place.

Exercise 2.3

1. (a) $\dfrac{1}{4}+\dfrac{2}{3}=\dfrac{3}{12}+\dfrac{8}{12}=\dfrac{11}{12}$

(b) $\dfrac{3}{5}+\dfrac{5}{3}=\dfrac{9}{15}+\dfrac{25}{15}=\dfrac{34}{15}$

(c) $\dfrac{12}{14}-\dfrac{2}{7}=\dfrac{6}{7}-\dfrac{2}{7}=\dfrac{4}{7}$

(d) $\dfrac{3}{7}-\dfrac{1}{2}+\dfrac{2}{21}=\dfrac{18}{42}-\dfrac{21}{42}+\dfrac{4}{42}=\dfrac{1}{42}$

(e) $1\dfrac{1}{2}+\dfrac{4}{9}+\dfrac{3}{2}=\dfrac{4}{9}+\dfrac{27}{18}+\dfrac{8}{18}=\dfrac{35}{18}$

(f) $2\dfrac{1}{4}-1\dfrac{1}{3}+\dfrac{1}{2}=\dfrac{9}{4}-\dfrac{4}{3}+\dfrac{1}{2}$
$=\dfrac{27}{12}-\dfrac{16}{12}+\dfrac{6}{12}=\dfrac{17}{12}$

(g) $\dfrac{10}{15}-1\dfrac{2}{5}+\dfrac{8}{3}=\dfrac{10}{15}-\dfrac{7}{5}+\dfrac{8}{3}$
$=\dfrac{10}{15}-\dfrac{21}{15}+\dfrac{40}{15}=\dfrac{29}{15}$

(h) $\dfrac{9}{10}-\dfrac{7}{16}+\dfrac{1}{2}-\dfrac{2}{5}=\dfrac{72}{80}-\dfrac{35}{80}+\dfrac{40}{80}-\dfrac{32}{80}$
$=\dfrac{45}{80}=\dfrac{9}{16}$

2. (a) $\dfrac{7}{8}+\dfrac{1}{3}=\dfrac{21}{24}+\dfrac{8}{24}=\dfrac{29}{24}$

(b) $\dfrac{1}{2}-\dfrac{3}{4}=\dfrac{2}{4}-\dfrac{3}{4}=-\dfrac{1}{4}$

(c) $\dfrac{3}{5}+\dfrac{2}{3}+\dfrac{1}{2}=\dfrac{18}{30}+\dfrac{20}{30}+\dfrac{15}{30}=\dfrac{53}{30}=1\dfrac{23}{30}$

(d) $\dfrac{3}{8}+\dfrac{1}{3}+\dfrac{1}{4}=\dfrac{9}{24}+\dfrac{8}{24}+\dfrac{6}{24}=\dfrac{23}{24}$

(e) $\dfrac{2}{3}-\dfrac{4}{7}=\dfrac{14}{21}-\dfrac{12}{21}=\dfrac{2}{21}$

(f) $\dfrac{1}{11}-\dfrac{1}{2}=\dfrac{2}{22}-\dfrac{11}{22}=-\dfrac{9}{22}$

(g) $\dfrac{3}{11}-\dfrac{5}{8}=\dfrac{24}{88}-\dfrac{55}{88}=\dfrac{-31}{88}$

3. (a) $\dfrac{5}{2}$ (b) $\dfrac{11}{3}$ (c) $\dfrac{41}{4}$ (d) $\dfrac{37}{7}$ (e) $\dfrac{56}{9}$

(f) $\dfrac{34}{3}$ (g) $\dfrac{31}{2}$ (h) $\dfrac{55}{4}$ (i) $\dfrac{133}{11}$ (j) $\dfrac{41}{3}$

(k) $\dfrac{113}{2}$

4. (a) $3\dfrac{1}{3}$ (b) $3\dfrac{1}{2}$ (c) $3\dfrac{3}{4}$ (d) $4\dfrac{1}{6}$

Self-assessment question 2.4

1. The numerators are multiplied together to form the numerator of the product. The denominators are multiplied to form the denominator of the product.

Exercise 2.4

1. (a) $\dfrac{2}{3}\times\dfrac{6}{7}=\dfrac{2}{1}\times\dfrac{2}{7}=\dfrac{4}{7}$

(b) $\dfrac{8}{15}\times\dfrac{25}{32}=\dfrac{1}{15}\times\dfrac{25}{4}=\dfrac{1}{3}\times\dfrac{5}{4}=\dfrac{5}{12}$

(c) $\dfrac{1}{4}\times\dfrac{8}{9}=\dfrac{1}{1}\times\dfrac{2}{9}=\dfrac{2}{9}$

(d) $\dfrac{16}{17}\times\dfrac{34}{48}=\dfrac{1}{17}\times\dfrac{34}{3}=\dfrac{1}{1}\times\dfrac{2}{3}=\dfrac{2}{3}$

(e) $2\times\dfrac{3}{5}\times\dfrac{5}{12}=\dfrac{3}{5}\times\dfrac{5}{6}=\dfrac{3}{1}\times\dfrac{1}{6}=\dfrac{1}{2}$

(f) $2\dfrac{1}{3}\times1\dfrac{1}{4}=\dfrac{7}{3}\times\dfrac{5}{4}=\dfrac{35}{12}$

(g) $1\dfrac{3}{4}\times2\dfrac{1}{2}=\dfrac{7}{4}\times\dfrac{5}{2}=\dfrac{35}{8}$

(h) $\dfrac{3}{4}\times1\dfrac{1}{2}\times3\dfrac{1}{2}=\dfrac{3}{4}\times\dfrac{3}{2}\times\dfrac{7}{2}=\dfrac{63}{16}$

2. (a) $\dfrac{2}{3}\times\dfrac{3}{4}=\dfrac{1}{2}$ (b) $\dfrac{4}{7}\times\dfrac{21}{30}=\dfrac{6}{15}=\dfrac{2}{5}$

(c) $\dfrac{9}{10}\times80=72$ (d) $\dfrac{6}{7}\times42=36$

3. Yes, because $\dfrac{3}{4} \times \dfrac{12}{15} = \dfrac{12}{15} \times \dfrac{3}{4}$

4. (a) $-\dfrac{5}{21}$ (b) $-\dfrac{3}{8}$ (c) $-\dfrac{5}{11}$ (d) $\dfrac{10}{7}$

5. (a) $5\dfrac{1}{2} \times \dfrac{1}{2} = \dfrac{11}{2} \times \dfrac{1}{2} = \dfrac{11}{4}$

   (b) $3\dfrac{3}{4} \times \dfrac{1}{3} = \dfrac{15}{4} \times \dfrac{1}{3} = \dfrac{5}{4}$

   (c) $\dfrac{2}{3} \times 5\dfrac{1}{9} = \dfrac{2}{3} \times \dfrac{46}{9} = \dfrac{92}{27}$

   (d) $\dfrac{3}{4} \times 11\dfrac{1}{2} = \dfrac{3}{4} \times \dfrac{23}{2} = \dfrac{69}{8}$

6. (a) $\dfrac{3}{5} \times 11\dfrac{1}{4} = \dfrac{3}{5} \times \dfrac{45}{4} = \dfrac{27}{4}$

   (b) $\dfrac{2}{3} \times 15\dfrac{1}{2} = \dfrac{2}{3} \times \dfrac{31}{2} = \dfrac{31}{3}$

   (c) $\dfrac{1}{4} \times \left(-8\dfrac{1}{3}\right) = \dfrac{1}{4} \times \left(-\dfrac{25}{3}\right) = -\dfrac{25}{12}$

Self-assessment question 2.5

1. To divide one fraction by a second fraction, the second fraction is inverted (that is, the numerator and denominator are interchanged) and then multiplication is performed.

Exercise 2.5

1. (a) $\dfrac{3}{4} \div \dfrac{1}{8} = \dfrac{3}{4} \times \dfrac{8}{1} = \dfrac{3}{1} \times \dfrac{2}{1} = 6$

   (b) $\dfrac{8}{9} \div \dfrac{4}{3} = \dfrac{8}{9} \times \dfrac{3}{4} = \dfrac{2}{9} \times \dfrac{3}{1} = \dfrac{2}{3}$

   (c) $-\dfrac{2}{7} \div \dfrac{4}{21} = -\dfrac{2}{7} \times \dfrac{21}{4} = -\dfrac{2}{1} \times \dfrac{3}{4} = -\dfrac{3}{2}$

(d) $\dfrac{9}{4} \div 1\dfrac{1}{2} = \dfrac{9}{4} \div \dfrac{3}{2} = \dfrac{9}{4} \times \dfrac{2}{3} = \dfrac{3}{4} \times \dfrac{2}{1} = \dfrac{3}{2}$

(e) $\dfrac{5}{6} \div \dfrac{5}{12} = \dfrac{5}{6} \times \dfrac{12}{5} = \dfrac{1}{6} \times \dfrac{12}{1} = 2$

(f) $\dfrac{99}{100} \div 1\dfrac{4}{5} = \dfrac{99}{100} \div \dfrac{9}{5} = \dfrac{99}{100} \times \dfrac{5}{9}$

$= \dfrac{11}{100} \times \dfrac{5}{1} = \dfrac{11}{20}$

(g) $3\dfrac{1}{4} \div 1\dfrac{1}{8} = \dfrac{13}{4} \div \dfrac{9}{8} = \dfrac{13}{4} \times \dfrac{8}{9}$

$= \dfrac{13}{1} \times \dfrac{2}{9} = \dfrac{26}{9}$

(h) $\left(2\dfrac{1}{4} \div \dfrac{3}{4}\right) \times 2 = \left(\dfrac{9}{4} \times \dfrac{4}{3}\right) \times 2$

$= \left(\dfrac{3}{4} \times 4\right) \times 2 = 3 \times 2 = 6$

(i) $2\dfrac{1}{4} \div \left(\dfrac{3}{4} \times 2\right) = \dfrac{9}{4} \div \left(\dfrac{3}{4} \times \dfrac{2}{1}\right)$

$= \dfrac{9}{4} \div \dfrac{3}{2} = \dfrac{9}{4} \times \dfrac{2}{3} = \dfrac{3}{4} \times \dfrac{2}{1} = \dfrac{3}{2}$

(j) $6\dfrac{1}{4} \div 2\dfrac{1}{2} + 5 = \dfrac{25}{4} \div \dfrac{5}{2} + 5$

$= \dfrac{25}{4} \times \dfrac{2}{5} + 5 = \dfrac{5}{2} + 5 = \dfrac{15}{2}$

(k) $6\dfrac{1}{4} \div \left(2\dfrac{1}{2} + 5\right) = \dfrac{25}{4} \div \left(\dfrac{5}{2} + 5\right)$

$= \dfrac{25}{4} \div \dfrac{15}{2} = \dfrac{25}{4} \times \dfrac{2}{15} = \dfrac{5}{4} \times \dfrac{2}{3} = \dfrac{5}{6}$

Chapter 3

Self-assessment questions 3.1

1. Largest is 23.01; smallest is 23.0.
2. 0.1

Exercise 3.1

1. (a) $\dfrac{7}{10}$ (b) $\dfrac{4}{5}$ (c) $\dfrac{9}{10}$

2. (a) $\dfrac{11}{20}$ (b) $\dfrac{79}{500}$ (c) $\dfrac{49}{50}$ (d) $\dfrac{99}{1000}$

3. (a) $4\dfrac{3}{5}$ (b) $5\dfrac{1}{5}$ (c) $8\dfrac{1}{20}$ (d) $11\dfrac{59}{100}$

   (e) $121\dfrac{9}{100}$

4. (a) 0.697   (b) 0.083   (c) 0.517

Self-assessment questions 3.2

1. Writing a number to 2 (or 3 or 4 etc.) significant figures is a way of approximating the number. The number of s.f. is the maximum number of non-zero digits in the approximation. The approximation is as close as possible to the original number.

2. The digits after the decimal point are considered. To write to 1 d.p. we consider the first two digits; to write to 2 d.p. we consider the first three digits, and so on. If the last digit considered is 5 or greater, we round up the previous digit; otherwise we round down.

Exercise 3.2

1. (a) 6960   (b) 70.4   (c) 0.0123   (d) 0.0110
   (e) 45.6   (f) 2350
2. (a) 66.00   (b) 66.0   (c) 66   (d) 70
   (e) 66.00   (f) 66.0
3. (a) 10   (b) 10.0
4. (a) 65.456   (b) 65.46   (c) 65.5
   (d) 65.456   (e) 65.46   (f) 65.5   (g) 65
   (h) 70

Chapter 4

Self-assessment question 4.1

1. Converting fractions to percentages allows for easy comparison of numbers.

Exercise 4.1

1. 23% of 124 $= \dfrac{23}{100} \times 124 = 28.52$

2. (a) $\dfrac{9}{11} = \dfrac{9}{11} \times 100\% = \dfrac{900}{11}\% = 81.82\%$

   (b) $\dfrac{15}{20} = \dfrac{15}{20} \times 100\% = 75\%$

   (c) $\dfrac{9}{10} = \dfrac{9}{10} \times 100\% = 90\%$

   (d) $\dfrac{45}{50} = \dfrac{45}{50} \times 100\% = 90\%$

   (e) $\dfrac{75}{90} = \dfrac{75}{90} \times 100\% = 83.33\%$

3. $\dfrac{13}{12} = \dfrac{13}{12} \times 100\% = 108.33\%$

4. 217% of 500 $= \dfrac{217}{100} \times 500 = 1085$

5. New weekly wage is 106% of £400,

   106% of 400 $= \dfrac{106}{100} \times 400 = 424$

   The new weekly wage is £424.

6. 17% of $1200 = \dfrac{17}{100} \times 1200 = 204$

The debt is decreased by £204 to
£1200 − 204 = £996.

7. (a) $50\% = \dfrac{50}{100} = 0.5$

(b) $36\% = \dfrac{36}{100} = 0.36$

(c) $75\% = \dfrac{75}{100} = 0.75$

(d) $100\% = \dfrac{100}{100} = 1$

(e) $12.5\% = \dfrac{12.5}{100} = 0.125$

8. £204.80

9. £1125

10. percentage change

$= \dfrac{\text{new value} - \text{original value}}{\text{original value}} \times 100$

$= \dfrac{7495 - 6950}{6950} \times 100$

$= 7.84\%$

11. percentage change

$= \dfrac{\text{new value} - \text{original value}}{\text{original value}} \times 100$

$= \dfrac{399 - 525}{525} \times 100$

$= -24\%$

Note that the percentage change is negative
and this indicates a reduction in price. There
has been a 24% reduction in the price of
the washing machine.

Self-assessment question 4.2

1. True

Exercise 4.2

1. $8 + 1 + 3 = 12$. The first number is $\frac{8}{12}$ of 180,
that is 120; the second number is $\frac{1}{12}$ of 180,
that is 15; and the third number is $\frac{3}{12}$ of
180, that is 45. Hence 180 is divided into 120,
15 and 45.

2. $1 + 1 + 3 = 5$. We calculate $\frac{1}{5}$ of 930 to be 186
and $\frac{3}{5}$ of 930 to be 558. The length is divided
into 186 cm, 186 cm and 558 cm.

3. $2 + 3 + 4 = 9$. The first piece is $\frac{2}{9}$ of 6 m, that
is 1.33 m; the second piece is $\frac{3}{9}$ of 6 m, that is
2 m; the third piece is $\frac{4}{9}$ of 6 m, that is 2.67 m.

4. $1 + 2 + 3 + 4 = 10$: $\frac{1}{10}$ of 1200 = 120; $\frac{2}{10}$ of
1200 = 240; $\frac{3}{10}$ of 1200 = 360; $\frac{4}{10}$ of 1200
= 480. The number 1200 is divided into 120,
240, 360 and 480.

5. $2\frac{3}{4} : 1\frac{1}{2} : 2\frac{1}{4} = \dfrac{11}{4} : \dfrac{3}{2} : \dfrac{9}{4} = 11 : 6 : 9$

Now, $11 + 6 + 9 = 26$, so

$\dfrac{11}{26} \times 2600 = 1100$ \qquad $\dfrac{6}{26} \times 2600 = 600$

$\dfrac{9}{26} \times 2600 = 900$

Alan receives £1100, Bill receives £600 and
Claire receives £900.

6. 8 kg, 10.67 kg, 21.33 kg

7. (a) $1:2$ (b) $1:2$ (c) $1:2:4$ (d) $1:21$

8. 24, 84

Chapter 5

Self-assessment questions 5.1

1. 'Algebra' refers to the manipulation of symbols,
as opposed to the manipulation of numbers.

2. The product of the two numbers is written
as mn.

3. An algebraic fraction is formed by dividing
one algebraic expression by another algebraic
expression. The 'top' of the fraction is the
numerator; the 'bottom' of the fraction is the
denominator.

4. Superscripts and subscripts are located in different positions, relative to the symbol. Superscripts are placed high; subscripts are placed low.
5. A variable can have many different values; a constant has one, fixed value.

Self-assessment questions 5.2

1. In the expression a^x, a is the base and x is the power.
2. 'Index' is another word meaning 'power'.
3. $(xyz)^2$ means $(xyz)(xyz)$, which can be written as $x^2y^2z^2$. Clearly this is distinct from xyz^2, in which only the quantity z is squared.
4. $(-3)^4 = (-3)(-3)(-3)(-3) = 81$.
$-3^4 = -(3)(3)(3)(3) = -81$. Here, the power (4) has the higher priority.

Exercise 5.2

1. $2^4 = 16$; $(\frac{1}{2})^2 = \frac{1}{4}$; $1^8 = 1$; $3^5 = 243$; $0^3 = 0$.
2. $10^4 = 10000$; $10^5 = 100000$; $10^6 = 1000000$.
3. $11^4 = 14641$; $16^8 = 4294967296$; $39^4 = 2313441$; $1.5^7 = 17.0859375$.
4. (a) $a^4b^2c = a \times a \times a \times a \times b \times b \times c$
   (b) $xy^2z^4 = x \times y \times y \times z \times z \times z \times z$
5. (a) x^4y^2 (b) $x^2y^2z^3$ (c) $x^2y^2z^2$
   (d) $a^2b^2c^2$
6. (a) $7^4 = 2401$ (b) $7^5 = 16807$
   (c) $7^4 \times 7^5 = 40353607$
   (d) $7^9 = 40353607$ (e) $8^3 = 512$
   (f) $8^7 = 2097152$
   (g) $8^3 \times 8^7 = 1073741824$
   (h) $8^{10} = 1073741824$
   The rule states that $a^m \times a^n = a^{m+n}$; the powers are added.
7. $(-3)^3 = -27$; $(-2)^2 = 4$; $(-1)^7 = -1$; $(-1)^4 = 1$.
8. -4492.125; 324; -0.03125.
9. (a) 36 (b) 9 (c) -64 (d) -8
   $-6^2 = -36$, $-3^2 = -9$,
   $-4^3 = -64$, $-2^3 = -8$

Self-assessment question 5.3

1. An algebraic expression is any quantity comprising symbols and operations (that is, $+$, $-$, $\times$, $\div$): for example, x, $2x^2$ and $3x + 2y$ are all algebraic expressions. An algebraic formula relates two or more quantities and must contain an '$=$' sign. For example, $A = \pi r^2$, $V = \pi r^2 h$ and $S = ut + \frac{1}{2}at^2$ are all algebraic formulae.

Exercise 5.3

1. 60
2. 69
3. (a) 314.2 cm^2 (b) 28.28 cm^2
   (c) 0.126 cm^2
4. $3x^2 = 3 \times 4^2 = 3 \times 16 = 48$;
   $(3x)^2 = 12^2 = 144$
5. $5x^2 = 5(-2)^2 = 20$; $(5x)^2 = (-10)^2 = 100$
6. (a) 33.95 (b) 23.5225 (c) 26.75
   (d) 109.234125
7. (a) $a + b + c = 25.5$ (b) $ab = 46.08$
   (c) $bc = 32.76$ (d) $abc = 419.328$
8. $C = \frac{5}{9}(100 - 32) = \frac{5}{9}(68) = 37.78$
9. (a) $x^2 = 49$ (b) $-x^2 = -49$
   (c) $(-x)^2 = (-7)^2 = 49$
10. (a) 4 (b) 4 (c) -4 (d) 12
    (e) -12 (f) 36
11. (a) 3 (b) 9 (c) -1 (d) 36
    (e) -36 (f) 144
12. $x^2 - 7x + 2 = (-9)^2 - 7(-9) + 2$
    $= 81 + 63 + 2 = 146$
13. $2x^2 + 3x - 11 = 2(-3)^2 + 3(-3) - 11$
    $= 18 - 9 - 11 = -2$
14. $-x^2 + 3x - 5 = -(-1)^2 + 3(-1) - 5$
    $= -1 - 3 - 5 = -9$
15. 0
16. (a) 49 (b) 43 (c) 1
    (d) 5
17. (a) 21 (b) 27 (c) 0
    (d) $\dfrac{1}{6}$
18. (a) 3 (b) $3\dfrac{4}{5}$ (c) 23 (d) 23

19. (a) $-\dfrac{1}{2}$ (b) 4 (c) 32

20. (a) 0 (b) 16 (c) $40\dfrac{1}{2}$

    (d) $2\dfrac{1}{2}$

21. (a) 17 (b) -0.5 (c) 7
22. (a) 6000 (b) 2812.5
23. (a) 17151 (b) 276951

Chapter 6

Self-assessment questions 6.1

1. $a^m \times a^n = a^{m+n}$
   $\dfrac{a^m}{a^n} = a^{m-n}$
   $(a^m)^n = a^{mn}$
2. a^0 is 1.
3. x^1 is simply x.

Exercise 6.1

1. (a) $5^7 \times 5^{13} = 5^{20}$ (b) $9^8 \times 9^5 = 9^{13}$
   (c) $11^2 \times 11^3 \times 11^4 = 11^9$
2. (a) $15^3/15^2 = 15^1 = 15$ (b) $4^{18}/4^9 = 4^9$
   (c) $5^{20}/5^{19} = 5^1 = 5$
3. (a) a^{10} (b) a^9 (c) b^{22}
4. (a) x^{15} (b) y^{21}
5. $19^8 \times 17^8$ cannot be simplified using the laws of indices because the two bases are not the same.
6. (a) $(7^3)^2 = 7^{3\times2} = 7^6$ (b) $(4^2)^8 = 4^{16}$
   (c) $(7^9)^2 = 7^{18}$
7. $1/(5^3)^8 = 1/5^{24}$
8. (a) $x^5 y^5$ (b) $a^3 b^3 c^3$
9. (a) $x^{10} y^{20}$ (b) $81 x^6$ (c) $-27 x^3$
   (d) $x^8 y^{12}$
10. (a) z^3 (b) y^2 (c) 1

Self-assessment question 6.2

1. a^{-m} is the same as $\dfrac{1}{a^m}$. For example, 5^{-2} is the same as $\dfrac{1}{5^2}$.

Exercise 6.2

1. (a) $\dfrac{1}{4}$ (b) $\dfrac{1}{8}$ (c) $\dfrac{1}{9}$ (d) $\dfrac{1}{27}$ (e) $\dfrac{1}{25}$
   (f) $\dfrac{1}{16}$ (g) $\dfrac{1}{9}$ (h) $\dfrac{1}{121}$ (i) $\dfrac{1}{7}$

2. (a) 0.1 (b) 0.01 (c) 0.000001 (d) 0.01
   (e) 0.001 (f) 0.0001

3. (a) $\dfrac{1}{x^4}$ (b) x^5 (c) $\dfrac{1}{x^7}$ (d) $\dfrac{1}{y^2}$
   (e) $y^1 = y$ (f) $\dfrac{1}{y^1} = \dfrac{1}{y}$ (g) $\dfrac{1}{y^2}$
   (h) $\dfrac{1}{z^1} = \dfrac{1}{z}$ (i) $z^1 = z$

4. (a) $x^{-3} = \dfrac{1}{x^3}$ (b) $x^{-5} = \dfrac{1}{x^5}$ (c) $x^{-1} = \dfrac{1}{x}$
   (d) x^5 (e) $x^{-13} = \dfrac{1}{x^{13}}$ (f) $x^{-8} = \dfrac{1}{x^8}$
   (g) $x^{-9} = \dfrac{1}{x^9}$ (h) $x^{-4} = \dfrac{1}{x^4}$

5. (a) a^{11} (b) x^{-16} (c) x^{-18} (d) 4^{-6}
6. (a) 0.001 (b) 0.0001 (c) 0.00001
7. $4^{-8}/4^{-6} = 4^{-2} = 1/4^2 = \frac{1}{16}$
   $3^{-5}/3^{-8} = 3^3 = 27$

Self-assessment questions 6.3

1. $x^{\frac{1}{2}}$ is the square root of x. $x^{\frac{1}{3}}$ is the cube root of x.

2. 10, −10. Negative numbers, such as −100, do not have square roots, since squaring a number always gives a positive result.

Exercise 6.3

1. (a) $64^{1/3} = \sqrt[3]{64} = 4$ since $4^3 = 64$
   (b) $144^{1/2} = \sqrt{144} = \pm 12$
   (c) $16^{-1/4} = 1/16^{1/4} = 1/\sqrt[4]{16} = \pm\frac{1}{2}$
   (d) $25^{-1/2} = 1/25^{1/2} = 1/\sqrt{25} = \pm\frac{1}{5}$
   (e) $1/32^{-1/5} = 32^{1/5} = \sqrt[5]{32} = 2$ since $2^5 = 32$

2. (a) $(3^{-1/2})^4 = 3^{-2} = 1/3^2 = \frac{1}{9}$
   (b) $(8^{1/3})^{-1} = 8^{-1/3} = 1/8^{1/3} = 1/\sqrt[3]{8} = \frac{1}{2}$

3. (a) $8^{1/2}$ (b) $12^{1/3}$
   (c) $16^{1/4}$ (d) $13^{3/2}$ (e) $4^{7/3}$

4. (a) $x^{1/2}$ (b) $y^{1/3}$
   (c) $x^{5/2}$ (d) $5^{7/3}$

Exercise 6.4

1. (a) 743
   (b) 74300
   (c) 70
   (d) 0.0007

2. (a) 3×10^2
   (b) 3.56×10^2
   (c) 0.32×10^2
   (d) 0.0057×10^2

Self-assessment question 6.5

1. Scientific notation is useful for writing very large or extremely small numbers in a concise way. It is easier to manipulate such numbers when they are written using scientific notation.

Exercise 6.5

1. (a) $45 = 4.5 \times 10^1$
   (b) $45000 = 4.5 \times 10^4$
   (c) $-450 = -4.5 \times 10^2$
   (d) $90000000 = 9.0 \times 10^7$
   (e) $0.15 = 1.5 \times 10^{-1}$
   (f) $0.00036 = 3.6 \times 10^{-4}$
   (g) 3.5 is already in standard form.
   (h) $-13.2 = -1.32 \times 10^1$
   (i) $1000000 = 1 \times 10^6$
   (j) $0.0975 = 9.75 \times 10^{-2}$
   (k) $45.34 = 4.534 \times 10^1$

2. (a) $3.75 \times 10^2 = 375$
   (b) $3.97 \times 10^1 = 39.7$
   (c) $1.875 \times 10^{-1} = 0.1875$
   (d) $-8.75 \times 10^{-3} = -0.00875$

3. (a) 2.4×10^8 (b) 7.968×10^8
   (c) 1.044×10^{-4}
   (d) 1.526×10^{-1}
   (e) 5.293×10^2

Chapter 7

Exercise 7.1

1. (a) $-5p + 19q$ (b) $-5r - 13s + z$ (c) not possible to simplify
   (d) $8x^2 + 3y^2 - 2y$ (e) $4x^2 - x + 9$

2. (a) $-12y + 8p + 9q$ (b) $21x^2 - 11x^3 + y^3$
   (c) $7xy + y^2$ (d) $2xy$ (e) 0

Self-assessment questions 7.2

1. Positive
2. Negative

Exercise 7.2

1. (a) 84   (b) 84   (c) 84
2. (a) 40   (b) 40
3. (a) $14z$ (b) $30y$ (c) $6x$ (d) $27a$
   (e) $55a$ (f) $6x$
4. (a) $20x^2$ (b) $6y^3$ (c) $22u^2$ (d) $8u^2$
   (e) $26z^2$
5. (a) $21x^2$ (b) $21a^2$ (c) $14a^2$
6. (a) $15y^2$ (b) $8y$
7. (a) $a^3b^2c^2$ (b) x^3y^2 (c) x^2y^4
8. No difference; both equal x^2y^4.

9. $(xy^2)(xy^2) = x^2y^4$; $xy^2 + xy^2 = 2xy^2$
10. (a) $-21z^2$ (b) $-4z$
11. (a) $-3x^2$ (b) $2x$
12. (a) $2x^2$ (b) $-3x$

Exercise 7.3

1. (a) $4x + 4$ (b) $-4x - 4$ (c) $4x - 4$
   (d) $-4x + 4$
2. (a) $5x - 5y$ (b) $19x + 57y$ (c) $8a + 8b$
   (d) $5y + xy$ (e) $12x + 48$
   (f) $17x - 153$ (g) $-a + 2b$ (h) $x + \frac{1}{2}$
   (i) $6m - 12m^2 - 9mn$
3. (a) $18 - 13x - 26 = -8 - 13x$ (b) $x^2 + xy$
   will not simplify any further

4. (a) $x^2 + 7x + 6$ (b) $x^2 + 9x + 20$
   (c) $x^2 + x - 6$ (d) $x^2 + 5x - 6$
   (e) $xm + ym + nx + yn$
   (f) $12 + 3y + 4x + yx$ (g) $25 - x^2$
   (h) $51x^2 - 79x - 10$
5. (a) $x^2 - 4x - 21$ (b) $6x^2 + 11x - 7$
   (c) $16x^2 - 1$ (d) $x^2 - 9$
   (e) $6 + x - 2x^2$
6. (a) $\frac{29}{2}x - \frac{5}{2}y$ (b) $\frac{5}{4}x + \frac{5}{4}$
7. (a) $-x + y$ (b) $-a - 2b$ (c) $-\frac{3}{2}p - \frac{1}{2}q$
8. $(x + 1)(x + 2) = x^2 + 3x + 2$. So
   $(x + 1)(x + 2)(x + 3) = (x^2 + 3x + 2)(x + 3)$
   $= (x^2 + 3x + 2)(x) + (x^2 + 3x + 2)(3)$
   $= x^3 + 3x^2 + 2x + 3x^2 + 9x + 6$
   $= x^3 + 6x^2 + 11x + 6$

Chapter 8

Self-assessment question 8.1

1. To factorise an expression means to write it as a product, usually of two or more simpler expressions.

Exercise 8.1

1. (a) $9x + 27$ (b) $-5x + 10$ (c) $\frac{1}{2}x + \frac{1}{2}$
   (d) $-a + 3b$ (e) $1/(2x + 2y)$
   (f) $x/(yx - y^2)$
2. (a) $4x^2$ has factors $1, 2, 4, x, 2x, 4x, x^2, 2x^2, 4x^2$
   (b) $6x^3$ has factors $1, 2, 3, 6, x, 2x, 3x, 6x, x^2, 2x^2, 3x^2, 6x^2, x^3, 2x^3, 3x^3, 6x^3$
3. (a) $3(x + 6)$ (b) $3(y - 3)$ (c) $-3(y + 3)$
   (d) $-3(1 + 3y)$ (e) $5(4 + t)$ (f) $5(4 - t)$
   (g) $-5(t + 4)$ (h) $3(x + 4)$ (i) $17(t + 2)$
   (j) $4(t - 9)$
4. (a) $x(x^3 + 2)$ (b) $x(x^3 - 2)$
   (c) $x(3x^3 - 2)$ (d) $x(3x^3 + 2)$
   (e) $x^2(3x^2 + 2)$ (f) $x^3(3x + 2)$
   (g) $z(17 - z)$ (h) $x(3 - y)$ (i) $y(3 - x)$
   (j) $x(1 + 2y + 3yz)$

5. (a) $10(x + 2y)$ (b) $3(4a + b)$
   (c) $2x(2 - 3y)$ (d) $7(a + 2)$ (e) $5(2m - 3)$
   (f) $1/[5(a + 7b)]$ (g) $1/[5a(a + 7b)]$
6. (a) $3x(5x + 1)$ (b) $x(4x - 3)$
   (c) $4x(x - 2)$ (d) $3(5 - x^2)$
   (e) $5x^2(2x + 1 + 3y)$ (f) $6ab(a - 2b)$
   (g) $8b(2ac - ba + 3c)$

Self-assessment question 8.2

1. $2x^2 + x + 6$ cannot be factorised. On the other hand, $2x^2 + x - 6$ can be factorised as $(2x - 3)(x + 2)$.

Exercise 8.2

1. (a) $(x + 2)(x + 1)$ (b) $(x + 7)(x + 6)$
   (c) $(x + 5)(x - 3)$ (d) $(x + 10)(x - 1)$
   (e) $(x - 8)(x - 3)$ (f) $(x - 10)(x + 10)$
   (g) $(x + 2)(x + 2)$ or $(x + 2)^2$
   (h) $(x + 6)(x - 6)$ (i) $(x + 5)(x - 5)$
   (j) $(x + 1)(x + 9)$ (k) $(x + 9)(x - 1)$
   (l) $(x + 1)(x - 9)$ (m) $(x - 1)(x - 9)$
   (n) $x(x - 5)$

2. (a) $(2x+1)(x-3)$ (b) $(3x+1)(x-2)$
   (c) $(5x+3)(2x+1)$
   (d) $2x^2+12x+16$ has a common factor of
       2, which should be written outside a
       bracket to give $2(x^2+6x+8)$. The
       bracket can then be factorised to give
       $2(x+4)(x+2)$.
   (e) $(2x+3)(x+1)$ (f) $(3s+2)(s+1)$
   (g) $(3z+2)(z+5)$
   (h) $9(x^2-4)=9(x+2)(x-2)$
   (i) $(2x+5)(2x-5)$

3. (a) $(x+y)(x-y)=x^2+yx-yx-y^2$
       $=x^2-y^2$
   (b) (i) $(4x+1)(4x-1)$

   (ii) $(4x+3)(4x-3)$
   (iii) $(5t-4r)(5t+4r)$

4. (a) $(x-2)(x+5)$ (b) $(2x+5)(x-4)$
   (c) $(3x+1)(3x-1)$ (d) $10x^2+14x-12$
       has a common factor of 2, which is written
       outside a bracket to give $2(5x^2+7x-6)$.
       The bracket can then be factorised to give
       $2(5x-3)(x+2)$; (e) $(x+13)(x+2)$.
   (f) $(-x+1)(x+3)$

5. (a) $(10+7x)(10-7x)$
   (b) $(6x+5y)(6x-5y)$

   (c) $\left(\dfrac{1}{2}+3v\right)\left(\dfrac{1}{2}-3v\right)$ (d) $\left(\dfrac{x}{y}+2\right)\left(\dfrac{x}{y}-2\right)$

Chapter 9

Self-assessment questions 9.2

1. There are no factors common to both
   numerator and denominator and hence
   cancellation is not possible. In particular 3
   is not a factor of the denominator.

2. Same as Q1 – there are no factors
   common to numerator and denominator.
   In particular x is a factor neither of the
   denominator nor of the numerator.

3. $\dfrac{x+1}{2x+2}=\dfrac{x+1}{2(x+1)}=\dfrac{1}{2}$

   The common factor, $x+1$, has been
   cancelled.

Exercise 9.2

1. (a) $\dfrac{3x}{y}$ (b) $\dfrac{9}{x}$ (c) $3y$ (d) $3x$ (e) 9
   (f) 3

2. (a) $\dfrac{5x}{y}$ (b) $\dfrac{3x}{y}$ (c) $15y$ (d) 15 (e) $-x^2$

(f) $-\dfrac{1}{y^4}=-y^{-4}$ (g) $\dfrac{1}{y}=y^{-1}$ (h) y^{-7}

3. (a) $\dfrac{1}{3+2x}$ (b) $1+2x$ (c) $\dfrac{1}{2+7x}$
   (d) $\dfrac{x}{2+7x}$ (e) $\dfrac{x}{1+7x}$ (f) $\dfrac{1}{7x+y}$
   (g) $\dfrac{y}{7x+y}$ (h) $\dfrac{x}{7x+y}$

4. (a) $5x+1$ (b) $\dfrac{3(5x+1)}{3(x+2y)}=\dfrac{5x+1}{x+2y}$
   (c) $\dfrac{3}{x+2}$ (d) $\dfrac{3}{y+2}$ (e) $\dfrac{13}{x+5}$
   (f) $\dfrac{17}{9y+4}$

5. (a) $\dfrac{1}{3+2x}$ (b) $\dfrac{2}{x+7}$
   (c) $\dfrac{2(x+4)}{(x-2)(x+4)}=\dfrac{2}{x-2}$ (d) $\dfrac{7}{ab+9}$
   (e) $\dfrac{y}{y+1}$

6. (a) $\dfrac{1}{x-2}$ (b) $\dfrac{2(x-2)}{(x-2)(x+3)} = \dfrac{2}{x+3}$

(c) $\dfrac{1}{x+2}$ (d) $\dfrac{(x+1)(x+1)}{(x+1)(x-3)} = \dfrac{x+1}{x-3}$

(e) $\dfrac{2}{x-3}$ (f) $\dfrac{1}{x-3}$ (g) $\dfrac{1}{2(x-3)}$

(h) $\dfrac{2}{x-3}$ (i) $\dfrac{1}{2(x+4)}$ (j) $\dfrac{1}{2}$ (k) 2

(l) $x+4$ (m) $\dfrac{1}{x-3}$ (n) $\dfrac{1}{x+4}$ (o) $\dfrac{1}{2}$

(p) $\dfrac{x+4}{2x+9}$

Self-assessment question 9.3

1. True

Exercise 9.3

1. (a) $\dfrac{y}{6}$ (b) $\dfrac{z}{6}$ (c) $\dfrac{2}{5y}$ (d) $\dfrac{2}{5x}$ (e) $\dfrac{3x}{4y}$

(f) $\dfrac{3x^2}{5y}$ (g) $\dfrac{3x}{5y}$ (h) $\dfrac{7x}{16y}$ (i) $\dfrac{1}{4x}$ (j) $\dfrac{x}{4}$

(k) $\dfrac{1}{x}$ (l) $\dfrac{x}{9}$ (m) $\dfrac{1}{x}$ (n) $\dfrac{1}{9x}$ (o) $\dfrac{x}{2}$

2. (a) $\dfrac{1}{x}$ (b) $\dfrac{x}{4}$ (c) 1 (d) x (e) $\dfrac{1}{x}$

(f) $\dfrac{4}{x}$ (g) $\dfrac{6}{x}$

3. (a) $\dfrac{a}{20}$ (b) $\dfrac{5a}{4b}$ (c) $\dfrac{1}{2ab}$ (d) $\dfrac{6x^2}{y^3}$

(e) $\dfrac{3b}{5a^2}$ (f) $\dfrac{x}{4y}$ (g) $\dfrac{x}{3(x+y)}$ (h) $\dfrac{1}{3(x+4)}$

4. (a) $\dfrac{3y}{z^3}$ (b) $\dfrac{3+x}{y}$ (c) $\dfrac{x}{12}$ (d) $\dfrac{b}{c}$

5. (a) $\dfrac{1}{x+4}$ (b) $\dfrac{4(x-2)}{x}$

(c) $\dfrac{12ab}{5ef} \times \dfrac{f}{4ab^2} = \dfrac{3}{5eb}$

(d) $\dfrac{x+3y}{2x} \times \dfrac{4x^2}{y} = \dfrac{2x(x+3y)}{y}$ (e) $\dfrac{9}{xyz}$

6. $\dfrac{2}{x+3}$

7. $\dfrac{x+4}{x+3}$

Self-assessment question 9.4

1. The l.c.d. is the simplest expression that is divisible by all the given denominators. To find the l.c.d. first factorise all denominators. The l.c.d. is then formed by including the minimum number of factors from the denominators such that each denominator can divide into the l.c.d.

Exercise 9.4

1. (a) $\dfrac{5z}{6}$ (b) $\dfrac{7x}{12}$ (c) $\dfrac{6y}{25}$

2. (a) $\dfrac{x+2}{2x}$ (b) $\dfrac{1+2x}{2}$ (c) $\dfrac{1+3y}{3}$

(d) $\dfrac{y+3}{3y}$ (e) $\dfrac{8y+1}{y}$

3. (a) $\dfrac{10-x}{2x}$ (b) $\dfrac{5+2x}{x}$ (c) $\dfrac{9-x}{3x}$

(d) $\dfrac{2x-3}{6}$ (e) $\dfrac{9+x}{3x}$

4. (a) $\dfrac{3}{x} + \dfrac{4}{y} = \dfrac{3y}{xy} + \dfrac{4x}{xy} = \dfrac{3y+4x}{xy}$

(b) $\dfrac{3}{x^2} + \dfrac{4y}{x} = \dfrac{3}{x^2} + \dfrac{4xy}{x^2} = \dfrac{3+4xy}{x^2}$

(c) $\dfrac{4ab}{x} + \dfrac{3ab}{2y} = \dfrac{8aby}{2xy} + \dfrac{3abx}{2xy} = \dfrac{8aby+3abx}{2xy}$

(d) $\dfrac{4xy}{a} + \dfrac{3xy}{2b} = \dfrac{8xyb}{2ab} + \dfrac{3xya}{2ab}$

$= \dfrac{8xyb+3xya}{2ab}$

(e) $\dfrac{3}{x} - \dfrac{6}{2x} = \dfrac{3}{x} - \dfrac{3}{x} = 0$

(f) $\dfrac{3x}{2y} - \dfrac{7y}{4x} = \dfrac{6x^2}{4xy} - \dfrac{7y^2}{4xy} = \dfrac{6x^2 - 7y^2}{4xy}$

(g) $\dfrac{3}{x+y} - \dfrac{2}{y} = \dfrac{3y}{(x+y)y} - \dfrac{2(x+y)}{(x+y)y}$

$= \dfrac{3y - 2x - 2y}{(x+y)y} = \dfrac{y - 2x}{(x+y)y}$

(h) $\dfrac{1}{a+b} - \dfrac{1}{a-b}$

$= \dfrac{a-b}{(a+b)(a-b)} - \dfrac{a+b}{(a+b)(a-b)}$

$= \dfrac{a-b-a-b}{(a+b)(a-b)} = \dfrac{-2b}{(a+b)(a-b)}$

(i) $2x + \dfrac{1}{2x} = \dfrac{4x^2}{2x} + \dfrac{1}{2x} = \dfrac{4x^2 + 1}{2x}$

(j) $2x - \dfrac{1}{2x} = \dfrac{4x^2}{2x} - \dfrac{1}{2x} = \dfrac{4x^2 - 1}{2x}$

5. (a) $\dfrac{x}{y} + \dfrac{3x^2}{z} = \dfrac{xz}{yz} + \dfrac{3x^2y}{yz} = \dfrac{xz + 3x^2y}{yz}$

(b) $\dfrac{4}{a} + \dfrac{5}{b} = \dfrac{4b + 5a}{ab}$

(c) $\dfrac{6x}{y} - \dfrac{2y}{x} = \dfrac{6x^2 - 2y^2}{xy}$

(d) $3x - \dfrac{3x+1}{4} = \dfrac{12x}{4} - \dfrac{3x+1}{4}$

$= \dfrac{12x - 3x - 1}{4} = \dfrac{9x - 1}{4}$

(e) $\dfrac{5a}{12} + \dfrac{9a}{18} = \dfrac{15a + 18a}{36} = \dfrac{33a}{36} = \dfrac{11a}{12}$

(f) $\dfrac{x-3}{4} + \dfrac{3}{5} = \dfrac{5(x-3) + 12}{20}$

$= \dfrac{5x - 15 + 12}{20} = \dfrac{5x - 3}{20}$

6. (a) $\dfrac{2x+3}{(x+1)(x+2)}$ (b) $\dfrac{3x+1}{(x-1)(x+3)}$

(c) $\dfrac{4x+17}{(x+5)(x+4)}$ (d) $\dfrac{4x-10}{(x-2)(x-4)}$

(e) $\dfrac{5x+4}{(2x+1)(x+1)}$ (f) $\dfrac{x+1}{x(1-2x)}$

(g) $\dfrac{3x+7}{(x+1)^2}$ (h) $\dfrac{x}{(x-1)^2}$

Exercise 9.5

1. (a) $\dfrac{4}{x+2} + \dfrac{3}{x+3}$ (b) $\dfrac{5}{x+4} - \dfrac{3}{x+1}$

(c) $\dfrac{1}{x+6} - \dfrac{2}{2x+3}$ (d) $\dfrac{2}{x-1} + \dfrac{3}{x-4}$

(e) $\dfrac{5}{2x-1} - \dfrac{1}{x+2}$

2. $\dfrac{5}{2(3x-2)} - \dfrac{3}{2(2x+3)}$

3. (a) $\dfrac{4}{x+5} - \dfrac{3}{x-5}$ (b) $\dfrac{1}{x-3} - \dfrac{1}{(x-3)^2}$

(c) $\dfrac{3}{1-x} - \dfrac{2}{x+2}$ (d) $\dfrac{4}{3x-1} - \dfrac{1}{(3x-1)^2}$

Chapter 10

Self-assessment questions 10.1

1. The subject of a formula is a variable that appears by itself on one side of the formula. It appears nowhere else.

2. To change the subject of a formula you will have to transpose it.

3. Whatever is done to the left-hand side (l.h.s.) must be done to the right-hand side (r.h.s.). We can (a) add or subtract the same quantity

from both left- and right-hand sides,
(b) multiply or divide both sides by the same
quantity, (c) perform the same operation on
both sides.

Exercise 10.1

1. (a) $x = \dfrac{y}{3}$ (b) $x = \dfrac{1}{y}$ (c) $x = \dfrac{5+y}{7}$

(d) $x = 2y + 14$ (e) $x = \dfrac{1}{2y}$

(f) $2x + 1 = \dfrac{1}{y}$ so $2x = \dfrac{1}{y} - 1 = \dfrac{1-y}{y}$

finally, $x = \dfrac{1-y}{2y}$

(g) $y - 1 = \dfrac{1}{2x}$ so $2x = \dfrac{1}{y-1}$

and then $x = \dfrac{1}{2(y-1)}$

(h) $x = \dfrac{y+21}{18}$ (i) $x = \dfrac{19-y}{8}$

2. (a) $m = \dfrac{y-c}{x}$ (b) $x = \dfrac{y-c}{m}$

(c) $c = y - mx$

3. (a) $y = 13x - 26$ and so $x = \dfrac{y+26}{13}$

(b) $y = x + 1$ and so $x = y - 1$

(c) $y = a + tx - 3t$ and so $tx = y + 3t - a$

and then $x = \dfrac{y + 3t - a}{t}$

4. (a) $x = \dfrac{y-11}{7}$ (b) $I = \dfrac{V}{R}$ (c) $r^3 = \dfrac{3V}{4\pi}$

and so $r = \sqrt[3]{\dfrac{3V}{4\pi}}$ (d) $m = \dfrac{F}{a}$

5. $n = (l - a)/d + 1$
6. $\sqrt{x} = (m - n)/t$ and so $x = [(m - n)/t]^2$
7. (a) $x^2 = 1 - y$ and so $x = \pm \sqrt{1-y}$
   (b) $1 - x^2 = 1/y$ and so $x^2 = 1 - 1/y$,
   finally, $x = \pm \sqrt{1 - 1/y}$
   (c) Write $(1 + x^2)y = 1 - x^2$, remove
   brackets to get $y + x^2 y = 1 - x^2$,
   rearrange and factorise to give
   $x^2(y + 1) = 1 - y$ from which
   $x^2 = (1 - y)/(1 + y)$, finally
   $x = \pm \sqrt{(1 - y)/(1 + y)}$.

Chapter 11

Self-assessment questions 11.1

1. A root is a solution of an equation.
2. A linear equation can be written in the form
   $ax + b = 0$, where a and b are constants. In a
   linear equation, the variable is always to the
   first power only.
3. Adding or subtracting the same quantity
   from both sides; multiplying or dividing
   both sides by the same quantity; performing
   the same operation on both sides of the
   equation.
4. Equation: The two sides of the equation are
   equal only for certain values of the variable

involved. These values are the solutions of
the equation. For all other values, the two
sides are not equal. Formula: A formula is
essentially a statement of how two, or more,
variables relate to one another, as for
example in $A = \pi r^2$. The area of a circle,
A, is always πr^2, for any value of the
radius r.

Exercise 11.1

2. (a) $x = \frac{9}{3} = 3$ (b) $x = 3 \times 9 = 27$
   (c) $3t = -6$, so $t = \frac{-6}{3} = -2$ (d) $x = 22$

(e) $3x = 4$ and so $x = \frac{4}{3}$ (f) $x = 36$

(g) $5x = 3$ and so $x = \frac{3}{5}$ (h) $x + 3 = 6$
and so $x = 3$

(i) Adding the two terms on the left we obtain

$$\frac{3x + 2}{2} + 3x = \frac{3x + 2}{2} + \frac{6x}{2}$$

$$= \frac{3x + 2 + 6x}{2} = \frac{9x + 2}{2}$$

and the equation becomes $(9x + 2)/2 = 1$, therefore $9x + 2 = 2$, that is $9x = 0$ and so $x = 0$.

3. (a) $5x + 10 = 13$, so $5x = 3$ and so $x = \frac{3}{5}$
   (b) $3x - 21 = 2x + 2$ and so $x = 23$
   (c) $5 - 10x = 8 - 4x$ so that $-3 = 6x$
   finally $x = \frac{-3}{6} = -\frac{1}{2}$

4. (a) $t = 9$ (b) $v = \frac{17}{7}$
   (c) $3s + 2 = 14s - 14$, so that $11s = 16$ and so $s = \frac{16}{11}$

5. (a) $t = 3$ (b) $t = 5$ (c) $t = -5$ (d) $t = 3$
   (e) $x = \dfrac{12}{5}$ (f) $x = -9$ (g) $x = 24$
   (h) $x = 15$ (i) $x = 23$ (j) $x = -\dfrac{43}{19}$

6. (a) $x = \dfrac{1}{5}$ (b) $x = \dfrac{2}{5}$ (c) $x = -\dfrac{3}{5}$
   (d) $x = 1\dfrac{2}{5}$ (e) $x = -1$ (f) $x = -3/5$
   (g) $x = -1/2$ (h) $x = -1/10$

Exercise 11.2

2. (a) Adding the two equations gives

$$3x + y = 1$$
$$\underline{2x - y = 2}$$
$$5x \quad\;\; = 3$$

from which $x = \frac{3}{5}$. Substitution into either equation gives $y = -\frac{4}{5}$.

(b) Subtracting the given equations eliminates y to give $x = 4$. From either equation $y = 1$.

(c) Multiplying the second equation by 2 and subtracting the first gives

$$2x + 6y = 24$$
$$\underline{2x - y \;\; = 17} \quad -$$
$$7y = 7$$

from which $y = 1$. Substitute into either equation to get $x = 9$.

(d) Multiplying the second equation by -2 and subtracting from the first gives

$$-2x + y \;\; = -21$$
$$\underline{-2x - 6y = \;\; 28} \quad -$$
$$7y = -49$$

from which $y = -7$. Substitution into either equation gives $x = 7$.

(e) Multiplying the first equation by 3 and adding the second gives

$$-3x + 3y = -30$$
$$\underline{3x + 7y = \;\; 20} \quad +$$
$$10y = -10$$

from which $y = -1$. Substitution into either equation gives $x = 9$.

(f) Multiplying the second equation by 2 and subtracting this from the first gives

$$4x - 2y = \;\; 2$$
$$\underline{6x - 2y = \;\; 8} \quad -$$
$$-2x \quad\;\; = -6$$

from which $x = 3$. Substitution into either equation gives $y = 5$.

3. (a) $x = 7$, $y = 1$ (b) $x = -7$, $y = 2$
   (c) $x = 10$, $y = 0$ (d) $x = 0$, $y = 5$
   (e) $x = -1$, $y = -1$

Self-assessment questions 11.3

1. If $b^2 - 4ac > 0$ a quadratic equation will have distinct real roots.
2. If $b^2 - 4ac = 0$ a quadratic equation will have a repeated root.
3. If $b^2 - 4ac < 0$ the quadratic equation has complex roots. When this case arises we are faced with finding the square root of a negative number. This is handled by introducing the symbol i to stand for the square root of -1.

Exercise 11.3

1. (a) $(x + 2)(x - 1) = 0$ so that $x = -2$ and 1
   (b) $(x - 3)(x - 5) = 0$ so that $x = 3$ and 5
   (c) $2(2x + 1)(x + 1) = 0$ so that $x = -1$ and $x = -\frac{1}{2}$
   (d) $(x - 3)(x - 3) = 0$ so that $x = 3$ twice
   (e) $(x + 9)(x - 9) = 0$ so that $x = -9$ and 9
   (f) $-1, -3$ (g) $1, -3$ (h) $1, -4$
   (i) $-1, -5$ (j) $5, 7$ (k) $-5, -7$
   (l) $1, -\frac{3}{2}$ (m) $-\frac{3}{2}, 2$ (n) $-\frac{3}{2}, 5$
   (o) $1, -1/3$ (p) $-1/3, 5/3$ (q) $0, -\frac{1}{7}$
   (r) $-\frac{3}{2}$ twice

2. (a) $x = \dfrac{-(-6) \pm \sqrt{(-6)^2 - 4(3)(-5)}}{6}$

   $= \dfrac{6 \pm \sqrt{96}}{6}$

   $= 2.633$ and -0.633

(b) $x = \dfrac{-3 \pm \sqrt{9 - 4(1)(-77)}}{2}$

   $= \dfrac{-3 \pm \sqrt{317}}{2}$

   $= 7.402$ and -10.402

(c) $x = \dfrac{-(-9) \pm \sqrt{(-9)^2 - 4(2)(2)}}{4}$

   $= \dfrac{9 \pm \sqrt{65}}{4}$

   $= 4.266$ and 0.234

(d) $1, -4$ (e) $1.758, -0.758$
(f) $0.390, -0.640$ (g) $7.405, -0.405$
(h) $0.405, -7.405$
(i) The roots are complex numbers:

   $x = \dfrac{-1 \pm \sqrt{1^2 - 4(11)(1)}}{22}$

   $= \dfrac{-1 \pm \sqrt{-43}}{22}$

   $= \dfrac{-1 \pm \sqrt{43}i}{22}$

(j) $2.766, -1.266$

3. (a) $(3x + 2)(2x + 3) = 0$
      from which $x = -2/3$ and $-3/2$
   (b) $(t + 3)(3t + 4) = 0$
      from which $t = -3$ and $t = -4/3$
   (c) $t = (7 \pm \sqrt{37})/2 = 6.541$ and 0.459

Chapter 12

Self-assessment question 12.1

1. A sequence is a set of numbers written down in a specific order. If the sequence continues on for ever, it is an infinite sequence. Otherwise, if it stops, it is a finite sequence.

Exercise 12.1

1. (a) $3, 6, 9, 12, 15$ (b) $1, 4, 16, 64, 256$
   (c) $\frac{1}{2}, 1, \frac{3}{2}, 2, \frac{5}{2}$

(d) $-3, 4, 11, 18, 25$
(e) $0, -1, -2, -3, -4$

2. $2, 5, 13, 35$

Self-assessment questions 12.2

1. An arithmetic progression is a sequence in which each new term is found by adding a fixed amount, known as the common difference, to the previous term.

2. 4, 6, 8, 10 ... is an arithmetic progression with first term 4 and common difference 2.

3. 4, 6, 9, 13 ... cannot be an arithmetic progression because the difference between the first two terms is 2 whereas that between the second and third is 3. In an arithmetic progression the difference must be fixed.

Exercise 12.2

1. 1, 5, 9, 13, 17
2. 3, 2, 1, 0, −1, −2
3. 77
4. −55

Self-assessment questions 12.3

1. A geometric sequence is one in which each term after the first is found by multiplying the previous term by a fixed amount called the common ratio.
2. $12, 6, 3, \frac{3}{2}$ is a geometric sequence with common ratio $\frac{1}{2}$. The sequence $12, 6, 0, -6$ is not a geometric sequence. In fact it is an arithmetic sequence.
3. Each term is the same.
4. Each term has the same size as the first term, but the terms alternate in sign.

Exercise 12.3

1. 8, 24, 72, 216
2. $\frac{1}{4}, \frac{1}{5}, \frac{4}{25}, \frac{16}{125}$
3. (a) $4 \times 2^4 = 64$ (b) $4 \times 2^{10} = 4096$
4. $-\frac{1}{2}$

Self-assessment question 12.4

1. If the terms of a sequence get closer and closer to a fixed value as we move along the sequence, then that sequence is said to converge. The fixed value is called the limit of the sequence.

Exercise 12.4

1. (a) No limit   (b) No limit   (c) 0   (d) 0
   (e) No limit   (f) No limit   (g) 0   (h) 7
   (i) 1

Exercise 12.5

1. (a) $1^2 + 2^2 + 3^2 + 4^2 + 5^2$
   (b) $3^2 + 5^2 + 7^2 + 9^2$
   (c) $1^2 + 3^2 + 5^2 + 7^2 + 9^2$
2. (a) $\frac{1}{2} + \frac{2}{3} + \frac{3}{4} + \frac{4}{5}$ (b) $\frac{1}{2} + \frac{2}{3} + \frac{3}{4} + \frac{4}{5}$.
   Although the expressions are different they represent the same sum.
3. $\frac{-1}{1} + \frac{1}{2} + \frac{-1}{3}$, which equals $-1 + \frac{1}{2} - \frac{1}{3}$.

Self-assessment questions 12.6

1. An arithmetic series is found by adding the terms of an arithmetic sequence.
2. $S_n = \frac{n}{2}(2a + (n - 1)d)$

Exercise 12.6

1. 648
2. −63
3. 20
4. −3
5. $-\frac{1}{2}$
6. 1740

Self-assessment questions 12.7

1. A geometric series is found by adding the terms of a geometric sequence.
2. $S_n = \dfrac{a(1 - r^n)}{1 - r}$
3. Division by zero is not allowed.
4. When $r = 1$, each term is the same, namely a. The sum of n terms will be na.

Exercise 12.7

1. 728
2. 55924050
3. 129

Exercise 12.8

1. 3
2. $\frac{3}{5}$
3. $\frac{16}{7} = 2\frac{2}{7}$
4. $-10\frac{2}{3}$

Chapter 13

Self-assessment question 13.1

1. (a) an individual member of a set
   (b) a set which is contained within another set
   (c) the set comprising all elements of interest in a particular problem
   (d) sets which have exactly the same elements
   (e) the complement of X is the set whose elements belong to $\mathcal{E}$ but which are not in X

Exercise 13.1

1. (a) T   (b) T   (c) T   (d) F   (e) F
2. (a) $\overline{X} = \{10, 12\}$ (b) $\overline{Y} = \{2, 4, 12\}$

Self-assessment questions 13.2

1. (a) the set whose elements belong to both given sets
   (b) the set whose elements include all elements of the given sets
2. Means 'is an element of' or 'belongs to'

Exercise 13.2

1. $A = \{6, 7, 8, 9, 10\}$
   $B = \{7, 9, 11, 13, 15, 17, 19\}$
   $C = \{15, 30, 45\}$
2. (a) $\{0, 1, 2, 3, 4, 8\}$ (b) $\{1, 3, 5, 7, 9\}$
   (c) $\{5, 6, 7, 9\}$ (d) $\{2, 4, 6, 8\}$
   (e) $\{0, 1, 2, 3, 4, 5, 7, 8, 9\}$ (f) $\{1, 3\}$
   (g) $\{0, 2, 4, 5, 6, 7, 8, 9\}$ (h) $\{6\}$ (i) $\{1, 3\}$
   (j) $\{0, 1, 2, 3, 4, 5, 7, 8, 9\}$

   (e) and (j) are equal sets. (f) and (i) are equal sets.
3. (a) T   (b) F   (c) T   (d) F   (e) F
   (f) T   (g) F   (h) F   (i) T   (j) F   (k) T

Exercise 13.3

1. Figure Exercise 13.3 Q1 (a), (b), (c)

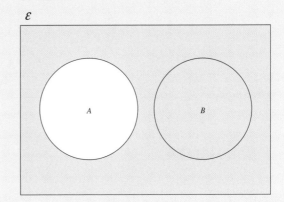

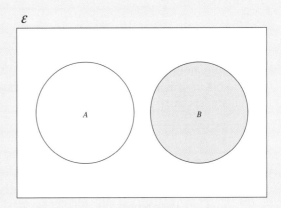

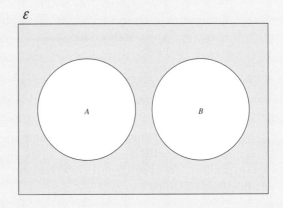

2. Figure Exercise 13.3 Q2

Self-assessment question 13.4

1. (a) a number of the form $\frac{p}{q}$, p and q being integers with $q \neq 0$.

   (b) any number which is not rational.

Chapter 14

Self-assessment question 14.1

1. Powers of 10

Self-assessment-questions 14.2

1. Repeatedly divide the decimal number by 2, carefully noting the remainder each time. The binary number is formed by writing the remainders in reverse order.

2. 2

3. 2

Exercise 14.2

1. (a) To convert to binary, the decimal number 19 is repeatedly divided by 2. Each time we note the remainder. The binary number is formed by writing the remainders in reverse order:

	Remainder
$19 \div 2 = 9$ r 1	1
$9 \div 2 = 4$ r 1	1
$4 \div 2 = 2$ r 0	0
$2 \div 2 = 1$ r 0	0
$1 \div 2 = 0$ r 1	1

   Reading the remainders from bottom to top: $19_{10} = 10011$.

   (b) 100100   (c) 1100100   (d) 1100011100
   (e) 1001110001000

2. (a) To convert a binary number to a decimal number we use the fact that each place is a power of 2: so

   $$111_2 = 1 \times 2^2 + 1 \times 2^1 + 1 \times 2^0$$
   $$= 1 \times 4 + 1 \times 2 + 1 \times 1$$
   $$= 4 + 2 + 1$$
   $$= 7$$

   (b) 21   (c) 57   (d) 113   (e) 255

3. (a) 3   (b) 7   (c) 15   (d) 31   (e) $2^N - 1$

4. The binary system is based on powers of 2. The examples in the text can be extended to the case of negative powers of 2 just as in the decimal system numbers after the decimal place represent negative powers of 10. So, for example, the binary number 11.101_2 is converted to decimal as follows:

   $$11.101_2 = 1 \times 2^1 + 1 \times 2^0 + 1 \times 2^{-1}$$
   $$+ 0 \times 2^{-2} + 1 \times 2^{-3}$$
   $$= 2 + 1 + \frac{1}{2} + \frac{1}{8}$$
   $$= 3\frac{5}{8}$$

   In the same way the binary equivalent of the decimal number 0.5 is 0.1.

Self-assessment questions 14.3

1. 8
2. 8

Exercise 14.3

1. (a) To convert to octal, the decimal number 971 is repeatedly divided by 8. Each time we note the remainder. The octal number is formed by writing the remainders in reverse order:

	Remainder
$971 \div 8 = 121$ r 3	3
$121 \div 8 = 15$ r 1	1
$15 \div 8 = 1$ r 7	7
$1 \div 8 = 0$ r 1	1

   Reading the remainders from bottom to top: $971_{10} = 1713_8$.

   (b) 5431 (c) 11626 (d) 23420 (e) 43006
2. (a) 59 (b) 671 (c) 3997 (d) 3519
   (e) 28661
3. (a) 7 (b) 63 (c) 511 (d) 4095
   (e) 32767 (f) $8^N - 1$

Self-assessment question 14.4

1. 16

Exercise 14.4

1. (a) The hexadecimal system is based on powers of 16. So

$$91_{16} = 9 \times 16^1 + 1 \times 16^0$$
$$= 144 + 1 = 145_{10}$$

   (b) $6C = 6 \times 16^1 + C \times 16^0 = 96 + C = 108$ (noting $C_{16} = 12_{10}$).
   (c) 2587 (d) 63956 (e) 43981
2. (a) To convert to hexadecimal, the decimal number 160 is repeatedly divided by 16. Each time we note the remainder. The hexadecimal number is formed by writing the remainders in reverse order:

	Remainder
$160 \div 16 = 10$ r 0	0
$10 \div 16 = 0$ r 10	10

   Note from Table 14.3 that 10 in hexadecimal is A, and reading the remainders from bottom to top: $160_{10} = A0$.

   (b) 18C (c) 1392 (d) 61A8 (e) F4240
3. (a) 15 (b) 255 (c) 4095 (d) 65535
   (e) $16^N - 1$

Chapter 15

Self-assessment question 15.1

1. A proposition is a statement that is true or false.

Exercise 15.1

1. (a), (c) and (e) are propositions.

Self-assessment question 15.2

1. (a) Conjunction corresponds to 'and' in English.

   (b) Implication corresponds to the 'If X then Y' structure.
   (c) Disjunction corresponds to 'or'.

Exercise 15.2

1. (a) The word has seven or more characters.
   (b) The program does not compile.
   (c) If the word has less than seven characters, then the program compiles.
   (d) The word has less than seven characters or the program does not compile.

(e) The program compiles and the word has less than seven characters.

2. (a) $\neg D$ (b) $V \rightarrow D$ (c) $\neg V \vee D$

Self-assessment questions 15.3

1. (a) A proposition that is always false is a contradiction.
   (b) A proposition that is always true is a tautology.

2. (a) 2^4 (b) 2^5 (c) 2^N

Exercise 15.3

1. (a)

P	Q	(i) $\neg(P \vee Q)$	(ii) $\neg P \wedge \neg Q$	(iii) $\neg(P \wedge Q)$	(iv) $\neg P \vee \neg Q$
T	T	F	F	F	F
T	F	F	F	T	T
F	T	F	F	T	T
F	F	T	T	T	T

(b) (i) and (ii) are logically equivalent; (iii) and (iv) are logically equivalent.

2. (a)

P	$\neg(\neg P)$
T	T
F	F

(b) and (c)

P	Q	R	(b) $(P \wedge Q) \vee R$	(c) $P \wedge (Q \vee R)$
T	T	T	T	T
T	T	F	T	T
T	F	T	T	T
T	F	F	F	F
F	T	T	T	F
F	T	F	F	F
F	F	T	T	F
F	F	F	F	F

3. The expressions are logically equivalent as they have the same truth table.

Chapter 16

Self-assessment questions 16.2

1. A function is a rule that receives an input and produces a single output.
2. The input to a function is called the independent variable; the output is called the dependent variable.
3. False. For example, if $f(x) = 5x$, then $f(\frac{1}{x}) = 5 \times \frac{1}{x} = \frac{5}{x}$ whereas $\frac{1}{f(x)} = \frac{1}{5x}$.
4. $f(x) = x^2 - 5x$

Exercise 16.2

1. (a) Multiply the input by 10.
   (b) Multiply the input by -1 and then add 2; alternatively we could say subtract the input from 2.
   (c) Raise the input to the power 4 and then multiply the result by 3.
   (d) Divide 4 by the square of the input.
   (e) Take three times the square of the input, subtract twice the input from the result and finally add 9.
   (f) The output is always 5 whatever the value of the input.
   (g) The output is always zero whatever the value of the input.

2. (a) Take three times the square of the input and add to twice the input; (b) take three times the square of the input and add to twice the input. The functions in parts (a) and (b) are the same. Both instruct us to do the same thing.

3. (a) $f(x) = x^3/12$ – letters other than f and x are also valid.
   (b) $f(x) = (x + 3)^2$
   (c) $f(x) = x^2 + 4x - 10$
   (d) $f(x) = x/(x^2 + 5)$ (e) $f(x) = x^3 - 1$
   (f) $f(x) = (x - 1)^2$ (g) $f(x) = (7 - 2x)/4$
   (h) $f(x) = -13$

4. (a) $A(2) = 3$ (b) $A(3) = 7$ (c) $A(0) = 1$
   (d) $A(-1) = (-1)^2 - (-1) + 1 = 3$
5. (a) $y(1) = 1$
   (b) $y(-1) = [2(-1) - 1]^2 = (-3)^2 = 9$
   (c) $y(-3) = 49$ (d) $y(0.5) = 0$
   (e) $y(-0.5) = 4$
6. (a) $f(t + 1) = 4(t + 1) + 6 = 4t + 10$
   (b) $f(t + 2) = 4(t + 2) + 6 = 4t + 14$
   (c) $f(t + 1) - f(t) = (4t + 10) - (4t + 6)$
       $= 4$
   (d) $f(t + 2) - f(t) = (4t + 14) - (4t + 6)$
       $= 8$
7. (a) $f(n) = 2n^2 - 3$ (b) $f(z) = 2z^2 - 3$
   (c) $f(t) = 2t^2 - 3$
   (d) $f(2t) = 2(2t)^2 - 3 = 8t^2 - 3$
   (e) $f(1/z) = 2(1/z)^2 - 3 = 2/z^2 - 3$
   (f) $f(3/n) = 2(3/n)^2 - 3 = 18/n^2 - 3$
   (g) $f(-x) = 2(-x)^2 - 3 = 2x^2 - 3$
   (h) $f(-4x) = 2(-4x)^2 - 3 = 32x^2 - 3$
   (i) $f(x + 1) = 2(x + 1)^2 - 3$
       $= 2(x^2 + 2x + 1) - 3 = 2x^2 + 4x - 1$
   (j) $f(2x - 1) = 2(2x - 1)^2 - 3$
       $= 8x^2 - 8x - 1$
8. $a(p + 1) = (p + 1)^2 + 3(p + 1) + 1$
   $= p^2 + 2p + 1 + 3p + 3 + 1 = p^2 + 5p + 5$
   $a(p + 1) - a(p) = (p^2 + 5p + 5)$
   $-(p^2 + 3p + 1) = 2p + 4$
9. (a) $f(3) = 6$ (b) $h(2) = 3$
   (c) $f[h(2)] = f(3) = 6$
   (d) $h[f(3)] = h(6) = 7$
10. (a) $f[h(t)] = f(t + 1) = 2(t + 1) = 2t + 2$
    (b) $h[f(t)] = h(2t) = 2t + 1$. Note that
        $h[f(t)]$ is not equal to $f[h(t)]$.
11. (a) $f(0.5) = 0.5$ (first part)
    (b) $f(1.1) = 1$ (third part) (c) $f(1) = 2$
    (second part)

Self-assessment questions 16.3

1. Suppose $f(x)$ and $g(x)$ are two functions. If
   the input to f is $g(x)$ then we have $f(g(x))$.
   This is a composite function.
2. Suppose $f(x) = x$, $g(x) = \frac{1}{x}$. Then both $f(g(x))$
   and $g(f(x))$ equal $\frac{1}{x}$. Note that this is in
   contrast to the more general case in which
   $f(g(x))$ is not equal to $g(f(x))$.

Exercise 16.3

1. (a) $f(g(x)) = f(3x - 2) = 4(3x - 2)$
       $= 12x - 8$
   (b) $g(f(x)) = g(4x) = 3(4x) - 2 = 12x - 2$
2. (a) $y(x(t)) = y(t^3) = 2t^3$
   (b) $x(y(t)) = x(2t) = (2t)^3 = 8t^3$
3. (a) $r(s(x)) = r(3x) = \frac{1}{6x}$
   (b) $t(s(x)) = t(3x) = 3x - 2$
   (c) $t(r(s(x))) = t(\frac{1}{6x}) = \frac{1}{6x} - 2$
   (d) $r(t(s(x))) = r(3x - 2) = \frac{1}{2(3x - 2)}$
   (e) $r(s(t(x))) = r(s(x - 2)) = r(3(x - 2))$
       $= \frac{1}{6(x - 2)}$
4. (a) $v(v(t)) = v(2t + 1) = 2(2t + 1) + 1$
       $= 4t + 3$
   (b) $v(v(v(t))) = v(4t + 3) = 2(4t + 3) + 1$
       $= 8t + 7$
5. (a) $m(n(t)) = m(t^2 - 1) = (t^2)^3 = t^6$
   (b) $n(m(t)) = n((t + 1)^3) = (t + 1)^6 - 1$
   (c) $m(p(t)) = m(t^2) = (t^2 + 1)^3$
   (d) $p(m(t)) = p((t + 1)^3) = (t + 1)^6$
   (e) $n(p(t)) = n(t^2) = t^4 - 1$
   (f) $p(n(t)) = p(t^2 - 1) = (t^2 - 1)^2$
   (g) $m(n(p(t))) = m(t^4 - 1) = (t^4)^3 = t^{12}$
   (h) $p(p(t)) = p(t^2) = t^4$
   (i) $n(n(t)) = n(t^2 - 1) = (t^2 - 1)^2 - 1$
   (j) $m(m(t)) = m((t + 1)^3) = [(t + 1)^3 + 1]^3$

Self-assessment questions 16.4

1. The inverse of a function, say $f(x)$, is
   another function, denoted by $f^{-1}(x)$.
   When the input to f^{-1} is $f(x)$, the output is x,
   that is
   $$f^{-1}(f(x)) = x$$
   Thus f^{-1} reverses the process in f.
2. The function $f(x) = 4x^4$ produces the same
   output for various inputs, for example
   $f(1) = 4, f(-1) = 4$. If an inverse of f existed,
   then when the input is 4, there would have
   to be outputs of both 1 and -1. This cannot
   be the case, as a function must produce a
   unique output for every input.

Exercise 16.4

1. In each case let the inverse be denoted by g
   (a) $g(x) = x/3$ (b) $g(x) = 4x$
   (c) $g(x) = x - 1$ (d) $g(x) = x + 3$
   (e) $g(x) = 3 - x$ (f) $g(x) = (x - 6)/2$
   (g) $g(x) = (7 - x)/3$ (h) $g(x) = 1/x$
   (i) $g(x) = 3/x$ (j) $g(x) = -3/4x$

2. (a) $f^{-1}(x) = \frac{x}{6}$
   (b) $f^{-1}(x) = \frac{x-1}{6}$
   (c) $f^{-1}(x) = x - 6$
   (d) $f^{-1}(x) = 6x$
   (e) $f^{-1}(x) = \frac{6}{x}$

3. (a) $g^{-1}(t) = \frac{t-1}{3}$

   (b) We require $g^{-1}\left(\dfrac{1}{3t+1}\right) = t$

   Let $z = \dfrac{1}{3t+1}$

   so that $t = \dfrac{1}{3}\left(\dfrac{1}{z} - 1\right)$

   Then $g^{-1}(z) = \dfrac{1}{3}\left(\dfrac{1}{z} - 1\right)$

   so $g^{-1}(t) = \dfrac{1}{3}\left(\dfrac{1}{t} - 1\right)$

   (c) $g^{-1}(t) = t^{\frac{1}{3}}$
   (d) $g^{-1}(t) = \left(\frac{t}{3}\right)^{\frac{1}{3}}$
   (e) $g^{-1}(t) = \left(\frac{t-1}{3}\right)^{\frac{1}{3}}$

   (f) We require $g^{-1}\left(\dfrac{3}{t^3+1}\right) = t$

   Let $z = \dfrac{3}{t^3+1}$

   so that $t = \left(\dfrac{3}{z} - 1\right)^{\frac{1}{3}}$

   Then $g^{-1}(z) = \left(\dfrac{3}{z} - 1\right)^{\frac{1}{3}}$

   so $g^{-1}(t) = \left(\dfrac{3}{t} - 1\right)^{\frac{1}{3}}$

4. (a) $h^{-1}(t) = \frac{t-3}{4}$ (b) $g^{-1}(t) = \frac{t+1}{2}$

   (c) $g^{-1}(h^{-1}(t)) = g^{-1}\left(\dfrac{t-3}{4}\right)$

   $= \dfrac{\frac{t-3}{4} + 1}{2}$

   $= \dfrac{t+1}{8}$

   (d) $h(g(t)) = h(2t - 1) = 4(2t - 1) + 3$
   $= 8t - 1$
   (e) $h(g(t))^{-1} = \frac{t+1}{8}$. We note that $[h(g(t))]^{-1}$ is the same as $g^{-1}(h^{-1}(t))$.

Chapter 17

Self-assessment questions 17.1

1. The horizontal axis is a straight line with a scale marked on it. It represents the independent variable. The vertical axis is also a straight line with a scale marked on it; this scale may be different from that which is marked on the horizontal axis. These two axes are at 90° to each other. The point where they intersect is the origin.
2. x coordinate
3. (0, 0)

Exercise 17.1

1.

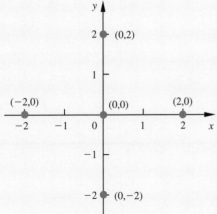

2. All y coordinates must be the same.
3. All x coordinates must be the same.

Self-assessment questions 17.2

1. (a) A closed interval is one that includes the two end-points.
   (b) An open interval is one that does not include the end-points.
   (c) A semi-closed interval is closed at one end and open at the other: that is, one end-point is included, the other is not.
2. (a) Square brackets, [ ], denote that the end-point is included. On a graph, a filled bullet, ●, is used to indicate this.
   (b) Round brackets, ( ), denote that the end-point is not included. On a graph, an open bullet, ○, is used to indicate this.

Exercise 17.2

1. (a) $\{x : x \in \mathbb{R}, 2 \leqslant x \leqslant 6\}$

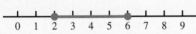

The interval [2, 6]

(b) $\{x : x \in \mathbb{R}, 6 < x \leqslant 8\}$

The interval (6, 8]
(c) $\{x : x \in \mathbb{R}, -2 < x < 0\}$

The interval (−2, 0)
(d) $\{x : x \in \mathbb{R}, -3 \leqslant x < -1.5\}$

The interval [−3, −1.5)
2. (a) T (b) T (c) T (d) T (e) F (f) T
   (g) F (h) T

Self-assessment questions 17.3

1. Dependent
2. Horizontal axis

Exercise 17.3

1.

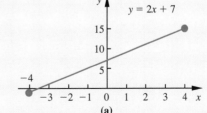

(a)

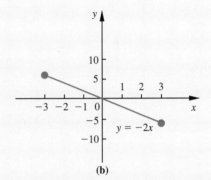

(b)

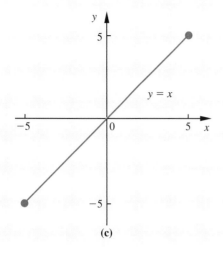

(c)

3. (a)

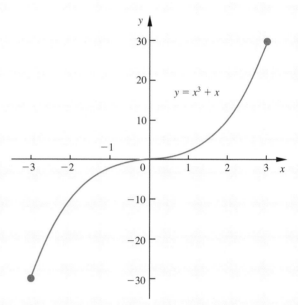

(b) $(1, 2)$ and $(-1, -2)$ lie on the curve.

4.

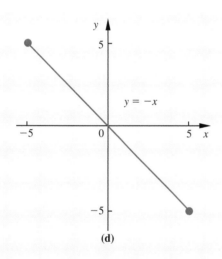

(d)

2.

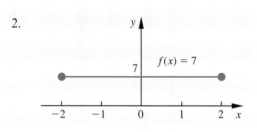

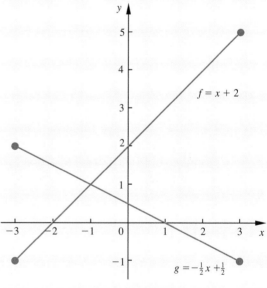

Graphs intersect at $(-1, 1)$

5.

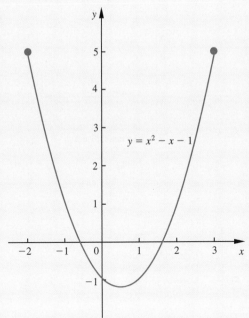

$$y = x^2 - x - 1$$

(a) The curve cuts the horizontal axis at
    $x = -0.6$ and $x = 1.6$.
(b) The curve cuts the vertical axis
    at $y = -1$.

6.

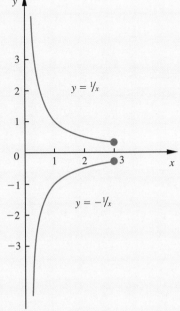

$y = \dfrac{1}{x}$

$y = -\dfrac{1}{x}$

7.

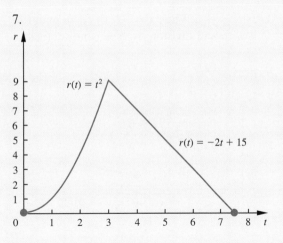

$r(t) = t^2$

$r(t) = -2t + 15$

8.

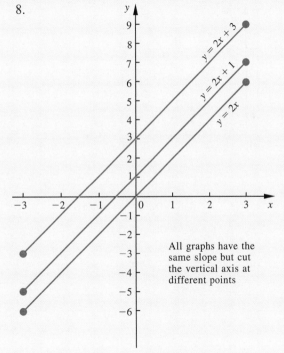

$y = 2x + 3$

$y = 2x + 1$

$y = 2x$

All graphs have the
same slope but cut
the vertical axis at
different points

Self-assessment questions 17.4

1. The domain is the set of values that the
   independent variable can take. The range is
   the set of values that the dependent variable
   can take.
2. When $x = 3$, the denominator of the fraction
   is 0. Division by 0 is not defined and so the
   value $x = 3$ must be excluded.

3. $f(x) = \frac{1}{x+2}$. When $x = -2$, the denominator is zero and division by zero is not defined.

Exercise 17.4

1.

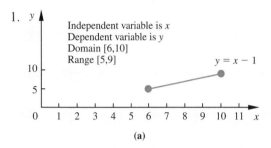

Independent variable is x
Dependent variable is y
Domain [6,10]
Range [5,9]

$y = x - 1$

(a)

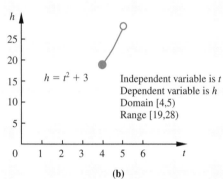

$h = t^2 + 3$

Independent variable is t
Dependent variable is h
Domain [4,5)
Range [19,28)

(b)

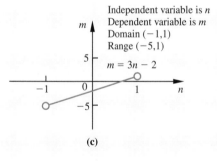

Independent variable is n
Dependent variable is m
Domain $(-1,1)$
Range $(-5,1)$

$m = 3n - 2$

(c)

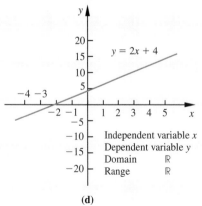

$y = 2x + 4$

Independent variable x
Dependent variable y
Domain $\mathbb{R}$
Range $\mathbb{R}$

(d)

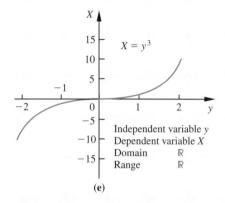

$X = y^3$

Independent variable y
Dependent variable X
Domain $\mathbb{R}$
Range $\mathbb{R}$

(e)

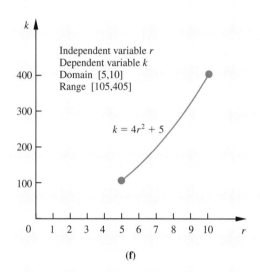

Independent variable r
Dependent variable k
Domain [5,10]
Range [105,405]

$k = 4r^2 + 5$

(f)

2.

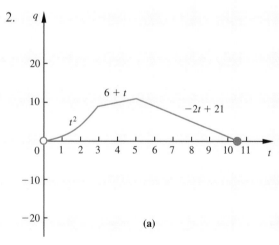

$6 + t$

t^2

$-2t + 21$

(a)

(b) Domain $(0, 10.5]$ (c) range $= [0, 11]$
(d) $q(1) = 1$, $q(3) = 9$, $q(5) = 11$, $q(7) = 7$

Self-assessment question 17.5

1. (a) The intersection of $f(x)$ with the x axis gives the solutions to $f(x) = 0$.
   (b) The intersection of $f(x)$ with the straight line $y = 1$ gives the solutions to $f(x) = 1$.

Exercise 17.5

1.

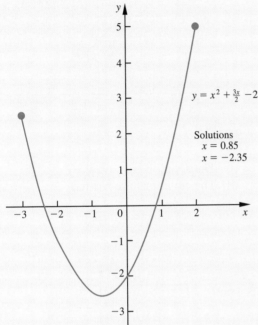

$y = x^2 + \frac{3x}{2} - 2$

Solutions
$x = 0.85$
$x = -2.35$

2.

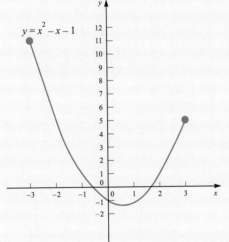

$y = x^2 - x - 1$

From the graph the roots of $x^2 - x - 1 = 0$ are $x = -0.6$ and $x = 1.6$.

3.

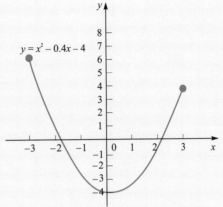

$y = x^2 - 0.4x - 4$

The required roots are $x = -1.8$ and $x = 2.2$.

4. (a)

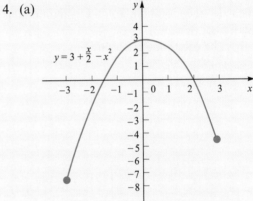

$y = 3 + \frac{x}{2} - x^2$

   (b) From the graph the roots of $x^2 - \frac{x}{2} - 3 = 0$ are $x = -1.5$ and $x = 2$.

Self-assessment questions 17.6

1. Points of intersection provide the solutions to simultaneous equations.
2. Three.
3. There are no solutions.

Exercise 17.6

1. (a) We write the equations in the form

$$y = -\frac{3}{2}x + 2, \qquad y = x - 3$$

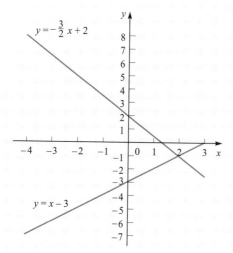

$$y = -\frac{3}{2}x + 2$$

$$y = x - 3$$

The point of intersection is $x = 2$, $y = -1$, which is the solution to the simultaneous equations.

(b) The equations are written as

$$y = -2x - 2, \qquad y = \frac{x}{4} + \frac{5}{2}$$

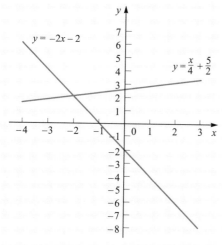

$$y = -2x - 2$$

$$y = \frac{x}{4} + \frac{5}{2}$$

From the graph the solution is $x = -2$, $y = 2$.

(c) The equations are written in the form

$$y = -2x + 4, \qquad y = \frac{4}{3}x + \frac{7}{3}$$

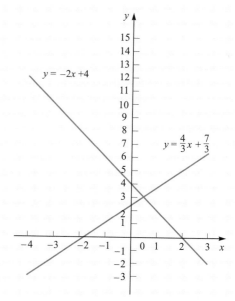

$$y = -2x + 4$$

$$y = \frac{4}{3}x + \frac{7}{3}$$

The solution is $x = \frac{1}{2}$, $y = 3$.

(d) The equations are written

$$y = \frac{x}{4} + \frac{7}{4}, \qquad y = 2x + 7$$

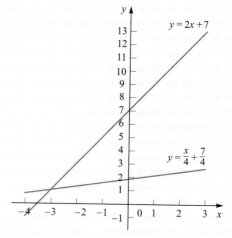

$$y = 2x + 7$$

$$y = \frac{x}{4} + \frac{7}{4}$$

The root is $x = -3$, $y = 1$.

(e) The equations are written

$$y = -\frac{x}{2} + \frac{1}{2}, \qquad y = -\frac{x}{10} - \frac{1}{10}$$

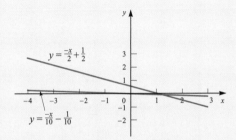

The root is $x = 1.5$, $y = -0.25$.

2.

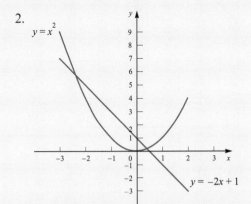

From the graph the required roots are approximately $x = -2.4$, $y = 5.8$ and $x = 0.4$, $y = 0.2$.

3.

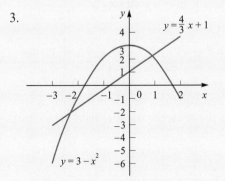

The roots of the simultaneous equations are approximately $x = -2.2$, $y = -2$ and $x = 0.9$, $y = 2.2$.

4.

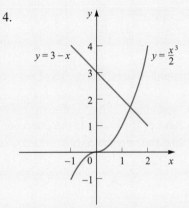

From the point of intersection we see the required root is approximately $x = 1.4$, $y = 1.5$.

5.

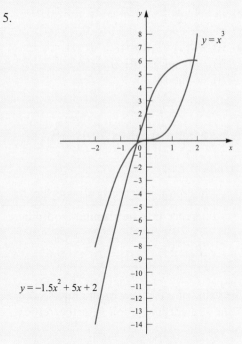

The points of intersection, and hence the required roots, are approximately $x = -0.4$, $y = -0.2$ and $x = 1.8$, $y = 6.1$.

Chapter 18

Self-assessment questions 18.1

1. $y = mx + c$
2. m is the gradient of the line. c is the intercept of the line with the vertical axis.

Exercise 18.1

1. (a), (b), (c) and (e) are straight lines.
2. (a) $m = 9$, $c = -11$ (b) $m = 8$, $c = 1.4$
   (c) $m = \frac{1}{2}$, $c = -11$ (d) $m = -2$, $c = 17$
   (e) $m = \frac{2}{3}$, $c = \frac{1}{3}$ (f) $m = -\frac{2}{5}$, $c = \frac{4}{5}$
   (g) $m = 3$, $c = -3$ (h) $m = 0$, $c = 4$
3. (a) (i) -1 (ii) 6   (b) (i) 2 (ii) -1
   (c) (i) 2 (ii) -1.5 (d) (i) $\frac{3}{4}$ (ii) 3
   (e) (i) -6 (ii) 18

Self-assessment questions 18.2

1.   gradient $= m =$

$$\frac{\text{difference in } y \text{ (vertical) coordinates}}{\text{difference in } x \text{ (horizontal) coordinates}}$$

$$m = \frac{y_2 - y_1}{x_2 - x_1}$$

2. The value of c is given by the intersection of the straight line graph with the vertical axis.

Exercise 18.2

1. $y = 2x + 5$

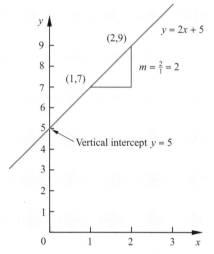

2. $y = 6x - 10$
3. $y = 2$
4. $y = x$
5. $y = -x$
6. $y = 2x + 8$
7. (b), (c) and (d) lie on the line.
8. The equation of the line is $y = mx + c$. The gradient is $m = -1$ and so

$$y = -x + c$$

When $x = -3$, $y = 7$ and so $c = 4$. Hence the required equation is $y = -x + 4$.
9. The gradient of $y = 3x + 17$ is 3. Let the required equation be $y = mx + c$.
   Since the lines are parallel then $m = 3$ and so $y = 3x + c$. When $x = -1$, $y = -6$ and so $c = -3$. Hence the equation is $y = 3x - 3$.
10. The equation is $y = mx + c$.
    The vertical intercept is $c = -2$ and so

$$y = mx - 2$$

When $x = 3$, $y = 10$ and so $m = 4$. Hence the required equation is $y = 4x - 2$.

Self-assessment questions 18.3

1. A tangent is a straight line that touches a curve at a single point.
2. The gradient of a curve at a particular point is the gradient of the tangent at that point.

Exercise 18.3

1.

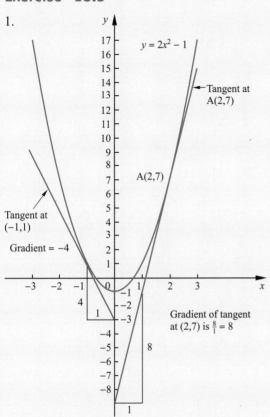

2.

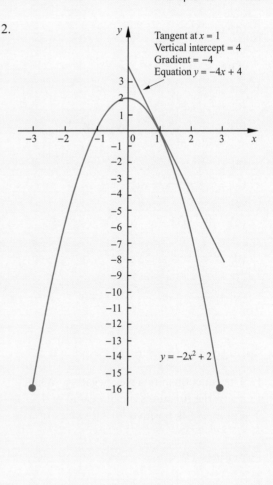

Chapter 19

Exercise 19.1

1. (a) 9.9742   (b) 6.6859   (c) 0.5488
   (d) 0.8808   (e) -8.4901

2. (a) $e^2.e^7 = e^{2+7} = e^9$ (b) $\frac{e^7}{e^4} = e^{7-4} = e^3$
   (c) $\frac{e^{-2}}{e^3} = e^{-2-3} = e^{-5}$ or $\frac{1}{e^5}$
   (d) $\frac{e^3}{e^{-1}} = e^{3-(-1)} = e^4$
   (e) $\frac{(4e)^2}{(2e)^3} = \frac{4^2 e^2}{2^3 e^3} = \frac{16 e^{2-3}}{8} = 2e^{-1} = \frac{2}{e}$

   (f) $e^2(e + \frac{1}{e}) - e = e^2 e + \frac{e^2}{e} - e$
   $= e^3 + e - e = e^3$
   (g) $e^{1.5} e^{2.7} = e^{1.5+2.7} = e^{4.2}$
   (h) $\sqrt{e}\sqrt{4e} = e^{0.5}(4e)^{0.5} = e^{0.5} 4^{0.5} e^{0.5}$
   $= 2e^1 = 2e$

3. (a) $e^x e^{-3x} = e^{x-3x} = e^{-2x}$
   (b) $e^{3x} e^{-x} = e^{3x-x} = e^{2x}$
   (c) $e^{-3x} e^{-x} = e^{-3x-x} = e^{-4x}$

(d) $(e^{-x})^2 e^{2x} = e^{-2x} e^{2x} = e^0 = 1$

(e) $\frac{e^{3x}}{e^x} = e^{3x-x} = e^{2x}$

(f) $\frac{e^{-3x}}{e^{-x}} = e^{-3x-(-x)} = e^{-2x}$

4. (a) $\frac{e^t}{e^3} = e^{t-3}$ (b) $\frac{2e^{3x}}{4e^x} = \frac{1}{2}e^{3x-x} = \frac{e^{2x}}{2}$

(c) $\frac{(e^x)^3}{3e^x} = \frac{e^{3x}}{3e^x} = \frac{e^{3x-x}}{3} = \frac{e^{2x}}{3}$

(d) $e^x + 2e^{-x} + e^x = 2e^x + 2e^{-x}$
$= 2(e^x + e^{-x})$

(e) $\frac{e^x e^y e^z}{e^{(x/2)-y}} = e^{x+y+z-((x/2)-y)} = e^{(x/2)+2y+z}$

(f) $\frac{(e^{\frac{t}{3}})^6}{(e^t + e^t)} = \frac{e^{(\frac{t}{3}) \times 6}}{2e^t} = \frac{e^{2t}}{2e^t} = \frac{e^t}{2}$

(g) $\frac{e^t}{e^{-t}} = e^{t-(-t)} = e^{2t}$

(h) $\frac{1+e^{-t}}{e^{-t}} = (1 + e^{-t})e^t = e^t + e^{-t}e^t$
$= e^t + e^0 = e^t + 1$

(i) $(e^z e^{-z/2})^2 = (e^{z-z/2})^2 = (e^{z/2})^2 = e^z$

(j) $\frac{e^{-3+t}}{2e^t} = \frac{e^{-3+t-t}}{2} = \frac{e^{-3}}{2} = \frac{1}{2e^3}$

(k) $\left(\frac{e^{-1}}{e^{-x}}\right)^{-1} = (e^{-1-(-x)})^{-1} = (e^{x-1})^{-1}$
$= e^{(x-1)(-1)} = e^{1-x}$

Self-assessment questions 19.2

1. The functions are never negative. When $x = 0$ the functions have a value of 1.
2. False; e^{-x} is positive for all values of x.

Exercise 19.2

1.

x	-1	-0.8	-0.6	-0.4	-0.2
e^{3x}	0.0498	0.0907	0.1653	0.3012	0.5488

x	0	0.2	0.4	0.6	0.8	1
e^{3x}	1	1.8221	3.3201	6.0496	11.0232	20.0855

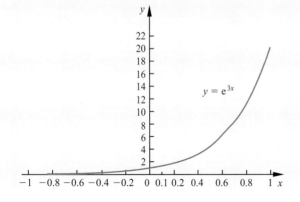

The graph has similar shape and properties to that of $y = e^x$.

2. (a)

t	0	1	2	3	4	5
P	5	8.1606	9.3233	9.7511	9.9084	9.9663

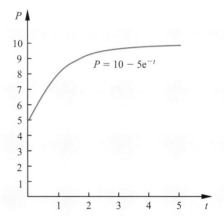

(b) Maximum population is 10, obtained for large values of t.

3. (a)

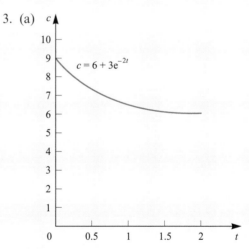

(b) 6

4. (a) $\cosh x + \sinh x$

$$= \frac{e^x + e^{-x}}{2} + \frac{e^x - e^{-x}}{2}$$

$$= \frac{e^x + e^{-x} + e^x - e^{-x}}{2} = e^x$$

(b) $\cosh x - \sinh x$

$$= \frac{e^x + e^{-x}}{2} - \frac{(e^x - e^{-x})}{2}$$

$$= \frac{e^x + e^{-x} - e^x + e^{-x}}{2} = e^{-x}$$

5. $3 \sinh x + 7 \cosh x$

$$= 3\left(\frac{e^x - e^{-x}}{2}\right) + 7\left(\frac{e^x + e^{-x}}{2}\right)$$

$$= 5e^x + 2e^{-x}$$

6. $6e^x - 9e^{-x} = 6(\cosh x + \sinh x)$

$$-9(\cosh x - \sinh x)$$

$$= 15 \sinh x - 3 \cosh x$$

Exercise 19.3

1. (a)

x	0	0.5	1	1.5	2	2.5	3
$15 - x^2$	15	14.75	14	12.75	11	8.75	6
e^x	1	1.65	2.72	4.48	7.39	12.18	20.09

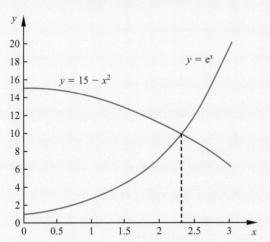

(b) Solving $e^x + x^2 = 15$ is equivalent to solving $e^x = 15 - x^2$. Approximate solution is $x = 2.3$.

2. (a)

x	-1	-0.75	-0.5	-0.25
$12x^2 - 1$	11	5.75	2	-0.25
$3e^{-x}$	8.15	6.35	4.95	3.85

x	0	0.25	0.5	0.75	1
$12x^2 - 1$	-1	-0.25	2	5.75	11
$3e^{-x}$	3	2.34	1.82	1.42	1.10

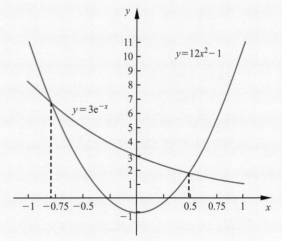

(b) Approximate solutions are $x = -0.80$ and $x = 0.49$.

3.

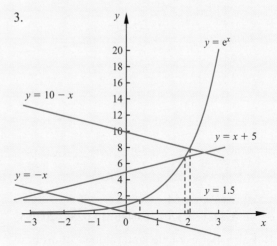

(a) The equation $e^x + x = 0$ is equivalent to $e^x = -x$. Hence draw $y = -x$. Intersection of $y = e^x$ and $y = -x$ will give solution to $e^x + x = 0$. Approximate solution: $x = -0.57$.

(b) The equation $e^x - 1.5 = 0$ is equivalent to $e^x = 1.5$. Hence also draw $y = 1.5$. This is a horizontal straight line. Intersection point gives solution to $e^x - 1.5 = 0$. Approximate solution: $x = 0.41$.

(c) The equation $e^x - x - 5 = 0$ can be written as $e^x = x + 5$. Hence draw $y = x + 5$. Approximate solution: $x = 1.94$.

(d) The equation

$$\frac{e^x}{2} + \frac{x}{2} - 5 = 0$$

can be written as

$$e^x = 10 - x$$

Hence also draw $y = 10 - x$. Approximate solution: $x = 2.07$.

(b) The t value corresponding to $y = 5$ on the graph is $t = 0.7$. Hence the root of $2 + 6e^{-t} = 5$ is $t = 0.7$.

(c) $y = t + 4$ is also shown. The point of intersection of $y = 2 + 6e^{-t}$ and $y = t + 4$ is a solution of

$$2 + 6e^{-t} = t + 4$$
$$6e^{-t} = t + 2$$

The solution is $t = 0.77$.

4. (a)

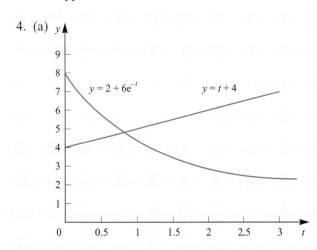

Chapter 20

Self-assessment question 20.1

1. False. For example, natural logarithms have base e (≈ 2.718).

Exercise 20.1

1. (a) 2.1761   (b) 5.0106   (c) −0.5003
   (d) −2.3026
2. (a) $\log_3 6561 = 8$ (b) $\log_6 7776 = 5$
   (c) $\log_2 1024 = 10$
   (d) $\log_{10} 100000 = 5$ (e) $\log_4 16384 = 7$
   (f) $\log_{0.5} 0.03125 = 5$
   (g) $\log_{12} 1728 = 3$ (h) $\log_9 6561 = 4$

3. (a) $1296 = 6^4$ (b) $225 = 15^2$
   (c) $512 = 8^3$ (d) $2401 = 7^4$ (e) $243 = 3^5$
   (f) $216 = 6^3$ (g) $8000 = 20^3$
   (h) $4096 = 16^3$ (i) $4096 = 2^{12}$

4. (a) 0.4771
   (b) $3 = 10^{0.4771}$

   $$3 \times 10^2 = 10^2 \times 10^{0.4771}$$

   $$300 = 10^{2.4771}$$

   $$\log 300 = 2.4771$$

(c)
$$3 = 10^{0.4771}$$
$$3 \times 10^{-2} = 10^{-2} \times 10^{0.4771}$$
$$0.03 = 10^{-1.5229}$$
$$\log 0.03 = -1.5229$$

5. (a) 7

(b)
$$\log 7 = 0.8451$$
$$7 = 10^{0.8451}$$
$$7 \times 10^2 = 10^2 \times 10^{0.8451}$$
$$700 = 10^{2.8451}$$
$$\log 700 = 2.8451$$

(c)
$$7 = 10^{0.8451}$$
$$10^{-2} \times 7 = 10^{-2} \times 10^{0.8451}$$
$$0.07 = 10^{-1.1549}$$
$$\log 0.07 = -1.1549$$

Exercise 20.2

1. (a) $\log_4 6 = \log 6/\log 4 = 1.2925$
   (b) $\log_3 10 = \log 10/\log 3 = 2.0959$
   (c) $\log_{20} 270 = \log 270/\log 20 = 1.8688$
   (d) $\log_5 0.65 = \log 0.65/\log 5 = -0.2677$
   (e) $\log_2 100 = \log 100/\log 2 = 6.6439$
   (f) $\log_2 0.03 = \log 0.03/\log 2 = -5.0589$
   (g) $\log_{100} 10 = \log 10/\log 100 = \frac{1}{2}$
   (h) $\log_7 7 = 1$

2.
$$\ln X = \frac{\log_{10} X}{\log_{10} e} = \frac{\log_{10} X}{0.4343} = 2.3026 \log_{10} X$$

3. (a) $\log_3 7 + \log_4 7 + \log_5 7$
$$= \frac{\log 7}{\log 3} + \frac{\log 7}{\log 4} + \frac{\log 7}{\log 5} = 4.3840$$

(b) $\log_8 4 + \log_8 0.25 = \dfrac{\log 4}{\log 8} + \dfrac{\log 0.25}{\log 8}$
$$= 0.6667 + (-0.6667)$$
$$= 0$$

(c) $\log_{0.7} 2 = \dfrac{\log 2}{\log 0.7} = -1.9434$

(d) $\log_2 0.7 = \dfrac{\log 0.7}{\log 2} = -0.5146$

Self-assessment question 20.3

1. $\log A + \log B = \log AB$
$$\log A - \log B = \log\left(\frac{A}{B}\right)$$
$$\log A^n = n \log A$$

Exercise 20.3

1. (a) $\log 5 + \log 9 = \log(5 \times 9) = \log 45$
   (b) $\log 9 - \log 5 = \log(\frac{9}{5}) = \log 1.8$
   (c) $\log 5 - \log 9 = \log(\frac{5}{9})$
   (d) $2 \log 5 + \log 1 = 2 \log 5 = \log 5^2$
   $= \log 25$
   (e) $2 \log 4 - 3 \log 2 = \log 4^2 - \log 2^3$
   $= \log 16 - \log 8 = \log(\frac{16}{8}) = \log 2$
   (f) $\log 64 - 2 \log 2 = \log(64/2^2) = \log 16$
   (g) $3 \log 4 + 2 \log 1 + \log 27 - 3 \log 12$
   $= \log 4^3 + 2(0) + \log 27 - \log 12^3$
   $$= \log\left(\frac{4^3 \times 27}{12^3}\right)$$
   $= \log 1 = 0$

2. (a) $\log 3x$ (b) $\log 8x$ (c) $\log 1.5$
   (d) $\log T^2$ (e) $\log 10X^2$

3. (a) $3 \log X - \log X^2 = 3 \log X - 2 \log X$
   $= \log X$
   (b) $\log y - 2 \log \sqrt{y} = \log y - \log(\sqrt{y})^2$
   $= \log y - \log y = 0$
   (c) $5 \log x^2 + 3 \log\dfrac{1}{x} = 10 \log x - 3 \log x$
   $= 7 \log x$
   (d) $4 \log X - 3 \log X^2 + \log X^3$
   $= 4 \log X - 6 \log X + 3 \log X$
   $= \log X$

(e) $3 \log y^{1.4} + 2 \log y^{0.4} - \log y^{1.2}$
$= 4.2 \log y + 0.8 \log y - 1.2 \log y$
$= 3.8 \log y$

4. (a) $\log 4x - \log x = \log(4x/x) = \log 4$

(b) $\log t^3 + \log t^4 = \log(t^3 \times t^4) = \log t^7$

(c) $\log 2t - \log \left(\dfrac{t}{4} \right) = \log \left(2t \div \dfrac{t}{4} \right)$

$= \log \left(2t \times \dfrac{4}{t} \right) = \log 8$

(d) $\log 2 + \log \left(\dfrac{3}{x} \right) - \log \left(\dfrac{x}{2} \right)$

$= \log \left(2 \times \dfrac{3}{x} \right) - \log \left(\dfrac{x}{2} \right)$

$= \log \left(\dfrac{6}{x} \div \dfrac{x}{2} \right)$

$= \log \left(\dfrac{6}{x} \times \dfrac{2}{x} \right)$

$= \log \left(\dfrac{12}{x^2} \right)$

(e) $\log \left(\dfrac{t^2}{3} \right) + \log \left(\dfrac{6}{t} \right) - \log \left(\dfrac{1}{t} \right)$

$= \log \left(\dfrac{t^2}{3} \times \dfrac{6}{t} \right) - \log \left(\dfrac{1}{t} \right)$

$= \log(2t) - \log \left(\dfrac{1}{t} \right)$

$= \log \left(\dfrac{2t}{1/t} \right)$

$= \log 2t^2$

(f) $2 \log y - \log y^2 = \log y^2 - \log y^2 = 0$

(g) $3 \log \left(\dfrac{1}{t} \right) + \log t^2 = \log \left(\dfrac{1}{t} \right)^3 + \log t^2$

$= \log \left(\dfrac{1}{t^3} \right) + \log t^2$

$= \log \left(\dfrac{t^2}{t^3} \right)$

$= \log \left(\dfrac{1}{t} \right) = -\log t$

(h) $4 \log \sqrt{x} + 2 \log \left(\dfrac{1}{x} \right)$

$= 4 \log x^{0.5} + \log \left(\dfrac{1}{x} \right)^2$

$= \log x^2 + \log \left(\dfrac{1}{x^2} \right)$

$= \log \left(x^2 \, \dfrac{1}{x^2} \right)$

$= \log 1 = 0$

(i) $2 \log x + 3 \log t = \log x^2 + \log t^3$
$= \log(x^2 t^3)$

(j) $\log A - \dfrac{1}{2} \log 4A = \log A - \log(4A)^{1/2}$

$= \log \left[\dfrac{A}{(4A)^{1/2}} \right]$

$= \log \left(\dfrac{A^{1/2}}{2} \right)$

(k) $\dfrac{\log 9x + \log 3x^2}{3} = \dfrac{\log(9x.3x^2)}{3}$

$= \dfrac{\log(27x^3)}{3}$

$= \dfrac{\log(3x)^3}{3}$

$= \dfrac{3\log(3x)}{3}$

$= \log(3x)$

(l) $\log xy + 2\log\left(\dfrac{x}{y}\right) + 3\log\left(\dfrac{y}{x}\right)$

$= \log xy + \log\left(\dfrac{x^2}{y^2}\right) + \log\left(\dfrac{y^3}{x^3}\right)$

$= \log\left(xy\,\dfrac{x^2}{y^2}\dfrac{y^3}{x^3}\right)$

$= \log\left(\dfrac{x^3y^4}{x^3y^2}\right)$

$= \log y^2$

(m) $\log\left(\dfrac{A}{B}\right) - \log\left(\dfrac{B}{A}\right) = \log\left(\dfrac{A}{B} \div \dfrac{B}{A}\right)$

$= \log\left(\dfrac{A}{B} \times \dfrac{A}{B}\right) = \log\left(\dfrac{A^2}{B^2}\right)$

(n) $\log\left(\dfrac{2t}{3}\right) + \dfrac{1}{2}\log 9t - \log\left(\dfrac{1}{t}\right)$

$= \log\left(\dfrac{2t}{3}\right) + \log(9t)^{1/2} - \log t^{-1}$

$= \log\left[\dfrac{2t}{3}(9t)^{1/2}\right] + \log t$

$= \log\left[\dfrac{2t}{3}(9t)^{1/2}t\right]$

$= \log\left(\dfrac{2t}{3}3t^{1/2}t\right)$

$= \log(2t^{2.5})$

5. $\log_{10} X + \ln X = \log_{10} X + \dfrac{\log_{10} X}{\log_{10} e}$

$= \log_{10} X + 2.3026\log_{10} X$

$= 3.3026\log_{10} X$

$= \log X^{3.3026}$

6. (a) $\log(9x - 3) - \log(3x - 1)$

$= \log\left(\dfrac{9x - 3}{3x - 1}\right) = \log 3$

(b) $\log(x^2 - 1) - \log(x + 1)$

$= \log\left(\dfrac{x^2 - 1}{x + 1}\right) = \log(x - 1)$

(c) $\log(x^2 + 3x) - \log(x + 3)$

$= \log\left(\dfrac{x^2 + 3x}{x + 3}\right) = \log x$

Exercise 20.4

1. (a) $\log x = 1.6000$

$x = 10^{1.6000} = 39.8107$

(b) $10^x = 75$

$x = \log 75 = 1.8751$

(c) $\ln x = 1.2350$

$x = e^{1.2350} = 3.4384$

(d) $e^x = 36$

$x = \ln 36 = 3.5835$

2. (a) $\log(3t) = 1.8$

$3t = 10^{1.8}$

$t = \dfrac{10^{1.8}}{3} = 21.0319$

(b) $10^{2t} = 150$

$2t = \log 150$

$t = \frac{1}{2}\log 150 = 1.0880$

(c) $\ln(4t) = 2.8$

$4t = e^{2.8}$

$t = \frac{1}{4}e^{2.8} = 4.1112$

(d) $e^{3t} = 90$

$3t = \ln 90$

$t = \frac{1}{3}\ln 90 = 1.4999$

3. (a) $\log x = 0.3940$

$x = 10^{0.3940} = 2.4774$

(b) $\ln x = 0.3940$

$x = e^{0.3940} = 1.4829$

(c) $10^y = 5.5$

$y = \log 5.5 = 0.7404$

(d) $e^z = 500$

$z = \ln 500 = 6.2146$

(e) $\log(3v) = 1.6512$

$3v = 10^{1.6512}$

$v = \frac{10^{1.6512}}{3} = 14.9307$

(f) $\ln\left(\frac{t}{6}\right) = 1$

$\frac{t}{6} = e^1 = e$

$t = 6e = 16.3097$

(g) $10^{2r+1} = 25$

$2r + 1 = \log 25$

$2r = \log 25 - 1$

$r = \frac{\log 25 - 1}{2} = 0.1990$

(h) $e^{(2t-1)/3} = 7.6700$

$\frac{2t-1}{3} = \ln 7.6700$

$2t - 1 = 3\ln 7.6700$

$2t = 3\ln 7.6700 + 1$

$t = \frac{3\ln 7.6700 + 1}{2} = 3.5560$

(i) $\log(4b^2) = 2.6987$

$4b^2 = 10^{2.6987}$

$b^2 = \frac{10^{2.6987}}{4} = 124.9223$

$b = \sqrt{124.9223} = \pm 11.1769$

(j) $\log\left(\frac{6}{2+t}\right) = 1.5$

$\frac{6}{2+t} = 10^{1.5}$

$6 = 10^{1.5}(2+t)$

$\frac{6}{10^{1.5}} = 2+t$

$t = \frac{6}{10^{1.5}} - 2 = -1.8103$

(k) $\ln(2r^3 + 1) = 3.0572$

$2r^3 + 1 = e^{3.0572}$

$2r^3 = e^{3.0572} - 1$

$r^3 = \frac{e^{3.0572} - 1}{2} = 10.1340$

$r = 2.1640$

(l) $\ln(\log t) = -0.3$

$\log t = e^{-0.3} = 0.7408$

$t = 10^{0.7408} = 5.5058$

(m) $10^{t^2-1} = 180$

$t^2 - 1 = \log 180$

$t^2 = \log 180 + 1 = 3.2553$

$t = \sqrt{3.2553} = \pm 1.8042$

(n) $10^{3r^4} = 170000$

$3r^4 = \log 170000$

$r^4 = \dfrac{\log 170000}{3} = 1.7435$

$r = \pm 1.1491$

(o) $\log 10^t = 1.6$

$t \log 10 = 1.6$

$t = 1.6$

(p) $\ln(e^x) = 20000$

$x \ln e = 20000$

$x = 20000$

4. (a) $e^{3x}e^{2x} = 59$

$e^{5x} = 59$

$5x = \ln 59$

$x = \dfrac{\ln 59}{5} = 0.8155$

(b) $10^{3t}.10^{4-t} = 27$

$10^{4+2t} = 27$

$4 + 2t = \log 27$

$2t = \log 27 - 4$

$t = \dfrac{\log 27 - 4}{2} = -1.2843$

(c) $\log(5-t) + \log(5+t) = 1.2$

$\log(5-t)(5+t) = 1.2$

$\log(25 - t^2) = 1.2$

$25 - t^2 = 10^{1.2}$

$t^2 = 25 - 10^{1.2}$

$= 9.1511$

$t = \pm 3.0251$

(d) $\log x + \ln x = 4$

$\log x + \dfrac{\log x}{\log e} = 4$

$\log x \left(1 + \dfrac{1}{\log e}\right) = 4$

$3.3026 \log x = 4$

$\log x = \dfrac{4}{3.3026} = 1.2112$

$x = 10^{1.2112} = 16.2619$

Self-assessment questions 20.5

1. Both $\log x$ and $\ln x$ tend to infinity as x tends to infinity. Both have a value of 0 when $x = 1$.
2. True. The graph of the logarithm function increases without bound, albeit very slowly.

Exercise 20.5

1. (a)

x	0.5	1	2	3	4
$\log x^2$	−0.6021	0	0.6021	0.9542	1.2041

x	5	6	7	8	9	10
$\log x^2$	1.3979	1.5563	1.6902	1.8062	1.9085	2

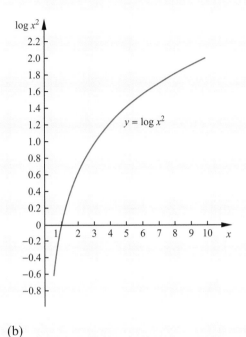

2. (a)

x	0.1	0.5	1	1.5	2	2.5	3
$\log x$	-1	-0.3010	0	0.1761	0.3010	0.3979	0.4771

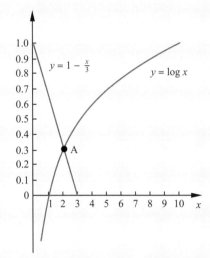

(b)

x	0.1	0.2	0.5	0.75	1
$\log(1/x)$	1	0.6990	0.3010	0.1249	0

x	2	3	4	5
$\log(1/x)$	-0.3010	-0.4771	-0.6021	-0.6990

x	6	7	8	9	10
$\log(1/x)$	-0.7782	-0.8451	-0.9031	-0.9542	-1

(b) $y = 1 - \frac{x}{3}$ is a straight line passing through $(0, 1)$ and $(3, 0)$.

(c) The graphs intersect at A. Approximate solution: $x = 2.06$.

3. (a)

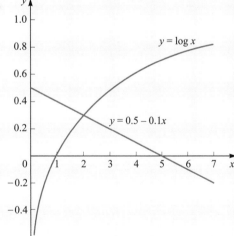

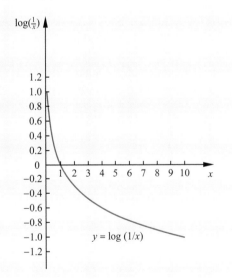

(b) The line $y = 0.5 - 0.1x$ is also shown on the figure.

(c) Solution $x = 2$.

Chapter 21

Exercise 21.2

1. 16 cm = 160 mm = 0.16 m
2. 356 mm = 35.6 cm = 0.356 m
3. 156 m = 0.156 km
4. (a) Note from Table 21.2 that 1 μm = 10^{-6} m. Therefore 150 μm = $150 \times 10^{-6} = 1.5 \times 10^{-4}$ m or 0.00015 m.
   (b) Using part (a) and the fact that 1 m = 100 cm, then 0.00015 m = 0.00015 $\times$ 100 = 0.015 cm.
   (c) 0.015 cm = 0.15 mm
5. (a) Given 1 m = 100 cm then 0.0046 m = 0.0046 $\times$ 100 cm = 0.46 cm.
   (b) 0.46 cm = 4.6 mm.
   (c) Given 1 m = $10^6 \mu$m then 0.0046 m = 0.0046 $\times 10^6$ μm = 4600 μm.
6. 0.0013 cm = 13 μm.

Exercise 21.3

1. (a) Because 1 $m^2 = 10^4$ cm^2 it follows that 10^{-4} m^2 = 1 cm^2. Then 25 cm^2 = 25×10^{-4} m^2 = 0.0025 m^2.
   (b) Because 1 cm^2 = 100 mm^2, then 25 cm^2 = 25 $\times$ 100 = 2500 mm^2.
2. Because 1 m$^2 = 10^6$ mm^2, then 0.005 m^2 = 0.005 $\times 10^6$ mm^2 = 5000 mm^2.
3. (a) 1 ml = $\frac{1}{1000}$ l. So 5 ml = 5×10^{-3} l = 0.005 l.
   (b) 1000 cm^3 = 1 l. So 0.005 l = 0.005 $\times$ 1000 = 5 cm^3. Alternatively, from the Key point on page 248 we note 1 ml = 1 cm^3, and so immediately 5 ml = 5 cm^3.
4. Forty teaspoons.
5. 500 cm^3 is the same as 500 ml. So 100 5 ml spoonfuls can be extracted from the bottle.
6. 1 l = 1000 cm^3 so 20 l = 20000 cm^3.

Self-assessment questions 21.4

1. A degree is a measurement of angle. One degree is $\frac{1}{360}$ of a complete revolution. Radians provide an alternative measure of angle. 2π radians represent a complete revolution. In a circle of radius 1 unit, an arc of length of 1 unit subtends an angle of 1 radian.
2. 360° = 2π radians. Equivalently, 180° = π radians.

Exercise 21.4

1. (a) 240° = $\pi/180 \times 240 = 4.19$
   (b) 300° = $\pi/180 \times 300 = 5.24$
   (c) 400° = $\pi/180 \times 400 = 6.98$
   (d) 37° = $\pi/180 \times 37 = 0.65$
   (e) 1000° = $\pi/180 \times 1000 = 17.45$
2. (a) $\pi/10 = 180/10 = 18°$
   (b) $2\pi/9 = \frac{2}{9} \times 180 = 40°$
   (c) 2 = $180/\pi \times 2 = 114.59°$
   (d) 3.46 = $180/\pi \times 3.46 = 198.24°$
   (e) 1.75 = $180/\pi \times 1.75 = 100.27°$
3. (a) 720° (b) 630° (c) 286.5° (d) 89.4°
4. (a) 2.8π (b) 1.2π (c) $\frac{7\pi}{3}$ (d) 0.7π
   (e) 1.8π

Exercise 21.5

1. $A = 14 \times 7 = 98$ cm^2
2. $A = 17.5$ m^2, cost = £87.33
3. $A = 2 \times 8 = 16$ m^2
4. $s = (a + b + c)/2 = 17$
   $A = \sqrt{s(s - a)(s - b)(s - c)}$
   $= \sqrt{2295} = 47.9$ cm^2
5. $A = \frac{1}{2}(a + b)h = 22$ cm^2
6. $A = \pi r^2 = 1017.9$ cm^2
7. $A = 78.5$ cm^2
8. $A = 26 = \pi r^2$ and so $r^2 = 26/\pi = 8.276$, then $r = 2.877$ cm
9. 13430 cm^2 = 1.3430 m^2
10. $V = \pi r^2 h = 1099.6$ cm^3 = 1.0996 litres
11. $V = \frac{4}{3}\pi r^3 = 2145$ cm^3
12. $86 = \frac{4}{3}\pi r^3$, therefore $r^3 = (3 \times 86)/(4 \times \pi) = 20.53$, then $r = 2.74$ cm
13. $A = \pi r^2 \times \theta/360 = 39.27$ cm^2

14. $V = \pi r^2 h = 28.274 \text{ m}^3$, now
    1000 litres = 1 m^3
    and so 28.274 m^3 = 28274 litres

15. (a) length of arc = $r\theta = 15(2) = 30$ cm
    (b) length of arc = $r\theta = 15(3) = 45$ cm
    (c) length of arc = $r\theta = 15(1.2) = 18$ cm
    (d) length of arc = $r\theta = 15(100 \times \pi/180)$
        = 26.18 cm
    (e) length of arc = $r\theta = 15(217 \times \pi/180)$
        = 56.81 cm

16. (a) angle subtended
    $$= \theta = \frac{\text{arc length}}{\text{radius}} = \frac{12}{12} = 1$$

    (b) $\theta = \dfrac{\text{arc length}}{\text{radius}} = \dfrac{6}{12} = 0.5$

    (c) $\theta = \dfrac{\text{arc length}}{\text{radius}} = \dfrac{24}{12} = 2$

    (d) $\theta = \dfrac{\text{arc length}}{\text{radius}} = \dfrac{15}{12} = 1.25$

    (e) $\theta = \dfrac{\text{arc length}}{\text{radius}} = \dfrac{2}{12} = \dfrac{1}{6}$

17. (a) Arc length = $r\theta = 18 \times 1.5 = 27$ cm
    (b) 19.8 cm   (c) 18.85 cm   (d) 31.68 cm

18. (a) Angle = $\dfrac{\text{arc length}}{\text{radius}} = \dfrac{18}{24} = 0.75$

    (b) 1.25   (c) 0.8333   (d) 2.5

19. (a) area of sector = $\dfrac{r^2\theta}{2} = \dfrac{9^2(1.5)}{2}$
        = 60.75 cm^2

    (b) area of sector = $\dfrac{r^2\theta}{2} = \dfrac{9^2(2)}{2} = 81$ cm^2

    (c) area of sector = $\dfrac{r^2\theta}{2} = \dfrac{9^2}{2}\left(100 \times \dfrac{\pi}{180}\right)$
        = 70.69 cm^2

    (d) area of sector = $\dfrac{r^2\theta}{2} = \dfrac{9^2}{2}\left(215 \times \dfrac{\pi}{180}\right)$
        = 151.97 cm^2

20.   area of sector = $\dfrac{r^2\theta}{2}$
    $$\theta = \frac{2 \times \text{area}}{r^2}$$

    When $r = 16$ we have
    $$\theta = \frac{\text{area}}{128}$$

    (a) Angle = $\theta = \frac{100}{128} = 0.78$
    (b) angle = $\frac{5}{128} = 0.039$
    (c) angle = $\frac{520}{128} = 4.06$

21. (a) Area = $\dfrac{r^2\theta}{2} = \dfrac{(18)^2(1.5)}{2} = 243$ cm^2
    (b) 356.4 cm^2
    (c) $120° = \frac{2\pi}{3}$ and so
    $$\text{Area} = \frac{(18)^2(2\pi/3)}{2} = 339.3 \text{ cm}^2$$

    (d) $217° = \dfrac{217}{180}\pi = 3.7874$
    $$\text{Area} = \frac{(18)^2(3.7874)}{2} = 613.6 \text{ cm}^2$$

22. Angle subtended by arc AB = $\frac{17}{25} = 0.68$
    $$\text{Area} = \frac{r^2\theta}{2} = \frac{(25)^2(0.68)}{2} = 212.5 \text{ cm}^2$$

23. Area = $\dfrac{r^2\theta}{2}$
    $$370 = \frac{(12)^2\theta}{2}$$
    $$\theta = \frac{2(370)}{12^2} = 5.1389$$

    Arc length = $r\theta = 12(5.1389) = 61.67$ cm

24. Arc length = $r\theta$, so $r = \dfrac{16}{1.2} = 13.33$ cm
    $$\text{Area} = \frac{r^2\theta}{2} = \frac{(13.33)^2(1.2)}{2} = 107 \text{ cm}^2$$

Exercise 21.6

1. (a) 12 kg = 12000 g
   (b) 12 kg = 12000000 mg
2. 168 mg = 0.168 g = 0.000168 kg or 1.68×10^{-4} kg
3. 0.005 tonnes = 5 kg
4. 875 tonnes = 875000 kg = 8.75×10^5 kg
5. 3500 kg = 3.5 tonnes

Exercise 21.8

1. Area has dimensions L×L = L^2.
2. ML^2T^{-2}.
3. Both sides have dimensions of time T.

Chapter 22

Self-assessment questions 22.1

1. In a right-angled triangle the side opposite the right angle is called the hypotenuse. For either of the other angles, the sine of that angle is given by the ratio of the lengths of the opposite side and the hypotenuse; the cosine is the ratio of the lengths of the adjacent side and the hypotenuse; the tangent is the ratio of the lengths of the opposite side and the adjacent side.
2. Sine, cosine and tangent are ratios of the lengths of two sides of a triangle. For example, if two lengths are 1 m and 3 m, the ratio is $\frac{1}{3}$. As a ratio it has no units. As another way of thinking of this – whatever units are used for the side lengths – the ratio of the lengths is the same. For example, lengths of 100 cm and 300 cm have the ratio $\frac{100}{300} = \frac{1}{3}$ as before.

hence
$\sin \theta = BC/AC$ and
$\cos(90° - \theta) = BC/AC$,
so $\sin \theta = \cos(90° - \theta)$
   (d) In Figure 22.1
   $\cos A = \cos \theta = AB/AC$
   $\sin C = \sin(90° - \theta) = AB/AC$
   So
   $\cos \theta = \sin(90° - \theta)$
3. (a) $\sin \theta = BC/AC = 3/\sqrt{18}$
   (b) $\cos \theta = AB/AC = 3/\sqrt{18}$
   (c) $\tan \theta = BC/AB = \frac{3}{3} = 1$
4. (a) $\sin \theta = \frac{4}{\sqrt{116}} = 0.3714$
   (b) $\cos \theta = \frac{10}{\sqrt{116}} = 0.9285$
   (c) $\tan \theta = \frac{4}{10} = 0.4$
   (d) $\sin \alpha = \frac{10}{\sqrt{116}} = 0.9285$
   (e) $\cos \alpha = \frac{4}{\sqrt{116}} = 0.3714$
   (f) $\tan \alpha = \frac{10}{4} = 2.5$

Exercise 22.1

1. (a) 0.5774 (b) 0.3420 (c) 0.2588
   (d) 0.9320 (e) 0.6294 (f) 1
2. (a) $\sin 70° = 0.9397$, $\cos 20° = 0.9397$
   (b) $\sin 25° = 0.4226$, $\cos 65° = 0.4226$
   (c) In Figure 22.1, $A = \theta$ and $C = 90° - \theta$,

Exercise 22.2

1. (a) 53.13° (b) 78.46° (c) 52.43°
2. (a) 0.6816 (b) 1.3181 (c) 1.1607
3. (a) 26.57° (b) 66.80° (c) 53.13°
4. (a) 0.1799 (b) 0.9818 (c) 0.7956

Chapter 23

Self-assessment question 23.1

1.

Second quadrant	First quadrant
sin is positive	sin is positive
cos is negative	cos is positive
tan is negative	tan is positive
Third quadrant	Fourth quadrant
sin is negative	sin is negative
cos is negative	cos is positive
tan is positive	tan is negative

Exercise 23.1

1. When $\sin \alpha > 0$ then α must be in the first or second quadrant. When $\cos \alpha < 0$ then α must be in the second or third quadrant. To satisfy both conditions, α must be in the second quadrant.
2. β can be in the third or fourth quadrant.
3. Second and third quadrants.
4. Third and fourth quadrants.
5. (a) $\sin \theta = \dfrac{1}{\sqrt{10}}$, $\cos \theta = -\dfrac{3}{\sqrt{10}}$, $\tan \theta = -\dfrac{1}{3}$

   (b) $-\dfrac{5}{\sqrt{29}}$, $-\dfrac{2}{\sqrt{29}}$, $\dfrac{5}{2}$

   (c) $-\dfrac{3}{\sqrt{18}} = -\dfrac{1}{\sqrt{2}}$, $\dfrac{3}{\sqrt{18}} = \dfrac{1}{\sqrt{2}}$, -1

Self-assessment questions 23.2

1. Any of the following: they are periodic functions – that is, they have a basic pattern that repeats over and over again; each has a period of 360°; each has a maximum value of 1 and a minimum value of −1.

2. The period of $\tan \theta$ is 180°; $\tan \theta$ has a jump (also called a discontinuity) whereas $\sin \theta$ and $\cos \theta$ can be drawn without lifting your pen from the paper – they are continuous; both $\sin \theta$ and $\cos \theta$ lie between −1 and 1, whereas $\tan \theta$ increases (and decreases) without bound.

3. For $\sin \theta$ the maximum is 1 and the minimum is −1. The maximum value occurs when $\theta = \ldots - 270°, 90°, 450° \ldots$. The minimum value occurs when $\theta = \ldots -90°, 270°, 630° \ldots$. Study the graphs in §23.2 to see this. For $\cos \theta$ the maximum is 1 and the minimum is −1. The maximum value occurs when $\theta = \ldots - 360°, 0°, 360° \ldots$. The minimum value occurs when $\theta = \ldots -180°, 180° \ldots$. Study the graphs in §23.2 to see this.

4. The cosine graph is symmetrical about the vertical axis. This means that the cosine of a negative angle is the same as the cosine of the corresponding positive angle – that is, $\cos(-\theta) = \cos \theta$. On the other hand the sine of a negative angle is the negative of the sine of the corresponding positive angle, that is $\sin(-\theta) = -\sin \theta$.

Exercise 23.2

1. (a) $\sin 0° = 0$, $\cos 0° = 1$
   (b) 1, 0   (c) 0, −1   (d) −1, 0   (e) 0, 1

2. (a)

θ	0	30	60	90	120	150	180
$\sin(\theta + 45°)$	0.71	0.97	0.97	0.71	0.26	−0.26	−0.71

θ	210	240	270	300	330	360
$\sin(\theta + 45°)$	−0.97	−0.97	−0.71	−0.26	0.26	0.71

sin $(\theta + 45)$

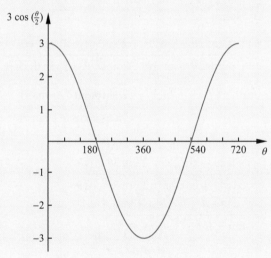

(c)

θ	0	30	60	90	120	150	180
$2 \tan(\theta + 60°)$	3.46	–	−3.46	−1.15	0	1.15	3.46

θ	210	240	270	300	330	360
$2 \tan(\theta + 60°)$	–	−3.46	−1.15	0	1.15	3.46

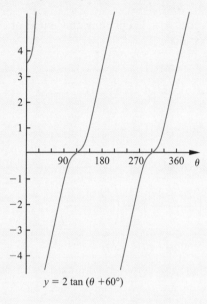

$y = 2 \tan (\theta + 60°)$

(b)

θ	0	60	120	180	240	300	360
$3 \cos(\theta/2)$	3	2.60	1.5	0	−1.5	−2.60	−3

θ	420	480	540	600	660	720
$3 \cos(\theta/2)$	−2.60	−1.5	0	1.5	2.60	3

$3 \cos \left(\frac{\theta}{2}\right)$

Chapter 24

Exercise 24.1

1. (a) $\theta = 50°$; $\theta = 180 - 50 = 130°$
   (b) $\theta = 180 - 40 = 140°$ and also
       $\theta = 180 + 40 = 220°$
   (c) $\theta = 20°$; $\theta = 180 + 20 = 200°$
   (d) $\theta = 180 + 70 = 250°$;
       $\theta = 360 - 70 = 290°$
   (e) $\theta = 10°$; $\theta = 360 - 10 = 350°$
   (f) $\theta = 180 - 80 = 100°$;
       $\theta = 360 - 80 = 280°$
   (g) $\theta = 0°$; $\theta = 180°, 360°$

(h) $\theta = 90°$; $\theta = 360 - 90 = 270°$

2. (a) $20°$, $340°$ (b) $190°$, $350°$
   (c) $140°$, $320°$

3. $\cos 2A = \cos^2 A - \sin^2 A$
   $$= \cos^2 A - (1 - \cos^2 A)$$
   $$= 2 \cos^2 A - 1$$

4. $\sin^2 A + \cos^2 A = 1$. Dividing this equation by $\cos^2 A$ gives

   $$\frac{\sin^2 A}{\cos^2 A} + \frac{\cos^2 A}{\cos^2 A} = \frac{1}{\cos^2 A}$$

   $$\tan^2 A + 1 = \frac{1}{\cos^2 A}$$

5. $\cos 2\theta + \cos 8\theta$

   $$= 2 \cos \left(\frac{2\theta + 8\theta}{2} \right) \cos \left(\frac{2\theta - 8\theta}{2} \right)$$

   $$= 2 \cos 5\theta \cos(-3\theta)$$

   Recall that $\cos(-x) = \cos x$ and so

   $$\cos 2\theta + \cos 8\theta = 2 \cos 5\theta \cos 3\theta$$

   and so

   $$\frac{\cos 2\theta + \cos 8\theta}{2 \cos 3\theta} = \cos 5\theta$$

6. (a) $\cos(270° - \theta)$
      $$= \cos 270° \cos \theta + \sin 270° \sin \theta$$
      $$= -\sin \theta$$
   (b) $\cos(270° + \theta)$
      $$= \cos 270° \cos \theta - \sin 270° \sin \theta$$
      $$= \sin \theta$$
   (c) $\sin(270° + \theta)$
      $$= \sin 270° \cos \theta + \sin \theta \cos 270°$$
      $$= -\cos \theta$$
   (d) $\tan(135° + \theta) = \dfrac{\tan 135° + \tan \theta}{1 - \tan 135° \tan \theta}$

      $$= \frac{-1 + \tan \theta}{1 + \tan \theta}$$

      $$= \frac{\tan \theta - 1}{\tan \theta + 1}$$

(e) $\sin(270° - \theta)$
   $$= \sin 270° \cos \theta - \cos 270° \sin \theta$$
   $$= -\cos \theta$$

7. $\dfrac{1 - \tan A}{1 + \tan A} = \dfrac{\tan 45° - \tan A}{1 + \tan 45° \tan A}$
   $$= \tan(45° - A)$$

8. Note that

   $$\cos 2\theta = \cos^2 \theta - \sin^2 \theta = 1 - 2 \sin^2 \theta$$

   using $\cos^2 \theta = 1 - \sin^2 \theta$

   and so $1 - \cos 2\theta = 2 \sin^2 \theta$.
   Also note that

   $$\cos 2\theta = \cos^2 \theta - \sin^2 \theta = 2 \cos^2 \theta - 1$$

   using $\sin^2 \theta = 1 - \cos^2 \theta$

   and so $1 + \cos 2\theta = 2 \cos^2 \theta$.
   Finally we note that $\sin 2\theta = 2 \sin \theta \cos \theta$.
   Hence

   $$\frac{1 - \cos 2\theta + \sin 2\theta}{1 + \cos 2\theta + \sin 2\theta} = \frac{2 \sin^2 \theta + 2 \sin \theta \cos \theta}{2 \cos^2 \theta + 2 \sin \theta \cos \theta}$$

   $$= \frac{2 \sin \theta (\sin \theta + \cos \theta)}{2 \cos \theta (\cos \theta + \sin \theta)}$$

   $$= \tan \theta$$

9. $\dfrac{\sin 4\theta + \sin 2\theta}{\cos 4\theta + \cos 2\theta} = \dfrac{2 \sin 3\theta \cos \theta}{2 \cos 3\theta \cos \theta} = \tan 3\theta$

10. $\dfrac{\tan A}{\tan^2 A + 1} = \dfrac{\sin A / \cos A}{\left(\dfrac{\sin A}{\cos A} \right)^2 + 1}$

   Multiplying numerator and denominator by $\cos^2 A$ gives

   $$\frac{\sin A \cos A}{\sin^2 A + \cos^2 A} = \sin A \cos A = \frac{1}{2} \sin 2A$$

11. (a) $\dfrac{1}{\cos A} - \cos A = \dfrac{1 - \cos^2 A}{\cos A}$

   $$= \frac{\sin^2 A}{\cos A} = \sin A \tan A$$

(b) $\dfrac{1}{\sin A} - \sin A = \dfrac{1 - \sin^2 A}{\sin A}$

$\qquad\qquad = \dfrac{\cos^2 A}{\sin A} = \dfrac{\cos A}{\tan A}$

12. $\sin 3\theta = \sin(2\theta + \theta)$

$\qquad = \sin 2\theta \cos \theta + \sin \theta \cos 2\theta$

$\qquad = (2 \sin \theta \cos \theta)\cos \theta$

$\qquad\quad + \sin \theta(\cos^2 \theta - \sin^2 \theta)$

$\qquad = 2 \sin \theta \cos^2 \theta + \sin \theta \cos^2 \theta - \sin^3 \theta$

$\qquad = 3 \sin \theta \cos^2 \theta - \sin^3 \theta$

$\qquad = 3 \sin \theta(1 - \sin^2 \theta) - \sin^3 \theta$

$\qquad = 3 \sin \theta - 4 \sin^3 \theta$

Self-assessment question 24.2

1. Consider for example $\sin 30°$, which equals 0.5. Similarly $\sin 150° = 0.5$. So if we seek the inverse sine of 0.5, that is the angle whose sine is 0.5, there will be more than one result, namely $30°$, $150°$ and others too.

Exercise 24.2

1. (a) Using a calculator, we see
   $\sin^{-1}(0.9) = 64.16°$.
   Also $\sin^{-1}(0.9) = 180 - 64.16 = 115.84°$.

   (b) Using a calculator we see
   $\cos^{-1}(0.45) = 63.26°$.
   Also
   $\cos^{-1}(0.45) = 360 - 63.26 = 296.74°$.

   (c) Using a calculator we see
   $\tan^{-1} 1.3 = 52.43°$.
   Also $\tan^{-1} 1.3 = 180 + 52.43 = 232.43°$.

   (d) Using a calculator we see
   $\sin^{-1} 0.6 = 36.87$. Hence
   $\sin^{-1}(-0.6) = 180 + 36.87 = 216.87°$
   and $360 - 36.87 = 323.13°$.

   (e) Using a calculator we see
   $\cos^{-1} 0.75 = 41.41°$. So
   $\cos^{-1}(-0.75) = 180 - 41.41 = 138.59°$
   and $180 + 41.41 = 221.41°$.

   (f) We have $\tan^{-1} 0.3 = 16.70°$ and so
   $\tan^{-1}(-0.3) = 180 - 16.70 = 163.30°$
   and $360 - 16.70 = 343.30°$.

   (g) $\sin^{-1} 1 = 90°$

   (h) $\cos^{-1}(-1) = 180°$

2. (a) $\sin(2\theta + 50°) = 0.5$

   $2\theta + 50° = (30), 150, 390, 510,$

   $\qquad\qquad 750 \ldots$

   $2\theta = (-20), 100, 340, 460,$

   $\qquad\qquad 700 \ldots$

   $\theta = 50, 170, 230, 350 \ldots$

   Solutions in the range $0°$ to $360°$ are

   $\theta = 50°, 170°, 230°, 350°$

   (b) $\cos(\theta + 110°) = 0.3$

   $\theta + 110° = (72.54), 287.46, 432.54$

   $\theta = 177.46°, 322.54°$

   (c) $\tan\left(\dfrac{\theta}{2}\right) = 1$

   $\dfrac{\theta}{2} = 45°, (225°)$

   $\theta = 90°$

   (d) $\sin\left(\dfrac{\theta}{2}\right) = -0.5$

   $\dfrac{\theta}{2} = 210, 330$

   $\theta = 420, 660$

   No solution between $0°$ and $360°$.

   (e) $\cos(2\theta - 30°) = -0.5$

   $2\theta - 30° = 120, 240, 480, 600$

   $2\theta = 150, 270, 510, 630$

   $\theta = 75°, 135°, 255°, 315°$

(f) $\tan(3\theta - 20°) = 0.25$

$$3\theta - 20° = 14.04, 194.04, 374.04,$$
$$554.04, 734.04, 914.04$$
$$3\theta = 34.04, 214.04, 394.04,$$
$$574.04, 754.04, 934.04$$
$$\theta = 11.35°, 71.35°, 131.35°,$$
$$191.35°, 251.35°, 311.35°$$

(g) $\sin 2\theta = 2\cos 2\theta$

$$\frac{\sin 2\theta}{\cos 2\theta} = 2$$
$$\tan 2\theta = 2$$
$$2\theta = 63.43, 243.43, 423.43, 603.43$$
$$\theta = 31.72°, 121.72°, 211.72°, 301.72°$$

3. (a) $\cos(\theta - 100°) = 0.3126$

$$\theta - 100° = \cos^{-1}(0.3126)$$
$$= -71.78°, 71.78°,$$
$$288.22° \dots$$
$$\theta = 28.22°, 171.78°$$

(b) $\tan(2\theta + 20°) = -1$

$$2\theta + 20° = \tan^{-1}(-1)$$
$$= 135°, 315°, 495°,$$
$$675° \dots$$
$$\theta = 57.5°, 147.5°, 237.5°,$$
$$327.5°$$

(c) $\sin\left(\dfrac{2\theta}{3} + 30°\right) = -0.4325$

$$\frac{2\theta}{3} + 30° = \sin^{-1}(-0.4325)$$
$$= 205.6°, 334.4° \dots$$
$$\theta = 263.4°$$

Chapter 25

Self-assessment questions 25.1

1. An equilateral triangle is one for which all sides have the same length. All angles will be equal (to 60°). An isosceles triangle has two sides of equal length. The angles opposite these two sides will be equal too. A scalene triangle has all sides of different length. All three angles will be different.
2. In any triangle the angles sum to 180°.
3. In a right-angled triangle, the side opposite the right angle is called the hypotenuse. Triangles that are not right angled will not possess a hypotenuse.
4. No. If a triangle contains an obtuse angle, this angle is greater than 90°. Because the sum of all the angles must be 180°, the sum of the remaining two angles must be less than 90°. Consequently the triangle cannot be right angled.

Exercise 25.1

1. $C = 180 - 70 - 42 = 68°$

2. $A + B + C = 180$

$$B + C = 180 - A = 180 - 42 = 138$$
$$2C + C = 138$$
$$C = 46°$$
$$B = 2C = 92°$$

3. $A + B + C = 180$

$A + B = 180 - C = 180 - 114 = 66$

Since A and B are equal then $A = B = 33°$.

4. In $\triangle ABC$, $A = 50°$, $B = 72°$ and $C = 58°$:
(a) the largest angle is B and so the longest side is b, that is, AC; (b) the smallest angle is A and so the shortest side is a, that is, BC.

5. There are two solutions.
(i) Suppose $A = 40°$ and $B = C$. Then

$A + B + C = 180°$

$B + C = 180 - 40 = 140$

$B = C = 70°$

The angles are $40°$, $70°$, $70°$.
(ii) Suppose $A = B = 40°$. Then $C = 100°$. The angles are $40°$, $40°$, $100°$.

Self-assessment questions 25.2

1. In a right-angled triangle, the square of the hypotenuse is equal to the sum of the squares of the two remaining sides.
2. Only right-angled triangles have a hypotenuse. The hypotenuse is the side opposite the right angle.

Exercise 25.2

1. $a^2 = c^2 - b^2 = 30^2 - 17^2 = 611$

$a = \sqrt{611} = 24.72$

The length of BC is 24.72 cm.

2. $d^2 = c^2 + e^2$

$= (1.7)^2 + (1.2)^2 = 4.33$

$d = \sqrt{4.33} = 2.08$

The length of CE is 2.08 m.

3. Let the length of $b =$ AC be k cm.
Then $AC/AB = b/c = k/c = \frac{2}{1}$ so $c = k/2$.
The length of $c =$ AB is $k/2$ cm.

Now,

$a =$ BC and $b^2 = a^2 + c^2$

$k^2 = a^2 + \left(\dfrac{k}{2}\right)^2 = a^2 + \dfrac{k^2}{4}$

$\dfrac{3}{4} k^2 = a^2$

$a = \sqrt{\dfrac{3}{4} k^2} = \dfrac{\sqrt{3}}{2} k$

So $AC : BC = k : (\sqrt{3}/2)k$. This is the same as $1 : \sqrt{3}/2$ or $2 : \sqrt{3}$.

4. Let the length of $y =$ XZ be k cm. The length of YZ is then $2k$ cm. Using Pythagoras' theorem we see that

$x^2 = y^2 + z^2$

$(2k)^2 = k^2 + 10^2$

$4k^2 = k^2 + 100$

$3k^2 = 100$

$k^2 = \dfrac{100}{3}$

$k = \sqrt{\dfrac{100}{3}} = 5.774$

$2k = 11.547$

The length of XZ is 5.774 cm and the length of YZ is 11.547 cm.

5. $m^2 = l^2 + n^2$

$30^2 = l^2 + 26^2$

$l^2 = 30^2 - 26^2 = 224$

$l = \sqrt{224} = 14.97$

The length of MN is 14.97 cm.

Self-assessment question 25.3

1. To solve a triangle means to find all its angles and the lengths of all of its sides.

Exercise 25.3

1. $B = 180 - 90 - 37 = 53°$

$$\tan A = \frac{BC}{AC} = \frac{a}{36}$$

$$\tan 37° = \frac{a}{36}$$

$$a = 36 \tan 37° = 27.13$$

$$\cos A = \frac{AC}{AB} = \frac{36}{c}$$

$$c = \frac{36}{\cos A} = \frac{36}{\cos 37°} = 45.08$$

The solution is

$A = 37°$ $a = BC = 27.13$ cm
$B = 53°$ $b = AC = 36$ cm
$C = 90°$ $c = AB = 45.08$ cm

2. Using Pythagoras' theorem

$$e^2 = c^2 + d^2$$

$$= 21^2 + 14^2 = 637$$

$$e = \sqrt{637} = 25.24$$

$$\tan C = \frac{21}{14} = 1.5$$

$$C = \tan^{-1}(1.5) = 56.31°$$

Finally, $D = 180 - 90 - 56.31 = 33.69°$. The solution is

$C = 56.31°$ $c = DE = 21$ cm
$D = 33.69°$ $d = CE = 14$ cm
$E = 90°$ $e = CD = 25.24$ cm

3. $Z = 180 - 90 - 26 = 64°$

$$\sin Y = \sin 26° = \frac{XZ}{YZ} = \frac{y}{45}$$

$$y = 45 \sin 26° = 19.73$$

and

$$\cos Y = \cos 26° = \frac{XY}{YZ} = \frac{z}{45}$$

$$z = 45 \cos 26° = 40.45$$

The solution is

$X = 90°$ $x = YZ = 45$ mm
$Y = 26°$ $y = XZ = 19.73$ mm
$Z = 64°$ $z = XY = 40.45$ mm

4. Using Pythagoras' theorem,

$$j^2 = i^2 + k^2$$

$$= 27^2 + 15^2 = 954$$

$$j = \sqrt{954} = 30.89$$

$$\tan I = \frac{i}{k} = \frac{27}{15} = 1.8$$

$$I = \tan^{-1} 1.8 = 60.95°$$

and

$$\tan K = \frac{15}{27}$$

$$K = \tan^{-1}\left(\frac{15}{27}\right) = 29.05$$

The solution is

$I = 60.95°$ $i = JK = 27$ cm
$J = 90°$ $j = IK = 30.89$ cm
$K = 29.05°$ $k = IJ = 15$ cm

5. $Q = 180 - 62 - 28 = 90°$ and so $\triangle PQR$ is a right-angled triangle.

$$\sin R = \frac{r}{q} = \frac{r}{22}$$

$$PQ = r = 22 \sin R = 22 \sin 28° = 10.33$$

and

$$\cos R = \frac{p}{q} = \frac{p}{22}$$

$$QR = p = 22 \cos R = 22 \cos 28° = 19.42$$

The solution is

$P = 62°$ $p = QR = 19.42$ cm

$Q = 90°$ $q = PR = 22$ cm

$R = 28°$ $r = PQ = 10.33$ cm

6. $T = 180 - 70 - 20 = 90°$ so $\triangle RST$ is a right-angled triangle.

$$\cos R = \cos 70° = \frac{s}{t} = \frac{12}{t}$$

$$RS = t = \frac{12}{\cos 70°} = 35.09$$

$$\tan R = \tan 70° = \frac{r}{s} = \frac{r}{12}$$

$$ST = r = 12 \tan 70° = 32.97$$

The solution is

$R = 70°$ $r = ST = 32.97$ cm

$S = 20°$ $s = RT = 12$ cm

$T = 90°$ $t = RS = 35.09$ cm

Self-assessment question 25.4

1. $\frac{a}{\sin A} = \frac{b}{\sin B} = \frac{c}{\sin C}$. The sine rule should be used when we know either (a) two angles and one side, or (b) two sides and a non-included angle.

Exercise 25.4

1. (a) $\dfrac{a}{\sin A} = \dfrac{b}{\sin B} = \dfrac{c}{\sin C}$

$$\frac{a}{\sin A} = \frac{24}{\sin B} = \frac{31}{\sin 37°}$$

So

$$\sin B = \frac{24 \sin 37°}{31} = 0.4659$$

$$B = \sin^{-1}(0.4659) = 27.77°$$

Hence $A = 180 - 37 - 27.77 = 115.23°$. Finally

$$a = \frac{31 \sin A}{\sin 37°} = \frac{31 \sin 115.23°}{\sin 37°} = 46.60$$

The solution is

$A = 115.23°$ $a = BC = 46.60$ cm

$B = 27.77°$ $b = AC = 24$ cm

$C = 37°$ $c = AB = 31$ cm

(b) Substituting the given values into the sine rule gives

$$\frac{a}{\sin A} = \frac{24}{\sin B} = \frac{19}{\sin 37°}$$

So

$$\sin B = \frac{24 \sin 37°}{19} = 0.7602$$

$$B = \sin^{-1}(0.7602) = 49.48° \text{ or } 130.52°$$

Both values are acceptable and so there are two solutions.

Case 1: $B = 49.48°$

$$A = 180 - 37 - 49.48 = 93.52°$$

$$a = \frac{19 \sin A}{\sin 37°} = \frac{19 \sin 93.52}{\sin 37°} = 31.51$$

Case 2: $B = 130.52°$

$$A = 180 - 37 - 130.52 = 12.48°$$

$$a = \frac{19 \sin A}{\sin 37°} = \frac{19 \sin 12.48°}{\sin 37°} = 6.82$$

The solutions are

$$A = 93.52° \quad a = BC = 31.51 \text{ cm}$$
$$B = 49.48° \quad b = AC = 24 \text{ cm}$$
$$C = 37° \quad c = AB = 19 \text{ cm}$$

and

$$A = 12.48° \quad a = BC = 6.82 \text{ cm}$$
$$B = 130.52° \quad b = AC = 24 \text{ cm}$$
$$C = 37° \quad c = AB = 19 \text{ cm}$$

(c) $\dfrac{17}{\sin 17°} = \dfrac{b}{\sin B} = \dfrac{c}{\sin 101°}$

$$c = \frac{17 \sin 101°}{\sin 17°} = 57.08$$

Also $B = 180 - 17 - 101 = 62°$. So

$$b = \frac{17 \sin B}{\sin 17°} = \frac{17 \sin 62°}{\sin 17°} = 51.34$$

The solution is

$$A = 17° \quad a = BC = 17 \text{ cm}$$
$$B = 62° \quad b = AC = 51.34 \text{ cm}$$
$$C = 101° \quad c = AB = 57.08 \text{ cm}$$

(d) $\dfrac{8.9}{\sin 53°} = \dfrac{b}{\sin B} = \dfrac{9.6}{\sin C}$

$$\sin C = \frac{9.6 \sin 53°}{8.9} = 0.8614$$

$$C = \sin^{-1}(0.8614) = 59.48° \text{ or } 120.52°$$

Both values are acceptable and so there are two solutions.

Case 1: $C = 59.48°$

$$B = 180 - 53 - 59.48 = 67.52°$$

$$b = \frac{8.9 \sin B}{\sin 53°} = \frac{8.9 \sin 67.52°}{\sin 53°} = 10.30$$

Case 2: $C = 120.52°$

$$B = 180 - 53 - 120.52 = 6.48°$$

$$b = \frac{8.9 \sin B}{\sin 53°} = \frac{8.9 \sin 6.48°}{\sin 53°} = 1.26$$

The solutions are

$$A = 53° \quad a = BC = 8.9 \text{ cm}$$
$$B = 67.52° \quad b = AC = 10.30 \text{ cm}$$
$$C = 59.48° \quad c = AB = 9.6 \text{ cm}$$

and

$$A = 53° \quad a = BC = 8.9 \text{ cm}$$
$$B = 6.48° \quad b = AC = 1.26 \text{ cm}$$
$$C = 120.52° \quad c = AB = 9.6 \text{ cm}$$

(e) $\dfrac{14.5}{\sin 62°} = \dfrac{b}{\sin B} = \dfrac{12.2}{\sin C}$

$$\sin C = \frac{12.2 \sin 62°}{14.5} = 0.7429$$

$$C = \sin^{-1}(0.7429) = 47.98°$$

Hence $B = 180 - 62 - 47.98 = 70.02°$.

$$b = \frac{14.5 \sin B}{\sin 62°} = \frac{14.5 \sin 70.02°}{\sin 62°} = 15.43$$

The solution is

$$A = 62° \quad a = BC = 14.5 \text{ cm}$$
$$B = 70.02° \quad b = AC = 15.43 \text{ cm}$$
$$C = 47.98° \quad c = AB = 12.2 \text{ cm}$$

(f) $\dfrac{a}{\sin A} = \dfrac{11}{\sin 36°} = \dfrac{c}{\sin 50°}$

$$c = \frac{11 \sin 50°}{\sin 36°} = 14.34$$

Also $A = 180 - 36 - 50 = 94°$ so that

$$a = \frac{11 \sin A}{\sin 36°} = \frac{11 \sin 94°}{\sin 36°} = 18.67$$

The solution is

$A = 94°$ $a = \text{BC} = 18.67$ cm

$B = 36°$ $b = \text{AC} = 11$ cm

$C = 50°$ $c = \text{AB} = 14.34$ cm

Self-assessment question 25.5

1. $a^2 = b^2 + c^2 - 2bc \cos A$ and the corresponding forms involving the other angles. This rule should be used when we know either (a) three sides, or (b) two sides and the included angle.

Exercise 25.5

1. (a) $a = \text{BC} = 86$, $b = \text{AC} = 119$, $c = \text{AB} = 53$

$$a^2 = b^2 + c^2 - 2bc \cos A$$

$$86^2 = 119^2 + 53^2 - 2(119)(53)\cos A$$

$$12614 \cos A = 119^2 + 53^2 - 86^2 = 9574$$

$$\cos A = \frac{9574}{12614} = 0.7590$$

$$= \cos^{-1}(0.7590) = 40.62°$$

$$b^2 = a^2 + c^2 - 2ac \cos B$$

$$119^2 = 86^2 + 53^2 - 2(86)(53)\cos B$$

$$9116 \cos B = 86^2 + 53^2 - 119^2 = -3956$$

$$\cos B = -\frac{3956}{9116} = -0.4340$$

$$B = \cos^{-1}(-0.4340) = 115.72°$$

$C = 180 - 40.62 - 115.72 = 23.66°$. The solution is

$A = 40.62°$ $a = \text{BC} = 86$ cm

$B = 115.72°$ $b = \text{AC} = 119$ cm

$C = 23.66°$ $c = \text{AB} = 53$ cm

(b) $b = \text{AC} = 30$, $c = \text{AB} = 42$, $A = 115°$

$$a^2 = b^2 + c^2 - 2bc \cos A$$

$$a^2 = 30^2 + 42^2 - 2(30)(42)\cos 115°$$

$$= 3728.9980$$

$$a = \sqrt{3728.998} = 61.07$$

Now using the cosine rule again,

$$b^2 = a^2 + c^2 - 2ac \cos B$$

$$30^2 = 3728.998 + 1764$$
$$- 2(61.07)(42)\cos B$$

$$5129.88 \cos B = 4592.998$$

$$\cos B = 0.8953$$

$$B = 26.45°$$

$C = 180 - 115 - 26.45 = 38.55°$. The solution is

$A = 115°$ $a = \text{BC} = 61.07$ cm

$B = 26.45°$ $b = \text{AC} = 30$ cm

$C = 38.55°$ $c = \text{AB} = 42$ cm

(c) $a = \text{BC} = 74$, $b = \text{AC} = 93$, $C = 39°$

$$c^2 = a^2 + b^2 - 2ab \cos C$$

$$c^2 = 74^2 + 93^2 - 2(74)(93)\cos 39°$$

$$= 3428.3630$$

$$c = 58.552$$

$$a^2 = b^2 + c^2 - 2bc \cos A$$

$$5476 = 8649 + 3428.3630$$
$$- 2(93)(58.552)\cos A$$

$$10890.672 \cos A = 6601.363$$

$$\cos A = 0.6061$$

$$A = 52.69°$$

$B = 180 - 52.69 - 39 = 88.31°$. The solution is

$A = 52.69°$ $a = \text{BC} = 74$ cm

$B = 88.31°$ $b = \text{AC} = 93$ cm

$C = 39°$ $c = \text{AB} = 58.55$ cm

(d) $a = BC = 3.6$, $b = AC = 2.7$,
   $c = AB = 1.9$

$$a^2 = b^2 + c^2 - 2bc \cos A$$

$$3.6^2 = 2.7^2 + 1.9^2 - 2(2.7)(1.9)\cos A$$

$$10.26 \cos A = -2.06$$

$$\cos A = -0.2008$$

$$A = 101.58°$$

$$b^2 = a^2 + c^2 - 2ac \cos B$$

$$2.7^2 = 3.6^2 + 1.9^2 - 2(3.6)(1.9)\cos B$$

$$13.68 \cos B = 3.6^2 + 1.9^2 - 2.7^2 = 9.28$$

$$\cos B = \frac{9.28}{13.68} = 0.6784$$

$$B = 47.28°$$

$C = 180 - 101.58 - 47.28 = 31.14°$. The solution is

$$A = 101.58° \quad a = BC = 3.6 \text{ cm}$$
$$B = 47.28° \quad b = AC = 2.7 \text{ cm}$$
$$C = 31.14° \quad c = AB = 1.9 \text{ cm}$$

(e) $a = BC = 39$, $c = AB = 29$, $B = 100°$.

$$b^2 = a^2 + c^2 - 2ac \cos B$$

$$= 39^2 + 29^2 - 2(39)(29)\cos 100°$$

$$= 2754.7921$$

$$b = 52.49$$

$$a^2 = b^2 + c^2 - 2bc \cos A$$

$$39^2 = 52.49^2 + 29^2 - 2(52.49)(29)\cos A$$

$$3044.42 \cos A = 52.49^2 + 29^2 - 39^2 = 2074.79$$

$$\cos A = \frac{2074.79}{3044.42} = 0.6815$$

$$A = 47.04°$$

$C = 180 - 100 - 47.04 = 32.96°$. The solution is

$$A = 47.04° \quad a = BC = 39 \text{ cm}$$
$$B = 100° \quad b = AC = 52.49 \text{ cm}$$
$$C = 32.96° \quad c = AB = 29 \text{ cm}$$

Chapter 26

Self-assessment questions 26.1

1. (a) A scalar has only magnitude.
   (b) A vector is a quantity with both magnitude and direction.
2. (a) The magnitude of a vector is the size, or length, of the vector.
   (b) A unit vector has a magnitude of 1.
   (c) The head of a vector is the 'end-point' of the vector.
   (d) The tail of a vector is the 'start-point' of the vector.

Exercise 26.2

1. (a) $2\mathbf{v}$ (b) $\frac{1}{4}\mathbf{v}$ (c) $-3\mathbf{v}$ (d) $\frac{1}{2}\mathbf{v}$ (e) $-\frac{1}{2}\mathbf{v}$
2. (a) $\frac{1}{5}\mathbf{a}$ (b) $-\frac{1}{5}\mathbf{a}$
3. (a) $12\mathbf{h}$ (b) $-\frac{1}{2}\mathbf{h}$

Self-assessment questions 26.3

1. To add $\mathbf{a}$ and $\mathbf{b}$, position the tail of $\mathbf{b}$ at the head of $\mathbf{a}$. Then $\mathbf{a} + \mathbf{b}$ is the vector from the tail of $\mathbf{a}$ to the head of $\mathbf{b}$.
2. (a) True (b) False (c) True
3. The sum $\mathbf{a} + \mathbf{b}$ is the resultant of $\mathbf{a}$ and $\mathbf{b}$.

Exercise 26.3

1. (a) $\overrightarrow{AC}$ (b) $\overrightarrow{AE}$ (c) $\overrightarrow{AE}$ (d) $\overrightarrow{DC}$
   (e) $\overrightarrow{AC}$ (f) $\overrightarrow{BA}$ (g) $\overrightarrow{EA}$

Self-assessment questions 26.4

1. (a) $\mathbf{i}$ (b) $\mathbf{j}$
2. (a) $-\mathbf{j}$ (b) $-\mathbf{i}$

3. (a) 90° (b) 0° (c) 180°
4. $\sqrt{x^2 + y^2}$
5. $6\mathbf{i}$
6. $-2\mathbf{j}$

Exercise 26.4

1. (a)$\sqrt{10}$ (b) 5 (c) 1 d) $\frac{1}{\sqrt{2}}$
2. (a) $3\mathbf{i} + 7\mathbf{j}$ (b) $2\mathbf{i} - 5\mathbf{j}$ (c) $-6\mathbf{i} - 4\mathbf{j}$
3. (a) $12\mathbf{i} - 8\mathbf{j}$ (b) $\mathbf{i} + \mathbf{j}$ (c) $\mathbf{i} - 15\mathbf{j}$
   (d) $\sqrt{13}$ (e) $4\sqrt{13}$ (f) $\sqrt{2}$

Self-assessment question 26.5

1. $\mathbf{a} \cdot \mathbf{b} = |\mathbf{a}||\mathbf{b}| \cos\theta$ and also $\mathbf{a} \cdot \mathbf{b} = a_1 b_1 + a_2 b_2$

Exercise 26.5

1. (a) 4 (b) −7 (c) −1
2. (a) 10.3° (b) 101.3° (c) 90° (d) 180°
4. (a) $\overrightarrow{AB} = 9\mathbf{i} + 6\mathbf{j}$ $\overrightarrow{BC} = -6\mathbf{i} - 10\mathbf{j}$
   (b) $\overrightarrow{AB}.\overrightarrow{BC} = -114$, $|AB| = \sqrt{117}$,
       $|BC| = \sqrt{136}$

$$\cos\theta = \frac{-114}{\sqrt{117}\sqrt{136}}$$

from which $\theta = 154.7°$

Chapter 27

Self-assessment questions 27.1

1. A matrix is a rectangular pattern of numbers.
2.
$$\begin{pmatrix} 6 \\ 2 \\ -1 \\ 0 \\ 9 \end{pmatrix}$$ is a 5 × 1 matrix.

   (6 2 −1 0 9) is a 1 × 5 matrix.

3. A square matrix has the same number of rows as columns.
4. The leading diagonal is that from the top left to the bottom right.
5. A diagonal matrix is a square matrix that is zero everywhere except on the leading diagonal. On this diagonal the elements can take any value.
6. An identity matrix is a diagonal matrix that has only 1s on its leading diagonal.
7. Yes

Exercise 27.1

1. A is 1 × 4, B is 4 × 2, C is 1 × 1, and D is 5 × 1.
2. A is square; B is neither diagonal nor square; C is square and diagonal (it is the 3 × 3 identity matrix); D and E are neither square nor diagonal; F is square and diagonal; G and H are square.
3. A 3 × 1 matrix has three elements, a 1 × 3 matrix has three elements. An $m \times n$ matrix has mn elements and an $n \times n$ matrix has n^2 elements.

Self-assessment questions 27.2

1. The two matrices must have the same order (that is, the same number of rows and same number of columns).
2. We have seen how to multiply a matrix by a number. In this case each element is multiplied by that number. The second sort of multiplication involves multiplying two matrices together. See §27.2.

3. The product will exist if $q = r$. Then AB will be $p \times s$. BA will exist if $s = p$ and the size of BA will be $r \times q$.

4. If A has order $p \times q$ then B must have order $q \times p$.

Exercise 27.2

1. $MN = \begin{pmatrix} 13 & 46 \\ 7 & 0 \end{pmatrix}$ $NM = \begin{pmatrix} 25 & 22 \\ 1 & -12 \end{pmatrix}$

   Note $MN \neq NM$.

2. (a) $3A = \begin{pmatrix} 3 & -21 \\ -3 & 6 \end{pmatrix}$

   (b) $A + B$ not possible   (c) $A + C$ not possible

   (d) $5D = \begin{pmatrix} 35 \\ 45 \\ 40 \\ 5 \end{pmatrix}$

   (e) BC not possible   (f) CB not possible
   (g) AC not possible   (h) CA not possible
   (i) $5A - 2C$ not possible

   (j) $D - E = \begin{pmatrix} 6.5 \\ 9 \\ 8 \\ 0 \end{pmatrix}$

   (k) BA not possible

   (l) $AB = \begin{pmatrix} 10 & 7 & 72 \\ -5 & -7 & -22 \end{pmatrix}$

   (m) $A^2 = AA = \begin{pmatrix} 8 & -21 \\ -3 & 11 \end{pmatrix}$

3. (a) $IA = \begin{pmatrix} 3 & 7 \\ -1 & -2 \end{pmatrix} = A$

   (b) $AI = A$;   (c) BI not possible

   (d) $IB = \begin{pmatrix} x \\ y \end{pmatrix} = B$

   (e) CI not possible
   (f) $IC = C$. Where it is possible to multiply by the identity matrix, the result is unaltered.

4. (a) $A + B = \begin{pmatrix} a+e & b+f \\ c+g & d+h \end{pmatrix}$

   (b) $A - B = \begin{pmatrix} a-e & b-f \\ c-g & d-h \end{pmatrix}$

   (c) $AB = \begin{pmatrix} ae+bg & af+bh \\ ce+dg & cf+dh \end{pmatrix}$

   (d) $BA = \begin{pmatrix} ea+fc & eb+fd \\ ga+hc & gb+hd \end{pmatrix}$

5. $\begin{pmatrix} -7x+5y \\ -2x-y \end{pmatrix}$

6. $\begin{pmatrix} 9a-5b \\ 3a+8b \end{pmatrix}$

7. (a) $BA = \begin{pmatrix} 10 & 3 & 1 \\ -2 & 9 & -5 \end{pmatrix}$

   (b) $B^2 = \begin{pmatrix} 7 & -2 \\ -3 & 10 \end{pmatrix}$

   (c) $B^2A = \begin{pmatrix} 6 & 21 & -9 \\ 34 & -9 & 13 \end{pmatrix}$

8. (a) $AB = \begin{pmatrix} 9 & 5 \\ 12 & 9 \end{pmatrix}$

   (b) $BA = \begin{pmatrix} 1 & -4 \\ 1 & 17 \end{pmatrix}$

   (c) $A^2 = \begin{pmatrix} -2 & 15 \\ -5 & 13 \end{pmatrix}$

   (d) $B^3 = \begin{pmatrix} -6 & -1 \\ 3 & -4 \end{pmatrix}$

10. $\begin{pmatrix} 3 & 0 \\ 0 & 2 \end{pmatrix}\begin{pmatrix} a & b \\ c & d \end{pmatrix} = \begin{pmatrix} 3a & 3b \\ 2c & 2d \end{pmatrix}$

   The effect of multiplying by $\begin{pmatrix} 3 & 0 \\ 0 & 2 \end{pmatrix}$ is to multiply the top row by 3 and the second row by 2.

Self-assessment questions 27.3

1. If A is a 2×2 matrix, its inverse is another matrix A^{-1} such that the products $A^{-1}A$ and AA^{-1} both equal the identity matrix. This inverse will exist if the determinant of A is not equal to zero.

2. A singular matrix is a square matrix with determinant equal to zero. It will not have an inverse.

3. $\begin{vmatrix} 7 & 0 \\ 0 & -9 \end{vmatrix} = -63$. Note that this is the

product of the elements on the leading diagonal.

Exercise 27.3

1. (a) -2 (b) -22 (c) 1 (d) 88
2. C and E are singular.

4. (a) $\dfrac{1}{-4}\begin{pmatrix} 10 & -4 \\ -6 & 2 \end{pmatrix} = \begin{pmatrix} \frac{-5}{2} & 1 \\ \frac{3}{2} & \frac{-1}{2} \end{pmatrix}$

(b) $\begin{pmatrix} 1 & 0 \\ 0 & 1 \end{pmatrix}$

(c) $\dfrac{1}{-8}\begin{pmatrix} 2 & -3 \\ -2 & -1 \end{pmatrix} = \begin{pmatrix} \frac{-1}{4} & \frac{3}{8} \\ \frac{1}{4} & \frac{1}{8} \end{pmatrix}$

(d) Has no inverse.

5. (a) $\begin{pmatrix} 1 & -2 \\ -4 & 9 \end{pmatrix}$

(b) $\dfrac{1}{2}\begin{pmatrix} 4 & -2 \\ -5 & 3 \end{pmatrix}$

(c) $\begin{pmatrix} -3 & -1 \\ -5 & -2 \end{pmatrix}$

(d) $\dfrac{1}{38}\begin{pmatrix} 5 & 2 \\ -9 & 4 \end{pmatrix}$

(e) $\dfrac{1}{2}\begin{pmatrix} 5 & -3 \\ -4 & 2 \end{pmatrix}$

Self-assessment question 27.4

1. $\begin{pmatrix} a & b \\ d & e \end{pmatrix}\begin{pmatrix} x \\ y \end{pmatrix} = \begin{pmatrix} c \\ f \end{pmatrix}$

Exercise 27.4

1. (a) Take $A = \begin{pmatrix} 1 & 1 \\ 2 & -1 \end{pmatrix}$

so that $A^{-1} = \begin{pmatrix} \frac{1}{3} & \frac{1}{3} \\ \frac{2}{3} & \frac{-1}{3} \end{pmatrix}$

Then $\begin{pmatrix} x \\ y \end{pmatrix} = \begin{pmatrix} \frac{1}{3} & \frac{1}{3} \\ \frac{2}{3} & \frac{-1}{3} \end{pmatrix}\begin{pmatrix} 17 \\ 10 \end{pmatrix} = \begin{pmatrix} 9 \\ 8 \end{pmatrix}$

so that $x = 9$ and $y = 8$.

(b) Take $A = \begin{pmatrix} 2 & -1 \\ 1 & 3 \end{pmatrix}$

so that $A^{-1} = \begin{pmatrix} \frac{3}{7} & \frac{1}{7} \\ \frac{-1}{7} & \frac{2}{7} \end{pmatrix}$

Then

$\begin{pmatrix} x \\ y \end{pmatrix} = \begin{pmatrix} \frac{3}{7} & \frac{1}{7} \\ \frac{-1}{7} & \frac{2}{7} \end{pmatrix}\begin{pmatrix} 10 \\ -2 \end{pmatrix} = \begin{pmatrix} 4 \\ -2 \end{pmatrix}$

so that $x = 4$ and $y = -2$.

(c) Take $A = \begin{pmatrix} 1 & 2 \\ 3 & -1 \end{pmatrix}$

so that $A^{-1} = \begin{pmatrix} \frac{1}{7} & \frac{2}{7} \\ \frac{3}{7} & \frac{-1}{7} \end{pmatrix}$

Then $\begin{pmatrix} x \\ y \end{pmatrix} = \begin{pmatrix} \frac{1}{7} & \frac{2}{7} \\ \frac{3}{7} & \frac{-1}{7} \end{pmatrix}\begin{pmatrix} 15 \\ 10 \end{pmatrix} = \begin{pmatrix} 5 \\ 5 \end{pmatrix}$

so that $x = y = 5$.

2. The matrix

$\begin{pmatrix} 2 & -3 \\ 6 & -9 \end{pmatrix}$

is singular and so has no inverse. Hence the matrix method fails to produce any solutions. The second equation is simply a multiple of the first – that is, we really only have one equation and this is not sufficient information to produce a unique value of x and y.

3. (a) $\begin{pmatrix} 3 & -1 \\ 1 & 1 \end{pmatrix}\begin{pmatrix} x \\ y \end{pmatrix} = \begin{pmatrix} 3 \\ 5 \end{pmatrix}$

$\begin{pmatrix} x \\ y \end{pmatrix} = \begin{pmatrix} \frac{1}{4} & \frac{1}{4} \\ -\frac{1}{4} & \frac{3}{4} \end{pmatrix}\begin{pmatrix} 3 \\ 5 \end{pmatrix}$

$= \begin{pmatrix} 2 \\ 3 \end{pmatrix}$

(b) $\begin{pmatrix} 2 & 3 \\ 1 & -2 \end{pmatrix}\begin{pmatrix} x \\ y \end{pmatrix} = \begin{pmatrix} 10 \\ -9 \end{pmatrix}$

$\begin{pmatrix} x \\ y \end{pmatrix} = \begin{pmatrix} \frac{2}{7} & \frac{3}{7} \\ \frac{1}{7} & -\frac{2}{7} \end{pmatrix}\begin{pmatrix} 10 \\ -9 \end{pmatrix}$

$= \begin{pmatrix} -1 \\ 4 \end{pmatrix}$

(c) $\begin{pmatrix} 1 & \frac{1}{2} \\ 4 & -3 \end{pmatrix} \begin{pmatrix} x \\ y \end{pmatrix} = \begin{pmatrix} 2 \\ 18 \end{pmatrix}$

$\begin{pmatrix} x \\ y \end{pmatrix} = \begin{pmatrix} \frac{3}{5} & \frac{1}{10} \\ \frac{4}{5} & -\frac{1}{5} \end{pmatrix} \begin{pmatrix} 2 \\ 18 \end{pmatrix}$

$= \begin{pmatrix} 3 \\ -2 \end{pmatrix}$

(d) $\begin{pmatrix} \frac{1}{2} & 3 \\ 2 & 5 \end{pmatrix} \begin{pmatrix} x \\ y \end{pmatrix} = \begin{pmatrix} 5 \\ 6 \end{pmatrix}$

$\begin{pmatrix} x \\ y \end{pmatrix} = \begin{pmatrix} -\frac{10}{7} & \frac{6}{7} \\ \frac{4}{7} & -\frac{1}{7} \end{pmatrix} \begin{pmatrix} 5 \\ 6 \end{pmatrix}$

$= \begin{pmatrix} -2 \\ 2 \end{pmatrix}$

Chapter 28

Self-assessment questions 28.1

1. Discrete data – the number of cars passing a given point during a given period of time. Continuous data – the length of time taken for an ambulance to arrive at the scene of an accident.
2. A tally chart is used to organise raw data into a frequency distribution.

Exercise 28.1

1. (a) discrete   (b) continuous   (c) discrete
   (d) continuous   (e) continuous
   (f) discrete   (g) continuous
   (h) continuous   (i) discrete
2. 1970   3
   1971   0
   1972   1
   1973   4
   1974   7
   1975   8
   1976   2
   Total   25

Self-assessment questions 28.2

1. When large amounts of data are involved it is often easier to analyse when it is grouped into classes.
2. Data is often recorded to the nearest value for the number of decimal places chosen. So for

example, if we are working to 1 decimal place, a value such as 14.2 represents a number in the interval 14.15–14.25. Class limits may be 14.0–14.1, 14.2–14.3, 14.4–14.5 etc., but theoretically these classes include values from 13.95 to 14.15, 14.15 to 14.35 etc. These latter values are the class boundaries.
3. The class width is the difference between the class boundaries.
4. The class midpoint can be found by adding half the class width to the lower class boundary.

Exercise 28.2

1. (a) See Table 1   (b) $\frac{1}{30} = 0.033$
   (c) $\frac{2}{30} = 0.067$
   (d) $\frac{4}{30} = 0.133 = 13.3\%$
2. (a) 5499.4 hours will appear in class 5000–5499; (b) 5499.8 hours will appear in class 5500–5999; (c) the class boundaries are 4999.5–5499.5, 5499.5–5999.5, 5999.5–6499.5, 6499.5–6999.5; (d) all classes have the same width, 5499.5 – 4999.5 = 500 hours; (e) the midpoints are 5249.5, 5749.5, 6249.5, 6749.5.
3. See Table 2.
4. The second class boundaries are 20.585–20.615. The class width is 20.615 – 20.585 = 0.03.

Table 1

Age	Tally	Frequency
15–19	JHT I	6
20–24	III	3
25–29	IIII	4
30–34	IIII	4
35–39	II	2
40–44	II	2
45–49	IIII	4
50–54		0
55–59	II	2
60–64	I	1
65–69		0
70–74	I	1
75–79	I	1
		30

Table 2

Mark	Frequency
0–9	0
10–19	4
20–29	3
30–39	2
40–49	6
50–59	15
60–69	10
70–79	5
80–89	11
90–99	4
	60

Self-assessment questions 28.3

1. In a vertical bar chart the length of a bar is proportional to the frequency. In a histogram it is the area of a column that is proportional to frequency.
2. Pictograms provide a readily accessible way of visualising statistical information. Pictograms can be drawn in a misleading way. Depending upon the amount of data available, pictograms may not provide an accurate representation of the data.
3. To avoid gaps on the horizontal axis.

Exercise 28.3

1.

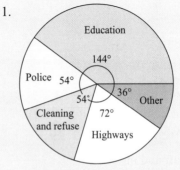

(a) Pie chart showing local council expenditure

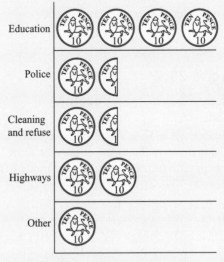

(b) Pictogram showing how each £1 spent by a local council is distributed

2.

Nitrogen 78%   Other gases 1%

280.8°

3.6°

75.6°

Oxygen
21%

Pie chart showing composition of the
atmosphere

3. (a) Class boundaries are 0–99.5, 99.5–199.5,
199.5–299.5, 299.5–399.5, 399.5–499.5.

(b) A claim for 199.75 would be placed in
the class 200–299.

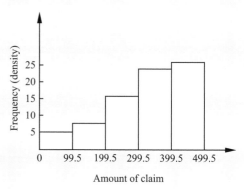

(c) Distribution of claim amounts

Chapter 29

Self-assessment questions 29.2

1. Mean, median and mode.
2. For example, suppose there are four
workers, earning £8000, £18,000, £19,000
and £21,000. The median salary is the mean
of the middle two, namely £18,500.

Exercise 29.2

1. Mean $= \bar{x} = \frac{50}{9} = 5.556$, median $= 5$,
mode $= 5$.
2. Mean $= \frac{262}{7} = 37.4$. The large, atypical value
of 256 distorts the mean; it may be more
appropriate to use the median or mode here.
3. Mean $= \frac{498}{7} = 71.1\%$, median $= 68\%$.
4. (a) $x_1 + x_2 + x_3 + x_4$
(b) $x_1 + x_2 + x_3 + x_4 + x_5 + x_6 + x_7$
(c) $(x_1 - 3)^2 + (x_2 - 3)^2 + (x_3 - 3)^2$
(d) $(2 - x_1)^2 + (2 - x_2)^2 + (2 - x_3)^2$
$+ (2 - x_4)^2$
5. (a) $\sum_{i=3}^{6} x_i$ (b) $\sum_{i=1}^{3}(x_i - 1)$

6. Mean $= 7.57/6 = 1.26$, median $= 1.32$, no
mode.
7. £320,800
8.

Midpoint x_i	Frequency f_i	$f_i \times x_i$
20.57	3	61.71
20.60	6	123.60
20.63	8	165.04
20.66	3	61.98
Total	20	412.33

Mean $= \bar{x} = 412.33/20 = 20.62$ (2 d.p.)

Self-assessment questions 29.3

1. An average gives no information about the
spread of a set of values. For example, the
sets 10,10,10 and 8,10,12 both have a mean
of 10 but the spread of the second set is
greater.

Class	f	Midpoint x_i	$f_i \times x_i$	$x_i - \bar{x}$	$(x_i - \bar{x})^2$	$f_i(x_i - \bar{x})^2$
20–24	2	22	44	−35.3	1246.09	2492.18
25–29	0	27	0	−30.3	918.09	0.0
30–34	1	32	32	−25.3	640.09	640.09
35–39	1	37	37	−20.3	412.09	412.09
40–44	3	42	126	−15.3	234.09	702.27
45–49	11	47	517	−10.3	106.09	1166.99
50–54	4	52	208	−5.3	28.09	112.36
55–59	3	57	171	−0.3	0.09	0.27
60–64	10	62	620	4.7	22.09	220.9
65–69	5	67	335	9.7	94.09	470.45
70–74	5	72	360	14.7	216.09	1080.45
75–79	1	77	77	19.7	388.09	388.09
80–84	3	82	246	24.7	610.09	1830.27
85–89	0	87	0	29.7	882.09	0.0
90–94	1	92	92	34.7	1204.09	1204.09
Total	50		2865			10720.50

Table for Exercise 29.3, Q4(a)

2. Find the mean of the set. Subtract the mean from each value, square and add the results. Divide the result by n, the number of data values. This is the variance. The standard deviation is the square root of the variance.

Exercise 29.3

1. (a) $\bar{x} = 20$, variance $= 0$, standard deviation $= 0$
   (b) $\bar{x} = 20$, variance $= 12.5$, standard deviation $= 3.54$
   (c) $\bar{x} = 20$, variance $= 200$, standard deviation $= 14.14$. The large standard deviation of the final set reflects the widely spread values.

2. Mean $= 5$, variance $= 9.2$, standard deviation $= 3.033$.

3. Jane's mean $= 51.4$, Tony's mean $= 51.6$. Jane's standard deviation $= 12.77$, Tony's standard deviation $= 17.40$. The students have a very close mean mark but Jane has achieved a more consistent set of results.

4. (a) See table above.
   (b) The modal class is 45–49, which is the most common class   (c) mean $= \frac{2865}{50} = 57.3$, variance $= \frac{10720.50}{50} = 214.41$, standard deviation $= 14.64$.

Chapter 30

Self-assessment questions 30.1

1. They must lie between 0 and 1 inclusive.
2. Two events are complementary if, when one can happen, the other cannot and vice versa.

Exercise 30.2

1. (a) $\frac{5}{6}$ (b) 0 (c) $\frac{1}{3}$ (d) 0
2. $\frac{12}{13}$
3. (a) $\frac{3}{10}$ (b) $\frac{3}{10}$ (c) $\frac{7}{10}$
4. There are 36 possibilities in all.
   (a) $\frac{1}{36}$ (b) 0 (c) 0 (d) $\frac{1}{36}$ (e) $\frac{26}{36}$ or $\frac{13}{18}$
5. $\frac{1}{30}$
6. (a) $\frac{1}{7}$ (b) $\frac{12}{35}$ (c) $\frac{2}{7}$ (d) $\frac{18}{35}$ (e) $\frac{6}{7}$ (f) $\frac{23}{35}$
7. If H represents head and T represents tail then the possible outcomes are: TTT, TTH, THT, THH, HTT, HTH, HHT, HHH. Note that all the outcomes have equal probability.
   (a) Number of outcomes with two heads and one tail is three. So P(two heads and one tail) $= \frac{3}{8}$.
   (b) Four outcomes have at least two heads. P(at least two heads) $= \frac{1}{2}$.
   (c) 'No heads' is equivalent to the outcome TTT. P(no heads) $= \frac{1}{8}$.

Self-assessment questions 30.3

1. If an experiment can be performed a large number of times it may be possible to use data from it to estimate the probabilities of different outcomes.
2. The final series should yield the best estimate of probability because the larger the number of trials, the better will be the estimate.

Exercise 30.3

1. (a) $\frac{143}{150}$ (b) $\frac{7}{150}$
2. 0.95
3. 89760
4. Number reached in less than one hour is $0.87 \times 17300 = 15051$.

5. P(component works) $= \frac{48700}{50000} = 0.974$.
   Number in batch working well $= 0.974 \times 3000 = 2922$.
   Number not working well $= 78$.

Self-assessment questions 30.4

1. Two events are independent if the occurrence of one does not affect the occurrence of the other.
2. The events are independent. Whichever card is selected from the first pack, this does not influence the selection from the second pack.
3. These are dependent. If the first card selected is a Spade, then there are fewer Spades remaining in the pack (12). If the first card selected is not a Spade there will be 13 Spades remaining when the choice of second card is made.

Exercise 30.4

1. $\frac{3}{6} \times \frac{1}{2} = \frac{1}{4}$
2. $\frac{1}{52} \times \frac{1}{52} = \frac{1}{2704} = 0.000370$
3. $(\frac{1}{2})^8 = 0.0039$
4. $(\frac{1}{6})^4 = 0.000772$
5. (a) 0.5 (b) 0.25
6. (a) P(all work well) $= (0.96)^4 = 0.8493$
   (b) P(none work well)
      $= (0.04)^4 = 2.56 \times 10^{-6}$
7. (a) P(passing three modules)
      $= (0.91)^3 = 0.7536$.
   (b) Suppose a student passes two modules; the possible outcomes are PPF, PFP, FPP. Each outcome has a probability of $(0.91)^2(0.09)$. Hence total probability is $3(0.91)^2(0.09) = 0.2236$.
   (c) Passing one module, the possible outcomes are FFP, FPF, PFF. Each outcome has probability $(0.09)^2(0.91)$. Hence total probability is $3(0.09)^2(0.91) = 0.0221$.
   (d) Passing no modules is equivalent to failing all three. This has a probability of $(0.09)^3 = 7.29 \times 10^{-4}$.

Chapter 31

Self-assessment questions 31.2

1. A scatter diagram is used as a preliminary tool for exploring possible associations between two variables.

2. When the points on a scatter diagram appear to cluster around a straight line which has a positive gradient, the two variables are said to show positive correlation. When the line has a negative gradient, the correlation is negative. If the points appear to be scattered randomly there is no correlation.

2. The data appears to exhibit weak positive correlation.

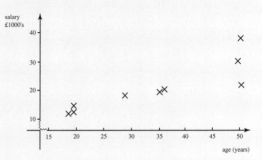

Exercise 31.2

1. (a) The data appears to exhibit fairly strong negative correlation.

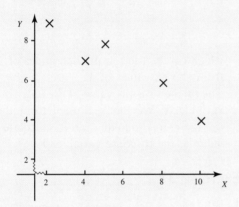

(b) The data appears to exhibit no correlation whatsoever.

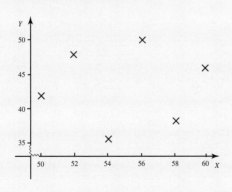

Self-assessment question 31.3

1. (a) perfect correlation (b) strong negative correlation (c) weak negative correlation (d) no correlation

Exercises 31.3

1.

x_i	y_i	x_i^2	y_i^2	x_iy_i
1	11	1	121	11
3	51	9	2601	153
4	96	16	9216	384
5	115	25	13225	575
13	273	51	25163	1123

$$r = \frac{4(1123) - (13)(273)}{\sqrt{4(51) - 13^2}\sqrt{4(25163) - 273^2}}$$

$$= \frac{943}{\sqrt{35}\sqrt{26123}} = 0.986$$

2.

x_i	y_i	x_i^2	y_i^2	$x_i y_i$
0	2	0	4	0
1	5	1	25	5
2	8	4	64	16
3	11	9	121	33
6	26	14	214	54

$$r = \frac{4(54) - (6)(26)}{\sqrt{4(14) - 6^2}\sqrt{4(214) - 26^2}}$$

$$= \frac{60}{\sqrt{20}\sqrt{180}} = 1$$

3. 0.8592

Self-assessment question 31.4

1. When, instead of being given specific data values, the information is provided as ranks or orderings.

Exercise 31.4

1. $\sum D^2 = 16$, $r = 0.5429$
2. $\sum D^2 = 80$, $r = 0.5152$

Chapter 32

Self-assessment questions 32.2

1. When drawing a scatter diagram for a regression problem the independent variable is plotted on the horizontal axis and the dependent variable is plotted on the vertical axis.
2. Given a set of values of variables X and Y consisting of pairs of data (x_i, y_i), the regression of Y on X is the equation of the straight line which best fits the data and which can be used to predict values of Y from chosen values of X.
3. The predictor is the value of the independent variable, X. We choose a predictor and then use the regression equation to calculate a value of the response variable, Y.

Exercise 32.2

1. (b)

student	x_i	y_i	$x_i y_i$	x_i^2
A	5	3	15	25
B	10	18	180	100
C	15	42	630	225
D	20	67	1340	400
E	25	80	2000	625
F	30	81	2430	900
	$\sum_{i=1}^{6} x_i = 105$	$\sum_{i=1}^{6} y_i = 291$	$\sum_{i=1}^{6} x_i y_i = 6595$	$\sum_{i=1}^{6} x_i^2 = 2275$

$$b = \frac{6(6595) - (105)(291)}{6(2275) - 105^2} = \frac{9015}{2625} = 3.434 \quad (3 \text{ d.p.})$$

$$a = \frac{291 - (3.434)(105)}{6} = -11.595 \quad (3 \text{ d.p.})$$

So $y = 3.43x - 11.60$ (2 d.p.).   (d) 29.56.

2. (b)

individual	x_i	y_i	$x_i y_i$	x_i^2
A	2	72	144	4
B	2.5	71	177.5	6.25
C	3	65	195	9
D	3.5	60	210	12.25
	$\sum_{i=1}^{4} x_i = 11$	$\sum_{i=1}^{4} y_i = 268$	$\sum_{i=1}^{4} x_i y_i = 726.5$	$\sum_{i=1}^{4} x_i^2 = 31.5$

$$b = \frac{4(726.5) - (11)(268)}{4(31.5) - 11^2} = \frac{-42}{5} = -8.4$$

$$a = \frac{268 - (-8.4)(11)}{4} = 90.1$$

So $y = 90.1 - 8.4x$.   (d) 63.22.

Chapter 33

Self-assessment questions 33.2

1. It is a function that provides a formula for calculating the gradient of another function.
2. $y' = nx^{n-1}$.
3. Derivative, derived function, first derivative, gradient function, rate of change.
4. $\frac{dy}{dx}$.
5. Zero. A constant function has a horizontal graph. Consequently its gradient is zero.
6. Radians.
7. The exponential function e^x.

Exercise 33.2

1. (a) $8x^7$ (b) $7x^6$ (c) $-x^{-2} = -1/x^2$
   (d) $-5x^{-6} = -5/x^6$ (e) $13x^{12}$
   (f) $5x^4$ (g) $-2x^{-3} = -2/x^3$

2. (a) $\frac{3}{2}x^{1/2}$ (b) $\frac{5}{2}x^{3/2}$ (c) $-\frac{1}{2}x^{-3/2}$
   (d) $(1/2)x^{-1/2} = 1/2\sqrt{x}$
   (e) $y = \sqrt{x} = x^{1/2}$ and so
      $y' = (1/2)x^{-1/2} = 1/2\sqrt{x}$
   (f) $0.2x^{-0.8}$

3. $y = 8$, $y' = 0$, the line is horizontal and so its gradient is zero.

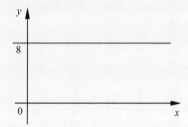

4. (a) $16x^{15}$ (b) $0.5x^{-0.5}$ (c) $-3.5x^{-4.5}$
   (d) $y = 1/x^{1/2} = x^{-1/2}$ and so $y' = -\frac{1}{2}x^{-3/2}$

(e) $y = 1/\sqrt{x} = 1/x^{1/2} = x^{-1/2}$ and so
$y' = -\frac{1}{2}x^{-3/2}$

5. $y' = 4x^3$:
   (a) at $x = -4$, $y' = -256$
   (b) at $x = -1$, $y' = -4$
   (c) at $x = 0$, $y' = 0$ (d) at $x = 1$, $y' = 4$
   (e) at $x = 4$, $y' = 256$. The graph is falling at the given negative x values and is rising at the given positive x values. At $x = 0$ the graph is neither rising nor falling.

6. (a) $y' = 3x^2$, $y'(-3) = 27$, $y'(0) = 0$,
      $y'(3) = 27$
   (b) $y' = 4x^3$, $y'(-2) = -32$, $y'(2) = 32$
   (c) $y' = \frac{1}{2}x^{-1/2} = 1/2\sqrt{x}$, $y'(1) = \frac{1}{2}$,
      $y'(2) = \frac{1}{2\sqrt{2}} = 0.354$
   (d) $y' = -2x^{-3}$, $y'(-2) = \frac{1}{4}$, $y'(2) = -\frac{1}{4}$

7. (a) $2t$ (b) $-3t^{-4}$ (c) $\frac{1}{2}t^{-1/2}$
   (d) $-1.5t^{-2.5}$

8. (a) $-7 \sin 7x$ (b) $\frac{1}{2} \cos(x/2)$
   (c) $-\frac{1}{2} \sin(x/2)$ (d) $4e^{4x}$ (e) $-3e^{-3x}$

9. (a) $1/x$ (b) $\frac{1}{3} \cos(x/3)$ (c) $\frac{1}{2}e^{x/2}$

10. $y' = \cos x$, (a) $y'(0) = 1$, (b) $y'(\pi/2) = 0$,
    (c) $y'(\pi) = -1$

11. (a) $7t^6$ (b) $3 \cos 3t$ (c) $2e^{2t}$ (d) $1/t$
    (e) $-\frac{1}{3} \sin(t/3)$

Self-assessment question 33.3

1. (a) The derivative of the sum of two functions is the sum of the separate derivatives.
   (b) The derivative of the difference of two functions is the difference of the separate derivatives. This means we can differentiate sums and differences term by term.
   (c) The derivative of $k f(x)$ is equal to k times the derivative of $f(x)$.

Exercise 33.3

1. (a) $y' = 6x^5 - 4x^3$
   (b) $y' = 6 \times (2x) = 12x$
   (c) $y' = 9(-2x^{-3}) = -18x^{-3}$ (d) $y' = \frac{1}{2}$
   (e) $y' = 6x^2 - 6x$
2. (a) $y' = 2 \cos x$
   (b) $y' = 3(4 \cos 4x) = 12 \cos 4x$

(c) $y' = -63 \sin 9x + 12 \cos 4x$
(d) $y' = 3e^{3x} + 10e^{-2x}$

3. The gradient function is $y' = 9x^2 - 9$.
   $y'(1) = 0$, $y'(0) = -9$, $y'(-1) = 0$. Of the three given points, the curve is steepest at $x = 0$ where it is falling.

4. When x^2 is differentiated we obtain $2x$. Also, because the derivative of any constant is zero, the derivative of $x^2 + c$ will be $2x$ where c can take any value.

5. $y' = 6 \cos 2t - 8 \sin 2t$
     $y'(2) = 6 \cos 4 - 8 \sin 4 = 2.1326$

6. $y' = 2 - 3x^2 + 2e^{2x}$
     $y'(1) = 2 - 3 + 2e^2 = 13.7781$

7. $y = (2x - 1)^2 = 4x^2 - 4x + 1$
      $y' = 8x - 4$
      $y'(0) = -4$

8. $f' = 2x - 2$
     $f'(1) = 0$

9. $y' = x^2 - 1$. We see that $y' = 0$ when $x^2 - 1 = 0$, that is when $x = -1$, 1.

10. $y' = 1 - \sin x$. When $y' = 0$ then $\sin x = 1$, and hence $x = \frac{\pi}{2}$.

Self-assessment questions 33.4

1. The second derivative of y is found by differentiating the first derivative of y.
2. y'' and, if y is a function of x, $\frac{d^2y}{dx^2}$.

Exercise 33.4

1. (a) $y' = 2x$, $y'' = 2$
   (b) $y' = 24x^7$, $y'' = 168x^6$
   (c) $y' = 45x^4$, $y'' = 180x^3$
   (d) $y' = \frac{3}{2}x^{-1/2}$, $y'' = -\frac{3}{4}x^{-3/2}$
   (e) $y' = 12x^3 + 10x$, $y'' = 36x^2 + 10$
   (f) $y' = 27x^2 - 14$, $y'' = 54x$
   (g) $y' = \frac{1}{2}x^{-1/2} - 2x^{-3/2}$,
      $y'' = -\frac{1}{4}x^{-3/2} + 3x^{-5/2}$
2. (a) $y' = 3 \cos 3x - 2 \sin 2x$,
      $y'' = -9 \sin 3x - 4 \cos 2x$
   (b) $y' = e^x - e^{-x}$, $y'' = e^x + e^{-x}$

3. $y' = 4 \cos 4x$, $y'' = -16 \sin 4x$, and so
   $16y + y'' = 16 \sin 4x + (-16 \sin 4x) = 0$ as
   required.

4. $y' = 4x^3 - 6x^2 - 72x$,
   $y'' = 12x^2 - 12x - 72$. The values of x where
   $y'' = 0$ are found from $12x^2 - 12x - 72 = 0$,
   that is $12(x^2 - x - 6) = 12(x + 2)(x - 3) = 0$
   so that $x = -2$ and $x = 3$.

5. (a) $y' = 6 \cos 3x - 8 \sin 2x$,
       $y'' = -18 \sin 3x - 16 \cos 2x$
   (b) $y' = k \cos kt$, $y'' = -k^2 \sin kt$
   (c) $y' = -k \sin kt$, $y'' = -k^2 \cos kt$
   (d) $y' = Ak \cos kt - Bk \sin kt$,
       $y'' = -Ak^2 \sin kt - Bk^2 \cos kt$

6. (a) $y' = 3e^{3x}$, $y'' = 9e^{3x}$
   (b) $y' = 3 + 8e^{4x}$, $y'' = 32e^{4x}$
   (c) $y' = 2e^{2x} + 2e^{-2x}$, $y'' = 4e^{2x} - 4e^{-2x}$
   (d) $y = e^{-x}$, $y' = -e^{-x}$, $y'' = e^{-x}$
   (e) $y = e^{2x} + 3e^x$, $y' = 2e^{2x} + 3e^x$,
       $y'' = 4e^{2x} + 3e^x$

7. $y' = 200 + 12x - 3x^2$, $y'' = 12 - 6x$
   $y'' = 0$ when $x = 2$

8. $y' = 2e^x + 2 \cos 2x$
   $y'' = 2e^x - 4 \sin 2x$
   $y''(1) = 2e - 4 \sin 2 = 1.7994$

9. $y' = 8 \cos 4t + 10 \sin 2t$
   $y'' = -32 \sin 4t + 20 \cos 2t$
   $y''(1.2) = -32 \sin 4.8 + 20 \cos 2.4 = 17.1294$

Self-assessment questions 33.5

1. Maximum, minimum and point of inflexion.
2. Suppose $x = a$ is a stationary point of $y(x)$.
   If y'' evaluated at $x = a$ is positive, the point
   is a minimum. If y'' evaluated at $x = a$ is
   negative that point is a maximum. If $y'' = 0$
   this test does not enable us to distinguish
   between the possible types of stationary
   point.
3. If $y'' = 0$.

Exercise 33.5

1. (a) $y' = 2x$, $y'' = 2$, $(0, 1)$ is a minimum;

(b) $y' = -2x$, $y'' = -2$, $(0, 0)$ is a
    maximum;   (c) $y' = 6x^2 + 18x$,
    $y'' = 12x + 18$, $(0, 0)$ is a minimum and
    $(-3, 27)$ is a maximum;
(d) $y' = -6x^2 + 54x$, $y'' = -12x + 54$, $(0, 0)$
    is a minimum and $(9, 729)$ is a
    maximum;
(e) $y' = 3x^2 - 6x + 3$, $y'' = 6x - 6$,
    stationary point by solving
    $3x^2 - 6x + 3 = 0$, at $(1, 2)$, second-
    derivative test fails, inspection of
    gradient on either side of $x = 1$ reveals a
    point of inflexion.

2. $y' = \cos x$. Solving $\cos x = 0$ gives $x = \frac{\pi}{2}, \frac{3\pi}{2}$.
   The maximum and minimum points occur at
   $x = \frac{\pi}{2}, \frac{3\pi}{2}$.
   We examine y''

   $$y'' = -\sin x$$

   $$y''\left(\frac{\pi}{2}\right) = -\sin \frac{\pi}{2} < 0 \quad \text{a maximum}$$

   $$y''\left(\frac{3\pi}{2}\right) = -\sin \frac{3\pi}{2} > 0 \quad \text{a minimum}$$

   When $x = \frac{\pi}{2}$, $y = 1$; when $x = \frac{3\pi}{2}$, $y = -1$. In
   summary, maximum at $(\frac{\pi}{2}, 1)$, minimum at
   $(\frac{3\pi}{2}, -1)$.

3. $y' = e^x - 1$. Solving $y' = 0$ gives $x = 0$. Now,
   $y'' = e^x$ and $y''(0) = 1 > 0$. When $x = 0$,
   $y = 1$, so there is a minimum at $(0, 1)$.

4. $y' = 6x^2 - 6x - 12 = 6(x + 1)(x - 2)$.
   Solving $y' = 0$ gives $x = -1, 2$. $y'' = 12x - 6$,
   $y''(-1) < 0$, a maximum, $y''(2) > 0$, a
   minimum.
   When $x = -1$, $y = 8$; when $x = 2$, $y = -19$.
   Hence there is a maximum at $(-1, 8)$, a
   minimum at $(2, -19)$.

5. $y' = -x^2 + 1$. Solving $y' = 0$ gives $x = -1, 1$.
   $y'' = -2x$, $y''(-1) > 0$, a minimum, $y''(1) < 0$,
   a maximum.
   When $x = -1$, $y = -\frac{2}{3}$, when $x = 1$, $y = \frac{2}{3}$. So
   there is a minimum at $(-1, -\frac{2}{3})$ and a
   maximum at $(1, \frac{2}{3})$.

6. $y' = 5x^4 - 5$
   $= 5(x^4 - 1)$
   $= 5(x^2 - 1)(x^2 + 1)$
   $= 5(x - 1)(x + 1)(x^2 + 1)$
   Solving $y' = 0$ gives $x = -1, 1$.

$y'' = 20x^3$, $y''(-1) < 0$, a maximum,
$y''(1) > 0$, a minimum.
When $x = -1$, $y = 5$, when $x = 1$, $y = -3$.
There is a maximum at $(-1, 5)$; a minimum at $(1, -3)$.

Chapter 34

Self-assessment question 34.2

1. The product rule states that if $y = uv$ then

$$\frac{dy}{dx} = \frac{du}{dx}v + u\frac{dv}{dx}$$

Exercise 34.2

1. (a) $\sin x + x \cos x$ (b) $2x \cos x - x^2 \sin x$
   (c) $\cos^2 x - \sin^2 x$ (d) $e^x + xe^x$
   (e) $e^{-x} - xe^{-x}$
2. (a) $2 \ln(3x) + 2$ (b) $3e^{3x} \sin x + e^{3x} \cos x$
   (c) $-e^{-\frac{x}{2}}(\frac{\cos 4x}{2} + 4 \sin 4x)$
   (d) $2 \cos 2x \cos 3x - 3 \sin 2x \sin 3x$
   (e) $e^x \ln(2x) + \frac{e^x}{x}$
3. $\frac{dy}{dx} = e^{-x}(\cos x - \sin x)$, $\frac{dy}{dx} (x = 1) = -0.1108$
4. $\frac{dy}{dx} = -2(\frac{\cos 2x + x \sin 2x}{x^3})$
5. $y' = 2(\sin x + x)(\cos x + 1)$
6. (a) $e^{-x}x(2 - x)$
   (b) $e^{-x}x(-x \sin x + 2 \sin x + x \cos x)$
7. $y' = x^2 e^x(3 + x)$, $y' = 0$ when $x = 0$ and $x = -3$

Self-assessment question 34.3

1. The quotient rule states that if $y = \frac{u}{v}$ then

$$\frac{dy}{dx} = \frac{v\frac{du}{dx} - u\frac{dv}{dx}}{v^2}$$

Exercise 34.3

1. (a) $\frac{xe^x}{(x+1)^2}$ (b) $\frac{1 - \ln x}{x^2}$ (c) $\frac{\ln x - 1}{(\ln x)^2}$
   (d) $-\frac{1}{\sin^2 t}$ (e) $\frac{2t^2 + 2t + 1}{(2t+1)^2}$

2. (a) $y' = \frac{1}{(x+2)^2}$, $y'(2) = \frac{1}{16}$
   (b) $y' = \frac{2(x+1)/x - 2 \ln(3x)}{(x+1)^2}$, $y'(2) = -0.0648$
   (c) $y' = \frac{2 \cos 2x \cos 3x + 3 \sin 2x \sin 3x}{(\cos 3x)^2}$,
       $y'(2) = -0.6734$
   (d) $y' = \frac{e^{3x} + 2e^{2x} - e^x}{(e^x + 1)^2}$, $y'(2) = 7.1791$
   (e) $y' = \frac{x^4 + 3x^2 - 2x}{(x^2 + 1)^2}$, $y'(2) = \frac{24}{25}$
3. $\frac{1}{\cos^2 \theta}$
4. $\frac{k}{\cos^2 k\theta}$

Self-assessment questions 34.4

1. $\frac{dy}{dt} = \frac{dy}{dx} \times \frac{dx}{dt}$
2. $\frac{dy}{dx} = \frac{dy}{dz} \times \frac{dz}{dx}$
3. When the input to a function is itself a function, we have a function of a function. So if, for example, the function $\theta = 7x^2$ is the input to the function $\sin \theta$, we obtain the function of a function $\sin(7x^2)$.
4. (a) Note that $\frac{1}{x}\cos x$ could be written as $\frac{\cos x}{x}$. So the function $\frac{1}{x}\cos x$ can be thought of as either a product of the functions $\frac{1}{x}$ and $\cos x$, or a quotient of $\cos x$ and x. Either the product rule or the quotient rule can be used.
   (b) $x \sin x$ is a product of the function x and $\sin x$.
   (c) $\sin(\frac{1}{x})$ is a function of a function. Note that if $y = \sin t$ and $t = \frac{1}{x}$ then $y = \sin \frac{1}{x}$.

Exercise 34.4

1. (a) $\frac{dy}{dt} = \cos x \times 2t = 2t \cos t^2$

   (b) $\frac{dy}{dt} = \frac{1}{2}x^{-1/2} \times 1$
   $= \frac{1}{2}x^{-1/2} = \frac{1}{2}(t+1)^{-1/2} = \frac{1}{2(t+1)^{1/2}}$

   (c) $\frac{dy}{dt} = e^x \times 4t = 4te^{2t^2}$

   (d) $\frac{dy}{dt} = \frac{15}{2}x^{3/2} \times (3t^2 + 2)$
   $= \frac{15}{2}(3t^2 + 2)(t^3 + 2t)^{3/2}$

2. (a) $\frac{dy}{dt} = -6t \sin(3t^2)$

   (b) $\frac{dy}{dt} = 2\cos(2t+1)$

   (c) $\frac{dy}{dt} = \frac{3}{3t-2}$

   (d) $\frac{dy}{dt} = (2t+3)\cos(t^2 + 3t + 2)$

3. (a) $\frac{dy}{dx} = 3\cos(3x+1)$

   (b) $\frac{dy}{dx} = 4e^{4x-3}$

   (c) $\frac{dy}{dx} = -2\sin(2x-4)$

   (d) $\frac{dy}{dx} = \frac{1}{2}(x^2 + 5x - 4)^{-1/2}(2x+5)$

   (e) $\frac{dy}{dx} = -\frac{1}{x^2 \cos^2(1/x)}$

4. (a) $\frac{dy}{dx} = \frac{4}{4x-3}$

   (b) $\frac{dy}{dx} = \frac{6x}{3x^2} = \frac{2}{x}$

   (c) $\frac{dy}{dx} = \frac{\cos x}{\sin x}$

   (d) $\frac{dy}{dx} = \frac{2x+3}{x^2 + 3x}$

Chapter 35

Self-assessment questions 35.2

1. An indefinite integral is found by reversing the process of differentiation. Such an integral will contain an arbitrary constant called the constant of integration.
2. During the process of differentiation any constants disappear. Consequently, when reversing this process we must allow for the possibility that a constant(s) was present.

Exercise 35.2

1. (a) $x^2/2 + c$ (b) $x^2 + c$ (c) $x^6/6 + c$
   (d) $x^8/8 + c$ (e) $-x^{-4}/4 + c$
   (f) $2x^{1/2} + c$ (g) $x + c$ (h) $4x^{3/4}/3 + c$
2. (a) $9x + c$ (b) $\frac{1}{2}x + c$
   (c) $-7x + c$ (d) $0.5x + c$. In general,
   $\int k \, dx = kx + c$.
3. (a) $-(\cos 5x)/5 + c$ (b) $-2\cos(x/2) + c$
   (c) $2\sin(x/2) + c$ (d) $-e^{-3x}/3 + c$
   (e) $-4\cos 0.25x + c$ (f) $-2e^{-0.5x} + c$
   (g) $2e^{x/2} + c$
4. (a) $\int \cos 5x \, dx = \frac{\sin 5x}{5} + c$

   (b) $\int \sin 4x \, dx = -\frac{\cos 4x}{4} + c$

   (c) $\int e^{6t} \, dt = \frac{e^{6t}}{6} + c$

   (d) $\int \cos(\frac{t}{3}) \, dt = 3 \sin(\frac{t}{3}) + c$

5. (a) $\int \cos 3y \, dy = \frac{\sin 3y}{3} + c$

   (b) $\int \frac{1}{\sqrt{x}} \, dx = \int x^{-1/2} \, dx = 2x^{1/2} + c$
   $= 2\sqrt{x} + c$

   (c) $\int \sin(\frac{3t}{2}) \, dt = -\frac{2}{3}\cos(\frac{3t}{2}) + c$

   (d) $\int \sin(-x) \, dx = \cos(-x) + c = \cos x + c$

Self-assessment question 35.3

1. (a) The integral of the sum of two functions is the sum of the separate integrals.
   (b) The integral of the difference of two functions is the difference of the separate integrals. This means we can integrate sums and differences term by term.
   (c) The integral of $k f(x)$ is equal to k times the integral of $f(x)$.

Exercise 35.3

1. (a) $x^3/3 + x^2/2 + c$
   (b) $x^3/3 + x^2/2 + x + c$ (c) $\frac{4}{7}x^7 + c$
   (d) $5x^2/2 + 7x + c$ (e) $\ln x + c$
   (f) $\int (1/x^2) \, dx = \int x^{-2} \, dx = x^{-1}/(-1) + c$
   $= -1/x + c$

(g) $3 \ln x - \frac{7}{x} + c$ (h) $\frac{2}{3} x^{3/2} + c$

(i) $8x^{1/2} + c$

(j) $\displaystyle\int \frac{1}{\sqrt{x}} \, dx = \int x^{-1/2} \, dx = 2x^{1/2} + c$

(k) $\dfrac{3x^4}{4} + \dfrac{7x^3}{3} + c$ (l) $\dfrac{2x^5}{5} - x^3 + c$

(m) $\dfrac{x^6}{6} - 7x + c$

2. (a) $3e^x + c$ (b) $\frac{1}{2} e^x + c$

   (c) $3e^x + 2e^{-x} + c$ (d) $5e^{2x}/2 + c$

   (e) $-\cos x - (\sin 3x)/3 + c$

3. (a) $x^3 - 2e^x + c$ (b) $x^2/4 + x + c$

   (c) $-\frac{2}{3} \cos 3x - (8 \sin 3x)/3 + c$

   (d) $-(3 \cos 2x)/2 + (5 \sin 3x)/3 + c$

   (e) $5 \ln x + c$

4. $5 \ln t - t^3/3 + c$

5. (a) $\int (3 \sin 2t + 4 \cos 2t) \, dt$
   $= -\frac{3}{2} \cos 2t + 2 \sin 2t + c$

   (b) $\int (-\sin 3x - 2 \cos 4x) \, dx$
   $= \frac{\cos 3x}{3} - \frac{\sin 4x}{2} + c$

   (c) $\int [1 + 2 \sin(\frac{x}{2})] \, dx = x - 4 \cos(\frac{x}{2}) + c$

   (d) $\int [\frac{1}{2} - \frac{1}{3} \cos(\frac{x}{3})] \, dx = \frac{x}{2} - \sin(\frac{x}{3}) + c$

6. (a) $\int [e^x(1 + e^x)] \, dx = \int e^x + e^{2x} \, dx$
   $= e^x + \frac{e^{2x}}{2} + c$

   (b) $\int [(x + 2)(x + 3)] \, dx = \int (x^2 + 5x + 6) \, dx$
   $= \frac{x^3}{3} + \frac{5x^2}{2} + 6x + c$

   (c) $\int (5 \cos 2t - 2 \sin 5t) \, dt$
   $= \frac{5}{2} \sin 2t + \frac{2}{5} \cos 5t + c$

   (d) $\int [\frac{2}{x}(x + 3)] \, dx = \int (2 + \frac{6}{x}) \, dx$
   $= 2x + 6 \ln x + c$

Self-assessment question 35.4

1. A definite integral usually has a specific numerical value. It does not contain a constant of integration.

Exercise 35.4

1. (a) $\left[\dfrac{7x^2}{2} \right]_{-1}^{1} = 0$

(b) $\left[\dfrac{7x^2}{2} \right]_{0}^{3} = \dfrac{63}{2} = 31.5$

(c) $\left[\dfrac{x^3}{3} \right]_{-1}^{1} = \dfrac{2}{3}$ (d) $\left[-\dfrac{x^3}{3} \right]_{-1}^{1} = -\dfrac{2}{3}$

(e) $\left[\dfrac{x^3}{3} \right]_{0}^{2} = \dfrac{8}{3}$

(f) $\left[\dfrac{x^3}{3} + 2x^2 \right]_{0}^{3} = 9 + 18 = 27$

(g) $[\sin x]_0^\pi = 0$

(h) $[\ln x]_1^2 = \ln 2 - \ln 1 = \ln 2 = 0.693$

(i) $[\cos x]_0^\pi = \cos \pi - \cos 0 = -1 - 1 = -2$

(j) $[13x]_1^4 = 52 - 13 = 39$

(k) $\left[\dfrac{x^3}{3} - \cos x \right]_{0}^{\pi}$

$= \left(\dfrac{\pi^3}{3} - \cos \pi \right) - (0 - \cos 0)$

$= \dfrac{\pi^3}{3} + 1 + 1 = \dfrac{\pi^3}{3} + 2 = 12.335$

(l) $\left[\dfrac{5x^3}{3} \right]_{1}^{3} = 45 - \dfrac{5}{3} = 43.333$

(m) $\left[-\dfrac{x^2}{2} \right]_{0}^{1} = -\dfrac{1}{2}$

(n) $\left[-\dfrac{1}{x} \right]_{1}^{3} = -\dfrac{1}{3} - \left(-\dfrac{1}{1} \right) = \dfrac{2}{3}$

(o) $\left[-\dfrac{\cos 3t}{3} \right]_{0}^{1}$

$= \left(-\dfrac{\cos 3}{3} \right) - \left(-\dfrac{\cos 0}{3} \right)$

$= 0.6633$

2. (a) $\int_0^2 (3e^x + 1)\,dx = [3e^x + x]_0^2 = 21.17$

(b) $\int_0^1 (2e^{2x} + \sin x)\,dx$

$= [e^{2x} - \cos x]_0^1$

$= (e^2 - \cos 1) - (e^0 - \cos 0)$

$= 6.8488$

(c) $\int_1^2 (2\cos 2t - 3\sin 2t)\,dt$

$= \left[\sin 2t + \dfrac{3}{2}\cos 2t \right]_1^2$

$= \left(\sin 4 + \dfrac{3}{2}\cos 4 \right) - \left(\sin 2 + \dfrac{3}{2}\cos 2 \right)$

$= -2.0223$

(d) $\int_1^3 \left(\dfrac{2}{x} + \dfrac{x}{2} \right)\,dx$

$= \left[2\ln x + \dfrac{x^2}{4} \right]_1^3$

$= \left(2\ln 3 + \dfrac{9}{4} \right) - \left(2\ln 1 + \dfrac{1}{4} \right)$

$= 4.1972$

3. (a) $\int_0^2 [x(2x+3)]\,dx$

$= \int_0^2 (2x^2 + 3x)\,dx$

$= \left[\dfrac{2x^3}{3} + \dfrac{3x^2}{2} \right]_0^2$

$= \dfrac{34}{3}$

(b) $\int_{-1}^1 \left(3e^{2x} - \dfrac{3}{e^{2x}} \right)\,dx$

$= \int_{-1}^1 (3e^{2x} - 3e^{-2x})\,dx$

$= \left[\dfrac{3e^{2x}}{2} + \dfrac{3e^{-2x}}{2} \right]_{-1}^1$

$= 0$

(c) $\int_0^\pi \left[2\sin\left(\dfrac{t}{2} \right) - 4\cos 2t \right]\,dt$

$= \left[-4\cos\left(\dfrac{t}{2} \right) - 2\sin 2t \right]_0^\pi$

$= \left(-4\cos\dfrac{\pi}{2} - 2\sin 2\pi \right)$

$\quad - (-4\cos 0 - 2\sin 0)$

$= 4$

(d) $\int_1^2 \left(\sin 3x - \dfrac{3}{x} \right)\,dx$

$= \left[-\dfrac{\cos 3x}{3} - 3\ln x \right]_1^2$

$= \left(-\dfrac{\cos 6}{3} - 3\ln 2 \right)$

$\quad - \left(-\dfrac{\cos 3}{3} - 3\ln 1 \right)$

$= -2.7295$

4. (a) $\displaystyle\int_1^2 \left[\frac{2}{x}\left(3+\frac{1}{x}\right)\right] dx$

$= \displaystyle\int_1^2 \left(\frac{6}{x}+\frac{2}{x^2}\right) dx$

$= \left[6\ln x - \dfrac{2}{x}\right]_1^2$

$= (6\ln 2 - 1) - (6\ln 1 - 2)$

$= 5.1589$

(b) $\displaystyle\int_0^1 [e^{-2x}(e^x - 2e^{-x})]\, dx$

$= \displaystyle\int_0^1 (e^{-x} - 2e^{-3x})\, dx$

$= \left[-e^{-x} + \dfrac{2e^{-3x}}{3}\right]_0^1$

$= \left(-e^{-1} + \dfrac{2e^{-3}}{3}\right) - \left(-e^0 + \dfrac{2e^0}{3}\right)$

$= -0.001355$

(c) $\displaystyle\int_0^{0.5} (\sin \pi t + \cos \pi t)\, dt$

$= \left[\dfrac{-\cos \pi t + \sin \pi t}{\pi}\right]_0^{0.5}$

$= \dfrac{2}{\pi}$

(d) $\displaystyle\int_0^1 (2x - 3)^2\, dx$

$= \displaystyle\int_0^1 (4x^2 - 12x + 9)\, dx$

$= \left[\dfrac{4x^3}{3} - 6x^2 + 9x\right]_0^1$

$= \dfrac{13}{3}$

Self-assessment question 35.5

1. Given $y(x)$, suppose the area required lies between $x = a$ and $x = b$. When this area lies entirely above the horizontal axis its value is $\int_a^b y(x)\, dx$. When an area lies entirely below the axis, definite integration will produce a negative value. The size of this area is found by ignoring the negative sign. If parts lie above and below the axis, each part should be evaluated separately.

Exercise 35.5

1. 26
2. Evaluate $\int_{-1}^2 (4 - x^2)\, dx$ and $\int_2^4 (4 - x^2)\, dx$ separately to get 9 and $-10\frac{2}{3}$ respectively. Total area is $19\frac{2}{3}$.
3. $[e^{2t}/2]_1^4 = e^8/2 - e^2/2 = 1486.78$
4. $[\ln t]_1^5 = \ln 5 = 1.609$
5. $[\frac{2}{3}x^{3/2}]_1^2 = \frac{2}{3}2^{3/2} - \frac{2}{3} = 1.219$
6. $\text{Area} = \displaystyle\int_1^4 \frac{1}{x}\, dx = [\ln x]_1^4 = 1.3863$
7. Note that all the area is above the t axis.

$\text{Area} = \displaystyle\int_0^{0.5} \cos 2t\, dt$

$= \left[\dfrac{\sin 2t}{2}\right]_0^{0.5}$

$= \dfrac{\sin 1}{2} - \dfrac{\sin 0}{2}$

$= 0.4207$

Chapter 36

Self-assessment questions 36.2

1. The product of functions $f(x)$ and $g(x)$ is found by multiplying them together to give $f(x) \times g(x)$, usually written as $f(x)g(x)$ or even simply fg.

2.
$$\int u \frac{dv}{dx} \, dx = uv - \int v \frac{du}{dx} \, dx$$

Exercise 36.2

1. (a) $\int x \cos x \, dx = x \sin x - \int \sin x \, dx$
   $= x \sin x + \cos x + c$

   (b) $\int 3te^{2t} \, dt = 3t \frac{e^{2t}}{2} - \int 3 \frac{e^{2t}}{2} \, dt =$
   $\frac{3te^{2t}}{2} - \frac{3e^{2t}}{4} + c$

   (c) $\int xe^{-3x} \, dx = x\frac{e^{-3x}}{-3} - \int \frac{e^{-3x}}{-3} \, dx =$
   $-\frac{1}{3}xe^{-3x} - \frac{1}{9}e^{-3x} + c$

   (d) $\int 4z \ln z \, dz = 2z^2 \ln z - \int 2z^2 \frac{1}{z} \, dz =$
   $2z^2 \ln z - z^2 + c$

2. $\int x^2 \sin x \, dx = -x^2 \cos x - \int 2x(-\cos x) \, dx$

   $= -x^2 \cos x + \int 2x \cos x \, dx$

   $= -x^2 \cos x + 2x \sin x - \int 2 \sin x \, dx$

   $= -x^2 \cos x + 2x \sin x + 2 \cos x + c$

Exercise 36.3

1. (a) $\int_0^1 xe^x \, dx = [xe^x]_0^1 - \int_0^1 e^x \, dx =$
   $[xe^x]_0^1 - [e^x]_0^1 = e^1 - e^1 + e^0 = 1$

   (b) $\int_1^2 xe^x \, dx = [xe^x]_1^2 - \int_1^2 e^x \, dx =$
   $[xe^x]_1^2 - [e^x]_1^2 = 2e^2 - e^1 - e^2 + e^1 = e^2$

   (c) $\int_1^2 xe^{-x} \, dx = [-xe^{-x}]_1^2 - \int_1^2 -e^{-x} \, dx =$
   $[-xe^{-x}]_1^2 - [e^{-x}]_1^2 = -2e^{-2} + e^{-1} - e^{-2} +$
   $e^{-1} = -3e^{-2} + 2e^{-1}$

   (d) $\int_{-1}^1 xe^{-x} \, dx = [-xe^{-x}]_{-1}^1 - \int_{-1}^1 -e^{-x} \, dx =$
   $[-xe^{-x}]_{-1}^1 - [e^{-x}]_{-1}^1 = -e^{-1} - e^1 - e^{-1} +$
   $e^1 = -2e^{-1}$

(e) $\int_1^2 1 \cdot \ln x \, dx = [x \ln x]_1^2 - \int_1^2 x \frac{1}{x} \, dx =$
$[x \ln x]_1^2 - [x]_1^2 = 2 \ln 2 - 2 + 1 =$
$2 \ln 2 - 1$

(f) $\int_0^\pi x \cos 2x \, dx = [x \frac{\sin 2x}{2}]_0^\pi - \int_0^\pi \frac{\sin 2x}{2} \, dx =$
$[x \frac{\sin 2x}{2}]_0^\pi + [\frac{\cos 2x}{4}]_0^\pi = \frac{1}{4} - \frac{1}{4} = 0$

(g) $\int_0^{\pi/2} x \cos 2x \, dx = [x \frac{\sin 2x}{2}]_0^{\pi/2} -$
$\int_0^{\pi/2} \frac{\sin 2x}{2} \, dx = [x \frac{\sin 2x}{2}]_0^{\pi/2} + [\frac{\cos 2x}{4}]_0^{\pi/2} =$
$-\frac{1}{4} - \frac{1}{4} = -\frac{1}{2}$

(h) $\int_{-\pi/2}^{\pi/2} x \sin x \, dx = [-x \cos x]_{-\pi/2}^{\pi/2} -$
$\int_{-\pi/2}^{\pi/2} -\cos x \, dx = 0 + [\sin x]_{-\pi/2}^{\pi/2} =$
$1 - (-1) = 2$

Exercise 36.4

1. (a) $I = \int (4x + 3)^3 \, dx$. Make the substitution $y = 4x + 3$, $\frac{dy}{dx} = 4$. Hence

   $$I = \int y^3 \frac{1}{4} \, dy$$

   $$= \frac{1}{4} \left[\frac{y^4}{4} \right] + c$$

   $$= \frac{(4x + 3)^4}{16} + c$$

   (b) $I = \int \sqrt{3x - 1} \, dx$. Make the substitution $y = 3x - 1$ from which $\frac{dy}{dx} = 3$. Hence

   $$I = \int \sqrt{y} \frac{1}{3} \, dy = \frac{2}{9} y^{\frac{3}{2}} + c = \frac{2}{9}(3x - 1)^{\frac{3}{2}} + c$$

   (c) $I = \int \frac{2}{1-x} \, dx$. Make the substitution $y = 1 - x$ from which $\frac{dy}{dx} = -1$. Hence

   $$I = \int \frac{2}{y}(-1) \, dy = -2 \ln y + c$$

   $$= -2 \ln(1 - x) + c$$

   (d) $I = \int \sin(6x + 5) \, dx$. Make the substitution $y = 6x + 5$ from which $\frac{dy}{dx} = 6$. Hence

   $$I = \int \sin y \frac{1}{6} \, dy = -\frac{1}{6} \cos y + c$$

   $$= -\frac{\cos(6x + 5)}{6} + c$$

(e) $I = \int \cos(\frac{x}{2} + 3)\, dx$. Make the substitution $y = \frac{x}{2} + 3$ from which $\frac{dy}{dx} = \frac{1}{2}$. Hence

$$I = \int \cos y (2)\, dy = 2 \sin y + c$$

$$= 2 \sin\left(\frac{x}{2} + 3\right) + c$$

2. (a) $I = \int x e^{(x^2)}\, dx$. Let $y = x^2$ so $\frac{dy}{dx} = 2x$. Hence I may be written as

$$I = \frac{1}{2} \int e^y\, dy = \frac{1}{2} e^y + c = \frac{1}{2} e^{(x^2)} + c$$

3. $I = \int \frac{e^{(1/x)}}{x^2}\, dx$. Let $y = \frac{1}{x}$ so $\frac{dy}{dx} = -\frac{1}{x^2}$. Hence I may be written as

$$I = \int e^y (-1)\, dy = -e^y + c = -e^{1/x} + c$$

4. (a) Let $I = \int (\frac{\theta}{3} - 1)^5\, d\theta$ and let $y = \frac{\theta}{3} - 1$ from which $\frac{dy}{d\theta} = \frac{1}{3}$. So I may be written as

$$I = \int y^5\, 3\, dy = \frac{y^6}{2} + c = \frac{(\frac{\theta}{3} - 1)^6}{2} + c$$

(b) Let $I = \int t^2 \sqrt{t^3 - 1}\, dt$, $y = t^3 - 1$ and so $\frac{dy}{dt} = 3t^2$. Hence

$$I = \int \sqrt{y}\, \frac{1}{3}\, dy = \frac{2}{9} y^{3/2} + c$$

$$= \frac{2}{9}(t^3 - 1)^{3/2} + c$$

(c) Let $I = \int \cos(\frac{y+1}{2})\, dy$, $\theta = \frac{y+1}{2}$ from which $\frac{d\theta}{dy} = \frac{1}{2}$. Hence we have

$$I = \int \cos \theta (2)\, d\theta = 2 \sin \theta + c$$

$$= 2 \sin\left(\frac{y+1}{2}\right) + c$$

(d) Let $I = \int \sin(10 - x)\, dx$, $y = 10 - x$ from which $\frac{dy}{dx} = -1$. So

$$I = \int \sin y (-1)\, dy = \cos y + c$$

$$= \cos(10 - x) + c$$

(e) Let $I = \int \frac{1}{e^{2x-1}}\, dx = \int e^{1-2x}\, dx$. Let $y = 1 - 2x$, $\frac{dy}{dx} = -2$ and so

$$I = \int -e^y \frac{1}{2}\, dy = -\frac{e^y}{2} + c = -\frac{e^{1-2x}}{2} + c$$

5. (a) $I = \int_2^6 (\frac{x}{2} - 1)^4\, dx$. With the substitution $y = \frac{x}{2} - 1$, I becomes

$$I = \int_0^2 y^4\, 2\, dy = \left[\frac{2y^5}{5}\right]_0^2 = \frac{64}{5}$$

(b) $I = \int_{-0.5}^1 \cos(4\theta + 2)\, d\theta$. With the substitution $y = 4\theta + 2$, I becomes

$$I = \int_0^6 \cos y \frac{1}{4}\, dy = \left[\frac{\sin y}{4}\right]_0^6 = \frac{\sin 6}{4} = -0.070$$

(c) $I = \int_{\pi/2}^{\pi} 2 \sin(\pi - 2\theta)\, d\theta$. With the substitution $y = \pi - 2\theta$, I becomes

$$I = \int_0^{-\pi} -\sin y\, dy = [\cos y]_0^{-\pi} = -2$$

(d) $I = \int_0^1 x e^{(-x^2)}\, dx$. With the substitution $y = -x^2$, I becomes

$$I = \frac{1}{2} \int_0^{-1} -e^y\, dy = \frac{1}{2}[-e^y]_0^{-1} = 0.3161$$

(e) $I = \int_0^5 \sqrt{10 - 2x}\, dx$. With the substitution $y = 10 - 2x$, I becomes

$$I = \int_{10}^0 -\frac{\sqrt{y}}{2}\, dy = -\left[\frac{y^{3/2}}{3}\right]_{10}^0 = 10.54$$

Exercise 36.5

1. $$\frac{x+6}{x^2 + 6x + 8} = \frac{x+6}{(x+2)(x+4)} = \frac{2}{x+2} - \frac{1}{x+4}$$

So

$$\int \frac{x+6}{x^2 + 6x + 8}\, dx = \int \frac{2}{x+2} - \frac{1}{x+4}\, dx$$

$$= 2\ln(x+2) - \ln(x+4) + c$$

2. Using standard techniques of partial fractions we find

$$\frac{10x - 1}{4x^2 - 1} = \frac{10x - 1}{(2x-1)(2x+1)} = \frac{3}{2x+1} + \frac{2}{2x-1}$$

So

$$\int \frac{10x - 1}{4x^2 - 1}\, dx = \int \frac{3}{2x+1} + \frac{2}{2x-1}\, dx$$

$$= \frac{3}{2}\ln(2x+1) + \ln(2x-1) + c$$

3. $\dfrac{3x-2}{9x^2-6x+1}=\dfrac{3x-2}{(3x-1)^2}=\dfrac{1}{3x-1}-\dfrac{1}{(3x-1)^2}$

So

$\displaystyle\int\dfrac{3x-2}{9x^2-6x+1}dx=\int\dfrac{1}{3x-1}-\dfrac{1}{(3x-1)^2}dx$

$\qquad\qquad=\dfrac{1}{3}\ln(3x-1)+\dfrac{1}{3(3x-1)}+c$

4. $\dfrac{3x^2+6x+2}{x^3+3x^2+2}=\dfrac{3x^2+6x+2}{x(x+1)(x+2)}$

$\qquad\qquad=\dfrac{1}{x}+\dfrac{1}{x+1}+\dfrac{1}{x+2}$

Hence

$\displaystyle\int_1^2\dfrac{3x^2+6x+2}{x^3+3x^2+2x}\,dx$

$\qquad=\displaystyle\int_1^2\dfrac{1}{x}+\dfrac{1}{x+1}+\dfrac{1}{x+2}\,dx$

$\qquad=[\ln(x)+\ln(x+1)+\ln(x+2)]_1^2$

$\qquad=\ln 2+\ln 3+\ln 4-\ln 1-\ln 2-\ln 3$

$\qquad=1.3863$

Chapter 37

Self-assessment questions 37.1

1. A function of two variables is a rule that operates upon two variables, chosen independently, and produces a single output value known as the dependent variable.

2. (a) In $z=f(x,y)$, y is an independent variable. The user is free to choose a value for y.

   (b) In $y=f(x)$, y is the dependent variable. Once a value for x has been chosen, then y is determined automatically – we have no choice.

3. Given $P=f(V,T)=\dfrac{RT}{V}$, P is the dependent variable. V and T are independent variables.

4. In $y=f(x,t)$, x and t are independent variables and y is the dependent variable.

Exercise 37.1

1. $f(8,2)=7(8)+2(2)=56+4=60$
2. (a) $f(2,3)=-11(2)+3=-19$
   (b) $f(11,1)=-11(11)+1=-121+1$
       $=-120$
3. $z(1,1)=3e^1-2e^1+1=3.718$

4. $w=7-(-3)(-9)=7-27=-20$
5. $\sin(50°)=0.7660$
6. $f(0.5,3)=e^{(2)(0.5)(3)}=e^3=20.086$

Self-assessment question 37.2

1. Your sketch should look like Figure 37.3 in the main text, with the axes ordered in the same way.

Exercise 37.2

1. (a) $z=-12$ (b) -18 (c) 15

Self-assessment questions 37.3

1. In each case, zero. The derivative of a constant is always zero.
2. Zero. When performing partial differentiation with respect to x, y is treated as a constant and consequently its derivative is zero.
3. $\dfrac{\partial z}{\partial x}$. Always remember to use the curly d.
4. $\dfrac{\partial z}{\partial y}$. Always remember to use the curly d.
5. $\dfrac{\partial y}{\partial x}$, $\dfrac{\partial y}{\partial t}$.

Exercise 37.3

1. (a) $\frac{\partial z}{\partial x} = 5, \frac{\partial z}{\partial y} = 11$ (b) $-14, -7$ (c) 8, 0
   (d) 0, -5 (e) 3, 8 (f) $-3, 2$ (g) 0, 0
   (h) 0, -3 (i) $4x, -7$ (j) $7, -9y^2$ (k) $-9, 9$
   (l) 9, 9

2. (a) $\frac{\partial z}{\partial x} = y, \frac{\partial z}{\partial y} = x$ (b) $3y, 3x$
   (c) $-9y, -9x$ (d) $2xy, x^2$ (e) $18xy, 9x^2$
   (f) $8y^2, 16xy$

3. $\frac{\partial z}{\partial x} = 9$ and $\frac{\partial z}{\partial y} = 2y$. Hence at $(4, -2)$ these
   derivatives are, respectively, 9 and -4.

4. (a) $\frac{\partial z}{\partial x} = 2e^{2x}$ and $\frac{\partial z}{\partial y} = 0$

   (b) $\frac{\partial z}{\partial x} = 0$ and $\frac{\partial z}{\partial y} = 5e^{5y}$

   (c) $\frac{\partial z}{\partial x} = ye^{xy}$ and $\frac{\partial z}{\partial y} = xe^{xy}$

   (d) $\frac{\partial z}{\partial x} = 0$ and $\frac{\partial z}{\partial y} = 8e^{2y}$

5. $\dfrac{\partial y}{\partial x} = \sin t, \dfrac{\partial y}{\partial t} = x \cos t$

Exercise 37.4

1. (a) $\dfrac{\partial z}{\partial x} = ye^x(x+1), \dfrac{\partial z}{\partial y} = xe^x$

   (b) $ye^y, xe^y(y+1)$ (c) $e^{xy}(xy+1), x^2e^{xy}$
   (d) $y^2e^{xy}, e^{xy}(xy+1)$
   (e) $x^2y \cos(xy) + 2x \sin(xy), x^3 \cos(xy)$
   (f) $-y^2 \sin(xy), -yx \sin(xy) + \cos(xy)$
   (g) $1 + \ln(xy), \frac{x}{y}$ (h) $3y^3e^x(x+1), 9xy^2e^x$

Self-assessment question 37.5

1. $\dfrac{\partial^2 z}{\partial y \partial x}$ means differentiate $\dfrac{\partial z}{\partial x}$ with respect
   to y. $\dfrac{\partial^2 z}{\partial x \partial y}$ means differentiate $\dfrac{\partial z}{\partial y}$ with
   respect to x. It is usually the case that the
   result is the same either way.

Exercise 37.5

In all solutions $\dfrac{\partial^2 z}{\partial x \partial y}$ and $\dfrac{\partial^2 z}{\partial y \partial x}$ are identical.

1. (a) $\dfrac{\partial^2 z}{\partial x^2} = 0, \dfrac{\partial^2 z}{\partial y \partial x} = 1, \dfrac{\partial^2 z}{\partial y^2} = 0$

(b) $\dfrac{\partial^2 z}{\partial x^2} = 0, \dfrac{\partial^2 z}{\partial y \partial x} = 7, \dfrac{\partial^2 z}{\partial y^2} = 0$

(c) $\dfrac{\partial^2 z}{\partial x^2} = 0, \dfrac{\partial^2 z}{\partial y \partial x} = 0, \dfrac{\partial^2 z}{\partial y^2} = 0$

(d) $\dfrac{\partial^2 z}{\partial x^2} = 0, \dfrac{\partial^2 z}{\partial y \partial x} = 16y, \dfrac{\partial^2 z}{\partial y^2} = 16x$

(e) $\dfrac{\partial^2 z}{\partial x^2} = -4y^3, \dfrac{\partial^2 z}{\partial y \partial x} = -12y^2 x,$

   $\dfrac{\partial^2 z}{\partial y^2} = -12yx^2$

(f) $\dfrac{\partial^2 z}{\partial x^2} = 0, \dfrac{\partial^2 z}{\partial y \partial x} = 0, \dfrac{\partial^2 z}{\partial y^2} = 0$

2. (a) $\dfrac{\partial^2 z}{\partial x^2} = \dfrac{2}{x^3}, \dfrac{\partial^2 z}{\partial y \partial x} = 0, \dfrac{\partial^2 z}{\partial y^2} = 0$

(b) $\dfrac{\partial^2 z}{\partial x^2} = \dfrac{2y}{x^3}, \dfrac{\partial^2 z}{\partial y \partial x} = -\dfrac{1}{x^2}, \dfrac{\partial^2 z}{\partial y^2} = 0$

(c) $\dfrac{\partial^2 z}{\partial x^2} = 0, \dfrac{\partial^2 z}{\partial y \partial x} = -\dfrac{1}{y^2}, \dfrac{\partial^2 z}{\partial y^2} = \dfrac{2x}{y^3}$

(d) $\dfrac{\partial^2 z}{\partial x^2} = \dfrac{2}{x^3}, \dfrac{\partial^2 z}{\partial y \partial x} = 0, \dfrac{\partial^2 z}{\partial y^2} = \dfrac{2}{y^3}$

3. (a) $\dfrac{\partial^2 z}{\partial x^2} = 0, \dfrac{\partial^2 z}{\partial y \partial x} = \cos y, \dfrac{\partial^2 z}{\partial y^2} = -x \sin y$

(b) $\dfrac{\partial^2 z}{\partial x^2} = -y \cos x, \dfrac{\partial^2 z}{\partial y \partial x} = -\sin x, \dfrac{\partial^2 z}{\partial y^2} = 0$

(c) $\dfrac{\partial^2 z}{\partial x^2} = 4ye^{2x}, \dfrac{\partial^2 z}{\partial y \partial x} = 2e^{2x}, \dfrac{\partial^2 z}{\partial y^2} = 0$

(d) $\dfrac{\partial^2 z}{\partial x^2} = ye^{-x}, \dfrac{\partial^2 z}{\partial y \partial x} = -e^{-x}, \dfrac{\partial^2 z}{\partial y^2} = 0$

4. (a) $\dfrac{\partial^2 z}{\partial x^2} = 8y^2e^{xy}, \dfrac{\partial^2 z}{\partial y \partial x} = 8e^{xy}(xy+1),$

   $\dfrac{\partial^2 z}{\partial y^2} = 8x^2e^{xy}$

(b) $\dfrac{\partial^2 z}{\partial x^2} = -3e^x \sin y, \dfrac{\partial^2 z}{\partial y \partial x} = -3e^x \cos y,$

   $\dfrac{\partial^2 z}{\partial y^2} = 3e^x \sin y$

(c) $\dfrac{\partial^2 z}{\partial x^2} = -4e^y \cos x,\ \dfrac{\partial^2 z}{\partial y \partial x} = -4e^y \sin x,$

$\dfrac{\partial^2 z}{\partial y^2} = 4e^y \cos x$

5. (a) $\dfrac{\partial^2 z}{\partial x^2} = -\dfrac{1}{x^2},\ \dfrac{\partial^2 z}{\partial y \partial x} = 0,\ \dfrac{\partial^2 z}{\partial y^2} = 0$

(b) $\dfrac{\partial^2 z}{\partial x^2} = 0,\ \dfrac{\partial^2 z}{\partial y \partial x} = 0,\ \dfrac{\partial^2 z}{\partial y^2} = -\dfrac{1}{y^2}$

(c) $\dfrac{\partial^2 z}{\partial x^2} = -\dfrac{1}{x^2},\ \dfrac{\partial^2 z}{\partial y \partial x} = 0,\ \dfrac{\partial^2 z}{\partial y^2} = -\dfrac{1}{y^2}$

(d) $\dfrac{\partial^2 z}{\partial x^2} = 0,\ \dfrac{\partial^2 z}{\partial y \partial x} = \dfrac{1}{y},\ \dfrac{\partial^2 z}{\partial y^2} = -\dfrac{x}{y^2}$

(e) $\dfrac{\partial^2 z}{\partial x^2} = -\dfrac{y}{x^2},\ \dfrac{\partial^2 z}{\partial y \partial x} = \dfrac{1}{x},\ \dfrac{\partial^2 z}{\partial y^2} = 0$

Self-assessment question 37.6

1. First calculate $\dfrac{\partial w}{\partial z}$. Then partially differentiate this with respect to y. Finally, partially differentiate the result with respect to x.

Exercise 37.6

1. $8, -3, -4$
2. $2x, 3y^2, -10t$
3. $\dfrac{\partial w}{\partial x} = yz,\ \dfrac{\partial w}{\partial y} = xz,\ \dfrac{\partial w}{\partial z} = xy$

Index

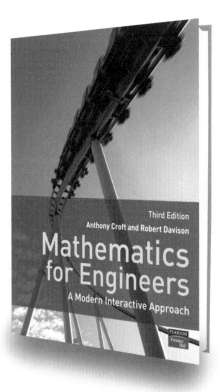